Advances in Solar Energy

An Annual Review of
Research and Development

Volume 16

Editorial Board

Advances in Solar Energy
An Annual Review of
Research and Development

Volume 16

Edited by

D. Yogi Goswami
University of Florida
Gainesville, FL

AMERICAN SOLAR ENERGY SOCIETY, INC.
Boulder, Colorado

London • Sterling, VA

First published by Earthscan in the UK and USA in 2005

American Solar Energy Society, Inc.
2400 Central Avenue, Suite A
Boulder, CO 80301

ISSN: 0731-8618
ISBN: 1-84407-244-4   hardback

Printed and bound in the UK by Cromwell Press, Trowbridge
Cover design by Yvonne Booth

A Continuation Order Plan is available for this series. A continuation order will bring
delivery of each new volume immediately upon publication. Volumes are billed only
upon actual shipment. For further information, and for a full list of publications please
contact:

Earthscan
8–12 Camden High Street
London, NW1 0JH, UK
Tel: +44 (0)20 7387 8558
Fax: +44 (0)20 7387 8998
Email: earthinfo@earthscan.co.uk
Web: **www.earthscan.co.uk**

22883 Quicksilver Drive, Sterling, VA 20166-2012, USA

Earthscan is an imprint of James & James (Science Publishers) Ltd and publishes
in association with the International Institute for Environment and Development

A catalogue record for this book is available from the British Library

Library of Congress Cataloging-in-Publication Data
Advances in solar energy — Vol. 1 (1982) —
-New York: American Solar Energy Society, c1983
   v. ill; 27 cm

   Annual
   ISSN 0731-8618 = Advances in solar energy.

   1. Solar energy — periodicals. I. American Solar Energy Society.
TJ809.S38   621.47'06-dc19   85-646250
                                                        AARC 2MARC-S

Library of Congress   (8603)

Printed on elemental chlorine-free paper

Foreword

As we move deeper into the 21st century it is becoming increasingly clear that human activity is changing the global climate – this is mainly caused by the energy resources we are predominately using. Fortunately, advancements in solar energy technologies are occurring at an increasing rate. This volume of *Advances in Solar Energy* covers both of these timely topics.

A majority of the technological advances covered in this volume relate to photovoltaics and solar thermal power. Both of these areas are expected to play a major role in our future electricity production. In addition, the topic of transportation using direct solar, wind, wave, and hydrogen fuel cells is also covered in this volume. Solar pond technologies and passive cooling are also reviewed in-depth in this volume. Since this year is the 50th anniversary of the International Solar Energy Society (ISES), we have included for the first time, a topic of historical importance – an innovative high temperature solar concentration system developed in the 19th century. Finally, in our continuing series on country-specific topics, this volume presents an article on renewable energy for the Russian economy. The authors of these invited reviews are world-renowned experts, researchers and practitioners in their respective fields.

The first chapter analyzes and answers the questions of whether climate is changing under the influence of human activity and whether these changes will cause serious negative socio-economic impacts in the future. Chapter author, Ganopolski, concludes that even though natural processes have contributed to global climate changes to some extent it is extremely unlikely that the scope of these changes during the last century could be solely attributed to natural factors. Results of model simulations suggest that crossing some currently known thresholds could trigger dramatic and possibly irreversible climate change.

Chapter 2 analyzes the potential contribution of photovoltaics in meeting part of European and US future electricity needs. Byrne and his co-authors conclude that in the forecast period of 2025 to 2070 the cumulative energy production from PV in Europe and USA will be higher than the domestic oil output in both regions. The chapter also examines the role of national energy policies in the two regions.

In Chapter 3, author Hamakawa reviews recent advances in PV technologies including improvements in efficiencies and reduction in costs. The paper also discusses a technological roadmap for PV up to the year 2030 and its future prospects. In addition, the paper considers the role of PV in future environmental issues.

Chapters 4 and 5 describe the advances in two different types of PV technologies. In Chapter 4, Yamaguchi and co-authors discuss the research and development of III-V compound multi-junction solar cells which have a potential for achieving conversion efficiencies of more than 50%. The authors explain that efficiencies of over 37% have been demonstrated under concentration of 200-suns. Chapter 5 gives an overview of the progress in crystalline silicon solar cells.

Chapter author Saitoh discusses the issue of long- and short-term reliability of crystalline silicon cells. Although these cells are considered highly reliable, the author presents ways of increasing their reliability even further.

Solar parabolic troughs are the lowest cost large-scale renewable energy (RE) power technology, primarily because of the field's development and experience associated with the large-scale power plants in California. In chapter 6 co-authors Kearney and Price and their team, review the current state-of-the-art of parabolic trough solar power technology and describe on-going research and development to further enhance its potential. The chapter also examines the economics of solar thermal power based on parabolic trough technology.

Chapter 7 reviews the advances in solar pond technology, especially during the past 20 years. The key advantage of solar ponds over any other RE technology is that energy collection and storage are integrated. Akbarzadeh and his co-authors describe many other advantages of solar ponds which include a wide variety of applications for industrial process heat, electrical power generation, mariculture and biogas production, desalination, and chemical production. The chapter describes key research activities around the world and identifies pivotal research and development areas.

Chapter 8 provides a review of passive cooling technologies for buildings. This topic has become extremely important because of the increasing penetration of mechanical air conditioning worldwide and the resulting adverse environmental impact. Author Santamouris reviews various passive cooling techniques, their present status and on-going research. The chapter also discusses the urban microclimate and possible improvements to urban thermal conditions by passive methods.

Chapters 9 and 10 deal with the use of renewable energy in transportation. Paul MacCready, the creator of the NASA "Helios" solar airplane, describes various ways of using direct solar radiation, wind power, waves, and etc. for transportation in air, land and water. He describes how natural and atmospheric effects such as up-currents, thermals, waves, winds, shears, turbulence and wakes are used by albatrosses and vultures and how these effects could be exploited by human-made machines. Chapter 10 deals with solar hydrogen fuel cell systems. In this chapter, authors Duffy and Aurora describe a solar hydrogen fuel cell system model, which will help understand the behavior of such systems and to aid in their optimal design.

Chapter 11 explores the potential of RE use in Russia and makes a strong case for its increased use in spite of this region's abundance of fossil fuels. Chapter author Tyukhov and coauthors describe some production activities of the Russian renewable energy equipment industry. They also describe a new policy initiative to significantly improve the contribution of RE to the Russian economy.

Finally, Chapter 12 describes an interesting historical topic – the early development of a solar concentrator. The "Pyreheliophoro" was actually invented by a Jesuit Priest (nicknamed Father Himalaya) in the 19th Century. The concentrator produced a sustained temperature of 3800°C, the highest temperature produced with solar energy at that time. Chapter author Collares-Pereira provides an interesting account of Fr.

Himalaya's life, his inventions, and his travels and travails to bring his work to the world.

I would like to thank the associate editors who helped in selecting topics, the authors for their chapters, and the reviewers who gave very valuable suggestions. Finally, I would like to thank Ms. Barbara Graham, Assistant Editor, and Ms. Allyson Haskell, Editorial Assistant, whose hard work and total dedication was essential in compiling Volume 16 of *Advances in Solar Energy*.

D. Yogi Goswami

Table of Contents

Chapter 1: Anthropogenic Global Warming: Evidence, Predictions and Consequences 1

Andrey Ganopolski
Potsdam Institute for Climate Impact Research
PO BOX 601203
Potsdam 14412, Germany
andrey@pik-potsdam.de

Chapter 2: Beyond Oil: A Comparison of Projections of PV Generation and European and U.S. Domestic Oil Production 35

John Byrne*, Lado Kurdgelashvili, Allen Barnett, Kristen Hughes and Aaron Thurlow
Center for Energy and Environmental Policy
278 Graham Hall
University of Delaware
Newark, DE 19716   USA
* jbbyrne@UDel.Edu,
86023@udel.edu, 84750@udel.edu, 10257@udel.edu, 15519@udel.edu

Yoshihiro Hamakawa
Advisor Professor to the Chancellor
Ritsumeikan University
1-1-1 Nojihigashi
Kusatsu Shiga 525-8577 Japan
hamakawa@se.ritsumei.ac.jp

Masafumi Yamaguchi*   Tatsuya Takamoto
Toyota Technological Institute   *Sharp Corporation*
2-12-1 Hisakata, Tempaku   *Hajikami, Shinjo*
Nagoya 468-8511, Japan   *Nara 639-2198, Japan*
*masafumi@toyota-ti.ac.jp

Kenji Araki
Daido Steel Corporation
Daido-cho, Minami
Nagoya 457-8545, Japan

Chapter 5: Progress of Highly Reliable Crystalline Si Solar Devices and Materials ... 129

Tadashi Saitoh
Department of Electrical and Electronics Engineering
Tokyo University of Agriculture and Technology
Koganei, Tokyo 184-8588, Japan
tadashi@cc.tuat.ac.jp

Chapter 6: Recent Advances in Parabolic Trough Solar Power Plant Technology 155

David Kearney*   *Henry Price*
Kearney and Associates   *National Renewable Energy Laboratory*
P.O. Box 2568   *1617 Cole Boulevard*
Vashon WA 98070   *Golden, CO 80401*
dkearney@attglobal.net   *henry_price@nrel.gov*

Contributors
Nikolaus Benz, Dan Blake, Avi Brenmiller, Douglas Brosseau, Robert Cable, Gilbert Cohen, Fritz-Dieter Doenitz, Robert Fimbres, Scott Frier, Randy Gee, Michael Geyer, Karen Gummo, Mary Jane Hale, Cheryl Kennedy, Ulf Herrmann, Eckhard Lüpfert, Rod Mahoney, Eli Mandelberg, Kenneth May, Luc Moens, Paul Nava, Robert Pitz-Paal, Nicholas Potrovitza, Rainer Tamme, Eduardo Zarza

Aliakbar Akbarzadeh*, John Andrews
Energy Conservation and
  Renewable Energy Group
School of Aerospace, Mechanical
  and Manufacturing Engineering
RMIT University
 Bundoora East Campus P.O. Box 71
Bundoora, Victoria
Australia 3083
* akbar@rmit.edu.au
andrews.john@rmit.edu.au

Peter Golding
 Department of Metallurgical and
  Materials Engineering
University of Texas at El Paso
500 W University Avenue
El Paso, TX 79968 USA
pgolding@utep.edu

Matheos Santamouris
Group Building Environmental Studies
Physics Department
University of Athens
Panepistimiopolis, Zografou, 157 84
Athens, Greece
msantam@cc.uoa.gr

Paul B. MacCready
AeroVironment Inc.
825 S. Myrtle
Monrovia, CA 91016-3424 USA
maccready@aerovironment.com

John J. Duffy*   *Peter H. Aurora*
Energy Engineering Program   *Engineering Reliability*
University of Massachusetts Lowell   *and Product Development*
One University Avenue   *RWE Schott Solar Inc.*
Lowell, MA 01854, USA   *4 Suburban Park Drive*
John_Duffy@uml.edu   *Billerica, MA 01821 USA*
                                       peter.aurora@rweschottsolar.us

Pavel P. Bezrukih
Energy Ministry of Russian Federation
7 Kitaigorodsky pr.
Moscow, 103074 Russia
degvv@mte.gov.ru

Dmitry S. Strebkov, Igor I. Tyukhov*
The All-Russian Research Institute for
  Electrification of Agriculture (2 VIESH)
1-st Veshnyakovsky proezd
Moscow, 109456 Russia
* ityukhov@yahoo.com
viesh@dol.ru

Manuel Collares Pereira
Instituto Nacional de Engenharia, Tecnologia e Inovação (INETI)
  Renewable Energies Department
Edifício G, Az. dos Lameiros
1649-038 Lisbon, Portugal
collares.pereira@ineti.pt

David R. Mills* and Christopher J. Dey
School of Physics, A28
University of Sydney
2006 Sydney, New South Wales
Australia
*d.mills@physics.usyd.edu.au

David R. Mills* and Christopher J. Dey
School of Physics, A28
University of Sydney
2006 Sydney, New South Wales
Australia
*d.mills@physics.usyd.edu.au

About the Authors

Aliakbar Akbarzadeh

Aliakbar Akbarzadeh is Professor of Mechanical Engineering at RMIT University, and Associate Editor of the Journals *Applied Thermal Engineering* and *Solar Energy*. He has been working in the area of salinity-gradient solar ponds since 1976. His R&D work on solar ponds covers the physics of solar ponds, their stability, maintenance, and performance, as well as their applications as a source of low grade heat. He has lectured in thermal science and engineering for 30 years, and supervised over 20 Ph.D. students. He has published more than 50 journal and conference papers with one of these being chosen as the ASME Best Paper of the Year on solar energy (1993-1994). He is the author of two books. His research on industrial waste heat recovery led to his winning of an Australian National Energy Award in 1996. He believes that for the sustainability of our planet to become a reality we need to consume less, conserve more, and rely substantially on renewable energy.

John Andrews

Dr. John Andrews is a Research Fellow on Sustainable Energy in the Energy Conservation and Renewable Energy Group, School of Aerospace, Mechanical and Manufacturing Engineering, at RMIT University, Melbourne. He has been working in this field for over 25 years, both on research and development, and commercialisation and policy-related activities. He has written three books, 15 major reports, and 22 journal and conference papers. He played a pioneering role in assessing the potential of, developing suitable technologies for, and setting up an appropriate policy environment to encourage the utilisation of wind energy for electricity generation in Australia. He initiated the first thorough investigation of the wind energy potential in the state of Victoria while Policy Manager at the Victorian Solar Energy Council in 1984. The book he coauthored, *Living Better with Less: How to Cut Waste and Improve our Lives* (Penguin, 1981) was one of the first works to propose sustainable development and demonstrate what it would involve in the Australian context. More recently he has been the project manager and lead researcher in two major R&D projects conducted by RMIT in conjunction with industry partners: a three-year project funded by the Energy Research Development Corporation to develop, demonstrate and prepare for commercialisation heat pipe heat exchangers for industrial waste heat recovery; and a two-year project to develop and demonstrate solar ponds for industrial process heating supported by the Australian Greenhouse Office.

Kenji Araki

Dr. Kenji Araki is a senior researcher of Daido Steel Col, Ltd. and is currently a visiting researcher at the Toyota Technological Institute. Dr. Araki is researching concentrator modules which use III-V solar cells (NEDO project). In the past his research has involved silicon cell concentrator modules and the development of high efficiency high power light bulbs.

Peter H. Aurora

Peter H. Aurora got his bachelor's degree in Mechanical Engineering at the Universidad Nacional de Ingenieria (Lima, Peru) in December 2000. He obtained a MS. Eng. in Energy Engineering in December 2003 at University of Massachusetts Lowell. He has been a member of the PerUML Project of the University of Massachusetts Lowell (Huarmey, Peru) since 2001. During his time as an employee of RWE Schott Solar, Inc (RSS), he served as a member of the Product Development and Reliability group. Major research areas include fuel cell, hydrogen production using renewable energy, hydrogen storage and sustainable development. Currently is he a doctoral student at University of Michigan in the department of Mechanical Engineering.

Allen M. Barnett

Dr. Allen Barnett is senior policy fellow for solar power projects at the Center for Energy and Environmental Policy and professor of electrical and computer engineering at the University of Delaware. Prior to re-joining the university, he was President and CEO of AstroPower, for many years the world's largest independent manufacturer of solar electric power products. Dr. Barnett serves on the Advisory Board of the U.S. Department of Energy's National Center for Photovoltaics, the Board of the Solar Energy Industries Association (U.S.), and the IEEE Photovoltaic Specialist Conference Committee. He is a Fellow of the Institute of Electrical and Electronic Engineers, recipient of the IEEE William R. Cherry Award and recipient of the Karl W. Böer Solar Energy Award. He has authored or co-authored more than 200 papers and has been awarded 24 U.S. patents.

Nikolaus Benz

Dr. Nikolaus Benz is responsible for R&D in the Solar Thermal business unit of the SCHOTT Corporation. Over the last several years he has led the development of low temperature tube collectors and of an improved thermal receiver for parabolic trough collectors. Prior to working at SCHOTT he was a group manager for solar thermal activities at the Bavarian Centre for Applied Energy Research in Munich. He has a more than 15 years experience in development of components for solar applications, system analysis and testing. Nikolaus Benz has his Diploma and Ph.D. in Physics from the Ludwig-Maximilians-University of Munich in Germany.

Pavel P. Bezrukih

Dr. Pavel P. Bezrukikh is a Head of Department of Science and Technology at the Ministry of Fuel and Energy of Russia, Chairman of Russian Renewable Energy Committee. Dr. Bezrukikh has more than 25 years of experience in the field of Russian Renewable of Russia. He is an honorary power-engineering specialist of Russia. He developed a number of the state level programs for promotion renewable energy in Russia including "Development of Untraditional Power Engineering of Russia for 2001-2005".

He is the author of 160 papers and patents. Dr. Bezrukikh teaches students of the Moscow Power Engineering Institute (part time). He is co-author of the book "Resources and Efficiency of the Use of Renewable Sources of Energy in Russia"(Saint-Petersburg, Nauka, 2002). He is member of the International Solar Energy Society (ISES).

Daniel Blake

Daniel M. Blake, Ph.D. is a Principal Scientist at the National Renewal Energy Laboratory, NREL, Golden, Colorado. Prior to joining NREL (then SERI) in 1986 he held positions as Senior Scientist and Manager of the Chemical Process Development Group for ARCO Metals Company, Tucson, AZ, and Professor of Chemistry at the University of Texas at Arlington. He has a Ph.D. in chemistry from Washington State University and a B.S. in chemistry from Colorado State University. His areas of expertise include inorganic, photo, environmental, and materials chemistry; process development; and renewable resources and energy. He has directed projects on destruction of hazardous chemicals and microorganisms, new heat transfer and thermal storage fluids, materials for solar hydrogen production reactors, chemistry in liquid desiccant systems, and development of catalytic processes for renewable energy and environmental applications. He has been awarded two US patents on metal oxide production and a photocatalytic reactor design and has two patents pending on air cleaning systems; has published more than 60 papers and book chapters; and made over 100 presentations at conferences, symposia, and professional meetings.

Avi Brenmiller

Avi Brenmiller is CEO & President of Solel Solar Systems Ltd., and has extensive hands-on engineering experience. He worked at LUZ from 1988-91 on a wide range of engineering and operational projects. From 1991-94, he served as a member of the management team at Elisra Electronic Systems Ltd., a major Israeli electronics concern and part of Koor Industries. In 1994 Mr. Brenmiller joined Solel. Since his appointment as CEO in 1997, he has been the driving force behind the company's development of new industrial and consumer solar collectors and solutions, and has overseen Solel's expansion into the global market as well as the signing of strategic partnerships in Europe, the U.S. and the Far East. Mr. Brenmiller is widely renowned as an expert in solar thermal applications and is a frequent guest speaker at international energy conferences. Mr. Brenmiller studied at the Technikum, in Givataim, Israel, and holds a Degree in ME from Ohio State University and a certificate in business administration from the Israel Management Center.

Douglas Brosseau

Doug Brosseau is a Senior Member of Technical Staff at Sandia National Laboratories. Mr. Brosseau has been employed at Sandia since 1985 and is a degreed Mechanical Engineer. A relative newcomer to the concentrating solar power arena, he joined his colleagues at the Sandia National Solar Thermal Test Facility in Albuquerque, New Mexico about 2-1/2 years ago. Mr. Brosseau works closely with SunLab partners at the National Renewable Energy Laboratory on DOE-funded activities related to parabolic trough power plant technologies and R&D. He has been involved with molten salt studies and material durability testing of thermal energy storage components in molten salt operating environments and supports a broad range of testing activities at the test site. Mr. Brosseau has a broad range of prior experience at Sandia and was employed in the electric power industry prior to joining Sandia.

John Byrne

John Byrne is director of the Center for Energy and Environmental Policy (CEEP) and professor of energy and environmental policy in the Graduate School of Urban Affairs & Public Policy, University of Delaware. He is co-executive director of the Joint Institute for a Sustainable Energy and Environmental Future (headquartered in Seoul, South Korea), a founding member of *Solar Cities* (a pioneering program sponsored by the International Solar Energy Society), editor of the book series *Energy and Environmental Policy* (Transaction Publishers), associate editor of the *Bulletin of Science, Technology and Society*, and a contributing author to the second and third assessment reports of the United Nations-sponsored Intergovernmental Panel on Climate Change. His work has received support from UNDP, UNEP, the World Bank, the W. Alton Jones Foundation, the U.S. Department of Energy, the National Renewable Energy Laboratory, and the U.S. Environmental Protection Agency. He has published 12 books and over 150 articles.

Robert Cable

Robert Cable is currently Senior Engineering Analyst for Solargenix Energy with headquarters in Raleigh, NC. As such, he assists with the design, performance and cost modeling, and project management of solar thermal projects. Previously, Mr. Cable was a Performance Engineer for LUZ Engineering and KJC Operating Company at the SEGS plants located in the California Mojave desert, for a combined total of over 10 years. During this time, he developed in-depth knowledge of the engineering, performance, modeling, operation, and maintenance of large scale concentrating solar thermal power plants. Mr. Cable holds a B.S. in Mechanical Engineering from California State Polytechnic University, Pomona, 1990.

Gilbert Cohen

Gilbert E. Cohen is Vice President of Engineering & Operations at Solargenix Energy. Prior to working at Solargenix Mr. Cohen was the Technical Services Manager of KJC Operating Company, the operator of the most successful Solar Electric Generating Systems (SEGS) in the Mojave Desert (California). His services at the SEGS spanned more than 14 years. Mr. Cohen's interest in solar energy began 25 years ago as a staff engineer of the Israeli Scientific Research Foundation, where he was involved in various R&D activities. He has authored numerous technical papers on the development, operation and maintenance on the parabolic trough technology. Mr. Cohen is an active member of the Solar Energy Division of the American Society of Mechanical Engineers (ASME), and of the American Solar Energy Society (ASES), where he served two terms on the ASES Board of Directors. In 2002 Mr. Cohen received the prestigious ASES Hoyt Clarke Hottel Award, which honors individuals who have made significant contributions to technology in the solar energy field.

Manuel Pedro Ivens Collares-Pereira

Electrotechnical Engineer (IST- Technical University of Lisbon) , Ph.D. in Physics (Univ. of Chicago). Presently he his Coordinator of Research at INETI (National Institute for Engineering and Technology in Industry)- Lisbon, Renewable Energies Department and Professor in the Physics Department (IST, Technical University of Lisbon-Lisbon). He his also responsible for the R&D company SOLPOWER, Ltd in association with AO SOL, Energias Renováveis, Ltd, manufacturer of CPC type collectors, under a special agreement with INETI to that effect. Founder of SUN CO, Companhia de Energia Solar, S.A., producer of solar cookers. He has done extensive research in Solar Energy as a specialist in Optics and Thermodynamics, with a large number of publications in areas like Non Imaging Optics applied to Solar Thermal, Photovoltaics, Illumination and Photocatalysis, Solar Irradiation Statistics, Solar Thermal Systems Design and Engineering, Domestic Hot Water and Heating and Cooling Systems, Absorption and Adsorption Cooling Equipment, Solar Drying, Solar Ponds, Solar Cooking , Greenhouses Heating. He has also been founder and twice President of CCE- Centro para a Conservação de Energia (the then acting Portuguese Energy Agency), founder and twice President of SPES- Sociedade Portuguesa de Energia Solar, the Portuguese Section of ISES, and President of ACTD- Associação para a Ciência, Tecnologia e Desenvolvimento.

Fritz-Dieter Doenitz

Dr. Fritz-Dieter Doenitz is a Consultant for Solar Thermal Power Technology in Mitterteich, Bavaria (Germany). Up to his retirement in May 2004, he was the head of the Process & Product Development with Schott Rohrglas, the glass tubing division of the international SCHOTT group. In this function he was responsible for the planning and execution of the German PARASOL project for the development of a new generation of parabolic trough receivers. Dr. Doenitz is a physicist with experience in glass science and technology. He studied at the University of Jena, Thuringia (Germany) where he received a Ph.D. in Mineralogy. During the 1970-80 period he was engaged in materials research, leading a laboratory for new glass ceramics at the university. He started his industrial career with Schott Rohrglas in the early 1990's. Amongst other developments he initiated the qualification of traditional glass tubing for solar purposes. In 2004 Dr. Doenitz received the German Solar Award for his outstanding work in the field of evacuated solar receivers.

John J. Duffy

John Duffy is a professor in the Mechanical Engineering Department of the University of Massachusetts Lowell, the coordinator of the graduate program in solar engineering, and director of the Center for Sustainable Energy. He has over 70 papers in engineering, environmental science, and engineering education. In recent years he has focused on engineering, installing, and testing solar systems appropriate in rural areas in developing countries.

Robert Fimbres

Robert Fimbres is the senior operations manager for SEGS III-VII located in Kramer Junction, California. He has over 18 years of operational experience with parabolic trough collectors, and has participated in a number of R&D projects aimed at improving solar field and power block performance sponsored by Sandia National Laboratories, NREL, and KJC Operating Company. He has done extensive monitoring and testing of key components necessary for parabolic trough SEGS operation, which resulted in important improvements in the quality and durability of solar field components. Mr. Fimbres has contributed to a number of publications focused on the cost reduction of SEGS operation and new developments in parabolic trough technology.

Scott Frier

Scott Frier, Executive Vice President – Operations and Maintenance of Solargenix Energy, has long experience with parabolic trough power projects. As KJC Operating Company's Chief Operating Officer, he was primarily responsible for Company operations since 1992, heading the operation of 150 MW of trough SEGS (Solar Electricity Generation Systems) facilities. Previous to that, he was the General Manager of the 44 MW Daggett SEGS facility, comprised of the SEGS I and II ... the world's first utility-scale, commercial solar thermal power plant. Mr. Frier became acquainted with solar energy when he joined LUZ Engineering (a subsidiary of LUZ Industries International, the developer of the existing nine SEGS) in 1985, after having come from a construction management background in the petrochemical industry.

Andrey Ganopolski

Dr. Andrey Ganopolski is senior research scientist at the Potsdam Institute for Climate Impact Research. Dr. Ganopolski is an expert in the field of climate modeling, past climate changes and future climate predictions. He is the author of 40 papers in prestigious scientific journals, including Nature and Science, four book chapters, and a contributing author of the Third Assessment Report of the Intergovernmental Panel on Climate Change (IPCC). He has given invited lectures on climate change and past climate modeling at the discussion meeting of the Royal Society and at numerous Universities and research centers.

Randy Gee

Randy Gee is the Director of Research, Development and Deployment for Solargenix Energy LLC, the major U.S. developer of parabolic trough systems. He has worked in the field of solar thermal energy for over twenty-five years. Mr. Gee's solar career began as a mechanical engineer with Honeywell's Energy Resources Center, and he then joined the Solar Energy Research Institute (now NREL) as a research engineer. Mr. Gee co-founded Industrial Solar Technology, a small business that developed and commercialized a parabolic trough solar collector system for process heat applications. Now as Solargenix Energy R, D & D Director, he continues his focus on technology advancement of solar thermal technology. Mr. Gee is a registered Professional Engineer in Colorado, Illinois and Arizona.

Michael Geyer

Dr.-Ing. Michael Geyer obtained his Physics Diploma at the University of Tübingen in 1981 and holds a Ph. D. in Mechanical Engineering from the University of Essen. Since 1981 his professional activities have been dedicated to the concentrating solar power systems, including positions as researcher and deputy director at the German Aerospace Agency (DLR) departments in Stuttgart and PSA (1981-1989), Manager for System Engineering, Research and Development at Flachglas Solartechnik and ABB (1989-1993), Professor for Energy-, Power Plant Technology and Process Technology at the Polytechnic University of Regensburg (1993-1995) and Head of DLR's Division at PSA from 1995 until 2001. Since July 2001 he is the Director of the Spanish Office of Flagsol GmbH and responsible for the solar thermal project development of the Solar Millennium Group in Spain. In September 2000 he was elected as the new

Executive Secretary of the IEA SolarPACES Implementing Agreement, which represents the international Solar Thermal Community of R&D institutions. He is author or co-author of more than 60 publications in the mentioned fields, including technical books, journals, conference papers, studies and reports.

Peter Golding

Peter Golding is Associate Professor of Metallurgical & Materials Engineering at The University of Texas at El Paso. Peter has served on various national committees of the Australian and New Zealand Solar Energy Society [ANZSES], including Victorian Chair and Editor of *Solar Progress*, the Journal of the ANZSES, Chaired the Ocean Thermal Energy Conversion & Solar Ponds Technical Committee of the Solar Energy Division of the American Society of Mechanical Engineers [ASME], been a member of the International Solar Energy Society, the American Solar Energy Society, and the American Society of Heating, Refrigeration & Air Conditioning Engineers [ASHRAE]. He has published articles on solar pond technologies and applications in journals and conference proceedings, co-authored an authoritative textbook on solar ponds, and led technical program committees for three international conferences and workshops on solar pond science and engineering.

Carin Gummo

Carin Gummo is currently the performance engineer for KJC Operating Company, providing operations and maintenance service to the five SEGS plants (III-VII) located in Kramer Junction, Calif. In 2001, she graduated from California Polytechnic State University in San Luis Obispo with a B.S. in Mechanical Engineering, and holds a P.E. license in the state of California. While her tenure with parabolic trough technology has been brief, she has worked extensively with plant modeling and co-authored two papers on new developments in solar field component technology.

Mary Jane Hale

Mary Jane Hale is a Senior Engineer at the National Renewable Energy Laboratory, Golden, Colorado, where she has worked since 1991. She has extensive experience in performance and economic analysis for solar industrial and solar thermal electric technologies, including solar absorption cooling, molten-salt central receiver and parabolic-trough technologies. Ms. Hale has authored or coauthored approximately 15 papers for professional journals and for conferences. She also has served two terms as chair of the American Society of Mechanical Engineer's Solar Energy Division Testing and Measurement Committee. Ms. Hale holds M.S. and B.S. degrees in Mechanical Engineering from the University of Utah, and is a licensed Professional Engineer in Colorado.

Yoshihiro Hamakawa

Dr. Yoshihiro Hamakawa is currently an Advisor Professor to the Chancellor and was a Professor of the Department of Photonics in the Ritsumeikan University and the Vice Chancellor of the Ritsumeikan 1998-2000. He is also an Emeritus Professor of Osaka University since 1996. He had been with the Osaka University since his completion of Ph.D. since 1964. He has performed a number of remarkable research achievements in the fields of Semiconductor Physics, Optoelectronics, and Solar Photovoltaic Conversion, particularly famous for the development of "modulation spectroscopy" which now becomes a useful tool for the characterization of band structure of solids, inventions of "the Multi-band gap Tandem type Solar Cells" (US Patent 4,271,328, March 20, 1979 and Japanese Patent Sho-54-32993 for example), success of Valency Electron Control of Amorphous Silicon Carbide Films and their applications to High Efficiency Solar Cells etc.. He has authored 50 books, 540 papers, 182 invited review papers and 78 Patents.

In addition to these scientific activities, Professor Hamakawa has made many key contributions to the promotion of PV technology developments around the world. He has been as a director of the International Solar Energy Society (ISES) from 1981-1983, a member of the International Advisory Committee of the IEEE PVSC since 1978, and the EU-PVSEC since 1981. He organized the Photovoltaic Science and Engineering Conference (PVSEC) in 1984 as the Chairman of the PVSEC-1. He has also been an active leader in the industrial development of PV solar energy as an initiator of the Sunshine Project, and has been a Board Member for the Sunshine Project since 1973. He has served the Chairman of the Solar Energy Division in the New Energy Section of the Industrial Development Council, AIST, MITI Japan. For his distinguished services, he received IEEE W. Cherry Award in 1994, 1995 Purple Ribbon Prize from Japanese Emperor (Heisei), 1st WCPEC Award (First World Conference on Photovoltaic Energy Conversion Award) in 2003, and 11 other awards.

Ulf Herrmann

Dr. Ulf Herrmann is the Chief Engineer at FLAGSOL GmbH in Cologne, Germany. FLAGSOL GmbH is an engineering company for the development, planning, and construction of solar power plants. He has a broad background in the development and evaluation of numerous aspects of solar thermal technology. Within FLAGSOL, Dr. Herrmann is responsible for the design and engineering of parabolic trough solar fields and solar boilers. He holds a Ph.D. in Mechanical Engineering from the University of Aachen.

Kristen Hughes

Kristen Hughes is a research associate and Ph.D. candidate at the Center for Energy and Environmental Policy (CEEP), University of Delaware. She received her master's and baccalaureate degrees from the University of Texas at Austin. Her research interests include renewable energy, electricity restructuring, and comparative studies of U.S. and European energy policy. She has previously published papers on U.S. policy initiatives to promote international sustainable development. Prior to joining CEEP, she held positions in state (Texas) and federal government (including work for the White House) in the U.S.

David Kearney

Dr. David Kearney consults internationally in the commercial development and project implementation of solar thermal systems, focusing on parabolic trough solar electric power plants and R&D projects to advance the technology. He has over forty years mechanical engineering experience in the fields of thermal and power engineering, with 25 years specialization in solar and other renewable and energy-efficient thermal energy systems. Dr. Kearney has led or been a key member on solar thermal electric power plant evaluations for New Mexico, California, Mexico, Hawaii, India, Spain, Morocco, and Crete. Manager of the Solar Thermal and Wind Division at SERI in its early days (now NREL), he also was U.S. VP of Advanced Technology with LUZ International Ltd., which developed and operated 354 MWe of solar parabolic trough plants in California in the late 1980's. Dr. Kearney holds a B.S. from the Univ. of Rochester (1959), an M.S. from Union College (1966), and a Ph.D. from Stanford University (1970), all in Mechanical Engineering.

Bruce Kelly

Bruce Kelly is a Senior Consultant in the Energy Delivery and Management Group at Nexant, Inc. in San Francisco, California. Mr. Kelly has spent 25 years in the technical and economic analysis of solar thermal technologies for electric power generation. Recent activities include the final design and startup of the 10 MWe Solar Two nitrate salt central receiver project in Barstow, California, the preliminary design of the 880 MWht nitrate salt thermal storage system for the 55 MWe AndaSol parabolic trough project in Spain, and the preliminary design for a 27 MWht thermocline storage system for the 1 MWe parabolic trough project sponsored by Arizona Public Service Company. Mr. Kelly has an MS and a BS in Mechanical Engineering from the University of California at Berkeley, and is a registered Mechanical Engineer in California.

Cheryl Kennedy

Cheryl Kennedy joined the National Renewable Energy Laboratory (NREL) in 1987, and has led the Advanced Materials Team since 2003. The team develops advanced reflector and absorber materials for use in low-cost, high-performance, and high-reliability systems that use concentrated sunlight to generate power, with an emphasis on large multi-megawatt parabolic trough systems and smaller kilowatt-scale concentrating photovoltaic systems. The work includes collaboration with solar manufacturers and interactions with the coatings industry. Ms. Kennedy is engaged in testing and analyzing the performance and durability of optical materials, thin-film optical modeling, and thin-film deposition of reflective and barrier coatings. She has received two patents in these disciplines. Ms. Kennedy holds a B.A. Physics, University of Colorado at Denver, a B.S. Chemistry from Mary Washington, and an M.S. Material Science from the Colorado School of Mines.

Lado Kurdgelashvili

Lado Kurdgelashvili is a research associate and Ph.D. candidate at the Center for Energy and Environmental Policy (CEEP), University of Delaware. He interned during summer 2003 at the Washington DC Projects Office of the Lawrence Berkeley National Laboratory. He received an MS in Renewable Energy and Environment from the University of Reading, UK, and an MS in Environmental Sciences and Policy from Central European University. Prior to joining CEEP's doctoral program, he served as assistant to the president of Georgian Gas International Corporation (GIC), and was a consultant for GeoEngineering (Tbilisi, Georgia).

Eckhard Lüpfert

Dr. Eckhard Lüpfert, located at the German Aerospace Center (DLR) at Plataforma Solar de Almeria in Spain, is research area manager for Qualification of Solar Thermal Systems in the DLR Institute of Technical Thermodynamics. He has been on the DLR staff in Cologne and Almeria since 1992, working on scientific projects in the area of solar parabolic troughs, solar tower systems, and solar chemistry. He has broad experience in technology development in these areas, as well as planning, management and controlling of R&D projects in the field of concentrating solar thermal power technologies, and dissemination and education activities regarding the application of solar thermal energy. Dr. Lüpfert holds a degree as Mechanical Engineer (Energy Technology) and a Ph.D. from the Technical University of Aachen, Germany.

Paul MacCready

Dr. Paul MacCready has a B.S., Physics, Yale University 1947, and a Ph.D., Aeronautics, California Institute of Technology, 1952. His professional activities have covered various fields, and five of his air and ground vehicles have been acquired by the Smithsonian Institution. The last two decades he has emphasized efficient use of energy, especially with the energy not being generated by the consumption of oil, natural gas, or coal, all of which produce CO2. AeroVironment's products, of which he has always been a part, include in air: the giant solar-powered Helios which reached

96,863' (2 miles higher than any other plane has ever achieved continuous flight), and many small, battery-powered drones; on land: the GM Sunraycer and IMPACT car which became GM's EV-1; in water: various vehicles that operate on or under water; plus quick-charge electric forklifts and airport vehicles. He is a member of the Natural Academy of Engineering, the American Academy of Arts & Sciences, and the American Philosophical Society, and has fifteen patents and five honorary degrees. He has received over 40 prestigious awards, including the Collier Trophy from NAA, the Reed Aeronautical Award from AIAA, the Ingenieur of the Century Gold Medal from ASME, the Lindbergh Award, Time Magazine's "The Century's Greatest Minds", the Chrysler Award for Innovation in Design, the Guggenheim Medal from AIAA, SAE, and ASME, and in 2003 the Heinz Award for Technology, Economy and Employment, and the Bower Award and Prize in Scientific Achievement from The Franklin Institute.

A. Roderick Mahoney

A. Roderick (Rod) Mahoney is a Principal Member of Technical Staff at Sandia National Laboratories in Albuquerque, New Mexico. Mr. Mahoney's current assignment is as Project Leader in the Alternating Current (AC) Laboratory at the Primary Standards Laboratory (PSL) at Sandia, where his responsibilities include pulsed high voltage, pulsed current, and impedance measurements and consultation for Nuclear Weapons Complex customers. Prior to the PSL assignment he was the lead parabolic trough thermal performance measurement engineer at the National Solar Thermal Test Facility, located at Sandia, Concentrating Solar Power liaison for technical support and consultation to the SEGS plants in California, and the DOE Solar Buildings program manager at Sandia. In addition to these activities, Mr. Mahoney is a recognized expert in the field of optical/thermal radiative properties and solar materials R&D, and holds several patents for both environmental protective coatings and solar absorber materials. Mr. Mahoney has a BS in Mechanical Engineering from the University of New Mexico and an AS in Electrical Engineering from Kansas State University – Salina.

Eli Mandelberg

Eli Mandelberg is the Executive Vice President, Engineering and Solar Products of Solel Solar Systems Ltd. He earned a B.S. in Mechanical Engineering from Tel-Aviv University, where he also undertook additional Business Management studies. Mr. Mandelberg served from 1981-86 as a scientist and engineer of the Photovoltaic R&D group in Tel - Aviv University. Joining LUZ Industries Israel in 1987, he was the Project Manager of the solar selective coating line from 1987-89, and from 1989-91 served as the head of Product Engineering. He was an independent consultant in energy and thin-film coating engineering from 1991-96, and joined Solel in 1996.

E. Kenneth May

Kenneth May is the President and co-founder of Industrial Solar Technology Corporation. IST is a leading designer, manufacturer and installer of parabolic trough solar collector systems for large-scale applications, operating trough systems throughout the US and several countries overseas. Prior to starting IST in 1983, Mr. May was employed as a senior engineer at the Solar Energy Research Institute (now NREL). Before that, Mr. May worked for UOP Inc., commissioning petroleum and petrochemical plants throughout the world. Mr. May received the ASES Hoyt Clarke Hottel Award 2000 for his "unique contribution to make solar thermal technology a commercial reality". Mr. May has an MS in Chemical Engineering from the University of Colorado, Boulder, and a BS in ChEng from Imperial College of Science, Technology and Medicine, University of London, England and is a registered professional engineer in Colorado.

Luc Moens

Dr. Luc Moens is a senior scientist at the National Renewable Energy Laboratory in Golden, Colorado. His work within the DOE Concentrating Solar Power program focuses on the development of advanced heat transfer fluids that have extremely high thermal stability under very high temperatures, and that can serve as new thermal storage media for use in next-generation solar parabolic trough systems for power production. Dr. Moens received a Ph.D. in synthetic organic chemistry at the University of California, Santa Barbara, and a BS in Agrochemical Engineering at the State University of Ghent, Belgium.

Paul Nava

Paul Nava is Managing Director of FLAGSOL GmbH in Cologne, Germany. FLAGSOL has designed the AndaSol parabolic trough plants in southern Spain that include 510,120 m² of SKAL-ET solar field and 8 hours of molten salt storage. FLAGSOL is responsible for engineering of the solar thermal power plants and delivery of key components for the solar field. Mr. Nava has been active in parabolic trough technology since 1987, when the German mirror supplier FLABEG and LUZ Industries Israel started joint project development activities. He has led a number of technical and economic studies for solar thermal parabolic trough collector power projects in Brazil, United States, Israel, Spain, Morocco, Greece, Namibia, and Iran. Mr. Nava is a graduated engineer for nuclear power plants (State College of Engineering at Juelich), and holds an M.S. degree in computer science from the University of Bonn.

Robert Pitz-Paal

Dr. Robert Pitz-Paal is head of the Solar Research Division at the German Aerospace Centre supervising a staff of approximately 60 people. In addition he is Professor for Solar Technology at the Technical University in Aachen, Germany. He has been working as scientist for more than 15 years in the field of concentrating solar technologies. He serves as the ASME associate editor on solar thermal power for the International Journal of Solar Energy Engineering. He was the Task III operating agent and currently is the national ExCo Member for the IEA SolarPACES implementing agreement. Robert Pitz-Paal has a Diploma in Physics from the Ludwig Maximillians University in Munich, Germany and a Ph.D. in Mechanical Engineering from the Ruhr University in Bochum, Germany.

Nicholas Potrovitza

Nicholas Potrovitza is the Technical Services Manager for KJC Operating Company, a leading provider of maintenance and operation services for the parabolic-trough based Solar Energy Generating System (SEGS) projects located at Kramer Junction in California. He has eighteen years of experience in all phases of concentrated solar power plant development from feasibility study through design, construction, start-up, operation and maintenance. Mr. Potrovitza has an MS degree in Quality Assurance from California State University (Dominguez Hills) and a BS in Mechanical Engineering from Polytechnic University, Timisoara, Romania.

Henry Price

Henry Price is a senior engineer at the National Renewable Energy Laboratory in Golden, Colorado. A recognized expert in the field of parabolic trough technology and applications, he is the DOE project manager for parabolic-trough R&D and is responsible for directing efforts at NREL, Sandia National Laboratory, and contracted R&D with industry. Mr. Price is a systems analyst with experience modeling the cost and performance of Concentrating Solar Power technologies. Prior to working at NREL he was the lead performance engineer for the LUZ SEGS plants in the California Mojave Desert. Mr. Price has an MS in Engineering from the University of Wisconsin – Madison, and a BS in Environmental Resource Engineering from Humboldt State University in California, and is a registered mechanical engineer in California.

Tadashi Saitoh

Dr. Tadashi Saitoh received BE and ME degrees from Hokkaido University, Sapporo, in 1962 and 1964. In 1978, he received a Doctoral Degree in electrical engineering from Osaka University. He was Professor of Electrical and Electronics Engineering at Tokyo University of Agriculture and Technology (1993-2004) and was chair of the department (1995-1996). Before he moved to the university in 1989, he had worked with Central Research Laboratory, Hitachi, Ltd. for 24 years. Since 1973, he has been involved in solar cell research using crystalline Si and InP semiconductors. In particular, from 1982, he has played a leading role in the research and development of high-efficiency, low-cost crystalline silicon solar cells in Japan.. His current interests are directed to research on gettering process of lifetime killer impurities for single and multi-crystalline Si solar cells, development of high-quality multicrystalline Si ingots and suppression of light-induced lifetime degradation in CZ Si solar cells. He is the author of about 200 papers in various journals and international conferences. He organized several international workshops on crystalline Si materials and solar cells and also chaired international conferences, sessions and workshops in photovoltaics. He is the IEEE Fellow and Advisory member of International conferences.

Matheos Santamouris

Mat Santamouris is an Associate Professor of Energy Physics at the University of Athens. He is Associate Editor of the Solar Energy Journal and Member of the Editorial Board of the International Journal of Solar Energy and the Journal of Ventilation. He is Editor of the Series of Book on Buildings, Energy and Solar Technologies published by James and James Science Publishers in London. He has published 9 books on topics related to solar energy and energy conservation in buildings. He is guest editor of six special issues of various scientific journals. He has coordinated many international research programs and he is author of almost 120 scientific papers published in international scientific journals. He is visiting professor at the Metropolitan University of London.

Dmitry S. Strebkov

Dmitry S. Strebkov is a Director of the All Russian Research Institute for Electrification of Agriculture. He is an academician of the Russian Academy of Agricultural Sciences, Professor of Chair "Fundamentals of Radio Engineering and Television", Chair-holder of UNESCO Chair "Renewable Energy and Rural Electrification", Chair-holder of the Chair at Moscow Agricultural Engineering University "Renewable Energy and Electrification of Agriculture". His public position is a Chairman of ISES—Russia and Deputy Chairman of Russian Renewable Energy Committee.

He graduated from Lomonosov Moscow State University, Department of Mathematics and from the Moscow Agricultural Engineering University, Department of Electrical Engineering for Rural Electrification. He is a Doctor of Science (Technology). Professor Strebkov is an expert in photovoltaics and renewable energy utilization in agriculture. He is the author of 300 papers and 150 inventions, including 16 U.S. patents. He is a member of the Board of Editors of the Journal of Solar Engineering, published in Tashkent Uzbekistan, Journal "Agricultural Engineering," Czech Republic, Journal "Electric Power Sources", Belgrade, Serbia, "Renewable Energy" (Moscow), "Proceedings of Russian Academy of Agricultural Science" (Moscow).

A. I. Budsko gold medal award of Russian Academy of Agriculture for the achievements in solar cells and solar PV energy conversion, 2003.

Tatsuya Takamoto

Dr. Tatsuya Takamoto, a researcher of the solar systems development center, joined SHARP Corporation in 2001 after working for 14 years at Japan Energy Corporation on III-V solar cells including InP homo-junction cells, CdTe thin film cells and InGaP/GaAs tandem cells. He is responsible for developing high efficiency multijunction cells for space use and terrestrial concentrator applications. Dr. Takamoto received his B.Sc.in Material Physics from Tsukuba University, Tsukuba, Japan in 1987 and his Ph.D. in Material Engineering from Toyota Technological Institute, Nagoya, Japan in 1999.

Ranier Tamme

Dr. Rainer Tamme is head of the Thermal Process Technology division at the Institute of Technical Thermodynamics of the German Aerospace Center in Stuttgart, Germany. He is a recognized expert in the field of thermal energy storage, thermo-chemical conversion and fuel processing technologies and has many years of experience in managing and coordinating numerous German and European projects involving a wide range of industrial companies and research institutes. His work in energy storage has included development, manufacturing and testing of thermal and chemical energy storage materials, and modelling and design of TES systems for new power generation systems and for the industrial process heat sector. Dr. Tamme is a Diplom Chemiker in chemistry and chemical engineering from the Technische University Hannover, and has a Ph.D. in solid state and inorganic chemistry. Prior to DLR, he worked 7 years as assistant professor at University of Hannover.

Aaron R. Thurlow

Aaron Thurlow is a research associate and master's student at the Center for Energy and Environmental Policy, University of Delaware. He received his BA in Urban Studies and Planning from the University of California, San Diego. His current research is focused on an analysis of state renewable portfolio standards in the U.S., which he presented at the American Solar Energy Society conference in July 2004 in Portland, Oregon.

Igor I. Tyukhov

Dr. Igor I. Tyukhov is Deputy Chair-holder UNESCO Chair "Renewable Energy and Rural Electrification" at the All-Russian Research Institute for Electrification of Agriculture, a docent (Assoc. Prof.) of the V. A Fabricant Physics Department at the Moscow Power Engineering Institute, Deputy-chair Holder of the Chair at the Moscow Agricultural Engineering University "Renewable Energy and Electrification of Agriculture", and member of the International Solar Energy Society (ISES).

He graduated from the Moscow Power Engineering Institute, Physics Department. He is kandidat tekhnicheskikh nauk (Ph. D). Dr. Tyukhov was visiting scholar at George Mason University (JFDP, 1999/2000 academic year), at the University of Oregon and at the Oregon Institute of Technology (Fulbright Program, 2002/2003 academic year). Dr. Tyukhov is an expert in the field of photovoltaics and renewable energy. He is the author more than 100 papers (including published in international conference proceedings) and inventions.

Masafumi Yamaguchi

Dr. Yamaguchi is a Principal Professor at Toyota Technological Institute, where his research into high efficiency 3-junction PV cells has resulted with world record efficiencies. He has also contributed in developing high-efficiency outdoor 3-junction solar cell modules in collaboration with industry. During the 1960-80s, Dr. Yamaguchi's research on blue light emitting diodes, III-V compound solar cells and materials and photonic functional devices, found superior radiation-resistance of InP materials and solar cells, and their great potential for space applications. His group also proposed a double-hetero structure tunnel junction for realizing a high performance and stable multi-junction cell interconnection. Dr. Yamaguchi has served as a chairman at three of the World Conference on Photovoltaic Energy Conversion (WCPEC). He received the Vacuum Science Paper Award for the study of Si MBE in 1981, the Irving Weinberg Award for contributions to Space Photovoltaics in 1997, and the Alexander Becquerel Prize for contribution to development of super-high-efficiency multi-junction and concentrator cells in 2004. Yamaguchi received the B.S. and Ph.D. degrees from Hokkaido University in 1968 and 1978, respectively.

Eduardo Zarza

Eduardo Zarza is an industrial engineer with a Ph.D. from the University of Seville (Spain) and 20 years experience in parabolic-trough collector solar systems. He is a member of the scientific staff of the Plataforma Solar de Almería (PSA), and leader of the PSA medium temperature solar energy applications R&D Unit. From 1996 through 2001, he was the Director of the European DISS (Direct Solar Steam) project which investigated the feasibility of direct steam generation in the absorber pipes of parabolic-trough collectors. Since 1985, he has actively participated in many international projects dealing with the parabolic-trough collector technology.

Chapter 1

Anthropogenic Global Warming: Evidence, Predictions and Consequences

by

Andrey Ganopolski
Potsdam Institute for Climate Impact Research,
PO Box 601203, Potsdam 14412, Germany

Abstract

The question of whether climate is changing under the influence of human activity, and whether these changes will cause serious negative socio-economic impact in the foreseeable future is of fundamental importance. The general consensus among the vast majority of scientists studying different aspects of climate is summarized in the recent IPCC report. The report states that the rate of change in the climate has been unusual during the past century compared to that of the last millennium. It is highly probable that the main cause of these changes can be attributed to anthropogenic activity, primarily to the emission of carbon dioxide and other greenhouse gases. There is a growing body of evidence supporting statistically significant trends of different climate characteristics on the global and regional scales, which are consistent with model simulations of human-induced global warming. Although considerable uncertainties remain in prediction of future climate change both due to uncertainties in future economic development and insufficient understanding of climate processes, the majority of climate experts believe, that the problem of global warming is a real one, and it deserves serious attention.

At the same time, the issue of global warming has frequently been attacked in recent years by opponents, who have pointed to controversy in the reconstruction of recent climate changes and apparent discrepancies between observed and simulated climate trends. In particular, claims have been made that observed climate changes could be entirely explained by variations in solar activity, and thus anthropogenic influence plays no role in climate change. Although a number of problems still remain to be

---

* Corresponding Author. E-mail: andrey@pik-potsdam.de

tackled, many of these apparent discrepancies were either found to be inconclusive or have been successfully resolved by invoking previously overlooked processes. It is now generally accepted that variation of solar insolation on decadal to centennial time scales is indeed an important factor of climate variability, but comprehensive analysis of climate data and results of model simulations suggest that it is extremely unlikely that climate change during the 20th century could be solely attributed to the natural factors.

Apart from relatively linear trends in temperature and other climate characteristics, which can be expected as a response to a gradual build up of greenhouse gases, there is a growing concern among climate experts that the response of the climate system to the anthropogenic influence could be strongly non-linear. Results of some model simulations suggest that crossing some currently unknown thresholds could trigger dramatic and possibly irreversible climate change.

Among such processes are the destabilizations of the West Antarctic Ice Sheet and the shutdown of the Atlantic thermohaline circulation. Paleoclimatic records present numerous examples of rapid climate changes in the past, though they cannot be considered as a direct analogue for future climate changes. So far it is impossible to quantify the probability of rapid climate changes, but even if the probability of such events is low, they deserve serious attention due to their extremely high potential socio-economic impact.

1. Introduction

Very few scientific ideas attract as much public attention as the concept of anthropogenic global warming. The problem of global warming is in the center not only of scientific, but also of public and political debates, especially in relation to the ratification of the Kyoto protocol. There are a number of questions, which proponents and opponents of immediate actions to mitigate global warming answer in opposite ways. Do we have reliable evidences of considerable increase of temperature during the 20th century? Is the rate of observed climate change unusual for the recent climate history? Can we attribute these changes to the anthropogenic emission of carbon dioxide and other greenhouse gases with sufficient degree of confidence? How confident are we that future climate changes will have a large socio-economic impact, which would justify a considerable cost of mitigation of greenhouse gases emission?

While proponents of climate protection are confident that they have scientifically sound answers to all these questions, the opponents, or so-called "climate skeptics," on the contrary believe that there is no reason to worry about climate change. Such different opinions reflect not only different political or corporate interests, but also the enormous complexity of the Earth system and insufficient knowledge about many important processes and mechanisms.

Nonetheless, it is undeniable that during the past decade considerable progress has been made in the understanding of different aspects of the climate change problem, and the growing body of evidence supports the concept of anthropogenic climate change. The comprehensive analysis of all available sets of historic observations provides new evidences of internally consistent climate trends over the 20th century. Longer reconstructions of climate history emerged, showing that the magnitude of the 20th century warming is likely to be the largest of any of the centuries during the last millennium, and the 1990s were the warmest decade of the 20th century. Positive temperature trends are observed not only in surface air temperatures, but also in sea surface and subsurface ocean temperature, and the heat content of the ocean steadily increased during the past decades. Positive precipitation trends have been recorded in many regions and discharge of some rivers has increased considerably. The shrinking of snow and ice cover areas in the Northern Hemisphere, and the retreat of the majority of glaciers worldwide also reflect the effect of global warming.

Still some controversy remains. In particular, over the past two decades the balloon and satellite records show significantly smaller lower-tropospheric warming than that observed at the surface. Another apparent discrepancy is the absence of warming over the large part of Antarctica, while climate models predict that warming in Antarctica should be larger than the global mean. These problems were analysed in a number of recent papers, and some apparent contradictions have been resolved. It is also important to realise that some observed time series are still too short and too contaminated by natural variability to detect climate changes with confidence.

Based on the thorough review of all available climate data a large group of climate experts stated in the Third Assessment Report of the Intergovernmental Panel on Climate Change (IPCC, 2001) that although the unequivocal attribution of observed climate changes to the anthropogenic influence is still not possible, it is highly unlikely that these changes can be explained solely by the natural causes. At the same time, the opponents of global change argue that an additional increase of CO_2 concentration in the atmosphere due to fossil fuels combustion will add a little to the natural greenhouse effect, and therefore will have a minor climate impact. A number of claims have been made that both recent and past climate changes are results of solar variability, and the greenhouse effect of carbon dioxide is not important for climate. This conclusion was solely based on a relatively good correlation between solar variability and climate trends, but a more scrupulous analysis shows that this correlation did not hold for the recent decades. This implies that although natural factors might have played an important role during the first half of the 20th century, the acceleration of global warming since 1950s would be impossible to explain without invoking the anthropogenic factors.

Much more is known now about past climate changes. Reliable paleoclimatic records reveal numerous abrupt climate changes in the recent history of climate, thus providing important evidence of the strongly non-linear behaviour of the climate system. Whether such abrupt past climate changes can be considered as a direct analogue for the future climate changes is unclear, but the fact that the climate system can shift from one state to another within just several years, clearly indicates that more attention should be paid to the possibility of "climate surprises". The experts in climate impact analysis believe that abrupt climate changes could have even more devastating socio-economic consequences than gradual changes because the latter are more predictable and this is why they are more easy to cope with. Moreover, it is possible that abrupt climate changes could be irreversible, whic implies that after the crossing of some threshold no measures to combat global warming will reverse a chain of dangerous events.

Considerable progress has been made in recent years in the development and the application of climate models of a growing degree of complexity. These models have been extensively validated against the present climate state and were applied to simulate past climate changes. Compared to previous generations of climate models, the newest models simulate both the mean climate state and the natural climate variability much more accurately. Moreover, thanks to increased spatial resolution and improved description of climate processes, the growing number of climate models no longer employ unphysical corrections of the fluxes between the ocean and the atmosphere to simulate realistic present day climate. The necessity of such correction in the previous generation of climate models has often been used by critics to state that results of climate models are unreliable. The new generation of climate models confirms in all major aspects the results of previous climate change predictions, thus proving that global warming is not an artifact of the previous generation of climate models.

At the same time, in spite of improved skill in simulation of the present day climate, it was not possible to considerably narrow the uncertainty range of future climate projections. Different models still simulate quite different global and regional climate response to the changes in the concentration of greenhouse gases. This is why considerable efforts are directed towards a better understanding of the reasons for discrepancies between different climate models, and to provide potential users of climate change scenarios with the probabilistic assessment of the range of possible uncertainties in climate predictions. It also becomes apparent that accurate simulations of climate change require invoking additional components and processes, such as biosphere and biogeochemistry, and their interaction with the anthropogenic component. The first step in this direction has been made already in the form of so-called integrated models of climate change and considerable further progress can be expected over the next decades.

2. Fundamentals of Climate Change

2.1 Climate Change

Although the word "climate" is broadly used in common language, it is difficult to give a strict scientific definition of climate. It is equally problematic to find the precise meaning and a clear distinction between terms "climate variability" and "climate change". These problems reflect the enormous complexity of the climate system and our limited knowledge about its internal dynamics.

Initially the term climate referred to the atmosphere only and meant the averaged weather conditions, such as the mean July temperature or annual precipitation. For many applications it is important to know not only the averaged characteristics, like monthly averaged temperature, but also different measures of variability (statistics), like the interannual variability of the temperature.

More recently the term climate has been extended to the whole climate system, which includes the atmosphere, ocean, land surface and the cryosphere. This is why climate is characterized more accurately as "the statistical description (of the climate system) in terms of the mean and variability of relevant quantities" (IPCC, 2001). In this sense, climate is a comprehensive statistical description of the given climatic state.

Gradual evolution of the climate state is named climate change. It is known that climate has changed in the past and will change in the future under the influence of natural factors such as solar luminosity, orbital parameters of the Earth, volcanoes, and many others. The term "anthropogenic climate change" refers to the part of climate change attributed to all aspect of human activity, such as emission of greenhouse gases and aerosols, land cover changes and others.

The climate change observed during the 20th century is a combination of natural and anthropogenic climate changes, and because the magnitude of natural variability is not accurately known, it is difficult to separate natural and anthropogenic components of the climate change with confidence. However, if the model predictions of the future climate changes are correct, the anthropogenic signal will soon considerably exceed the level of natural variability, and the discussions about the unequivocal attribution of climate change will cease.

2.2 Greenhouse Effect

The idea that the growing concentration of greenhouse gases in the earth's atmosphere will cause global warming is far from being new. In the 1860s British scientist John Tyndall discovered the greenhouse effect and proposed that carbon dioxide plays an important role in climate. Several years later Swedish scientist Svante Arrhenius calculated the warming effect of the doubling of CO_2 concentration quite accurately. However, it was not until direct measurements confirmed the steady growth of CO_2 concentration in the atmosphere and the first computer models of climate were developed in the 1960s that a concern about possible anthropogenic impact on climate gained a solid scientific basis.

Still, in the 1970s when global warming made a pause, possible global cooling due to an increase of atmospheric aerosols, and potential climate consequences of nuclear war ("nuclear winter"), were the dominant concern in the climate community. Even now some scientists continue to claim that the Earth is on the verge of a new ice age and thus global warming is welcome.

The situation changed in the 1980s when global warming resumed, and climate models were considerably improved. Since then the analysis of climate data and results of model simulations consistently reinforce the concept of anthropogenic global warming caused by an increase of atmospheric concentration of carbon dioxide, methane, nitrous oxide, and other greenhouse gases.

In the light of numerous discussions about global warming it may appear that the greenhouse effect is still nothing more than a scientific hypothesis. However, the mechanism of the greenhouse effect is well understood and is based on accurate physical measurements. The greenhouse effect stems from the fact that certain gases in the earth atmosphere are transparent to the incoming short-wave solar radiation but trap the outgoing infrared terrestrial radiation. As a result, for a given surface temperature the earth emits less energy into space than it would if the atmosphere was transparent for infrared radiation. Natural greenhouse gases include water vapour, carbon dioxide, methane, nitrous oxide and ozone.

The role of the greenhouse effect for the earth temperature can be readily estimated. Assuming that the atmosphere is fully transparent for terrestrial radiation, and taking the reflectivity of the earth measured from the satellites, one can calculate that the equilibrium globally averaged temperature of the earth's surface would be -18ºC. However, direct measurements show that the globally averaged surface temperature is +14ºC. The 32ºC difference between these two values represents the natural greenhouse effect. The fact that climate models accurately simulate observed surface temperature indicates that they correctly account for the natural greenhouse effect.

Opponents of anthropogenic global warming often argue that the additional increase of CO_2 concentration in the atmosphere due to fossil fuel consumption will add only a little to the natural greenhouse effect, and thus will have only a minor impact on climate. It is important to realize that the radiative properties of the atmospheric gases are accurately known from direct measurements, but calculation of the radiative fluxes in the real atmosphere requires some simplifications and empirical parameterizations.

Still, the direct effect of CO_2 on radiation balance of the earth computed with different climate models is rather similar. In accordance with IPCC (2001), the globally averaged radiative forcing due to the doubling of CO_2 concentration is 4 W/m². This means that if the concentration of CO_2 doubles while the atmospheric temperature and humidity are kept constant, the outgoing long-wave radiation at the top of the atmosphere drops by 4 W/m². Such change in the radiative balance alone would cause an increase of the surface temperature by less than 1ºC, which is below the range of global temperature response to a doubling of CO_2 concentration simulated by different climate models (1.5-4.5ºC). This additional warming arises from a number of positive feedbacks operating in the climate system.

The most important of them is the so-called water vapor feedback. Water vapor is the most important atmospheric greenhouse gas. Unlike other greenhouse gases its concentration is controlled by climate. An increase of atmospheric temperature due to direct effect of CO_2 will lead to an increase of water vapor content, and thus to an additional warming. Results of model simulations suggest that the water vapor feed-back approximately doubles the direct effect of CO_2, although the uncertainties in the magnitude of water vapor feedback are large.

Another important positive feedback is the snow/ice albedo feedback. This feed-back is explained by the fact that warming reduces the area covered by snow and sea ice. Since snow and sea ice have a much higher albedo (reflectivity) than snow-free land and open water, any reduction in snow and ice cover will decrease the amount of radiation reflected by the earth, thus amplifying warming. The global effect of albedo feedbacks is probably smaller than the effect of water vapor feedback but in the high latitudes this feedback plays a very important role and explains much stronger warming in the high latitudes compared to the tropics.

Apart from these two positive feedbacks there are several others which region-ally could be very important, but on the global scale even the sign of these feedbacks is not known. One of them is cloud feedback, an extremely complicated and uncertain feedback. The impact of clouds on climate strongly depends on the latitude, season, and type of clouds. In the low latitudes clouds have a predominantly cooling effect, while in the high latitudes they warm the earth surface. Another feedback is related to the change in the vertical thermal structure of the atmosphere, the so-called lapse rate feedback. Results of model simulations show that the vertical temperature gradient in the atmosphere decreases in the tropics and increases in the high latitudes as a result of global warming. This leads to attenuation of the surface warming in the tropics, and its amplification in the high latitudes.

The combined global effect of all aforementioned feedbacks is positive, which explains why global warming caused by doubling of CO_2 should be much larger than 1°C. At the same time, the combination of all uncertainties associated with the climate feedbacks explains the broad range of simulated climate response to doubling of CO_2.

3.   Empirical Evidences of Anthropogenic Climate Change

3.1 Observed Climate Changes During the 20th Century

3.1.1 Observed Temperature Changes

Historical data based on meteorological observations over the last century clearly shows a warming trend on the global and hemispheric scales (Figure 1) as well as over most regions and for all seasons. In accordance with such reconstructions (Jones et al., 2001; Folland et al., 2001) the globally averaged surface temperature has in-creased by about 0.6°C during the 20th century. In some regions, especially over the continents in the middle and high latitudes of the Northern Hemisphere, the tempera-ture has risen much faster than the globally averaged, but there are several regions where no warming or even some cooling is observed.

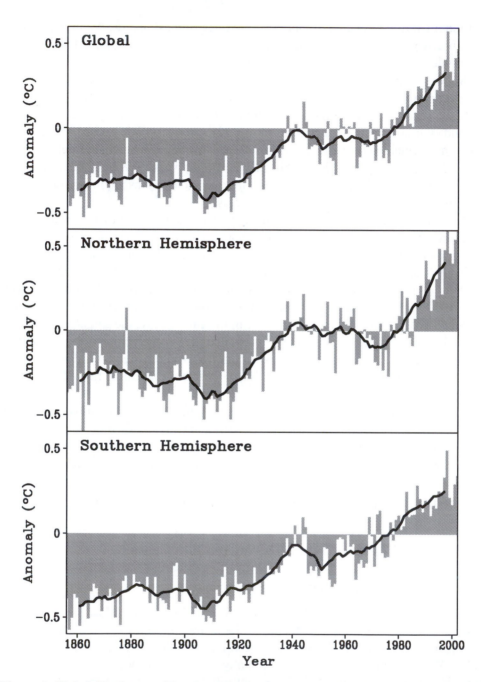

Figure 1. Global, Northern and Southern Hemisphere combined land surface air and sea surface temperatures anomalies over the last 140 years. Grey bars represent annual data, solid lines show ten-year running mean. Temperature anomalies are calculated relative to the averaging period 1950-1979. The data are from the HadCRUT2 database (http://www.cru.uea.ac.uk/cru).

Not only the surface ocean temperature, but also the volume averaged temperature of the ocean has risen considerably since the 1950s (Levitus et al., 2000). All these reconstructions reveal a consistent temporal pattern of temperature changes: the onset of warming at the beginning of the 20th century interrupted by cooling in the middle of the century, and resumption of warming after the 1970s. Such similarity supports the credibility of individual temperature reconstructions.

Indeed, each reconstruction has a number of uncertainties, related to changes in the number and location of observation, changes in observational techniques, etc. For example, the measurements of sea surface temperature in the 1940s were systematically affected because the observations mainly came from ship engine intake, while before this period they were made using wooden and canvas buckets where temperature was systematically underestimated due to heat losses caused by evaporation.

Similarly, the measurements of surface air temperature in a number of meteorological stations have been affected by the so-called "urban heat islands" effect. This effect has been suspected as being partially responsible for the observed increases in land air temperatures over the last few decades, in particular, because observed data show a clear trend in reduction of diurnal amplitude of temperature variations, which is typical for the urban areas. Nevertheless, extensive tests have shown that the urban heat island effect is responsible only for a small portion of the measured temperature rise. Moreover, the consistency between the land and sea surface temperature trends clearly indicate that the warming over land is not an artifact of growing urbanization.

While the surface air and sea surface temperature data clearly shows consistent warming during the 20th century, the analysis of above surface temperatures (low and high troposphere) measured from balloons and satellites shows a much less pronounced warming trend. In particular, over the last two decades the mean tropospheric temperature hardly shows any significant warming. This is apparently at odds with the results of model simulations, which predict that the troposphere should warm with approximately the same rate as the earth surface. The discrepancy between the surface and tropospheric temperature trends casts some doubts on the accuracy of the surface temperature reconstructions and was used by climate skeptics to argue that observed warming is not real. However, a careful analysis shows that such a conclusion is at least premature, because of the large uncertainties in the estimates of troposheric temperature trends.

The main source of these uncertainties is the limited coverage of radiosond data and changes in the measurement techniques in recent decades. Although satellite measurements have a global coverage, intercalibration (consistency between different measurements) remains a serious problem (Santer et al., 1999). Due to all these factors, estimated tropospheric changes are subject to large uncertainties. Recent analysis performed by Santer et al. (2003) shows that satellite measurements could be successfully reconciled with the results of model simulations.

Although some unexplained differences remain, this problem does not impose a direct challenge to the concept of global warming. Satellite and radiosond measurements show a pronounced cooling trend in the upper atmosphere, which is fully consistent with the results of model simulations.

3.1.2 Observed Changes in the Hydrological Cycle

Although observed temperature rise represents the most direct manifestation of climate change during the last century, other climate characteristics related to the hydrological cycle, cryosphere, and extreme weather events, also have experienced detectable changes. Theses changes are not only important as additional indicators of global warming, but also because they affect natural and anthropogenic components of the Earth system.

It is expected that global warming will lead to an increase in atmospheric water content and, as a result, an increase in precipitation. Indeed, such a trend is observed over most land areas (e.g. Jones and Hulme, 1996). The observed data shows that the total precipitation over the land area has increased by about 2% during the 20th century, and at high latitudes in the Northern Hemisphere this increase was as large as 5-10%.

However, the increase of precipitation was not uniform during the past several decades. For example, northern Africa suffered long and devastating droughts. A negative trend in precipitation has also been observed in Southern Europe while precipitation has increased in north-western Europe. Such patterns of precipitation changes can be partly attributed to the changes in North-Atlantic Oscillations. Apart from precipitation changes, an increase in river discharge has been observed in several regions. For example, the total discharge of Siberian rivers into the Arctic Ocean has increased by 7% during the 20th century (Peterson et al., 2002).

3.1.3 Observed Changes in the Cryosphere

If global warming is real, one should expect gradual "shrinking" of the cryosphere, that is, a reduction of snow and sea ice cover. Indeed, the data show a decrease in snow and sea ice areas in the Northern Hemisphere during the 20th century (Vinnikov et al., 1999). Not only is sea ice area decreasing, but the ice layer is becoming considerably thinner (Rothrock et al., 1999). These trends are statistically significant and consistent with the results of model simulations. Direct measurements of sea ice area in the Southern Hemisphere show little changes over past two decades. This is not surprising due to the short duration of these time series and large interannual variability of sea ice area in the Southern Hemisphere. At the same time, an indirect evidence from the Antarctic ice core shows a considerable retreat of the sea ice in the Southern Hemisphere since the 1950s (Curran et al., 2003). However, a longer period of time would be needed to detect a long-term trend in the Southern Hemisphere ice for a high degree of confidence.

Mountain glaciers are probably the most sensitive to climate changes. For several reasons they are able to respond vigorously and rapidly even to relatively small variations in climate. It is known that the glaciers are retreating worldwide since the beginning of the 20th century (IPCC, 2001). Because the mass balance of glaciers is affected both by temperature and precipitation, in some areas, where increase of precipitation dominated, some glaciers advanced in recent decades, but the overall number of retreating glaciers is much higher than the advancing glaciers. Although ice sheets have a much larger time scale of response as compared to glaciers, and changes in their mass balance are more difficult to monitor, several worrisome signals have been reported recently (e.g. Krabill et al., 1999). Results of remote sensing show surprisingly rapid thinning of the Greenland ice sheet and a part of the Antarctic Peninsula. Most of Antarctic Peninsula ice shelf has collapsed during recent years. Although this ice shelf is relatively small compared to the major Antarctic ice shelves, it is an important indication of the warming in the Southern Hemisphere.

3.1.4 Observed Changes in the Frequency of Extreme Events

Apart from trends in the averaged climate characteristics, changes in the frequency of extreme weather events attract much attention, because of their potential impact on the natural systems and economy. Some extreme events reveal statistically significant trends. For example, the number of frosty days has decreased during last decades, while the number of heat waves has increased (Frich et al., 2001). The latest example is the record hot summer of the year 2003 in Western Europe. At the same time, other dangerous extreme events such as hurricanes, haven't yet shown any signs of increase (Karl et al., 1995). It is important to note that extreme events are rare and longer time series is required to get sufficient statistics.

A short conclusion can be made based on the data analysis discussed above. There is a growing body of evidence showing statistically significant trends in different climate characteristics consistent with the concept of global warming. Some previously revealed problems which cast doubts on the validity of these findings have been at least partially explained. At the same time, it is important to realize that most observations are not conducted over a sufficiently long time, and the anthropogenic signal is still considerably masked by the natural climate variability.

3.2 Attribution of Climate Change

The problem of attribution of observed climate trends remains the focus of scientific debates. Most climate experts attribute the recent warming trend to the increased concentration of CO_2 and other greenhouse gases, caused by the anthropogenic factors (fossil fuel combustion, deforestation), while climate skeptics insist that observed climate change is caused solely by natural factors.

Because the anthropogenic signal is still relatively weak, the detection of the anthropogenic component of climate change remains a difficult problem, which leaves room for alternative interpretations. Most of the alternative explanations invoke the sun as the major driver of observed climate change. A number of hypotheses have been proposed in recent years claiming a strong correlation between solar activity and climate variations on different time scales. While the physical mechanisms of solar influence on climate are still not well understood, the authors of this hypothesis made their case based on the similarity between temperature changes and solar activity on one hand, and the dissimilarity between temperature and CO_2 changes on the other. Indeed, while CO_2 concentration increased monotonously during the past century, both solar activity and surface temperature experienced marked decline between the 1940s and 1970s.

Friis-Christensen and Lassen (1991), using the available global temperature and solar activity data, have claimed that the recent temperature changes could be entirely explained by solar variations. While the similarity between solar activity and temperature looks convincing at first glance, this similarity held only until the 1980s. During past two decades solar activity was almost constant, while temperature has risen considerably. This is a typical example of a premature conclusion based on the short time series. In fact, ten years later the same authors came to a different conclusion by using extended time series. They concluded that the temperature trend since the 1980s cannot be explained by solar variations and requires invoking the anthropogenic factors. This conclusion is consistent with model simulations of climate change over the 20th century performed with both natural and anthropogenic forcing. These results indicate that a considerable portion of climate variability during the first half of the 20th century is due to natural factors, but rapid warming at the end of the century cannot be explained by natural variability alone (Meehl et al., 2003).

Another attempt to attribute climate changes during the 20th century solely to the natural factors was undertaken by Svensmark and Friis-Christensen (1997). They presented evidence that global cloud cover is strongly correlated to galactic cosmic rays which are correlated to the solar activity. Since clouds are known to be an important climate factor, such strong correlation could in principle explain how the solar activity could cause large climate variations. The weakness of this hypothesis is a lack of clear physical mechanism explaining how the galactic rays could affect clouds, especially lower clouds. Moreover, as it was shown by Laut (2003), the high correlation between galactic cosmic rays and temperatures is apparently the result of using two internally inconsistent time series for global cloud cover. Furthermore, the correlation itself does not imply any causal relationship. Both variations in solar luminosity and galactic cosmic rays are known to be closely related to the solar activity. This is why the correlation between cloudiness and galactic rays can be readily explained by the fact that both processes are correlated with the variations in solar luminosity, and they are not directly related to each other.

Because there were very few direct temperature measurements prior to the 20th century, we do not know the magnitude of natural climate variability for the centenial time scale. This is why it is extremely important to reconstruct climate trends over a longer period of time. Should such time series show that warming observed during the 20th century exceeds the range of temperature variations during previous millennium, then it is likely that the observed recent warming is the result of anthropogenic influence.

A number of recent temperature reconstructions spanning the last millennium show that the 20th century warming is indeed unusual, and the temperatures observed at the end of the 20th century are well above temperatures reconstructed for the previous centuries (Mann et al., 1999). Bauer et al. (2002) simulated the temperature variations during the last millennium using climate models which take into account both natural and anthropogenic forcing including solar luminosity, volcanic activity, and greenhouse gases. The results of these simulations give additional credence to the hypothesis that anthropogenic factors caused a significant amount of warming over the 20th century.

Figure 2 illustrates a very good correspondence between reconstructed data by Mann et al. (1999) and simulated Northern Hemisphere temperature variations over the past millennium. It is important that both reconstructed and simulated temperatures not only agree with each other, but are also in agreement with historical temperature records available for the last 150 years. Temperature series reconstructed by Mann et al. (1999) clearly show that the warming during the 20th century is truly exceptional at least for the last 1000 years, however, this result is still ambiguous.

Another reconstruction of the temperature trend (Esper et al., 2002) shows much larger amplitudes of climate variations. In accordance with this reconstruction, the temperatures observed during the 20th century were not unusual for the last millennium. The reason for discrepancy between different reconstructions is still debated. At the same time, it is important to realize that any large-scale climate reconstruction beyond the 20th century is based on indirect climate proxies, such as tree rings and corrals, calibration and validation of which are rather complicated. Moreover, the magnitude of natural climate variability during the past millennium is of the same order of magnitude as the accuracy of paleoclimate reconstructions.

4. Reliability of Climate Models as Tools for Future Climate Predictions

Scenarios of the future climate changes are solely based on the simulations performed with climate models. That is why, whether climate models are a credible tool for climate predictions is the central question in the discussions about anthropogenic climate change and its consequences. In spite of considerable progress made in recent decades in the development of climate models, they are still imperfect and disagree in some important aspects of the future climate change (IPCC, 2001). This fact has often been used by the climate skeptics to portray model predictions of future climate changes as unreliable. However, the climate modellers believe that an extensive evaluation of climate models for present and past climates (McAvaney et al., 2001) clearly show that they provide credible simulations of climate and climate change.

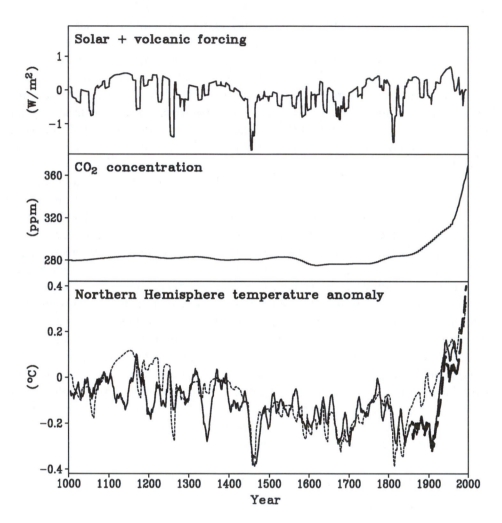

Figure 2. Reconstructed climate forcing and the Northern Hemisphere surface air temperature anomalies over the past millennium. Upper panel shows the combined globally averaged solar and volcanic forcing at the top of the atmosphere based on Lean et al. (1995) and Crowley (2000). Middle panel shows the temporal evolution of CO_2 based on the ice core data and direct measurements. The lower panel shows the Northern Hemisphere temperature anomalies. Thin solid line represents reconstructed temperature anomalies (Mann et al., 1999); thin dashed line is the simulation with the CLIMBER-2 model forced by prescribed solar, volcanic and CO_2 variations (Bauer et al., 2003); thick dashed line represents instrumental temperature record.

A broad spectrum of climate models ranging from very simple to extremely complex is now available for climate studies. The most sophisticated and comprehensive climate models, known as the coupled atmosphere-ocean general circulations models (GCMs), have been under intensive development for the past 30 years. These types of climate models are based on a physically sound approach and describe a large set of processes in the ocean and atmosphere. Compared to other types of climate models, coupled GCMs are based more extensively on the basic physical principles but the description of many processes still relies heavily on semi-empirical parameterizations.

Among such processes are turbulence, clouds, hydrology, sea ice and biophysical processes. It is generally accepted that these parameterizations represent a major source of discrepancies and deficiencies among the models. Therefore, development of better parameterizations is crucial for improvement of the climate models.

Another way to make climate models more realistic is better spatial resolution. It has been shown that doubling of spatial resolution leads to a considerable improvement in simulations of mean climate state and internal climate variability. Unfortunately, the extent of achievable spatial resolution is limited by the capability of contemporary computers. For example, a doubling of spatial resolution leads to a ten-fold increase in the computational cost, and since it takes about a decade to increase the computer performance by an order of magnitude, the improvement of model resolution is a rather slow process. Moreover, an increase of spatial resolution without appropriate improvement of model physics can only lead to a partial success.

Another important direction in the development of climate models is accounting for more climate processes. For example, the treatment of land surface processes and vegetation cover in climate models was very rudimentary until recently, but an understanding of the importance of these processes for climate has led to considerable improvements in the surface parameterizations employed in climate models.

As a result of the aforementioned improvements, modern climate models simulate both averaged climate characteristics and internal climate variability much more accurately than the climate models developed in the 1980s. In the past, most climate models had to use a correction for the fluxes between the atmosphere and the ocean to simulate climate realistically and to avoid a spurious climate drift. This fact has often been used by climate skeptics to challenge the results of future climate predictions. A growing number of new climate models do not employ this correction. Experiments performed with the climate models, which do not employ flux correction, demonstrate that flux correction has no systematic effect on climate response to the increase of greenhouse gases. Moreover, the newest models confirm all major aspects of climate change projections performed with the early climate models.

A model has to be validated before applying it for prediction. Since climate models contain physical constants as well as a number of free parameters, it would seem natural to calibrate climate models for one climate state, and then validate them against another. Unfortunately, this standard technique of model validation is not applicable to the climate models. The reason is that the only climate known with sufficient accuracy is the modern one. Simulations of paleoclimates are very useful for testing climate models, but for obvious reasons reconstructions of paleoclimates are not as detailed as the direct observations.

Climate trends observed during the 20th century represent another important opportunity to test climate models "in action" but the uncertainties in the natural and anthropogenic climate forcing remain large. That is why the validation of the climate models in the strict sense is not possible so far.

5. What Can We Learn From Past Climate Changes?

5.1 Past Climates as a Test-ground for Climate Models

Past climate changes represent an extremely useful test-ground for climate models and our understanding of how the climate system works. However, the problem is that direct observations of the climate system are almost entirely limited to the last century.

Although paleoclimate proxies are indirect and cannot provide as accurate and comprehensive information as the direct measurements, they can provide us with a reliable picture of the most important aspects of past climates when used in concert with direct measurements. Because past climates differ considerably from the modern one, the ability of the climate models to simulate past climates realistically enhances the credibility of their future projections (McAvaney et al., 2001).

Experiments performed with a number of state-of-the art climate models for past climates, such as the Last Glacial Maximum (21,000 years before present), show that the climate models are able to simulate some important aspects of past climates in good agreement with paleoclimate reconstructions. This is especially important because a large portion of the glacial age cooling has to be attributed to a lowering of CO_2 concentration, and that is why successful simulations of glacial climate confirm that the climate models have a realistic sensitivity to CO_2 changes. Similarly, a number of model simulations performed for the geological epochs, when CO_2 concentration was much higher than the current one, show no indications that climate models overestimate the response of climate to CO_2 changes as stated by some climate skeptics.

5.2 Climate and CO_2

Apart from the testing of climate models, paleoclimate records provide us with very rich information about temporal evolution of climate on different time scales and its interrelation with CO_2 concentration. For the very distant past, reconstructions of climate and CO_2 are not very reliable, nonetheless, it is generally accepted that extremely warm climates of past geological epochs cannot be explained without invoking high CO_2 levels. However, some controversy exists.

Veizer et al. (2000) have found that over the last half billion years reconstructed temperature variations in the tropics and CO_2 concentration were poorly correlated. This finding challenges the paradigm of the leading role of CO_2 in climate change but in the light of all uncertainties related to the reconstructions of temperature and CO_2 in such a distant past, it is premature to derive a conclusion based on a single finding.

Moreover, for the last 100 million years for which paleoclimatic reconstructions are much more reliable, there is a good correlation between temperature and CO_2. Unlike previous geological epochs, direct measurements of CO_2 concentration from the air bubbles trapped in the ice cores are available for the last 400,000 years (Petit et al., 1999). This data shows quasi-regular oscillations of CO_2 concentration with an amplitude of about 100 ppm (i.e. one third of its preindustrial level). These CO_2 variations are closely correlated with the reconstructed Antarctic temperature (Figure 3a), which implies a very tight link between CO_2 and climate and supports the results of simulations, showing that a considerable portion of temperature changes during glacial times can be explained by variations in CO_2 concentration.

As mentioned above the climate models reveal a rather broad range (1.5-4.5ºC) for the globally averaged surface air temperature sensitivity to a doubling of CO_2 (the so-called "climate sensitivity"). For accurate predictions of future climate change it is vital to narrow this range of uncertainties. A number of attempts have been made to constrain the climate sensitivity using measured temperature records over the last century. Unfortunately, these attempts gave a range even broader than that estimated with the climate models (e.g. Forest et al., 2002; Knutti et al., 2002).

The main reason why historical climate data cannot help to reduce the range of climate sensitivity is the large uncertainties in the estimates of radiative forcing, especially the aerosol forcing. In this respect, paleoclimate data represents a good alternative, since past climate changes were much larger than those observed during the last century, and the relative uncertainties in the major climate forcing are smaller.

By using the temperature and CO_2 history derived from the Antarctic ice core, Lorius et al. (1990) estimated the possible range for the climate sensitivity as 3-4°C. Hoffert and Covey (1992) extended this analysis by combining temperature reconstructions for the last glacial maximum and the warm mid-Cretaceous climate (100 million years ago). They found an average climate sensitivity of 2.3ºC. Recently, Cuffey and Brook (2000) used the same technique and improved paleotemperature reconstructions to arrive at a much higher estimate for the climate sensitivity of 3.9ºC, which falls towards the upper end of the climate model's sensitivity. These results indicate a combination of model simulations with paleoclimatic reconstructions may help eventually to narrow down the range of climate sensitivity with a high degree of confidence.

5.3 Abrupt Climate Changes

It has long been proposed that the climate system may respond to external or internal forcings in a strongly non-linear manner, which could manifest, for instance, in the existence of multiple equilibria and rapid transitions between different modes of operation, but only recently have paleoclimate records presented convincing proof of this notion. Analysis of Greenland ice cores and North Atlantic sediments performed

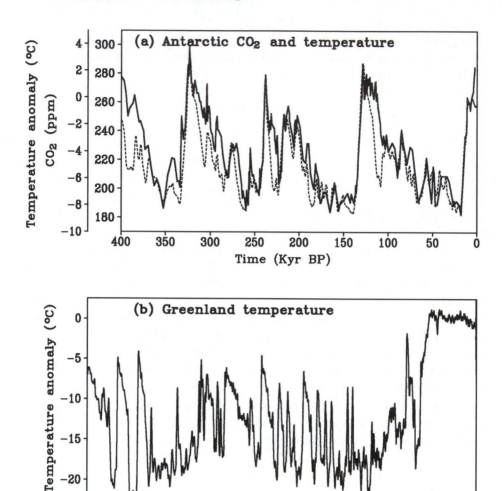

Figure 3. Past climate changes recorded in Antarctic and Greenland ice cores. (a) Variations of CO_2 concentration (solid line) and Antarctic temperature anomalies relative to the modern (dashed line) during the last 400,000 years (Petit et al., 1999); (b) Greenland temperature anomalies reconstructed from the GRIP ice core covering the last 80,000 years (Dansgaard et al., 1982).

in the early 1990s has shown that climate changes in this region were everything but gradual.

Reconstruction of Greenland temperature during the last glacial cycle shown in Figure 3b reveals an important instability of the climate system. Most of this temperature record, with a notable exception for the last 10,000 years, is punctuated by numerous rapid warming events, known as Dansgaard-Oeschger events (Dansgaard et al., 1982). These events are characterized by extremely rapid warming (up to ten degrees during one decade). The amplitude of warming during several Dansgaard-Oeschger events exceeds half of the glacial-interglacial temperature variations observed in Greenland.

Although the strongest Dansgaard-Oeschger events were recorded in Greenland and around the Northern Atlantic, synchronous warming events have been found in many other locations around the world. Analysis of paleoclimatic data and model simulations indicates that these abrupt warming events are related to the rapid reorganizations of the Atlantic thermohaline circulation (e.g. Ganopolski and Rahmstorf, 2001), which was apparently very unstable during the glacial time.

Another type of abrupt climate change discovered in the paleoclimatic records, the so-called Heinrich event, is related to the episodes of massive iceberg discharge into the Atlantic Ocean from surrounding ice sheets (Broecker et al., 1992). Recent analysis of sea level changes during the last glacial cycle (Chappell, 2002) shows a rapid increase of sea level by 10 metres or more synchronized with Heinrich events. Although the North American ice sheet, which vanished completely some 10,000 years ago, was the primary source of iceberg armadas during Heinrich events, there are indications that two currently existing ice sheets, Antarctic and Greenland, have also experienced abrupt and vigorous ice surges during the glacial age.

Due to the very different climate conditions of the glacial age compared to the present interglacial climate state, abrupt climate changes observed in the past cannot be considered as a direct analogue for the greenhouse world, but they are important evidence of the potential instability of two components of climate system—the thermohaline ocean circulation and ice sheets, the components which are suspected for the unpleasant surprises in the future greenhouse world.

6. Prediction of Future Climate Change

6.1 Uncertainties of Future Climate Predictions

So far, a large number of future climate change simulations have been performed with the whole spectrum of climate models and scenarios for greenhouse gases and aerosols. A comparison of different results is summarized in IPCC (2001), which shows a number of important similarities. However, there are also considerable differences, which demonstrates that the uncertainties in prediction of future climate changes remain large.

One of the major sources of these uncertainties is related to the prescription of anthropogenic and natural forcing. Simulations of future climates are based on the emission scenarios for different greenhouse gases and aerosols, which, in turn, are based on the scenarios of the socio-economic development in the 21st century. The latter depend critically on the assumptions made about future demographic and technological developments, and the measures which will be undertaken to mitigate greenhouse gas emissions (e.g. Hoffert et al., 1998). As an example, in the latest set of emission scenarios prepared by IPCC (SRES family of emission scenarios), the CO_2 emissions in the year 2100 differ by a factor of six between the most pessimistic and the most optimistic scenarios, leading to the large differences in corresponding CO_2 concentrations.

Another important source of uncertainties is a poor understanding of the radiative properties of aerosols. Analyses compiled in IPCC (2001) show that several different types of aerosol could contribute considerably to climate changes. Among them are sulphate, organic carbon, black carbon (soot), and mineral dust. Some of these aerosols cause cooling, while others cause warming of the earth's surface. In addition, several types of aerosols affect the optical properties of clouds and hydrological processes in the atmosphere (so-called indirect effect of aerosols). Uncertainties in the direct and indirect effects of aerosols remain very large. Furthermore, unlike the well-mixed greenhouse gases, spatial and temporal aerosol distribution is extremely inhomogeneous.

It is believed that another effect of atmospheric aerosols is cooling, thus aerosols to some extent counteract the effect of greenhouse gases, especially in the regions with high concentration of sulphur aerosol. In the first simulations of future climate change performed in the 1980s the effect of aerosols was not taken into account, which led to an overestimation of the observed rate of warming. Accounting for the direct aerosol effect helps to bring the model results into a better agreement with observations (Mitchell et al., 1995). Inclusion of the aerosol effect also reduces the projected warming for the 21st century. At the same time, it is important to note that the relative role of aerosols in climate change will decline with time because the emission of aerosols is not expected to grow as fast as the emission of greenhouse gases. Moreover, unlike CO_2, aerosols have a very short life-span in the atmosphere. That is why their concentration in the atmosphere is proportional to their emission, whilst the concentration of CO_2 and several other greenhouse gases is the integral of their emission over a long period of time.

Another important factor affecting the climate is land-use. Simulations show that deforestation in the high latitudes leads to strong regional cooling, while deforestation in the tropics, on the contrary, causes warming. A possibility of CO_2 sequestering by afforestation (planting new trees) in different regions is now widely discussed. In this respect it is interesting to note, that the net climate effect of afforestation in the middle latitude could be opposite to the expected one (Betts, 2000), because a reduction of warming due to sequestering of carbon dioxide will be overridden by warming, related to a lowering of surface albedo. Apart from that, it is still not well understood how natural vegetation will respond to the climate changes and CO_2 increase. Some model results suggest (e.g. Betts et al., 1997) that natural vegetation could, at least regionally, strongly amplify warming due to a reduction of evaporation and northward shift of tree line.

Another source of uncertainties is the internal and external (solar variations, volcanic aerosol) natural climate variability. It is important to realize that natural climate variability is unpredictable. This fact alone prevents accurate climate predictions even if all other sources of uncertainties are removed. Nonetheless, the range of uncertainties related to internal variability can be estimated by performing an ensemble of simulations with identical external forcing but different initial conditions (so-called Monte-Carlo simulations). Such experiments performed with climate models (e.g. Cubasch et al., 1994) have shown that internal variability significantly affects climate simulations at the early stage of global warming but with the increase of CO_2 concentration, the relative role of internal variability decreases and spatial patterns of climate change become more robust. Last, but not least, different climate models simulate substantially different responses to the same anthropogenic forcing. The globally averaged equilibrium surface temperature response to the doubling of CO_2 concentration is used as a benchmark for the climate model sensitivity to CO_2 changes. This characteristic, called "climate sensitivity," falls into the broad range between 1.5 and 4.5°C for different climate models, although most of them have climate sensitivity close to the median of this range.

The reason for such large differences in the climate sensitivity is primarily attributed to the uncertainties related to climate feedbacks, such as water vapor, cloud, and surface albedo feedbacks. For example, Colman (2003) has shown that the strength of water vapor feedback differs between climate models by as much as a factor of two, and even the sign of global cloud feedback is uncertain. Since it is not known which of the climate models is the most accurate for future climate prediction, the whole range of model results has to be used to assess the possible range of uncertainties.

6.2 Future Climate Change Projections

6.2.1 Predicted Temperature Changes

Although experiments with equilibrium climate response to doubling of CO_2 are useful for model intercomparison, they do not represent the true climate system response to a gradual increase in concentration of the greenhouse gases. Due to a very long response time of the deep ocean, the climate system needs hundreds of years to reach an equilibrium with the CO_2 concentration. That is why the transient response of climate to a gradual increase of CO_2 is weaker than the equilibrium response (Manabe et al., 1991). For example, in experiments where CO_2 concentration is growing at a rate of 1% per year, the global warming when CO_2 doubles is only 50-70% of its equilibrium value (IPCC, 2001). Moreover, the transient response differs from the equilibrium one in its spatial pattern. In equilibrium experiments temperature changes are relatively zonally uniform, while in the transient experiments the continents are warming much faster than the ocean. That is why, in the Southern Hemisphere where the ocean area is much larger than in the Northern Hemisphere, warming occurs at a lower rate. The Northern Atlantic and the Southern Atlantic, where the deep mixing occurs, and thus the thermal inertia of the ocean is the largest, show the weakest rate of warming (Figure 4a,c). At the same time, all of the climate models predict the highest rate of warming over land in winter at high latitudes and in the Arctic.

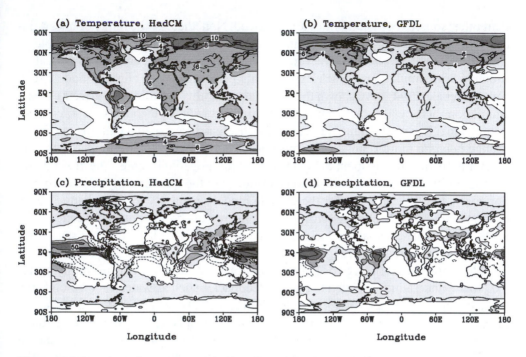

Figure 4. Mean annual temperature (a,b) and precipitation (c,d) changes around the year 2080 compared to the year 1990, simulated with the HadCM3 coupled climate model for the SRES A1F scenario (a,c) and with the GFDL coupled climate model for the SRES A2 scenario (b,d). Temperature changes are in °C, light shaded area represents warming above 2°C, dark shaded area corresponds to warming above 8°C. Precipitation changes are in millimetre per month. Shading represents the areas of increased precipitation. The data are from the IPCC Data Distribution Centre (http://ipcc-ddc.cru.uea.ac.uk/).

Global temperature rise will undoubtedly cause serious socio-economic impact. A direct consequence of global warming is the sea level rise discussed below. Another consequence is the inevitable increase in the frequency of extremely hot days (heat waves), which will have serious negative impact on health (see also Table 1). The unusually hot summer of 2003 in Western Europe is a worrisome example of such a development. Another potential health impact of global warming is related to the increase of the area where climate is suitable to malaria transmission. Currently malaria is limited to the tropics, but global warming could extend this area considerably leading to a large increase in the number of people exposed to malaria. As indicated by model simulations (Cox et al., 2000) an increase of temperature could cause a dieback of the Amazonian forest and a global increase of soil respiration, which will cause additional release of CO_2 into the atmosphere, and thus amplify global warming. In the high latitudes, warming will cause a degradation of the permafrost and an increase of methane release from the wetlands. Because methane is the second most important greenhouse gas, this will also amplify global warming.

Table 1. The most important trends in climate phenomena and extreme events related to climate change expected in the 21st century with the estimates of confidence level and examples of potentially important impacts. The table is based on IPCC WG1 and WG2 (Working Group 1 and 2) Third Assessment Report.

Phenomena/ extremes	confidence in changes	affected areas	possible impact
Averaged temperature increase	very high	entire earth	natural ecosystems, permafrost, health, crop
Maximum temperature increase, more hot days and heat waves	very high	most of land areas	health, crop, wildfires, droughts
Sea level rise	very high	coastal areas	coastal ecosystems, floods, coral reefs
More intense precipitation events	high	many land areas	floods, landslides, property, business interruption
Increase in frequency and intensity of tropical cyclones	high	some tropical areas	crop, natural ecosystems, human lives, property losses
Shutdown of the Atlantic thermohaline circulation	very low	North Atlantic and Western Europe	rapid regional climate changes, fishery, additional sea level rise
Disintegration of the West Antarctic Ice Sheet	very low	coastal areas	coastal ecosystems, floods, coral reefs

6.2.2 Predicted Changes in Hydrological Cycle

Another important aspect of global warming is change in the hydrological cycle. All climate models predict an increase of globally averaged precipitation due to global warming, however, simulated regional patterns of precipitation changes are much less robust and show a low correlation between different models (Figure 4b,d). This is related to a strong spatial variability of precipitation and the large number of factors affecting precipitation changes.

Still, some common features can be derived from model simulations. In particular, most of the climate models predict the largest increase in precipitation in the equatorial region and middle latitudes, while in the subtropics they predict the precipitation to remain unchanged or even to decrease. Due to the increased contrasts between land and ocean temperatures many models predict considerable intensification of the Asian and African monsoons.

At the same time, for many regions even the sign of changes varies from one model to another, for example, in the Amazonian region as shown in Figure 4. Despite the fact that global precipitation increases, an increase of temperature over most of the land area results in an increase of evaporation which overrides the increase of precipitation. This leads to a considerable decrease in soil moisture, especially in summer. In the simulations with the GFDL coupled climate model (Wetherald and Manabe, 2003) the soil moisture in the summers decreases by more than 10% over a large portion of Europe and United States in the middle of the 21st century. This could potentially cause a serious negative impact on natural vegetation and agriculture, and lead to an increase of forest fire frequency. At the same time, due to an increase in winter precipitation, annual river discharge in the high and mid latitudes can increase by more than 20%.

The combination of warming with changes in the hydrological cycle will have a serious impact on the water resources in many regions. Already, one third of the global population lives in the water-stress countries. Unmitigated global warming will considerably increase the number of people exposed to water stress (e.g. Arnell et al., 2002).

6.2.3 Sea Level Rise

Model simulations show that during the 21st century the sea level will rise several times faster than during the past century and by the year 2100 could rise by 20 cm to 1 m as compared to the present (Figure 5d). Sea level rise is caused by the thermal expansion of the ocean and melting of the ice sheets and glaciers. The thermal expansion of the ocean is the most important factor, at least on the time scales from decades to century. In accordance with model simulations, the thermal expansion will contribute more than a half of the expected global sea level rise. It is also important that unlike the contribution from ice sheets and glaciers, the thermal expansion is spatially inhomogeneous.

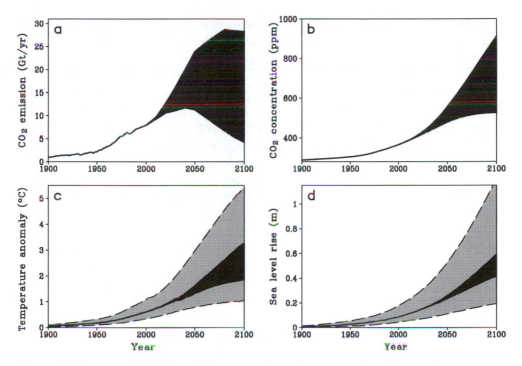

Figure 5. The range of SRES emission scenarios and corresponding climate change and sea level rise scenarios simulated with the CLIMBER-2 climate model. (a) Emission for the last 100 years and envelope of the SRES CO_2 emission scenarios for the next 100 years in Gt (1 Gt=10^9 metric tons) of carbon per year; (b) envelope of simulated CO_2 concentration scenarios corresponding to the envelop of SRES emission scenarios in ppm; (c) simulated range of global, annual surface air temperature changes; (d) globally averaged sea level rise in meters. Black areas in the panel (c) and (d) represent the range of uncertainties related to emission scenarios only simulated with the model version, which has climate sensitivity of 2.5ºC (medium climate sensitivity). Shaded areas represent the combined range of uncertainties related both to the emission scenarios and climate sensitivities ranging from 1.5 to 4.5ºC. Note that temperature and sea level anomalies shown here are computed relative to the pre-industrial climate state (year 1860), this is why simulated temperature anomalies and sea level rise for the year 2100 are higher than those given in IPCC (2001) where the reference year is 1995.

Although all models agree that spatial variations of the sea level rise can reach up to 100% of its globally averaged value, on the regional scale different models show little similarity (Gregory et al., 2001). The reason for such discrepancy is still not well understood. Nonetheless, it is clear that global sea level rise should not be considered as the maximum estimate for any individual location. Since thermal expansion is directly proportional to the average warming of the whole ocean, and the thermal inertia of the ocean is enormous, it will take at least a thousand years before the sea level will reach an equilibrium with the climate. This implies that sea level will continue to rise long after the CO_2 level stabilizes.

Long-term experiments performed with several climate models show that the final sea level rise after doubling of CO_2 could reach 1-2 meters, while at the time of CO_2 doubling it will be only 20-40 cm. This clearly demonstrates that global warming will have extremely long consequences, which are very difficult to prevent.

Apart from the thermal expansion, changes in the mass balance of ice sheet and glaciers will also contribute to sea level rise. It is expected that glaciers and small ice caps will predominantly melt, as it has already been observed during the 20th century, while the contribution of Greenland and the Antarctic ice sheet is less certain, because it depends on a combination of two opposite processes.

Increase of air temperature will enhance melting of the ice sheets, especially in Greenland. At the same time, an increase of snowfall simulated in the high latitudes by all climate models will have an opposite effect. Model simulations suggest that in Greenland the effect of melting will be the dominant one, and thus Greenland will contribute to sea level rise, whilst in Antarctica an increase of snowfall will dominate, thus Antarctic contribution to sea level will be negative. The combination of afore-mentioned uncertainties explains a wide range of sea level rise projections for the 21st century (Figure 5), with the most probable values of about 50 cm. Such sea level rise could have profound negative socio-economic impact by increasing the risk of coastal flooding and causing the loss in coastal wetlands. In particular, the estimates show that unmitigated global warming could increase the number of people in coastal areas affected by storm surges by a factor ten by the year 2080 (Arnell et al., 2002).

6.3 Unpleasant Surprises of Greenhouse

The aspects of climate change discussed above represent a relatively linear response of the climate system to the increasing concentration of greenhouse gases. Meanwhile there is a growing concern among climate experts that the response of the climate system to the anthropogenic global warming could be strongly non-linear and will trigger rapid and irreversible climate changes.

There are several candidates for such strongly non-linear behavior. Both paleoclimatic data and model simulations show that the Atlantic thermohaline circulation and continental ice sheets are subjected to rapid transitions caused by internal instability. The Atlantic thermohaline circulation is driven by the sinking of cold dense water in the northern part of the Atlantic. The thermohaline circulation is an important carrier of energy from the low to the high latitudes in the Northern Hemisphere and this is why it plays an extremely important role in the climate system. Global warming will reduce the density of surface water masses in high latitudes due to an increase of temperature, precipitation and river discharge. This could potentially lead to a cessation of the sinking of cold dense water in the high latitudes and, as a result, to complete shut-down of the Atlantic thermohaline circulation (Manabe and Stouffer, 1993). The latter will have a dramatic impact on the North Atlantic, will affect fishery and accelerate considerably regional sea level rise. Moreover, since the shutdown of the thermohaline ocean circulation will lead to a drastic reduction of the northward ocean heat transport, it could cause a sudden reversing of the gradual warming trend into a rapid cooling over the Northern Atlantic and North-Western Europe (Rahmstorf and Ganopolski, 1999). Such reverse of the temperature trend might be especially difficult for adaptation since it does not meet the expectations of people living in the greenhouse world (Hulme, 2003).

While most of the experts agree that shutdown of the thermohaline circulation will have a very serious negative socio-economic impact, neither likelihood nor timing of such events are possible to predict at the present state of knowledge. Most of the climate models predict a weakening of the themohaline circulation by gradual increase of CO_2 during the 21st century but very few experiments have been performed beyond the year 2100.

One such long-term experiment (Wood et al., 1999) shows that the HadCM3 model (unlike the GFDL model (Manabe and Stouffer, 1993)) does not simulate complete shutdown of the thermohaline circulation even for quadrupling of CO_2 concentration. It is likely that different climate models have different threshold values for CO_2 concentration above which the thermohaline circulation will collapse. It is also possible that this threshold value depends on the rate of CO_2 concentration changes (Stocker and Schmittner, 1997). At the same time, several climate models show either weak response or no response of the thermohaline ocean circulation to the global warming.

In the light of these uncertainties the complete shutdown of the thermohaline ocean circulation cannot be regarded as very probable, at least during the 21st century, but its probability will increase if CO_2 concentration continues to grow in accordance with business-as-usual scenarios. Due to potential huge socio-economic impact such phenomenon should be considered as very serious in spite of its low probability.

Another candidate for unpleasant surprises of the greenhouse world is the West Antarctic Ice Sheet. The West Antarctic Ice Sheet accumulates an amount of water enough to raise sea level by 6 m. If the West Antarctic Ice Sheet disintegrates and completely melts on the time scale of a century to a millennia, it will increase the rate of sea level rise by at least a factor of two.

The possibility of such a rapid disintegration of the West Antarctic Ice Sheet in the foreseeable future remains subject to debate and controversy (e.g. Oppenheimer, 1998). The major problem in the prediction of such type of event is a rather poor understanding of the mechanisms related to ice sheet instability, in particular, the processes at the bases of ice sheets and the interaction between grounded ice sheets and ice shelves. Paleoclimate data clearly show that during the glacial age the North American and European ice sheets experienced numerous rapid oscillations resulting in sea level rise by 10 meters and more in less than a thousand years. Such type of instability has been successfully simulated with the climate-ice sheet model (Calov et al., 2002).

Whether the West Antarctic Ice Sheet is subject to a similar type of instability remains unclear. Paleoclimate data indicate that during the past 10,000 years (current interglacial period) the West Antarctic Ice Sheet was relatively stable but an unprecedented warming could potentially destabilize it via an increase of bottom melting. Still, most of the experts believe that the probability of rapid disintegration of the entire West Antarctic Ice Sheet in the nearest future is rather low, and it is not expected that the West Antarctic Ice Sheet will make a significant contribution to the sea level rise at least over the next several centuries. On the other hand, because of the poor understanding of ice sheet dynamics and imperfectness of the existing model, it is premature to rule out the danger of such a catastrophic event.

7. Summary and Conclusions

It is undeniable that we have made a considerable step forward in our understanding of climate dynamics during the past decades, and it will benefit further from the progress in computer technologies, remote sensing and advances in theory. In particular, the concept of anthropogenically induced global warming in recent years was considerably reinforced by the analysis of available historical climate data and reconstructions of the recent and distant past climates.

Measurements show that CO_2 concentration has steadily risen during the past century and is currently the highest it has been in the last 400,000 years, indeed probably in the past several million years. It was shown that the 1990s was the warmest decade of the last century, and probably of the last millennium. The global average surface air temperature has increased over the 20th century by 0.6°C.

These facts alone are already serious enough to cause concern about the future consequences of unlimited combustion of fossil fuels. Moreover, predictions of climate changes in the 21st century performed with different climate models suggest that the globally averaged surface air temperature will increase by 1 to 4°C by the year 2100, which is an extremely high rate of change, even when compared to the most unstable climates of past epochs. Based on scrupulous analysis, the IPCC came to the conclusion that "the balance of evidence suggests a discernible human influence on global climate" (IPCC, 2001). Although it is recognized that a part of the climate variability during the 20th century should be attributed to natural causes, such as variations of solar insolation and volcanoes, IPCC concluded that it is very likely that the rapid warming observed since 1950s should be explained primarily by the increase in concentration of CO_2 and other greenhouse gases.

In spite of the considerable advance in the understanding of climate processes, there are still a number of problems related to the detection, attribution and prediction of future climate change. It appears that the climate system is more complex and difficult to model accurately than was believed several decades ago. That is why it is not yet possible to narrow down the large uncertainties in the assessment of future climate changes. However, most climate experts are confident that the results of future climate projections represent a useful basis for a comprehensive analysis of the possible socio-economic impacts of climate change.

Although there is no reason to expect that all aspects of climate change will have a negative impact on society, nonetheless, it is very likely that on a global scale and in many regions, this impact will be negative. Adaptation to rapidly changing environmental conditions could be too costly, especially for the poorest nations. Among the factors which could have a serious negative impact are the sea level rise, increase of extreme weather events, and the melting of the permafrost accompanied with additional release of methane. Another serious concern is caused by a possibility of the so-called "low probability-high impact" events such as the shutdown of the Atlantic thermohaline circulation and disintegration of the West Antarctic Ice Sheet.

Paleoclimate records present a great deal of evidence that the ocean thermohaline circulation and the large continental ice sheets are subject to instabilities and rapid transitions. Although past climate changes cannot be considered as direct analogues for the future, they give important support to the climate models, which are able to simulate such rapid transitions. At the current stage of knowledge it is impossible to quantify the probability of such events or predict the thresholds, the crossing of which could trigger rapid climate changes, but because of potentially enormous socio-economic impacts of abrupt climate changes, they deserve very serious attention.

8. References

Arnell, N.W., Cannell, M.G.R., Hulme, M., Kovats, R.S., Mitchell, J.F.B, Nicholls, R.J., Parry, M.L., Livermore, M.T.J., White, A., 2002. The consequences of CO_2 stabilisation for the impacts of climate change. Clim. Change, 53, pp. 413-446.

Bauer, E., Claussen, M., Brovkin, V., Huenerbein, A., 2003. Assessing climate forcings of the Earth system for the past millennium. Geophys. Res. Lett. 30, Art. No. 1276

Betts, R.A., 2000. Offset of the potential carbon sink from boreal forestation by decreases in surface albedo. Nature, 408, pp. 187-190.

Betts, R.A., Cox, P.M., Lee, S.E., Woodward, F.I., 1997. Contrasting physiological and structural vegetation feedbacks in climate change simulations. Nature, 387, pp. 796-799.

Broecker, W., Bond, G., Mieczyslawa, K., Clark, E., McManus, J., 1992. Origin of the northern Atlantic's Heinrich events. Clim. Dyn., 6, pp. 265-273.

Calov, R., Ganopolski, A., Petoukhov, V., Claussen, M., Greve, R., 2002. Large-scale instabilities of the Laurentide ice sheet simulated in a fully coupled climate-system model. Geophys. Res. Lett., 29, doi: 10.1029/2002GL016078.

Colman, R., 2003. A comparison of climate feedbacks in general circulation models. Clim. Dyn., 20, pp. 865-873

Cox, P.M., Betts, R.A., Jones, C.D., Spall, S.A., Totterdell, I.J., 2000. Acceleration of global warming by carbon cycle feedbacks in a 3D coupled model. Nature, 408, pp. 184-187.

Crowley, T.J., 2000. Causes of climate change over the past 1000 years. Science, 289, pp. 270–277.

Cubasch, U., Santer, B.D., Hellbach, A., Hegerl, G., Höck, H., Maier-Reimer, E., Mikolajewicz, U., Stössel, A., Voss, R., 1994. Monte Carlo climate change forecasts with a global coupled ocean-atmosphere model. Clim. Dyn., 10, pp. 1-19.

Cuffey, K.M., Brook, E.J., 2000. Ice sheets and the ice-core record of climate change. In: Jacobson, M.C., Charlson, R.J., Rodhe, H., Orians, G.H., eds. Earth System Science. From Biogeochemical Cycles to Global Change. Academic Press, New York, pp. 459-497

Curran, M.A.J., van Ommen, T.D., Morgan, V.I., Phillips, K.L., Palmer, A.S., 2003. Ice Core Evidence for Antarctic Sea Ice Decline Since the 1950s. Science 302, pp. 1203-1206.

Daansgard, W., Claussen, H.B., Gundestrup, N., Hammer, C.U., Johnsen, S.F., Kristindottir, P.M., Reeh, N., 1982. A new Greenland deep ice core. Science 218, pp. 1273-1277.

Esper, J., Cook, E.R., Schweingruber, F.H., 2002. Low-Frequency Signals in Long Tree-Ring Chronologies for Reconstructing Past Temperature Variability. Science, 295, pp. 2250-2253.

Folland, C.K., Rayner, N.A., Brown, S.J., Smith, T.M., Shen, S.S., Parker, D.E., Macadam, I., Jones, P.D., Jones, R.N., Nicholls, N., Sexton, D.M.H., 2000. Global temperature change and it uncertainties since 1861. Geophys. Res. Lett., 28, pp. 2621-2624.

Forest, C.E., Stone, P.H., Sokolov, A.P., Allen, M.R., Webster, M.D., 2002. Quantifying uncertainties in climate system properties with the use of recent climate observations. Science, 295, pp. 113-117.

Frich, P., Alexander, L.V., Della-Marta, P., Gleason, B., Haylock, M., Klein-Tank, A., Peterson, T., 2002. Observed coherent changes in climatic extremes during the second half of the 20th Century. Clim. Res., 19, pp. 193-212.

Friis-Christensen, E., Lassen, K., 1991. Length of the solar cycle: an indicator of solar activity closely associated with climate. Science, 254, pp. 698-700.

Ganopolski, A, Rahmstorf, S., 2001. Rapid changes of glacial climate simulated in a coupled climate model. Nature 409, pp. 153-158.

Gregory, J.M., Church, J.A., Boer, G.J., Dixon, K.W., Flato, G.M., Jackett, D.R., Lowe, J.A., O'Farrell, S.P., Roeckner, E., Russell, G.L., Stouffer, R.J., Winton, M., 2001. Comparison of results from several AOGCMs for global and regional sea-level change 1900-2100. Clim. Dym., 18, pp. 225-240.

IPCC, 2001: Climate Change 2001: The Scientific Basis. Contribution of Working Group I to the Third Assessment Report of theIntergovernmental Panel on Climate Change [Houghton, J.T.,Y. Ding, D.J. Griggs, M. Noguer, P.J. van der Linden, X. Dai, K. Maskell, and C.A. Johnson (eds.)]. Cambridge University Press, Cambridge, United Kingdom and New York, NY, USA, 881pp.

Hoffert M.I., Covey, C., 1992. Deriving global climate sensitivity from paleoclimate reconstructions. Nature, 360, pp. 573-576.

Hoffert M.I., Caldeira, K., Jain, A.K., Haites, E.F., Harvey, L.D.D., Potter, S.D., Schlesinger, M.E., Schneider, S.H., Watts, R.G., Wigley, T.M.L., Wuebbles, D.J., 1998. Energy implications of future stabilization of atmospheric CO_2 content. Nature, 395, pp. 881-884.

Hulme, M., 2003. Abrupt climate change: can society cope? Phil. Trans. Royal Soc., 361 pp. 2001-2019.

Jones, P.D., New, M., Parker, D.E., Martin, S., Rigor, I.G., 1999. Surface air temperature and its changes over the past 150 years. Rev.Geophys., 37, pp. 173-199.

Jones, P.D., Hulme, M., 1996. Calculating regional climatic time series for temperature and precipitation: methods and illustrations. Int. J. Climatol., 16, pp. 361-377.

Karl, T.R., Knight, R.W., Plummer, N., 1995. Trends in high-frequency climate variability in the twentieth century. Nature, 377, pp. 217-220.

Krabill, W., Frederick, E., Manizade, S., Martin, C., Sonntag, J., Swift, R., Thomas, R., Wright, W., Yungel, J., 1999. Rapid thinning of parts of the southern Greenland ice sheet. Science, 283, pp. 1522-1524.

Kristjansson, J.E., Kristiansen, J., 2000. Is there a cosmic ray signal in recent variations in global cloudiness and cloud radiative forcing? J. Geophys. Res., 105, pp. 11851-11863.

Knutti, R.F., Stocker, T.F., Joss, F., Plattner, G.K., 2002. Constraints on radiative forcing and future climate changes from observations and climate model ensembles. Nature, 416, pp. 719-723

Laut, P., 2003. Solar activity and terrestrial climate: an analysis of some purported correlations. J. Atmosospheric and solar-terrestrial physics, 65, pp. 801-812.

Lean, J., Beer, J., Bradley, R., 1995. Reconstruction of solar irradiance since 1610: Implications for climate change. Geophys. Res. Let., 22, pp. 3195-3198.

Levitus, S., Antonov, J., Boyer, T.P., Stephens, C., 2000. Warming of the World Ocean. Science, 287, pp. 2225-2229.

Lorius, C., Jouzel, J., Raynaud, D., Hansen, J., Le Treut, H., 1990. The ice-core record: climate sensitivity and future greenhouse warming. Nature, 347, pp. 139-145.

Manabe, S., Stouffer, R.J., Spelman, M.J., Bryan, K., 1991. Transient response of a coupled ocean-atmosphere model to gradual changes of atmospheric CO_2. Part I: Annual mean response. J. Climate, 4, pp. 785-818.

Manabe, S., Stouffer, R.J., 1993. Century-scale effects of increased atmospheric CO_2 on the ocean-atmosphere system. Nature, 364, pp. 215-218.

Mann, M.E., Bradley, R.S., Hughes, M.K., 1999. Northern Hemisphere Temperatures During the Past Millennium: Inferences, Uncertainties, and Limitations. Geophys. Res. Lett., 26, pp. 759-762.

McAvaney, B.J., Covey, C., Joussaume, S., Kattsov, V., Kitoh, A., Ogana, W., Pitman, A.J., Weaver, A.J., Wood, R.A., Zhao, Z.-C., 2001. Model evaluation. In: Houghton J.T. et al. (eds.) Climate change 2001: the scientific basis. Contribution of Working Group I to the 3rd Assessment Report of the Intergovernmental Panel on Climate Change. Cambridge University Press, Cambridge, p 471–523

Meehl, G.A., Washington, W.M., Wigley, T.M.L., Arblaster, J.M., Dai, A., 2003. Solar and Greenhouse Gas Forcing and Climate Response in the Twentieth Century. J. Climate, 16, pp. 426-444.

Mitchell, J.F.B., Johns, T.J., Gregory, J.M., Tett, S.B.F., 1995. Climate response to increasing levels of greenhouse gases and sulphate aerosols. Nature, 376, pp. 501-504.

Oppenheimer, M., 1998. Global warming and the stability of the West Antarctic Ice Sheet. Nature 393, pp. 325-332.

Peterson, B.J., Holmes, R.M., McClelland, J.W., Vörösmarty, C.J., Lammers, R.B., Shiklomanov, A.I. Shiklomanov, I.A., Rahmstorf, S., 2002. Increasing river discharge to the Arctic Ocean. Science, 298, pp. 2171-2173.

Petit, J.R., Jouzel, J., Raynaud, D., Barkov, N.I., Barnola, J.-M., Basile, I., Bender, M., Chappellaz, J., Davis, M., Delaygue, G., Delmotte, M., Kotlyakov, V.M., Legrand, M., Lipenkov, V.Y., Lorius, C., Pépin, L., Ritz, C., Saltzman, E., Stievenard, M., 1999. Climate and atmospheric history of the past 420,000 years from the Vostok ice core, Antarctica. Nature, 399, pp. 429-436.

Rahmstorf, S., Ganopolski, A., 1999. Long-term global warming scenarios computed with an efficient coupled climate model. Climatic Change, 43, pp. 353-367.

Rothrock, D.A., Yu, Y., Maykut, G.A., 1999. Thinning of the Arctic Sea-Ice Cover. Geophys. Res. Lett., 26, pp. 3469-3472.

Santer, B.D., Hnilo, J.J., Wrigley, T.M.L, Boyle, J.S., Doutriaux, C., Fiorino, M., Parker, D.E., Taylor, K.E., 1999. Uncertainties in observational based estimates of temperature change in the free atmosphere. J. Geophys. Res., 104, pp. 6305-6333.

Santer, B.D., Wigley, T.M.L., Meehl, G.A., Wehner, M.F., Mears, C., Schabel, M., Wentz, F.J., Ammann, C., Arblaster, J., Bettge, T., Washington, W.M., Taylor, K.E., Boyle, J.S., Brüggemann, W., Doutriaux, C., 2003. Reliability of Satellite Data Sets. Science, 300, pp. 1280-1284.

Stocker, T.F., Schmittner, A., 1997. Influence of CO_2 emission rates on the stability of the thermohaline circulation. Nature, 388, pp. 862-865.

Svensmark, H., 1998. Influence of cosmic rays on Earth's climate. Phys. Rev. Lett., 22, pp. 5027-5030.

Veizer, J., Godderis, Y., Francois, L.M., 2000. Evidence for decoupling of atmospheric CO_2 and global climate during the Phanerozoic eon. Nature, 408, pp. 698-701.

Vinnikov, K.Y., Robock, A., Stouffer, R.J., Walsh, J.E., Parkinson, C.L., Cavalieri, D.J., Mitchell, J.F.B, Garrett, D., Zakharov, V.F., 1999. Global warming and Northern Hemisphere sea ice extent. Science, 286, pp. 1934-1937.

Wetherald, R.T., Manabe, S., 2002. Simulation of hydrologic changes associated with global warming. J. Geophys. Res. 107, doi:10.1029/2001JD001195.

Wood, R.A., Keen, A.B., Mitchell, J.F.B, Gregory, J.M., 1999. Changing spatial structure of the thermohaline circulation in response to atmospheric CO_2 forcing in climate model. Nature, 399, pp. 572-575.

Chapter 2

Beyond Oil: A Comparison of Projections of PV Generation and European and U.S. Domestic Oil Production[1]

by

John M. Byrne*, Lado Kurdgelashvili, Allen Barnett,
Kristen Hughes and Aaron Thurlow
Center for Energy and Environmental Policy
278 Graham Hall, University of Delaware
Newark, DE 19716 USA

Abstract

The chapter analyzes the potential contribution of photovoltaics (PV) to meet part of European and U.S. future electricity needs. A logistic growth model based on historical trends of PV markets was utilized in order to forecast electricity generation from PV between 2005 and 2070. Historical growth of other energy sources (e.g., oil, natural gas) and other silicon-based commodities (e.g., cellular phones, personal computers) were also analyzed in order to assess whether the projected growth rates for PV capacity that result from the model can be considered realistic. The PV forecasts are also compared with projections of other researchers. The forecasted energy generated from PV is then compared in energy units (barrels of oil equivalent) with forecasted European and U.S. domestic oil production. Our conclusion is that cumulative energy production from PV will be higher than domestic oil output for both regions for the forecast period. The role of European and U.S. national energy policies promoting PV development is examined in light of this finding.

---

[1] An early version of parts of the argument appearing in this chapter is published in a recent issue of the journal *Energy Policy* (see Byrne et al., 2004). The earlier article included only an analysis of the U.S. case. Several additions and detailed improvements have been made to the argument and methodology that we believe further demonstrate the importance of comparing PV and domestic oil while expanding the scope of the analysis to Europe and considerably strengthening the empirical support for our findings.

* *Corresponding Author. Email: jbbyrne@UDel.Edu*

1. Introduction

To realize the potential of solar electric power, improvements in photovoltaic (PV) technology will be needed. While certain peak-shaving and so called "distributed utility" (see Weinberg et al., 1993) applications can be competitive at today's module prices (Byrne et al., 1996, 1997, 1998, 2000 and 2002, as discussed below), as can a full range of off-grid uses, wider market penetration of PV will depend on further technical and economic advances. Without considering the costs of externalities, current costs preclude PV from competing with fossil fuel-fired generation technology and wind energy and hydroelectric systems for the supply of low-cost bulk power to grid users.

Partly due to this fact, PV's future is often discussed, even by its promoters, as largely a matter of technology and markets. Forgotten in this treatment are three interrelated factors of equal or greater significance: 1) the major changes underway in the electricity sector; 2) an even larger set of changes in how societies assess energy services; and 3) the influence of past and future policies on PV's (and its competitors') development. Briefly, electricity technologies and economics are shifting from a nearly exclusive focus on the provision of large amounts of kWhs via centralized production systems to a service architecture that emphasizes modular development of electricity capacity when and where different needs arise (see, e.g., Hunt and Shuttleworth 1996). As discussed in section 4 of the chapter, comparison of PV to busbar costs of thermal plants using fossil fuels will become less and less relevant as this trend accelerates. Just as the lower per minute cost of land line telephone communication did not decide the fate of the cellular phone industry, it is unlikely that PV's prospects will rest upon competitive bulk power rates.

Possibly more significant is the trend in many societies to evaluate energy services according to new yardsticks that include environmental performance, health impacts, effects on local control and security implications. With the energy sector now understood to contribute to a variety of environmental ills from climate change to urban smog and acid rain, and with linkages now being drawn between its emissions and health problems (such as higher incidences of cardiovascular and respiratory diseases), there is a growing social consensus that changes are needed in the types of fuels used, the pricing of energy services and the general expectations of energy systems. With regard to the latter, the recent California energy crisis (in the wake of its electricity deregulation policy) and the persisting volatility in fossil fuel prices and international energy politics have likewise strengthened the interest of many communities in increasing the use of local renewable energy sources. Often, renewable energy generally, and PV specifically, reduces social risk and adds local control benefits (both for communities and companies) to energy service provision (Awerbuch, 1995; Lovins and Lovins, 1982).

These rising social concerns have led to a plethora of policy initiatives to incentivize renewable energy development, especially in the U.S. and Europe (discussed in some detail below). Not willing to let short term costs and benefits entirely determine their energy future, several communities in both jurisdictions have adopted regulations and employed policy tools to speed up the entry of renewables into the marketplace and, in some cases, have authorized taxes to discourage continued reliance on fossil fuels (see section 7 below). While the initiatives are embryonic, the new policy environment appears to seek change in the social structure of energy service, which if effective would raise doubts about the validity of benchmarking PV's performance against the energy *status quo*.

Thus, there are empirical reasons (discussed more fully below) why PV's future likely cannot be judged by existing technical and economic considerations alone. This should encourage research and policy analysis that canvasses wider opportunities for the technology than its ability to substitute for conventional systems. Yet there is very little research of this kind. In the interest of opening the PV research agenda to broader questions and in an effort to build a more complete analytical portrait of the technology's potential, we consider a question that is ordinarily neglected, namely, if and how PV could have the equivalent importance of a major energy source for the U.S. and Europe in the 21st century. Oil continues to be widely regarded as a major energy source for the American and European economies, while PV – both in the policies and investment patterns of the two jurisdictions[2] – is treated as a 'frontier' technology. Our question probes, rather than accepts, the validity of this treatment.

Thinking beyond conventional energy wisdom is actually grounded on a well-established feature of energy transitions. As with the contemporary cases of coal, large hydro, oil, natural gas and nuclear power, energy transitions are typically sudden and substantial (see, e.g., Smil 2000). The rapid change that defines them is closely linked to policy shifts, which often spur technical and economic change (Byrne and Rich, 1983). Therefore, understanding PV's potential will require us to ask not only technically and economically innovative questions, but policy questions as well.

2. Forecasting PV Electricity Generation

2.1 A logistic Growth Model

Energy forecasting is a well-established and widely used tool by private and public sector organizations to gauge market trends in this strategic sector. Typically, forecasts link physical (e.g., geology), economic and social information, often in mathematical models, in order to produce near-term (less then one year), short-term (1-5 years), medium-term (5-10 years) and long-term projections (more then 10 years) of energy supply and demand (EIA 2002b, IEA 2001, 2002a, 2002b, 2002c). For reasons

---

[2] As we discuss later in the chapter, recent changes in policy strategy and investments are evident in Europe and the U.S. If these changes continue to accumulate, it is possible that a shift in policy objectives–and energy investment outcomes–will result. See section 7 for details.

described below, we have adopted a logistic growth approach for forecasting PV market development in the U.S. and Europe. This methodology is commonly used to anticipate the entry of new technology generally (Fisher and Pry, 1971; Mignogna, 2001; Woodall, 2000) and new energy technologies specifically (e.g. EWEA, 1999; Roethle Group, 2002).

Logistic growth models have proven to be accurate tools for forecasting a wide range of phenomena, from human population growth (used by Belgian mathematician Pierre Verhulst in 1838) to oil development (Hubbert, 1962 and Laherrère, 2000). A logistic growth curve, according to Laherrère (2000) and Mignogna (2001), can be represented by the following equation:

$$Q = \frac{U}{1 + e^{-b(t-t_m)}} \qquad\qquad \text{(Eq. 1)}$$

where:

Q is the forecast variable (e.g., annual or cumulative energy production or the number of new users in the case of technologies like cellular phones and personal computers);
U is the saturation (maximum) level for Q;
b is the slope term, reflecting the initial growth rate;
t is time (in years); and,
t_m represents the midpoint of the logistic curve.

Variable Q in equation (1) can represent annual energy production in the case of renewable energy technologies, cumulative energy production in the case of oil and natural gas, and the number of users in the case of silicon–based technologies such as cellular phones or personal computers. U represents the maximum level of energy production from renewable energy (based on known or assumed economic, technical, and physical restrictions). Alternatively, U represents the physical resource limit for nonrenewable sources such as oil and natural gas (i.e., ultimate feasible production in a given geographic area), and the maximum number of potential customers in the case of cellular phones or personal computers. The slope term b is a constant and represents the rate of growth per unit of time. In equation (1), t_m is the time it takes for Q to reach the midpoint of its logistic growth trajectory.

Rearranging terms in Eq. 1 renders:

$$\frac{U-Q}{Q} = e^{-b(t-t_m)} \qquad\qquad \text{(Eq. 2)}$$

which offers an easy means of defining in functional form the parameters of interest. Taking the logarithm of both sides gives:

$$\ln(\frac{U-Q}{Q}) = -b \times t + b \times t_m \qquad \text{(Eq. 3)}$$

Grouping variables, we then obtain:

$$Y = \ln(\frac{U-Q}{Q})$$

This yields the familiar linear equation:

$$Y = \alpha + \beta \times X \qquad \text{(Eq. 4)}$$

where:

$X = t$

parameter $\alpha = b*t_m$

and parameter $\beta = -b$

Applying statistical regression methods to Eq. 4, the parameters α and β can be robustly estimated.

Noting that $t_m = -\alpha/\beta$ and $b = -\beta$, Eq.1 can be presented as:

$$Q = \frac{U}{1 + e^{\beta(t+\frac{\alpha}{\beta})}} \qquad \text{(Eq.5)}$$

Equation 5 and the linear regression method used to estimate parameters in Eq. 4 are consistent with the classic Fisher-Pry form of a logistic growth curve widely used to model technology diffusion (Fisher and Pry, 1971).[3]

In this way, a forecasting model can be built on empirical experience to date with the technology of interest (PV, in this case) and, equally important, the parameter values adopted for forecasting purposes can be benchmarked against experience with comparable technologies and market conditions. As discussed below, regression results enable the use of a growth rate in the forecast that is selected on objective grounds and can be compared to historical experience with other energy sources and technologies.

---

[3] As Mignogna (2001) has shown, the approach we have taken here is mathematically identical to the Fisher-Pry logistic growth model.

2.2 Validity of Using a Logistic Curve Model to Predict Growth in Use of Energy and Silicon-based Technologies

To validate the use of the logistic growth curve, historical data on growth patterns for selected energy and silicon based technologies were analyzed. Results from a regression analysis conducted for the cellular phone industry in the U.S. are shown for illustrative purposes in Figures 1 and 2. After obtaining regression parameters (Figure 1), a logistic growth curve is constructed for the period from market entry to diffusion of this technology (see Figure 2). The coefficient of determination for this regression (R^2=0.9942) indicates that over 99% of the variance in sales data for 1986-2003 is accurately predicted by the logistic growth model. The saturation level (U) is estimated to be 200 million subscribers.[4]

Similarly, an analysis of cumulative oil production shows the versatility of a logistic model to describe energy market development. Maximum oil reserves (U) were set at 216 billion bbl. This is the midpoint between the 210 billion bbl limit used by Laherrère (2000) and the 222 billion bbl maximum employed by Bartlett (2000). Using this (U) value, the regression for the U.S. cumulative oil production produced a quite high R^2 value (see Figure 3a).

Using estimates from the U.S. Geological Survey (1995; 1998) and Mineral Management Service (1996) for measured (proved) reserves and their estimates of "undiscovered" but economically recoverable resources[5] yields a higher estimate of maximum recoverable reserves (229 billion bbl). Figure 3b offers a logistic curve fit for historical oil production data (see US Census Bureau (1975) and Energy Information Administration (2002a)) based on this alternative U value. Inputting obtained regression parameters for the two scenarios yields actual versus estimated scatter plots (see Figures 4a and b).

Table 1 summarizes regression analysis results for the selected technologies and markets. All of the regressions use an underlying logistic growth curve and produce high R^2 values. Naturally, differences in units of the Q variable for each technology and energy option (i.e., subscriptions vs annual electricity generation vs cumulative production in tons or cubic meters) lead to differences in intercept values (the α's). Slope coefficient differences (the β's) reflect the particular economic, technical and physical factors influencing the development of these energy and technology alternatives. Notwithstanding these evident differences, Table 1 demonstrates the ability of a logistic growth model to accurately project market development. It is this feature that encouraged our use of the method.

---

[4] According to Telecompetition (2003), the maximum penetration rate of cellular phone subscriptions in the U.S. market is expected to be in the 70-75% range of the population, which is 281 million according to the last U.S. Census (2000). Applying a mid-point value of 72.5% yields 200 million. We use this value for the U term to create the regression results in Figure 1.

[5] See below for a discussion of these terms.

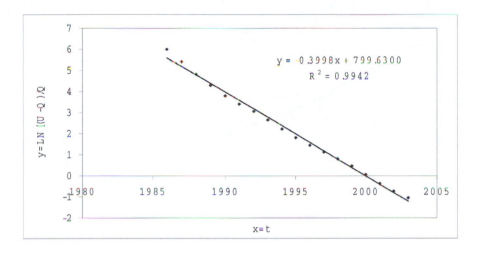

Figure 1. Regression analysis, using a logistic growth model, for cumulative U.S. cellular phone subscriptions with saturation level realized at 200 million subscriptions (Data sources: CTIA 2003).

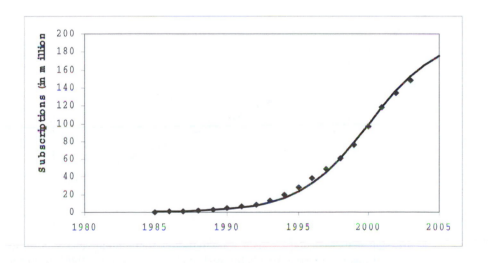

Figure 2. Resulting logistic curve for actual numbers of annual cellular phone subscriptions in the U.S. during 1985-2000 (Data source: CTIA 2003).

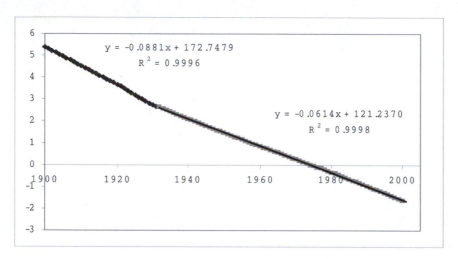

Figure 3a. Regression analysis, using a logistic growth model, for cumulative U.S. domestic oil production with ultimate oil reserves of 216 billion bbl - based on Bartlett (2000) and Laherrere (2000). (Data sources: US Census Bureau 1975, EIA 2002a).

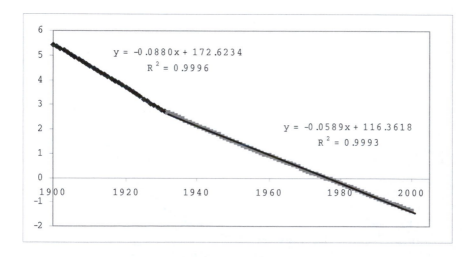

Figure 3b. Regression analysis, using a logistic growth model, for cumulative U.S. domestic oil production with ultimate oil reserves of 229 billion bbl - based on USGS (1995; 1998) and MMS (1995) (Data sources: US Census Bureau 1975, EIA 2002a).

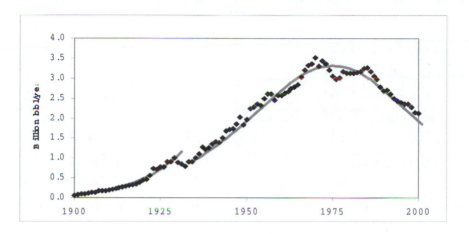

Figure 4a. U.S. domestic annual oil production (diamonds = actual and solid line = estimated) with ultimate oil reserves of 216 billion bbl.

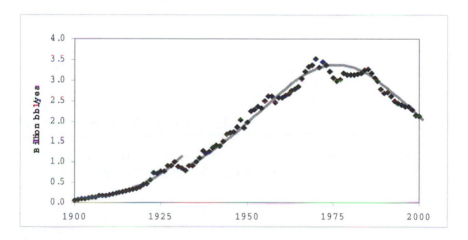

Figure 4b. U.S. domestic annual oil production (actual and estimated) with ultimate oil reserves of 229 billion bbl.

Table 1. Statistics for logistic curves fitting oil and natural gas production, cellular phone subscriptions, personal computer sales, and PV shipments.

Technology/Energy Option	α	β	R²
Cumulative oil production (1900-1931 period)*	172.75	-0.0881	~100%
Cumulative oil production (1932-2000 period)*	121.24	-0.0614	~100%
Cumulative oil production (1900-1931 period)**	172.62	-0.0880	~100%
Cumulative oil production (1932-2000 period)**	116.36	-0.0589	~100%
Cumulative natural gas production	141.13	-0.0711	~100%
Cellular Phone industry # of subscribers	799.63	-0.3998	99%
Cumulative PC sales	413.42	-0.2071	~100%
Cumulative PV shipments in the US at 10% cap	363.2004	-0.1783	~100%
Cumulative PV shipments in the US at 15% cap	363.5500	-0.1782	~100%
Cumulative PV shipments in the EU at 10% cap	365.3638	-0.1793	~100%
Cumulative PV shipments in the EU at 15% cap	365.7169	-0.1793	~100%

* A U value of 216 billion bbls is assumed (based on Bartlett (2000) and Laherrère (2000)).
** Based on the maximum reserves estimate of 229 billion bbls (assuming a $30 selling price see USGS, 1995; 1998 and MMS, 1996).

2.3 Forecast of PV Generation: 2000-2070

For the U.S. and EU,[6] we assumed initial PV domestic sales growth to average 20% per year. This is consistent with regression results for the period 1984-2002 for the U.S. and EU domestic PV markets (see Figures 5a and 5b).[7] We also assumed that the contribution of PV to European and U.S. electricity supply would grow until, in each case, it reached a level of 10-15% of total electricity supply.

These assumptions, when applied in a logistic growth model, mean that PV annual domestic sales are expected to be approximately 20% until 2020 or so, at which point slower rates of growth appear and by 2040 single digit increases are projected. As discussed in section 2.4, there are sound empirical reasons to expect such a growth path.

We assumed that annual EU and U.S. electricity generation growth would be 1.1% and 1.8%, respectively, for the period 2001-2020. The rates were chosen be-

---

[6] For data consistency, only the current 15 members of the European Union (Austria, Belgium, Denmark, Finland, France, Germany, Greece, Ireland, Italy, Luxembourg, Netherlands, Portugal, Spain, Sweden, United Kingdom) and Norway (a major European oil producer) are included in our analysis. Thus, the abbreviation "EU" hereinafter includes the original 15 members plus Norway.

[7] The slope terms in the regressions in Figure 5a and b can be converted, as follows, to gain the estimated initial annual growth rates in PV domestic generation: $e^\beta-1$= annual growth rate. For β=0.1783 (U.S.) and β=0.1793 (EU), initial annual growth in PV generation is approximately 20% for the 1984-2002 period.

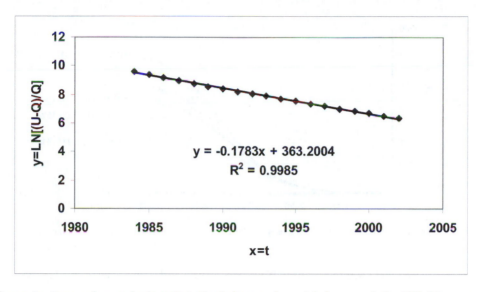

Figure 5a. Regression analysis, using a logistic growth model, for cumulative PV shipments in the U.S. during 1984-2002 (Data source: Maycock, 2002).

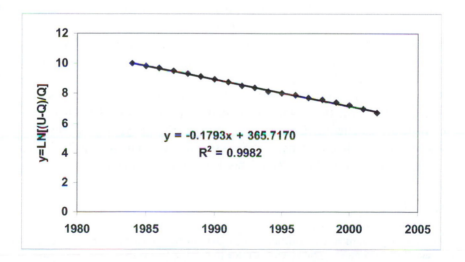

Figure 5b. Regression analysis, using a logistic growth model, for cumulative PV shipments in the EU during 1984-2002 (Data source: Maycock, 2002).

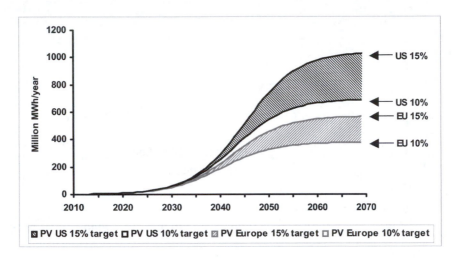

Figure 6. Potential U.S. and European PV supply at 10% and 15% of electricity supply target.

cause they are the ones used by IEA for the EU and EIA for the U.S. (IEA, 2002c and EIA, 2002b). We also assumed that both the EU and U.S. total electricity consumption would stabilize by 2050. Since the demand for PV generation is linked in our forecast to national/regional electricity demand, this has the effect of assuming that electricity demand growth declines gradually from 1.1% (for Europe) and 1.8% (for U.S.) in 2020 to 0% in 2050.[8]

Figure 6 provides our forecast of domestic PV market sales for the EU and U.S. that result from the assumptions described above. We expect PV generation into the EU and U.S. grids to remain quite small until 2025 when the level of output from domestic installations begins to be noticeable. By 2050, we project PV generation to diffuse into the electricity market and to represent a 10 to 15 % share of the generation mixes of both jurisdictions.

Because ours is a long-term forecast (70 years), many uncertainties are associated with the forecast. In the following section, we consider these uncertainties as part of an effort to benchmark our forecast with empirical experience from the energy and silicon-based technology markets. Specifically, we compare our PV forecast with actual diffusion patterns of oil, natural gas, cellular phone and personal computers.

---

[8] While some may regard a zero growth rate after 2050 to be unrealistic, it is useful to consider that substitutes for conventional grid electric service (including fuel cells and/or "knowledge improvements") may change the electricity services market over the next 50 years. But the real purpose of a zero growth scenario is to ensure a conservative forecast for PV, since continued growth in electricity demand, even with a cap on PV's share, would otherwise lead to *higher* forecasted growth in PV generation demand after 2050.

3. Reasonableness of the Forecasts

The annual compound growth rate for PV shipments from the U.S. and EU companies over the last two decades has averaged 20%.[9] As shown in Table 2, this growth level is not unusual for the range of technologies we examined. While fossil fuels diffused at slower rates, these energy options entered U.S. and European markets early in the 20th century, when modernization was still at an early stage. By contrast, silicon technologies have diffused much more rapidly. Naturally, the economic context of their diffusion is contemporary with PV, but they share additional features with PV of being service oriented and decentralized in their applications. By contrast, the fossil fuels are built on large-scale, manufacturing-focused applications.

Past experience shows that growth rates were quite stable for these diverse energy options and technologies until 1% of saturation levels were reached. At that point, growth rates gradually declined for all comparable cases examined here. Thus, our assumption of an initial annual growth rate of 20% is both statistically (Table 1) and empirically supported by technology and energy market experience (Table 2).

The assumption of a 10-15% limit on PV's contribution to future European and U.S. electricity supply is conservative. The technical limit for grid use of an intermittent source of energy is ordinarily thought to be around 30% (e.g., Kelly and Weinberg, 1993), and one group of researchers has suggested that for PV, specifically, it may be upwards of 20% (Perez et al., 1993). According to an EPIA (1996) study, PV on rooftops, without considering other PV applications, can provide more than 17% of electricity supply in the U.S. and 16% in the EU. Thus, our 10-15% cap is in the lower range of research projections and technical limits.

A 2050 peak for growth in PV installations in European and U.S. markets, as well as an assumption that, after 2050, growth is limited to replacement demand, are likewise conservative. Both anticipate slow market maturity for the technology when silicon-based technologies have tended to grow more rapidly than these assumptions allow (see Table 2). Studies of the diffusion of new energy technologies suggest that higher growth and market share could be assumed than we have in our analysis (e.g., Payne et al., 2001).

The reasonableness of our PV forecast can be additionally established by comparing historical experience of energy options and silicon technologies with our projection of growth in PV demand at strategic points in the logistic growth pathway. The early phase of logistic growth can be divided into two key segments—the growth pattern as an option reaches 1% of saturation and the growth pattern as it reaches 10% of saturation. Indicators in units of time and percent growth for both saturation levels are reported in Table 2 in order to establish a basis for comparison. An inspection of these data reveals our PV forecast anticipates conservative to mid-range performance for both segments of early logistic growth.

---

[9] The National Center for Photovoltaics of the U.S. Department of Energy's National Renewable Energy Laboratory uses a 15-25% range of annual growth for its American PV "roadmap."

Table 2. Market growth rates for different technologies.

Technology	Number of years for reaching 1% of saturation level	Annual growth rate	Number of years for reaching 10% of saturation level from the 1% level	Annual growth rate	Number of years for reaching 90% of saturation level from the 10% level
Cumulative U.S. oil production	50	9%	30	6%	~70
Cumulative U.S. natural gas production	40	8%	35	7%	~60
Cellular Phone industry # of subscribers	4	49%	6	43%	11
Cumulative PC sales	2	160%	4	21%	21
PV US* at 10% electricity cap	34	20%	13	18%	24
PV US* at 15% electricity cap	36	20%	13	18%	24
PV EU** at 10% electricity cap	31	20%	13	18%	24
PV EU** at 15% electricity cap	33	20%	13	18-20%	24

  * 1% of saturation to be reached in 2018-2020, and 10% in 2031-2033.
 **1% of saturation to be reached in 2015-2017 and 10% in 2028-2030.

PV is projected to take 30-35 years to reach 1% of saturation, a time interval consistent with the fossil fuels and significantly longer than the experience of silicon technologies such as the cellular phone and the PC. The rate of annual market growth for PV during this early period of development in our model (20%) is well below the empirically experienced range of the silicon options.

In the second segment of logistic growth, we project PV to take an additional 13 years to reach 10% of saturation, well within the 5-35 year range of the selected cases reported in Table 2. The decline in annual market growth during this period to 18% is consistent with the declines in growth experienced by oil and natural gas. Likewise, silicon technologies have experienced declines in the early phases of their diffusion. But it is notable that the projected growth for PV is quite a bit lower than its silicon-based counterparts and is only modestly faster than its energy competitors. As to the latter, it is important to consider that we have capped demand for PV-based generation at 10-15% of national/regional electricity demand and, further, we have assumed that national/regional electricity demand will reach an *absolute* peak in 2050. Both assumptions mean that PV is assumed *not* to realize annual energy demand levels as substantial as the fossil fuels. This can explain the difference in growth rates to some degree: PV is being forecasted to approach a comparatively small demand level, while oil and natural gas grew to reach a much larger eventual value. In this instance, it is not unreasonable for PV to be projected to experience more rapid early growth than the fossil fuels as it realizes a moderate level of energy demand.

A third method for assessing the reasonableness of our PV forecast is to compare it to the projections of other researchers. While the amount of research on this topic is limited, we were able to identify forecasts by organizations whose work is typically regarded as credible in the energy field. These include for the U.S., the National Renewable Energy Laboratory's National Center for Photovoltaics (NCPV), Tellus Institute and Brookhaven National Laboratory; for Europe, we relied on forecasts from the EU Energy Commission and consultants to that commission. Because these forecasts reported their projections in units of installed capacity, we converted our PV generation forecast into capacity units, using an average capacity factor of 18.5% for the U.S. and 16% for the EU (see Wenger et al., 1996 and Sinke, 2001 for support for our assumed capacity factors). Tables 3 and 4 summarize the results.

Clearly, our projected capacity additions are well within those of similar assessments by other reputable researchers for the initial 20 to 30 years of the forecast period. We were not able to find another rigorous long-term forecast of the length of our analysis in the research literature, except a study conducted by the International Institute for Applied Systems Analysis (IIASA) (Christiansson, 1995). That effort forecasts PV generation at the end of this century at 1,800 TWh for North America and at 1,293 TWh for Europe. Our forecast, for comparison, is between 694-1,042 TWh for the U.S and between 380-571 TWh for Europe. Supportive evidence of our shorter term forecast of PV-based electricity generation is found in the U.S. EIA's recent projection of distributed generation, which anticipates electricity from PV in the U.S. ranging from 1.0 to 7.0 billion kWh in 2010 (its reference case) and 2.0-16 billion kWh in 2020 (assuming a 40% tax credit for PV and advanced technological improvements-see EIA, 2000). In our own model, PV is forecasted to produce 1.7 billion kWh in 2010 and 10.3 billion kWh in 2020. Thus again, our forecast appears to be conservative.

4. Current and Competitive PV Prices

Because electricity generation via PV for grid service (i.e., the supply of kWhs to an electric grid) is currently expensive compared to thermal power plants, a projection such as that in Figure 6 may be judged by some to be unrealistic. Unless PV generation costs decline dramatically, skeptics might argue that the technology will continue to be uncompetitive and unable to realize even the modest amount forecasted by us. To address this issue, we consider evidence regarding the time frame in which PV is likely to reach its break-even price, such evidence speaks directly to the question of when PV can compete favorably in the energy market. We also consider the volume of PV domestic sales necessary to realize a break-even price and compare it to our forecast levels in order to evaluate our projection by this key economic criterion. We address both questions using experience curve analysis.

Table 3. Forecasted Installed Photovoltaic Capacity in U.S. (In GW$_p$).

	2010	2020	2025	2030
Our Forecast	1.0	6.4	15.5	36.8 – 37.9
NCPV Roadmap		15.0		
Tellus Institute – 5%	1.7			
Brookhaven National Laboratory	1.2	11.9	36.6	111.9

Table 4. Forecasted installed Photovoltaic Capacity in Europe (In GW$_p$).

	2010	2020	2030
Our Forecast	1.2	7.1	38.8 - 40.8
EC 1997 White Paper	3.0		
EurObserv'ER 2003, PV Barometer	1.4		
EPIA & Greenpeace	2.8	54.0	
Jäger-Waldau, "PV Status Report 2003"	3.0	15.0	30.0

Experience curve analysis is often used to evaluate market entry issues related to new technologies. In essence, the price of a new technology (or its services) is conceived to be an inverse function of production and application experience. As makers of the technology learn about factors that affect production and use, adjustments are made which improve efficiency in both dimensions and enable production and application volumes to be increased. Unit costs decline over time and applications are discovered at these lower costs that widen the market, thereby reinforcing trends toward lower prices and/or higher market value (see, e.g., Mansfield, 1993; IEA, 2000).

Experience curves can be described by the following equation:

$$\text{Price at year } t = P_0 * X^E \tag{Eq. 6}$$

where:

P_0 is the price of the first unit of cumulative shipments
X is cumulative shipments at year t
and E is the experience index (which sets the rate of change in the price-production relationship of the experience curve).

The experience index can be derived from what is termed a "progress ratio" (PR) (or vice versa) given that $PR = 2^E$ (IEA, 2000). With a reasonable value for E established (see below), the experience curve equation can be used to calculate the breakeven level of cumulative shipments necessary to bring the average selling price to a level that can be expected to be competitive with other options.

The average selling price for modules at which PV should become competitive has been extensively debated in the research literature. Forecasts vary from $0.50/$W_p$ to $2.50/$W_p$ (Neij, 1997; International Energy Agency, 2000; NCPV, 2001; Zwann and Rabl, 2003). We adopt $1.50/$W_p$ as a mid-range value. Using log-linear regression analysis for the period 1985-2001, a PR of 80% is statistically estimated (assuming a break-even price of $1.50/$W_p$). With a PR of 80%, the resulting break-even level of cumulative worldwide shipments is about 22,000 MW_p (Byrne et al., 2004). U.S. and European sales are expected by our forecast (see Figure 6) to reach 22 GW before 2025. Thus it would appear that the forecast in Figure 6 is easily achievable at a break-even price of $1.50/$W_p$. These results are not far from those of other researchers. For example Poponi (2003), who uses a PV *system* break-even price of $3.20/$W_p$ and market growth rates[10] of 15-30% anticipates the break-even year to occur between 2011 and 2017. And EIA (2000) expects that PV prices will reach a break-even price of $3.00 per Wp between 2010 and 2014.

But there is also a broader issue at stake, namely, whether PV is properly conceived as a fuel/technology substitute *within* the existing energy system; or whether PV is to be forecasted as an option that could significantly alter the architecture and function of future energy systems. If it is the former, then PV's price should be compared to existing energy prices. In this case, the current cost of $0.30 per kWh of PV-generated electricity compares unfavorably to the $0.03-$0.05 per kWh for power from natural gas or coal fired plants (Gross et al., 2002).[11] The break-even price for the above analysis would be approximately $0.10 per kWh, still twice that of thermal plants. While it is plausible that PV could find competitive opportunities in special grid service applications equivalent to the GWs forecast in Figure 6, realizing such a scale would be difficult and, in this instance, there would be grounds for skepticism.

However, if PV is instead regarded as a new type of service technology intended to facilitate the transition to a decentralized energy architecture, the above approach is ill-conceived. In this case, PV can be considered to compete in several markets. For example, PV can be used as a source of on-site generation with its output wholly or partly consumed by its owner. Where net metering is allowed,[12] PV costs should be compared to retail electricity prices (with an adjustment for avoided transmission and distribution (T&D) losses). Under this scenario, PV systems that sell for $3.00/Wp (i.e., module prices of $1.50/$W_p$) would be competitive in several U.S. and European markets (notably, Germany, Italy, California and New York). No longer a technology

---

[10] Typically, the PV component of an installed system (which additionally includes an inverter, power conditioning, a small amount of buffer storage, and wiring and array structure) represents 50% of system cost. Thus, for comparison to our module cost of $1.50 W_p, Poponi's system cost is the equivalent of $1.60 W_p for PV modules.

[11] This does not include the health and environmental costs of fossil fuel use. When these are taken into account, some argue that electricity from fossil fuel-fired plants is more expensive than PV (e.g., Hohmeyer, 1992).

[12] Currently, 36 states within U.S. require utilities to purchase generation surpluses from PV and other qualified renewable energy (UCS, 2003).

for specialized markets, PV would be economical as an option for retail electricity markets. Of course, the possibility of competing in retail markets depends upon policy interventions that require open access to customers. Further, it would be important that PV is properly credited for the benefits of avoided T&D losses that its use would create.

PV can also provide a range of services that large power plants cannot. These include: peak-shaving and load control (what has elsewhere been termed "energy management"–see Byrne et al., 1996; 1997; 1998 and 2000), emergency/back-up power (Byrne et al., 1996; 1997; 1998) T&D decongestion (see Letendre et al, 1996), environmental improvements (e.g., reducing pollution that leads to urban smog and acid rain, and mitigating carbon emissions that are associated with climate change) and fuel diversity (see, especially, Awerbuch, 1995). When the values of these services are incorporated, PV can compete in *new* markets not served (or only partly served) by the current energy system. In particular, these services coincide with the shift to a decentralized energy system in which the economics of modularity (Hoff and Herig, 1997; and Dunn 2000) replaces the traditional scale–based economies of fossil and nuclear fuels (Messing et al, 1979; Hunt and Shuttleworth, 1996) in guiding technology and market development of the energy sector. Again, policy frameworks are needed that enable social and market institutions to accurately evaluate these benefits.

In this respect, PV's future may be only partly gauged by cost and price comparisons with the existing energy system. A wider consideration of PV applications may be required to realize an accurate portrait of its potential. Furthermore, the role of policy in creating conditions for fair and full competition would need to be addressed.[13] If a broader approach is granted, then we are confident that our PV forecast for the U.S. and Europe is reasonable.

5. Forecasting Oil Production

Oil production forecasts for Europe and the U.S. are regularly prepared by government and independent research sources. We relied on forecasts from IEA country reports (2001, 2002a, 2002b) for Europe[14] through 2010 and on an EIA (2002b) forecast to 2010 for the U.S. For further projections to 2070, a logistic growth function

---

[13] The role of policy in shaping PV's future is discussed in detail in section 7 of this chapter.

[14] Because the major oil production area in Europe is the North Sea and no other significant petroleum deposits in Europe have been identified by research (see, e.g., USGS 2000), North Sea oil was the only known European deposit included in the analysis. Of course, an estimate of undiscovered reserves in Europe, prepared by the USGS, was added.

was applied to USGS (2000) estimates of remaining and "undiscovered" but economically recoverable oil reserves in the U.S. and Europe.[15] This method assumes that after maturation of known and expected (so-called "undiscovered") production areas, and long after passing peak production levels (1970 for U.S. and around 2000 for Europe–see below), annual oil production will decline exponentially.

Our approach follows the methodology pioneered by M.K. Hubbert (1962), whose well-known forecast for Shell Oil has proved to be quite accurate. It is also consistent with the methodological refinements developed by Laherrère (2000), a contemporary of Hubbert. To derive annual declining rates, production levels for the beginning of the forecast period (P) and potential remaining oil reserves (S) at that time are utilized. Potential remaining oil reserves (including "undiscovered" deposits) during the 2010-2070 forecast period are calculated by the following formula:

$$S=(R+X)-(H+F) \tag{Eq. 7}$$

Where:

S is the remaining potential reserve beyond 2010
R is the USGS estimate of remaining reserves
X is the USGS estimate of undiscovered oil reserves
H is historical cumulative production from 1995 to 2001
F is cumulative production from the early forecast period of 2002-2010.

It can be shown by the following mathematical manipulation that the rate of decline is:

$$r=P/S \tag{Eq. 8}$$

The proof is as follows:

$$P/(1+r)+P/(1+r)^2+ P/(1+r)^3+\ldots=S \tag{Eq. 9}$$

Multiplying both sides by $1/(1+r)$ gives:

$$P/(1+r)^2+ P/(1+r)^3+ P/(1+r)^4\ldots=S/(1+r) \tag{Eq. 10}$$

---

[15] The term "undiscovered" is used by the USGS to capture new on-shore oil findings that are not now known. It is predicated on experience throughout the 20th century in which new oil reserves were frequently found, even in extensively explored areas. In addition, it includes oil that could be found as a result of technology improvements. The concept is not without controversy and some researchers believe that it is being incorrectly used in current forecasts (Bentley, 2002). The USGS then divides its "undiscovered" oil estimates into "economically recoverable" and "uneconomical" deposits. The same method is adopted by MMS for the off-shore case. Our oil forecast includes on- and off-shore, economically recoverable "undiscovered" deposits, as well as the remaining oil in already known on- and off-shore reserves.

Subtracting (10) from (9) gives:

$P/(1+r)=S-S/(1+r)$ (Eq. 11)

Multiplying both sides by $(1+r)$ gives:

$S+S*r-S=P$ (Eq. 12)

After canceling S terms and dividing both sides by S, equation (8) results.

After obtaining r values, annual oil production forecasts for Europe and U.S. were projected to 2070.

Domestic oil production is expected to decline in the U.S. and Europe. IEA (2002c) has concluded that European oil production peaked in 2000, while historical data from EIA (2002a) indicate that American production peaked in 1970 and continues to decline. While undiscovered oil deposits are anticipated in both regions,[16] these will not be sufficient to reverse the decline in domestic oil production in either region. The controversial category of reserve growth was not included in our analysis.[17]

If we consider the projected energy production value of all current oilfields and potential, but as yet undiscovered, oil production areas, and fit a logistic curve to link them to 2010 IEA/EIA forecasts, we can find a plausible pathway for domestic oil supply through 2070.[18] Using the methodology described above, production forecasts were made for U.S. and Europe (Figures 7 and 8). By 2055 and by 2070, respectively, Europe and the U.S. are projected to have fully depleted their domestic reserves. Thus, from the perspective of domestic production, the oil era will have concluded by 2070 for both jurisdictions.

6. Comparison of the Energy Values of Future Oil Production and PV Systems

Forecasted energy generation from PV is converted into barrels of oil equivalent to facilitate the comparison of energy from PV with the oil production forecasts developed above. For our conversion, we made the following assumptions: a 30-year lifetime for PV systems; and 1 peak Watt of PV generates an average of 1.6 kWh[19] of electricity annually in the U.S. and 1.4 kWh in Europe (taking into account avoided

---

[16] USGS (2000) believes that 19.6 billion bbls of oil are still to be found in Europe and USGS (1998) and MMS (1996) believes that 37.4 billion bbls can be economically recovered at $30 per bbl in the U.S. from this category.

[17] Because this category of oil additions is disputed by many in the forecast community (Laherrère, 1999; Bentley 2002), we decided to exclude it from our analysis.

[18] Of course, forecasting future oil production is a risky analytical enterprise that, understandably, can yield widely varying estimates.

[19] According to Wenger et al., 1996, the capacity factor for U.S. is 17-20%, with a mid-point of 18.5%. According to Sinke (2001), for Europe the capacity factor is 8-24%, with a mid-point of 16%. These capacity factors yield 1.6 kWh and 1.4 kWh of annual generation per 1 Wp of installed PV, respectively for the U.S. and EU.

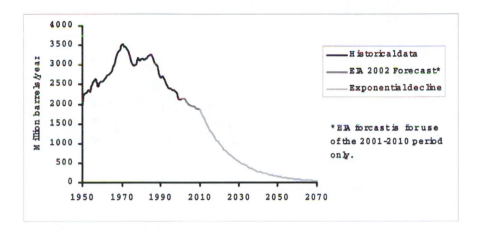

Figure 7. Historical and forecasted U.S. oil production (Data sources: EIA 2002a, EIA 2002b, USGS 2000).

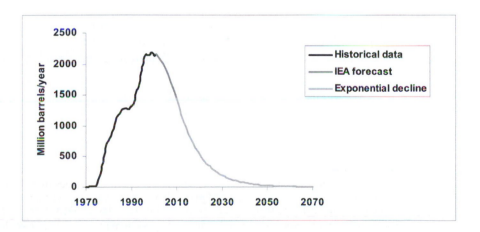

Figure 8. Historical and forecasted European oil production (Data sources: IEA 2001, IEA 2002a, IEA 2002b, USGS 2000).

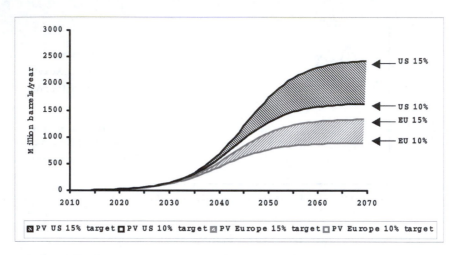

Figure 9. Potential U.S. and European PV supply at 10% and 15% of electricity supply target expressed in barrels of oil.

T&D losses). According to the Society of Petroleum Engineers (SPE, 1984) 1 barrel (bbl) of crude oil = 6.12×10^9 J. Using 1 kWh = $3.6 \ 10^6$ J and taking into account a typical efficiency of an internal combustion engine of approximately 25%,[20] 1 bbl of crude oil is equivalent to 425 kWh of electricity. Our PV forecast, converted to oil equivalent, is presented in Figure 9.[21]

By our calculations, U.S. cumulative domestic oil production for the period 2010-2070 could amount to 30.3 billion barrels.[22] PV energy supply in oil equivalent is forecasted to be between 45.0 and 62.2 billion barrels. Thus, the contribution of PV to U.S. energy supply for 2010-2070 is likely to be 1.5 to 2.0 times that from domestic oil (Figure 10). For Europe, cumulative domestic oil production could amount to 14.6 billion barrels, while PV energy supply in oil equivalent for the same period could grow to between 27.5 and 38.2 billion barrels. Thus, the contribution of PV to European energy supply for 2010-2070 would be at least twice that from domestic oil (Figure 11).

---

[20] Typical energy efficiency for internal combustion engines under test conditions ranges from 20% to 40%, with the higher end for diesel engines (Plint and Martyr, 1995), Under real operating conditions, depending on the type of engine, efficiency can range from 15% to 30% (Lumley, 1999). Consequently, an efficiency of 25% was assumed for the current research. ICC technology was selected for this conversion because the principal use of oil in the U.S. and EU is to power vehicles and stationary engines.

[21] Figure 9 is identical to Figure 6, except for the energy unit used to map PV market development.

[22] This estimate assumes the higher U value of 229 billion bbls–see section 5 above.

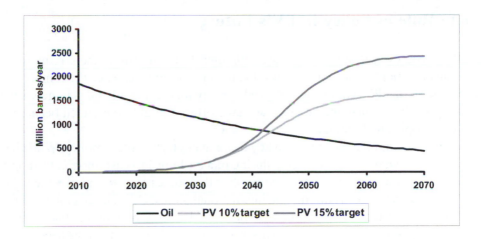

Figure 10. Comparison of forecasts of U.S. PV energy supply and U.S. oil production from existing domestic reserves.

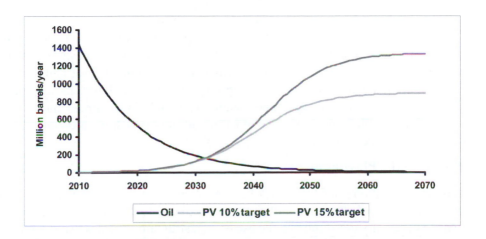

Figure 11. Comparison of forecasts of European PV energy supply and European oil production from domestic reserves.

7. The Role of Policy in PV's Future

Of course, even if our PV forecast is thought to be plausible, there remains an important objection to our analysis. Specifically, some might challenge the comparison itself, arguing that PV would compete in electricity markets, while oil is largely used for transport. In essence, the analysis we have offered might be rejected because it compares 'apples with oranges.'

We believe this objection misses a key factor, namely, the role of policy. While it is likely that PV and oil would compete during much of our forecast horizon to supply distinctly different services,[23] the ability of both sources to serve markets will be significantly dependent on national and international policy. Oil's status as an essential energy source for industrialized society comes with policy obligations that include national and global security commitments, subsidies to relieve users of the need to pay the full social and environmental costs of oil production and consumption, and increasingly favorable treatment for investments in oil extraction (because exploration and drilling will only increase in cost over the next 65-70 years). The oil industry has demonstrated an impressive capacity to obtain needed policy attention in all of these areas in order to sustain its market viability.

For PV to attract even modest policy support (beyond its present treatment as a 'frontier' technology deserving 'market priming' assistance), it must compete in the very important policy 'marketplace.' Comparing PV to oil is essential for PV's participation in this competition. National and international energy policy largely exists as a fuels policy, rather than a sectoral or user-focused policy. The findings of the research reported here can be used to 'level the playing field' of policy-making, and provide grounds for challenging the current pro-carbon bias in energy policy throughout Europe and the U.S. Hopefully, it may also encourage Europe and the U.S. to re-think the meaning of and strategic planning for energy security.

Thus, we think there are sound reasons for comparing the long-term prospects for oil and PV. Not only the changing economics of energy markets, but the highly important role of energy policy suggest that the comparison is warranted. That said, we now turn our attention to a discussion of energy policy in Europe and the U.S. in order to consider the sorts of tools that might be used to promote PV investment and uses. In this way, it is possible to consider if an operational policy path is emerging that, in effect, might "realize" our forecast.

7.1 Leveling the Playing Field

Historically, national and regional policies have greatly influenced energy development. The creation of state monopolies to shelter fossil fuels, hydropower, nuclear energy and electricity development, the use of infrastructure subsides to pay for pipe-

---

[23] Of course, there is the possibility that oil and PV could be direct competitors. If a hydrogen economy emerges by 2020 or so, or if breakthroughs in electric vehicle technology (especially with respect to storage) are realized, we could see direct competition between these sources.

lines, ports and T&D systems, and the authorization of a wide range of preferential tax treatments to favor investment in and consumption of selected energy sources and technologies are examples of how policies have shaped the current energy system. Globally, policy support for fossil fuels, hydropower and nuclear energy is estimated to total more than $240 billion annually (van Beers and de Moors, 2001), creating an obvious market advantage for the energy *status quo*.

While fossil fuels and, especially, oil receive disproportionate energy policy attention in Europe and the U.S. and garner the bulk of the yearly world energy subsidy of $240 billion, there is evidence that renewable energy is beginning to attract modest policy support. Below we describe the evolving policy contexts in both jurisdictions with respect to renewable energy generally and PV specifically. Policy will likely prove to be critical to PV's long-term market development. This is because energy market distortions of the magnitude that now exist can only be overcome by policy action.

But equally important, PV promises basic change in the nature and type of energy services available in a society (see above) and it promises to change the political and economic character of energy production and consumption. The environmentally unsustainable and socially inequitable features of the present system, as well as the "abundant energy" ideology that has justified it (Byrne and Rich, 1986) are all challenged by PV, which is environment-enhancing and inherently decentralized, thereby offering opportunities for democratic management of energy systems (Lovins, 1977; Byrne and Rich, 1992;Flavin, 1994). Even ardent advocates of the fossil fuel era have recognized that oil, natural gas and coal systems tend to be technocratic and centralized in character (Burton, 1980).

7.2 U.S. PV Policy

A variety of policies have been and will be important to the growth of PV use in the U.S. Besides net-metering (which has been adopted by 36 U.S. states), interconnection standards, tax credits, renewable portfolio standards and public benefit charges have been key to market development for PV in the U.S.

Interconnection remains a barrier to PV use in the U.S. because no national standard exists and requirements among utilities are diverse and inconsistent. The lack of a uniform standard is a burden for PV installers and creates additional transaction costs, increasing the overall cost of a PV system (Beck and Martinot, 2004). Adoption of a national standard would appear to be essential for long-term development of PV.

The federal energy investment tax credit, a product of the 1992 Energy Policy Act, provides a 10% tax subsidy to companies that purchase certain renewable energy systems including PV (DSIRE, 2003). The tax credit has been a significant incentive for installing PV systems. However, a limitation is that it only benefits companies with appropriate tax liability. Making the tax credit both tradable and available to all parties would create a wider incentive for the use of PV.

The policy mechanism providing the greatest potential for future growth of installed PV capacity in the U.S. is the Renewable Energy Portfolio Standard (RPS). An RPS requires that a certain percentage of all electricity generated in a jurisdiction is provided from qualified renewable sources of energy (typically, solar thermal and electric, wind, some forms of biomass, and geothermal). Sixteen states have adopted renewable portfolio standards, which will result in 14 GW of new renewables by 2012. But to date, a national RPS has not been passed (UCS, 2003).

Recently, two bills on a national RPS appeared in the U.S. Senate and one in the U.S. House. Senator Jim Jeffords' Renewable Energy and Energy Efficiency Investment Act called for renewable resources to account for 20% of U.S. electricity supply by 2020, while Senators Tom Daschle and Jeff Bingaman's Energy Policy Act included an RPS of 10% by 2020. Congressman Mo Udall's H.R. 1294 included an RPS of 20% by 2025. A national study conducted by Tellus Institute concluded that an RPS would have a significant impact on the growth of PV capacity in the U.S. Using the National Energy Modeling System (NEMS) created by the U.S. Energy Information administration, Tellus found that 5% of renewables needed to reach a national RPS of 10% by 2010 would come from solar (Bernow et al., 1997).

Another key policy tool widely used in the U.S. is the public benefit charge (PBC), which assesses a small fee per kWh sold by utilities (usually $0.001-$0.003 per kWh). Accrued funds from PBCs are then auctioned in a competitive bidding process to companies and non-profit organizations for the promotion of energy efficiency and renewable energy projects. Thirty-one states currently employ these charges and 14 have specific set asides, for renewable energy development. Through 2012, these 14 states are obliged to spend at least $4.3 billion for PV, wind and other qualifying renewables (see UCS, 2003).

To ensure that PV is able to enter and transform the U.S. energy market, national RPS and PBC policies, interconnection standards and investment tax credits are probably needed. States have already taken the lead in this effort to ensure both diversity and success in their energy policies. Realization of our U.S. PV forecast would undoubtedly be significantly advanced if such a policy paradigm were adopted at the national level.

7.3 European PV policy

The European Commission's 1997 White Paper called for a doubling of renewables in EU energy consumption, from 6 percent in 1997 to 12 percent by 2010. A 2001 directive called for 22.1 percent of electricity to be derived from renewables, also by 2010 (European Parliament, 2001). For photovoltaic applications, the specific target is 3 GW of installed capacity (which is a little less than 1% of current EU generation capacity) (European Commission, 1997 and 2003).

The EU has pursued its expressed goals for PV development through a number of measures, including the "Campaign for Take-Off" that set targets for the installation of 650 MWp of PV capacity in domestic markets by 2003 (EurObserv'ER, 2001).

Building on partnerships with industry, associations and public officials, the EU has also supported scientific research and demonstration projects in order to increase PV's technological competitiveness and market penetration. To better coordinate research and technology development among countries in the region, the European Commission in partnership with national government agencies and research organizations has established the "PV-EC-NET" project (PV-EC-NET, 2003).

These policies reflect growing momentum due to Europe-wide public support for the technology. Such support has resulted in impressive recent gains for PV. Thus, by the end of 2002, member states' installed PV capacity reached 392 MWp, more than doubling the EU's total of 188 MWp in 2000 (EPIA, 2002). However, of that total capacity, some 278 MWp were installed in Germany. This uneven installation pattern among member states demonstrates how Europe's overall success in PV development can largely be traced to particular policies, tied to specific and long-running support, at national levels.

Germany is the undisputed leader in European PV development. German support for PV has been operationalized through tax incentives and low investment loans, as well as a net metering program that rewards developers with sizable feed-in tariffs, recently equal to approximately $0.60 per kWh (Sijm, 2002). The country's "1,000 Roofs Program" started in 1991 to subsidize PV-electricity generation costs (IEA, 1998). The program's 1999 follow-up, the "100,000 Rooftops Solar Electricity Program," called for the installation of 300 MWp through low-interest loans for solar energy equipment (IEA, 2002d). The country is on track to meet its target (European Commission, 2003: 53). At the same time, the Renewable Energy Sources Act, passed in 2000, altered PV feed-in tariffs to include percentage reductions over time (IEA-PVPS, 2003). However, PV development should remain strong in Germany, as the Federal Environment Ministry and the Federal Ministry of Economics have agreed to provide additional payments (beyond basic feed-in tariffs) for PV systems installed on buildings and facades. These measures are specifically designed to "compensate for the expiry of the successful 100,000 roofs solar electricity program," which ended in June 2003 (Federal Environment Ministry, 2003).

The Netherlands, Italy, Spain, and France are also using feed-in tariffs and additional measures to stimulate investor interest in PV electricity generation (de Vries et al., 2003; ENER-IURE 2002a and 2002b). However, many of these programs are relatively new, having been implemented only in the last two to three years. For example, Italy's version of a solar PV rooftop program, adopted in 2001, provides investment subsidies to support the installation of 10,000 mainly small-scale PV units (EurObserv'ER, 2003 and European Union, 2001).

Goals and projections for installed PV capacity in Europe, while somewhat different in their results, point to continued and sometimes very significant growth in coming decades. In its April 2003 report, the EurObserv'ER Photovoltaic Barometer cautions that future growth in installed PV capacity in the EU remains "very much centered on Germany" (EurObserv'ER, 2003). Because the long-term contributions of policies in France, Italy, and Spain remain uncertain, the Barometer suggests that EU installed capacity may only reach 1,400 MW in 2010. By contrast, the European Photovoltaic Industry Association (EPIA), in a joint effort with Greenpeace, has projected annual growth rates of 27 percent until 2009 and 34 percent between 2010 and 2020. Their scenario holds that PV can supply 10 percent of Europe's electricity needs in 2020 (EPIA, 2001). But a European Commission report on global prospects for PV offers that the EPIA-Greenpeace target is overly ambitious, as it would require an estimated 54 GW of installed solar electrical capacity in the EU by 2020. Building on the 1997 White Paper's target of 3 GW installed capacity by 2010 with a growth scenario of 17 percent per year, Jäger-Waldau (leading author of the updated study) provides a more cautious projection of 15 GW of installed capacity in 2020 (Jäger-Waldau, 2003). Even so, this is twice the level that our forecast expected for 2020 (see Table 4). Thus, slower growth than the optimistic EPIA-Greenpeace scenario could nevertheless lead to PV playing a greater role in Europe's energy future than the region's oil reserves.

The diversity of targets in Europe suggests the need for robust implementation of existing policies in member states, as well as a need to increase policy support beyond Germany in order for the region to make use of the option. In particular, a regional Renewable Energy Portfolio Standard (RPS) implemented at the EU level would seem to be a worthwhile policy option. Under this framework, targets would be allocated among different types of renewable energy technologies, including PV, in order to ensure their market development and penetration while also encouraging "cost-reducing innovations" and a "genuine EU-wide level playing field" among renewable electricity producers (Jansen, 2003). Greater parity in installed capacity throughout the region could help the PV sector become less vulnerable to changes in particular markets and assume a stronger role in EU-wide electricity generation.

8. From Forecast to Policy

Energy debates are usually dominated by issues of technology and markets. This reflects an implied belief that incremental change will characterize the future. But energy change often is dramatic and sudden (e.g., MacKenzie, 1997). Moreover, energy choices can be highly affected by policy decisions. It is wise, therefore to evaluate policy alternatives that do not assume the energy *status quo,* in order to understand the true magnitude of policy choices that are at stake. The direct comparison of PV and oil is an example of a less constrained approach to energy policy analysis. Our findings from this comparison suggest that PV has a realistic potential of providing services in the U.S. that would be 1.5 to 2.0 times that of all U.S. domestic oil reserves. For Europe, the role may be even greater, possibly more than twice as much as the region's domestic oil reserves.

These findings highlight the disjuncture between energy potential and energy policy in our current situation. Indeed, energy policy in both the EU and the U.S. assumes that oil will be a major source for reliably and economically meeting social needs well into the future, while PV is presently regarded in policies of the two jurisdictions as something between an R&D project and a niche technology to fulfill specialized needs. Recent policies have taken tentative steps to broaden the role of PV in the energy systems of the two areas. But more will be needed to overcome the disjuncture between potential and existing market development.

While we are confident that the future will prove PV's skeptics to be wrong, we are nevertheless concerned that the chasm between the potential and current policy unnecessarily hinders the ability of the EU and the U.S. to move forward. In this regard, we worry that many technologists, economists and policy makers still seek an understanding of our energy future by an investigation of its past. Forecast and policy offer tools to avoid the error in this way of thinking. Through both, we can depict where we have been and also how we might arrive at a different destination, if we are willing to act beyond the limits of the *status quo*. All energy transformations, as far as we can tell, occurred when societies were able to do precisely this.

9. References

Awerbuch, S., Dillard, J., Mouck, T., Preston, A., 1996. Capital budgeting, technological innovation and the emerging competitive environment of the electric power industry. Energy Policy, 24 (2), pp. 195-202.

Awerbuch, S., 1995. New economic cost perspective for valuing solar technologies. In Advances in Solar Energy; Annual Review of Research and Development, Boer K. W. (ed), Vol 10, Boulder, Colorado: American Solar Energy Society.

Bartlett, A., 2000. An Analysis of U.S. and World Oil Production Patterns Using Hubbert-Style Curves. Mathematical Geology, 32 (1), pp. 1-17.

Beck, F., Martinot, E., 2004. Renewable Energy Policies and Barriers. Encyclopedia of Energy, Academic Press/Elsevier Science.

Bentley, R.W., 2002. Global oil & gas depletion: an overview. Energy Policy, 30 (3), pp. 189- 205.

Bernow, S., Dougherty, B., Duckworth, M., 1997. Quantifying the Impacts of a National, Tradable Renewables Portfolio Standard. The Electricity Journal, May 1997, pp. 42-52.

Bingham, J., Daschle, T., "S. 1766" 107th Congress. Available on the worldwide web at: http://thomas.loc.gov/cgi-bin/query/z?c107:S.1766:

Burton, D., 1980. The Governance of Energy, New York: Praeger.

Byrne, J., Kurdgelashvili, L., Poponi, D., Barnett, A., 2004. The potential of solar electricity for meeting future US energy needs: a comparison of projections of solar electricity energy generation and Arctic National Wildlife Refuge oil production. Energy Policy, 32 (2), pp. 289-297.

Byrne, J., Agbemabiese, L., Boo, K.J., Wang, Y-D., Alleng, G., 2000. An international comparison of the economics of building integrated PV in different resource, pricing and policy environments: the cases of the U.S. Japan and South Korea. Proceedings of the American Solar Energy Society Solar 2000 Conference, pp. 81-85.

Byrne, J., Letendre, S., Agbemabiese, L., Bouton, D., Kliesch, J., Aitken, D., 1998. Photovoltaics as an Energy Services Technology: A Case Study of PV Sited at the Union of Concerned Scientists Headquarters. Proceedings of the American Solar Energy Society Solar 98 Conference, Albuquerque, NM (June 15-17): pp. 131-136.

Byrne, J., Letendre S., Agbemabiese, L., Redin, D., Nigro, R., 1997. Commercial building integrated photovoltaics: market and policy implications. Proceedings of the 26th IEEE Photovoltaic Specialists Conference, pp. 1301- 1304.

Byrne, J., Letendre, S., Govindarajalu, C., Wang, Y-D., 1996. Evaluating the economics of photovoltaics in a demand-side management role. Energy Policy, 24 (2), pp. 177-185.

Byrne, J., Rich, D., 1992. Toward a Political Economy of Global Change. In Energy and Environment: the Policy Challenge, Byrne J., and Rich, D. (eds), New Brunswick, NJ and London: Transaction Publisher, pp. 269-302.

Byrne, J., Rich, D., 1986. In Search of the Abundant Energy Machine. In The Politics of Energy R&D, Byrne, J. and Rich, D. (eds.), New Brunswick, NJ and London: Transaction Publishers, pp. 141-160.

Byrne, J., Rich, D., 1983. The Solar Energy Transition as a Problem of Political Economy. In The Solar Energy Transition: Implementation and Policy Implications, Boulder, Rich, D., Barnett, A. M., Viegel, J. M. and Byrne, J. (eds) CO: Westview Press, pp. 163-186.

Cellular Telecommunications & Internet Association, 2003. CTIA's Semi-Annual Wireless Industry Survey Results: June 1985 – June 2003. CTIA, Washington, DC. Available on the worldwide web at: http://www.wow-com.com/industry/stats/surveys/

Christiansson, L., 1995. Diffusion and learning curves of renewable energy technologies. IIASA, WP-95-126, December 1995, Appendix A.

de Vries, H.J., Roos, C.J., Beurskens, L.W.M., Kooijmam-vam Dijk A.L., Uyterlinde, M.A., 2003. Renewable Electricity Policies in Europe: Country Fact Sheets 2003, Energy research Centre of the Netherlands, ECN-C—03-071, ECN Beleidsstudies.

Dunn, S., 2000. Micropower: the next electrical era. Worldwatch Paper 151. Worldwatch Institute, Washington, DC.

Energy Information Administration (EIA), 2003. Analysis of a 10-percent Renewable Portfolio Standard, SR/OIAF/2003-01.

Energy Information Administration (EIA), 2002a. Historical Petroleum Supply/Disposition Data. Available on the worldwide web at: http://www.eia.doe.gov/neic/historic/hpetroleum3.htm

Energy Information Administration (EIA), 2002b. Annual Energy Outlook 2001: with Projections to 2020. EIA, DOE/EIA-0383 Washington, DC, U.S.

Energy Information Administration (EIA), 2002c. North Sea. Available on the world-wide web at: http://www.eia.doe.gov/emeu/cabs/northsea2.html

Energy Information Administration (EIA), 2000. Modeling Distributed Electricity Generation in the NEMS Buildings Models. Available on the worldwide web at: http://www.eia.doe.gov/oiaf/analysispaper/pdf/distgen.pdf

ENER-IURE FASE III, 2002a. Analysis of the legislation regarding renewable energy sources in the E.U. Member States, France/Electricity, 5.

ENER-IURE FASE III, 2002b. Analysis of the legislation regarding renewable energy sources in the E.U. Member States, France/Financial Measures, 5.

EurObserv'ER, 2003. Photovoltaic Barometer, pp. 1-10.

EurObserv'ER, 2001. New and Renewable Energies/Photovoltaics: Objectives – Technology, pp. 1-2.

European Commission, 2003. PV Status Report 2003: Research, Solar Cell Production and Market Implementation in Japan, USA and the European Union, pp. 1-72.

European Commission, 1997. Energy for the Future: Renewable Sources of Energy – White Paper for a Community Strategy and Action Plan. COM (97) 599, pp. 1-27 and 41.

European Parliament and the Council of the European Union, 2001 Directive 2001/77/EC: "Renewable Energy: The Promotion of Electricity from Renewable Energy Sources," pp. 1-4.

European Photovoltaic Industry Association (EPIA), 2002. Photovoltaic in Europe: Capacity Installed in Europe: 2000, 2001, 2002, pp. 1-2.

European Photovoltaic Industry Association (EPIA), 1996. Photovoltaics in 2010: Photovoltaics. Available on the worldwide web at: http://www.agores.org/Publications/PV2010.htm

European Union, 2001. National RE Partnership: PV Roofs Programme, Objectives/Actions, pp. 1-2.

European Wind Energy Association, (EWEA), 1999. Wind Force 10: A Blueprint to Achieve 10% of the World's Electricity from Wind Power by 2020. Forum for Energy and Development, and Greenpeace International, London.

Federal Environment Ministry of Germany, November 11, 2003. Ministries Agree on the Further Development of the Renewable Energy Sources Act, Berlin, pp. 1-3.

Fhenakis, V., Morris, S., 2002. Markal Analysis of PV Capacity and CO_2 Emissions' Reduction in the U.S., Brookhaven National Laboratory.

Fisher, J.C., Pry, R.H., 1971. A Simple Substitution Model of Technological Change. Technological Forecasting and Social Change, 3 (1), pp. 75-88.

Flavin, C., Lenssen, N., 1994. Power Surge, New York: W. W. Norton.

Gross, R., Matthew, L., Bauen, A., 2002. Progress in Renewable Energy. Environmental Policy and Management Group (EPMG) Working Paper Series, Imperial College London, October 2002.

Hoff, T.E., Herig, C., 1997. Managing risk using renewable energy technologies. In The Virtual Utility: Accounting, Technology & Competitive Aspects of the Emerging

Industry, Awerbuch, S. and Preston, A. (eds.), Norwell, Massachusetts: Kluwer Academic Publishers.

Homeyer, O., 1992. The Social Cost of Electricity Generation: Wind and Photovoltaic vs. Fossil and Nuclear Energy. In Energy and Environment: The Policy Challenge, Byrne, J. and Rich, D. (eds.), New Brunswick, NJ and London: Transaction Publishers. pp 141-186.

Hubbert, M.K., 1962. Energy resources: report to the committee on natural resources, Publication 1000-D, Washington, DC: National Academy of Sciences and National Research Council.

Hunt, S., Shuttleworth, G., 1996. Competition and Choice in Electricity, Chichester, England: John Wiley & Sons.

International Energy Agency (IEA), 2003. Photovoltaic Power Systems Programme. (IEA-PVPS), In Trends in Photovoltaic Applications: Survey Report of Selected IEA Countries Between 1992 and 2002, Paris, France: OECD/IEA Publications. pp. 1-20.

International Energy Agency (IEA), 2002a. Energy Policy of IEA Countries; The United Kingdom 2002 Review, Paris, France: OECD/IEA Publications.

International Energy Agency (IEA), 2002b. Energy Policy of IEA Countries; Denmark 2002 Review, Paris, France: OECD/IEA Publications.

International Energy Agency (IEA), 2002c. Electricity information 2002; IEA Statistics, Paris, France: OECD/IEA Publications.

International Energy Agency (IEA), 2002d. Energy Policies of Germany: 2002 Review, Paris, France: OECD/IEA Publications.

International Energy Agency (IEA), 2001. Energy Policy of IEA Countries; Norway 2001 Review, Paris, France: OECD/IEA Publications.

International Energy Agency (IEA), 2000. Experience Curves for Energy Technology Policy, Paris, France: OECD/IEA Publications.

International Energy Agency (IEA), 1998. Energy Policies of Germany: 1998 Review, OECD/IEA Publications.

Jansen, J.C., 2003. Policy Support for Renewable Energy in the European Union: A Review of the Regulatory Framework and Suggestions for Adjustment. Energy research Center of the Netherlands, ECN- C—03-113, ECN Beleidsstudies.

Jeffords, J., "S. 1333" 107th Congress, 1st Session. Available on the worldwide web at: http://frwebgate.access.gpo.gov/cgi-bin/
useftp.cgi?IPaddress=162.140.64.89&filename=s1333is.pdf&directory=/disk2/
wais/data/107_cong_bills

Kelly, H., Weinberg, C.J., 1993. Utility strategies for using renewables. In Renewable Energy. Johansson, T. B., Kelly, H., Reddy, A. K. N., Williams R. H. and Burnham L. (eds), Washington, DC: Island Press.

Laherrère, J.H., 2000. The Hubbert curve: its strengths and weaknesses. Available on the worldwide web at http://dieoff.com/page191.htm

Laherrère, J.H., 1999. Reserve growth: technological progress, or bad reporting and bad arithmetic. Geopolitics of Energy, 22 (4), pp 7-16.

Letendre, S., Weinberg, C., Byrne, J., Wang, Y-D., 1998. Commercializing photovoltaics: the importance of capturing distributed benefits. Proceedings of the American Solar Energy Society Solar 98 Conference, pp. 231-237.

Lovins, A.B., 1997. Soft Energy Path, New York: Harper Collins.

Lovins, A.B., Lovins, H.L., 1982. Brittle Power: Energy Strategy for National Security, New York: Brick House Pub Co.

Lumley, J.L., 1999. Engines an Introduction, New York: Cambridge University Press.

MacKenzie, J.J., 1997. Oil as a Finite Resource: When is Global Production Likely to Peak. Washington, DC: World Resources Institute.

Mansfield, E., 1993. Innovation and the diffusion of new techniques. In Technical Change and the Rate of Initiation, Mansfield E. (Eds.), Bookfield, Vermont: E. Elgar Publishing, pp. 293-318.

Maycock, P.D., 2002. "The World Photovoltaic Market," Warrenton, Virginia: PV Energy Systems Inc.

Messing, M., Friesema, H.P., Morell, D., 1979. Centralized Power: The Politics of Scale in Electricity Generation, Cambridge, Gunn & Hain, Massachusetts, Oelgeschlager.

Mignogna, R.P., 2001. Introduction to technology trend analysis, part 2:' Technology/ Engineering Management Inc. e-Newsletter Available on the worldwide web at: http://www.temi.com/Newsletter/Feb01/TechTrends2.html

Mineral Management Service (MMS), 1996. An Assessment of the Undiscovered Hydrocarbon Potential of the Nation's Outer Continental Shelf. Minerals Management Service OCS Report, MMS 96-0034

National Center for Photovoltaics (NCPV), 2001. Solar Electric Power: The U.S. Photovoltaic Industry Roadmap. Washington, DC: Department of Energy, National Center for Photovoltaics,

Neij, L., 1997. Use of experience curves to analyse the prospects for diffusion and adoption of renewable energy technology. Energy Policy, 25 (13), pp. 1099-1107.

North Carolina Solar Center, 2003. Database of State Incentives for Renewable Energy. Available on the worldwide web at: http://www.dsireusa.org.

Payne, A., Duke, R., Williams, R.H., 2001. Accelerating residential PV expansion: supply analysis for competitive electricity markets. Energy Policy, 29 (10), pp. 787-800.

Perez, R., Seals, R., Stewart, R., 1993. Assigning the load matching capability of photovoltaics for US utilities based upon satellite derived isolation data. Proceedings of 23rd IEEE Photovoltaic Specialists Conference, pp. 1146-1151.

Plint, M., Martyr, A., 1995. Engine Testing: Theory and Practice, Oxford, England: Butterworth-Heinemann Ltd.

Poponi, D., 2003. Analysis of diffusion path for photovoltaic technology based on experience curves. Solar Energy, 74 (4) pp. 331-340.

PV-EC-NET, 2003. Towards an Increased Efficiency of the European PV RTD Programmes. Available on the worldwide web at: (http://www.pv-ec.net/).

Roethle Group, 2002. The Business of Fuel Cells: What's Happening with Fuel Cells? The Roethle Group, Inc.

Sijm, J.P.M., 2002. The Performance of Feed-in Tariffs to Promote Renewable Electricity in European Countries. Energy Research Center of the Netherlands, ECN-C-02-083, ECN Beleidsstudies.

Sinke, W.C., 2001. Key technical and non-technical challenges for mass deployment of photovoltaic solar energy (PV). Energy Research Center of the Netherlands, ECN-RX—01-070, ECN Beleidsstudies.

Smil, V., 2000. Energy in World History, Boulder, CO: Westview Press.

Society of Petroleum Engineers (SPE), 1984. The SI Metric System of Units and SPE Metric Standard, 2nd Printing.

Telecompetition, Inc., 2003. US Mobile Subscribers and Penetration Rates by Major Trading Area and Basic Trading Area 2002-2003, March.

U.S. Census Bureau, 1975. Historical Statistics of the United States, Washington, DC: U.S. Bureau of the Census.

U.S. Census Bureau, 1991-2000a. Statistical Abstract of the United States, Washington, DC: U.S. Bureau of the Census.

U.S. Census Bureau, 1994-2001b. Manufacturing Profiles, Washington, DC: U.S. Bureau of the Census.

U.S. Census Bureau, 2000. Census 2000. Washington, DC: U.S. Bureau of the Census.

U.S. Geological Survey, (USGS), 2000. World Petroleum Assessment 2000: Description and Results, Washington, DC: USGS Digital Data Series DDS-60 Multi Disc Set Version 1.0

U.S. Geological Survey (USGS), 1998. Economics and the 1995 National Assessment of United States Oil and Gas Resources. USGS Circular 1145, Washington, DC: USGS Printing Office.

U.S. Geological Survey (USGS), 1995. National Assessment of United States Oil and Gas Resources, Washington, DC: USGS Printing Office.

Udall, "H.R. 1294" 108th Congress, 1st Session. Available on the worldwide web at: http://frwebgate.access.gpo.gov/cgi-bin/ getdoc.cgi?dbname=108_cong_bills&docid=f:h1294ih.txt.pdf

Union of Concerned Scientists (UCS), 2003. Available on the worldwide web at: http://www.ucsusa.org/clean-energy/renewable-energy

van Beers, C., de Moor, A., 2001. Public Subsides and Policy Failures: How Subsidies Distort the Natural Environment, Equity and Trade, and how to Reform them, Cheltenham, UK: Edward Elgar Publishers.

Weinberg, C.J., Iannucci, J.J., Reading, M.M., 1993. The Distributed Utility: Technology, Customer, and Public Policy Changes Shaping the Electrical Utility of Tomorrow. Energy Systems and Policy, 15 (4), pp. 307-322.

Wenger, H., Herig, C., Taylor, R., Eiffert, P., Perez, R., 1996. Niche Markets for Grid-Connected Photovoltaics, Washington, D.C.: Proceedings of IEEE Photovoltaic Specialists Conference, May, pp. 13-17.

Woodall, P., 2000. Untangling E-Conomics: A Survey of the New Economy. The Economist, Sep., pp. 23-29.

Zwann, B., Rabl, A., 2003. Prospects for PV: a learning curve analysis. Solar Energy, 74 (1), pp. 19-31.

Chapter 3

Recent Advances in Solar Photovoltaic Technology and its New Roles to Contribute to Environmental Issues

by

Yoshihiro Hamakawa*
Advisor Professor to the Chancellor, Ritsumeikan University
1-1-1 Nojihigashi Kusatsu Shiga 525-8577
Japan

Abstract

The current state of the art in solar photovoltaic (PV) technology is overviewed, with a brief discussion of the relationship between the industrial revolution and forms of energy resources. Evolutions in energy resources from coal (solid) to oil (liquid), and to LNG (gas) in the past two centuries are examined in view of economy and environmental issues, with electric energy as a final form of the energy in the 21st century. Also discussed is the 3E Trilemma which is the most important issue for the 21st century. Secondly, a new strategy and action plan in Japan is introduced for renewable energy promotion aimed for the year 2010. Thirdly, recent research and development efforts for the improvement of solar cell efficiency are reviewed together with their device physics and recent achievements with innovative technologies introducing new cell structure and materials. Then transitions of the annual worldwide solar cell module productions are reported. Progress of the conversion efficiency in various types of solar cells is also summarized. In the final part of paper, the technological roadmap of PV up to 2030, and its future prospects are demonstrated. Possible new roles to contribute to the global environmental issues with PV system utilizations are proposed and discussed.

* Corresponding Author. Email: *hamakawa@se.ritsumei.ac.jp*

1. Introduction: Evolution of Industrial Development with the Energy Revolution

Developing clean energy technology alternatives to fossil fuels has become one of the most important tasks assigned to modern science and technology. The reason for this strong motivation is to stop air pollution resulting from the mass consumption of fossil fuels and to protect the ecological cycles of the biosystems on the earth. As well recognized after the Kyoto Protocol, much attention has been focused on the effects of global warming resulting from the mass consumption of fossil fuels. The 21st century is envisioned to be characterized as an age using a best mix of different energy forms.

As is well known, the industrial revolution was initiated by the invention of the steam engine by James Watt in 1765. However, technological maturity from research and development to practical applications with full market penetration normally takes 30-50 years. The steam ships and the railway network powered by steam engines operated by coal fuel were developed in the middle of 19th century. The same is true of the petroleum fuel age, that is, with an early invention of the internal combustion engine by Etienne Lenoir in 1860 (Hamakawa, 2000). Then, the petroleum age blossomed in the 20th century with gasoline engines powering the automobile industry and airplane traffic network.

Figure 1 shows changes in civilized life with forms of energy resource since the industrial revolution. Three main fossil fuel energy expenditures are plotted separately up to the year of 2000, showing their prospective future (as the dashed lines) by following a scenario promoting renewable energy. It is seen that the form of energy resources has changed from coal to oil and to gases including recent LNG and LPG. There are two important driving forces of these evolutions. One is the convenience of use and mass consumption of energy — that is, mass production technology, ease of transport, and distribution with storage. Considering the motivation of energy form transition, the underlying driving force for the energy revolution is based upon the so called Great Principle of Economy.

Other reasons are governed by the environmental load. As can be seen from Table 1 reported by the United States Solar Energy Industries Association Directory (SEIA, 1996), the pollutant emission factors for electricity generation in carbon equivalent gram per kWh are 322.8 for coal, 258.5 for oil, and 178.0 for LNG, respectively. According to a recent energy analysis survey (Hata, 2002) the electricity generation ratio to primary energy is more than 40% in developed countries, and will increase to more than 50% during the first quarter period of the 21st century as shown in Figure 2. Electrical energy is the most convenient energy form for the above-mentioned reasons, in respect to mass production, transport and distribution for modern civilized life. Solar photovoltaic (PV) technology is able to generate this electric power without any environmental load.

Table 1. Pollutant emission factors for electrical generation in units of g/kWh (after US. SEIA).

Energy Source	CO_2	NO_2	SO_x
Coal	322.8	1.8	3.400
Oil	258.5	0.88	1.700
Natural Gas	178.0	0.9	0.001
Nuclear	7.8	0.003	0.030
Photovoltaic	5.3	0.007	0.020
Biomass	0.0	0.6	0.140
Geothermal	51.5	TR	TR
Wind	6.7	TR	TR
Solar Thermal	3.3	TR	TR
Hydropower	5.9	TR	TR

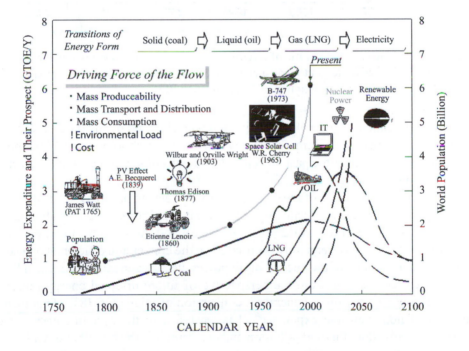

Figure 1. Change in civilized life with energy resource.

The history of PV goes back over 160 years to the late 1830s as shown Figure 1 E. Becquerel observed a photo-galvanic effect in a metal-electrolyte interface. A similar effect was observed in the thin gold coated selenium several decades later by W.G. Adams. However, the first practical device was invented by Chopin et al., at Bell Labs U.S.A. with Si p-n Junction in 1954, and was used as the space power source for the satellite Vangaurd 1 in 1958. Systematic investigations of PV technologies for

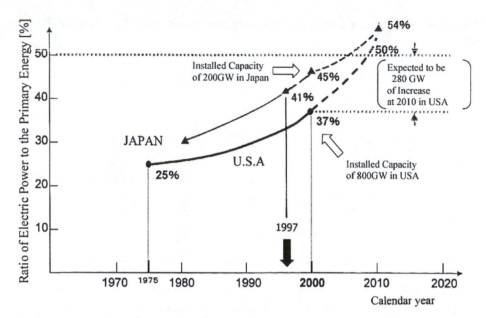

Figure 2. Rapid increase of the ratio of electric energy to the primary energy and its prospect in the United States and Japan electricity demand due to recent promotion of information technology (IT) penetration in the civilization life.

terrestrial applications started on a worldwide scale triggered by the onset of the first oil shock in October 1973. Remarkable advances in the technological achievements have since been seen over the past 30 years. The details of key issues will be given in the following sections.

Among the wide variety of renewable energy projects in progress, PV is the most promising as a future energy technology, because it is pollution-free, it is abundantly available everywhere in the world (even in space), and the system can be operated under diffuse light. It has frequently been said that the conversion efficiency of a solar cell is less than half that of the nuclear or oil power electric generation system. This is a nonsensical comparison from the perspective of future-oriented energy policies. Steam turbine electric power generation conversion efficiency is 38%. This means that 62% of nonrenewable expensive oil is wasted, and the system causes atmospheric contamination, thermal pollution and acid rain. On the other hand, with even only a 15% efficiency solar photovoltaic power generation is pollution-free plus the energy source is free. This is the basic difference between solar photovoltaic power generation and fossil fuel electric power generation.

2. 3E-Trilemma and Clean Energy Developments

In the discussions of the grand design for 21st century's civilized life, one of the most important issues to be solved was decided unanimously by the world community

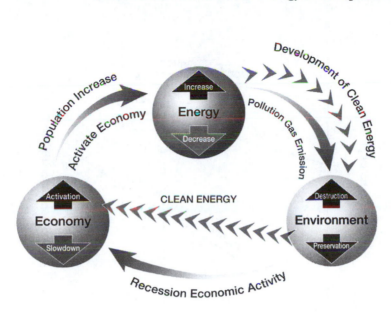

Figure 3. The 3-E trilemma is the most important task assigned to 21st century's civilization. The only way to solve this trilemma is to develop clean energy technology.

as the 3E-Trilemma. That is, for the economical development (E: *Economy*), we need an increase of the energy expense (E: *Energy*) with the present infrastructure in modern countries. However, it induces an environmental issue (E: *Environment*) by more emissions of pollutant gases. On the contrary, if the political option chooses a suppression of pollutant gas emissions, it inactivates the economical development. This is known as the 3E-Trilemma. Figure 3 shows an illustration of the cyclic correlation of Economy, Energy and Environment (Hamakawa, 2000). Here, the importance is placed on the change in the circuit from the infrastructure of fossil fuels energy supply to that of renewable energy developments and supply.

According to the result of world energy trends analysis, the energy consumption per capita in a country is directly proportional to the country's annual income per capita or its Gross National Product (GNP) (Hamakawa, 2000). On the other hand, the number of the world's inhabitants is steadily increasing as shown in Figure 4, and reached about 6.1 billion in the year 2000. It can be expected that worldwide energy demand will increase by multiples of population increase and another factor due to promotion of modernization. This increase in energy demand seems unavoidable in the near future even if energy conservation technologies under development are applied to a moderate degree. For example, the rate of energy consumption per production unit in the heavy industries in well-developed countries might decrease, but this will be completely compensated for by the rapid increase of energy demand in the newly advancing countries such as China, Malaysia and Thailand. As can be seen in Figure 5, energy consumption in East and South Asia is rapidly increasing with their economic

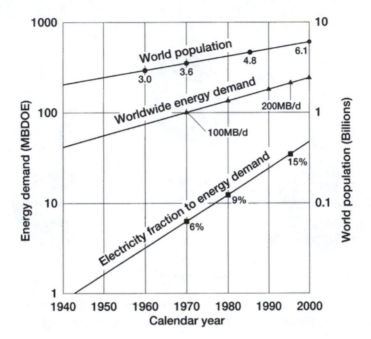

Figure 4. World population growth, worldwide energy demand and the ratio of electricity demand to total energy demand. Here, the unit of MBDOE stands for millions of barrels per day of oil equivalent.

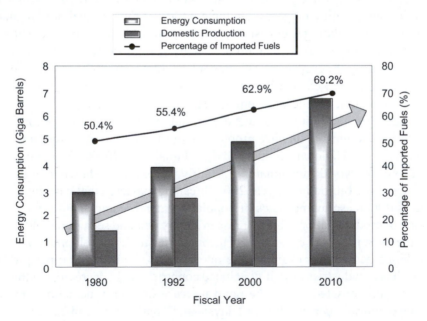

Figure 5. Transition of energy demand in eleven Asian countries. The percentages of imported fuels are increasing as the dotted line.

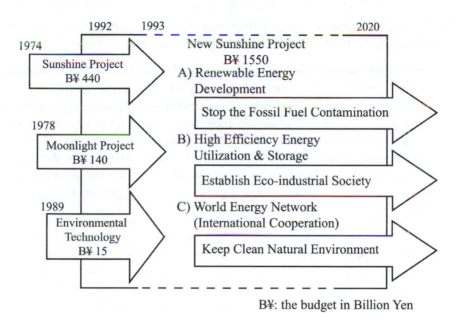

B¥: the budget in Billion Yen

Figure 6. Organizations and objectives of the New Sunshine Project, and their historical trajectory.

growth, and the percentage of imported fuels will reach almost 70% by the year 2010 (Hamakawa, 1998).

Considering the two-sided nature of the energy policy, that is, continuous growth of mass consumption of the limited fossil fuels on one side, and increasingly severe global environmental issues on the other side, the Agency of Industrial Science and Technology (AIST) in the Ministry of International Trade and Industry (MITI) in Japan has decided to establish the New Sunshine Program for the development of clean energy and environmental technologies. Figure 6 illustrates a comprehensive structure of the new program and its relation to the Sunshine Project, which was formulated in May 1973, prior to the first energy crisis and started in 1974. The Moonlight Project, which was initiated in 1978 for the development of energy conservation and environmental technologies, continued until 1989 (Miyazawa et al., 1997). Recent budgets in Billion Japanese Yen (¥) are also inserted under the projects: Renewable Energy Development Technology, High Efficiency Utilization of Fossil Fuels and Energy Storage, and International Energy Cooperation. This so-called World Energy Network (WENET) utilizes hydrogen fuel production technology by PV over a wide energy network nicknamed the Eco-Energy City.

The direct conversion of solar radiation to electricity by photovoltaics has a number of significant advantages as an electricity generator. They include:

1. *Clean and inexhaustible resource:* Solar photovoltaic conversion systems tap the inexhaustible resource of sun light which is free of charge and available anywhere in the world. The amount of energy supplied by the sun to the earth is more than five orders of magnitude larger than the world electric power consumption—more than enough to keep modern civilization going. Roofing tile photovoltaic generation, for example, saves excess thermal heat and conserves the local heat balance. This means that a considerable reduction of thermal pollution in densely populated city areas can be attained.

2. *Almost maintenance-free:* The uniqueness of photovoltaic conversion is based upon the photo-electric quantum effect in semiconductors and is totally different from thermal energy processes. It has no mechanically moving parts and needs no lubrication, which means power stations are easily operated and maintenance-free, as has already been proved in remote-controlled lighthouses, telecommunication relay stations, and space satellite power sources.

3. *Size-independent conversion efficiency:* Solar arrays consist of a number of solar cell modules and permit a wide range of application sizes and systems with essentially the same conversion efficiency and the same technology. For example, a 1 mW wrist watch solar cell and a 1MW photovoltaic power station could be operated with the same conversion efficiency.

4. *Modular mass production provides large scale merit:* The solar cell consists of semiconductor modules, and it can easily be mass-produced like transistors and semiconductor integrated circuits. Therefore, it is expected that mass production for expanding market size will induce cost reductions.

5. *Operation under diffuse light source:* Solar cells generate electric power under diffuse light. As it has already been shown in the consumer-oriented applications of electronics (i.e., wrist watches and pocket calculators) solar cells work in room light, even under fluorescent lamps.

3. High Conversion Efficiency Research and Development Efforts and Their Recent Achievements

In spite of the various advantages of photovoltaics as mentioned in section two, a big barrier impeding the expansion of the large-scale bulk power application using photovoltaic systems is the high price of solar cell modules, which was more than \$5/Wp (¥600/Wp) in 2000. Therefore, the cost reduction of solar cells is of prime importance. To achieve this objective, tremendous research and development efforts have been made in a wide variety of technical fields from solar cell material, cell structure,

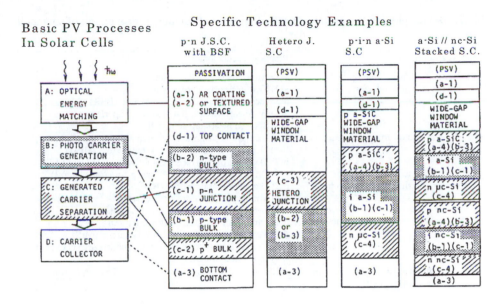

Figure 7. Four steps of basic processes in PV energy conversion and corresponding function in actual solar cells (a-Si: amorphous silicon, μc: microcrystalline, nc: nanocrystalline).

and mass production processes over the past twenty years. As a result, about an order of magnitude cost reduction was achieved in the 1990s. However, another one-fifth of cost reduction is required to levelize the present conventional electricity price according to the recent survey report (Izumina, 2003).

The photovoltaic energy conversion consists of two principal processes, namely the photocarrier generation by the interband or level-to-band absorption of light in semiconductors, and the separation of the generated hole-electron pairs by an internal electric field. This internal field is conventionally provided by the p-n junction of a semiconductor, the heterojunction, the p-i-n junction, and so on. In addition to these principal processes, the solar cell provides ohmic contacts for the collection of photogenerated carriers and to extract electric power. For the purpose of more efficient photon energy transfer into and also more efficient photon absorption in the semiconductor, an antireflective (AR) coating is commonly applied on the window side, to achieve efficient optical energy absorption by the semiconductor. A schematic representation of these processes and the corresponding structures of several typical solar cells is shown in Figure 7. In the figure, basic processes are expressed as A to D along the energy flow from photon to electron. For the improvement of solar cell efficiency, efforts should be made to increase the efficiency of each process and decrease energy losses.

Table 2 summarizes some concrete technologies to improve four basic processes in solar cell production. In process A, in order to efficiently guide photon energy to the

Table 2. Device physics and practical technologies for efficiency improvement of solar cells.

Process and Principle	Practical Technologies
(A) Better gulding of input optical energy to semiconductor and widening of special response	a-1) Antireflective (AR) coating a-2) Textured surface treatment a-3) Increase of back surface reflection (BSR) a-4) Wide gap window junction (HJ, superlattice) a-5) Stacked heterogap solar cell
(B) Efficient photo-carrier generations	b-1) Increase of μτ product (film quality) b-2) Decrease of surface recombination loss (HFJ) b-3) Decrease of interface recombination loss (graded J)
(C) Efficient generated-carrier collection (carrier confinement and drifting)	c-1) Drift type photovoltaic effect (p-I-n, graded gap J) c-2) Graded impurity profile (BSF) c-3) Minority carrier mirror effect (HJ) c-4) Decrease voltage factor loss (HJ, HFJ)
(D) Decrease of series resistance loss	d-1) Optimization of top electrode pattern d-2) Tunneling injection of minority carrier (MIS) d-3) Increase transparent electrode conductivity (ITO/SnO$_2$) d-4) Integrated cascade solar cell (TFSC)

HJ, heterojunction; HFJ, heteroface junction; MIS, metal-insulator-semiconductor; TFSC, thin-film solar cell; BSF back surface field

PV active layer (optical matching with the semiconductor), a number of materials such as SiO, TiO$_x$, Ta$_2$O$_5$, CeO$_2$, and SnO$_2$+In$_2$O$_3$ (abbreviated to ITO: indium tin oxide), have been investigated as AR-coating materials (Apfel, 1975). These films have been deposited by various technologies including sputtering, evaporation, and chemical vapor deposition (CVD). In the case of Si cells, SiO coatings cause a 20% increase in short-circuit current (Brandhorst, 1976). Textured surface treatment by preferential etching of a (100) surface is another technology which has been applied to amorphous silicon (a-Si), GaAs, and CdS solar cells (Baraona and Brandhorst, 1975). This treatment produces a micropyramidal surface by grooving via a difference in the chemical reaction between the various crystal surface orientations. The surface texturing improves optical confinement of photons in the semiconductor. The highest AM1.5 efficiency of 24.3% ever reported in a Si cell has recently been achieved with micro grooved passivated-emitter solar cells by Zhao, et al. (Zhao, 2001). AM1.5 refers to the optical conditions (angle of light incidence of air mass 1.5) at which the efficiency was measured.

One important remaining area for further improving solar cell efficiency in process A in Table 2 is more efficient collection of the carriers generated by longer wavelength photons, in particular those just above the band edge. For example, in the case of an a-Si solar cell, the penetration depth of photons with an energy of an a-Si:H solar cell is only 0.6 μm. A concept of efficient collection of long wavelength photon energy by a highly reflective random backside surface was first proposed by Den Boer and Van Strijp (Boer, 1982). This concept has been extended to the more efficient utilization of both optical and carrier confinement with multi-layered heterostructure junction (Hamakawa et al., 1982). Fujimoto et al. have developed a practical technology with a cell structure of ITO/n-μc-Si/i-p type-a-Si/TiO$_2$/Ag-plated semi textured stainless steel. With this treatment, short-circuit current density (Jsc) can be improved by more than 20% in the a-Si solar cell (Fujimoto et al., 1984).

As to B of Table 2, photocarrier generation by interband optical absorption is the principal process in the PV effect. Whereas the absorption coefficient is a kind of material constant, the quantum efficiency of photocarrier generation varies with imperfections in the semiconductor, including lattice defects, impurities, and grain boundaries in the crystalline base materials, and also the film quality in the case of a-Si films. Since there are still possibilities for improvement in a-Si film materials, a wide variety of efforts are in progress from high purity raw material gases to deposition technologies, such as the cross-field method, triode glow, and ECR (electron cyclotron resonance) plasma growth (Wallace et al., 1987). Therefore, the optimum design of the light absorbing region when matched with film quality is important for saving material, which will directly affect the cell cost. The size of this region is determined by the interband absorption coefficient, the minority carrier lifetime τ, the diffusion length, and the surface recombination velocity S at the front interface of the semiconductor. Many experimental device physics modelings and investigations have been conducted on crystalline Si, polycrystalline Si and GaAs basis compound semiconductors in these processes.

As to C of Table 2, an internal electric field is needed to separate the positive and negative photogenerated carriers, which induce the electromotive force. The open circuit voltage of the cell increases with increasing internal potential difference. The basic technologies to form graded p-i-n junctions, heterojunctions, and a back surface field (BSF) are well established.

There have been very important advances in increasing the open circuit voltage, V_{OC}, through the use of heterojunctions with wide gap window material as discussed by Hamakawa (Hamakawa, 1986) at the 1986 meeting of the Materials Research Society in Palo Alto, through the techniques of the super-lattice (Nakano et al., 1987), the BSF (Hamakawa, 1986), and the graded gap active layer (Arya et al., 1987).

Finally, as to D of Table 2, an illuminated solar cell on load suffers an internal resistance ohmic loss which reduces cell efficiency. Most of the series resistance consists of the top electrode and the very thin front region of the junction. An optimum design theory for the grid pattern of the top electrode has recently been investigated for application to crystalline p-n junctions and heterojunction solar cells (Takakura and Hamakawa, 1978). Recent aspects of the practical technologies in the field of high efficiency c-Si-basis solar cell have been reviewed by Green et al. (1987).

Another possibility for further improvement of amorphous solar cell efficiency is heterostructure stacked junctions utilizing internal carrier exchange effect through localized states at the heterojunction interface. This idea was first proposed by an Osaka University group, (Hamakawa et al., 1979) and was applied to the high voltage solar cell called HOMULAC (Horizontally Multi-Layered Photovoltaic Cell) (Hamakawa, 1979). This idea was extended to multiband gap stacking — that is, the stacked solar cell can convert electricity from a much wider solar radiation spectra

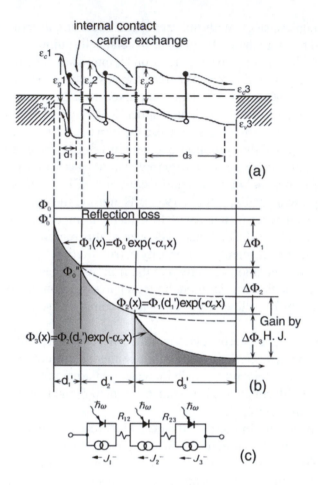

Figure 8. Band diagram explanation of the multi-band gap stacked solar cell (a), its photon flux penetration pattern (b), and equivalent circuit (c).

related to two or three different absorption spectra of the stacked materials. The right edge side of Figure 7 shows the basic PV processes in the two stacked solar cell. Figure 8 shows the energy band diagram and schematic representation of hole-electron exchange effects at the junction (a), and an equivalent circuit explanation of the HOMULAC (Horizontally Multi-Layered Cell) device (c). To obtain an efficient photon-energy collection with a multiband-gap stacked solar cell, there are some design rules for the selection of material combination for stacking. The optimum design theory selection rule and attainable conversion efficiency have already been done by several investigators (Fan and Palm, 1983; Takakura, 1992). In recent years many experimental approaches have also been done all over the world.

An improvement of cell efficiency is also directly connected to cost-reduction in photovoltaic systems. Therefore, a series of research and development efforts have

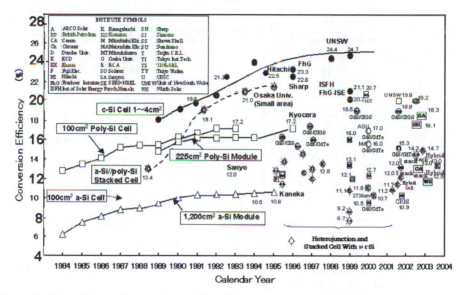

Figure 9. Transitions of research and development phase solar cell and sub module efficiencies for various material solar cells.

been made on each step of the photovoltaic process, including anti-reflective coating, non-reflective texture surface treatment, and heterojunctions with wide energy gap material. These became routine technology in the solar cell industry used to obtain more efficient optical energy collection. Some data of the highest solar cell efficiencies achieved in the research and development phase for various materials are summarized in Figure 9. Through these technological improvements, the efficiency of a single crystalline Si solar cell reaches 18-21% in mass-production. For example, Sanyo has recently announced 21% efficiency in the 10×10 cm^2 HIT cell (Heterojunction with Intrinsic Thin-layer) as shown in Figure 10 and module efficiency of 18.4% on the 80×120 cm^2 size module. This HIT cell has a double heterojunction sandwiched by *p* and *n* type a-Si (Tanaka et al., 2003). Sharp also reported 17.4% efficiency on the 5×5 inch square cell (Tomita, 2003). The polycrystalline and cast silicon solar cells show 15-18% on average. Figure 11 shows a comparison of a Kyocera 'd Blue cell and an ordinary poly-crystalline cell, and their output characteristics (OITDA, 2003a).

On the other hand, in the amorphous Si-based thin film cell field, tremendous research and development efforts have been focused on the a-Si/μc-Si (microcrystalline-silicon) and/or nc-Si (nanocrystalline-silicon) stacked solar cells as a next generation solar cell (Hamakawa and Takakura, 2000). In principle, this solar cell is able to collect a wide spectrum solar radiation using a combination of two to three kinds of stacked semiconductors. The Osaka University group showed 21% efficiency with a-Si/poly-Si combinations (Ma, 1993). Recently, they proposed a new combination of a-Si stacked with nc-Si. In this type, solar cell materials stacked both front and bottom could be fabricated by plasma CVD in mass production processing. Through com-

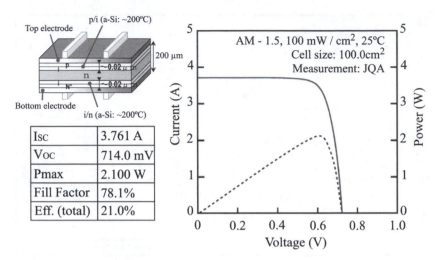

Isc	3.761 A
Voc	714.0 mV
Pmax	2.100 W
Fill Factor	78.1%
Eff. (total)	21.0%

Figure 10. The junction structure and output characteristics of HIT solar cells measurement by the Japanese Quality Assurance Organization (JQA).

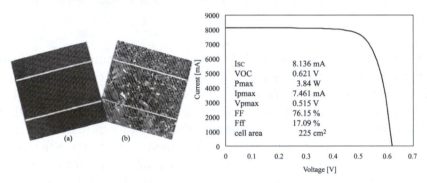

Figure 11. Left: Out look views of the d'-Blue cell (a) and that of ordinary poly-crystalline silicon solar cell (b). Right: Output characteristic of the d'-Blue cell.

puter simulation, they also recently showed a realistic attainable efficiency of 18.4% with only 8 μm total cell thickness (Matsui et al., 2001). Figure 12 shows this I-V characteristic calculated from the experimentally obtained design parameters. Recently, Kaneka group announced an experimental output characteristics of a small area 1×1 cm^2 cell having a 14.7% efficiency. The junction structure of the a-Si/μc-Si stacked cell initial efficiency is shown in Figure 13 (Yamamoto, 2003). Commercially produced moduleswith this stacked cell and their output characteristics are also shown in Figure 14 (Tawada, 2003). Canon Inc. has recently started to produce a three stacked thin film solar cell module by the roll to roll CVD production line. This module has an a-Si//nc-Si//nc-Si structure with a 356x239 mm^2 size and a module efficiency of 13.37% as shown in Figure 15 (Saito and Ogawa, 2003).

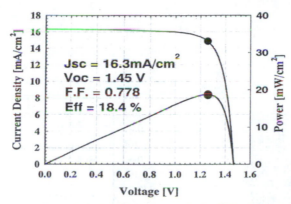

Figure 12. The output characteristic of a two terminal a-Si//n-Si double stacked solar cell obtained by computer simulation calculated by experimentally obtained realistic material constants.

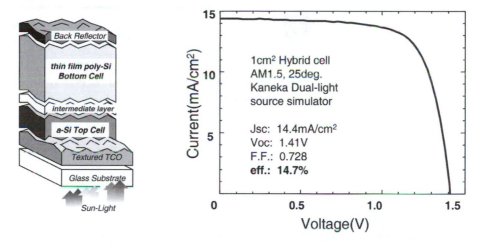

Figure 13. Schematic Device Structure and Highest I-V Performance of small sized a-Si//n-Si double stacked solar cell by Kaneka Solar Tech. Inc.

4. Accelerated Promotion of Photovoltaic Technology in Japan

Photovoltaic solar cells are made from the same semiconductor materials that are used in VLSIC (Very Large Scale Integrated Circuit) transistors. The cost of these transistors is dropping dramatically each year with the advances in manufacturing technology and mass production. In an effort to reduce the high manufacturing cost of solar cells, Japan has initiated solar cell mass production. The goal is to benefit from the economy of scale (sometimes referred to as "mass production scale merit"), thus reducing the cost of solar cells and breaking through the cost barrier.

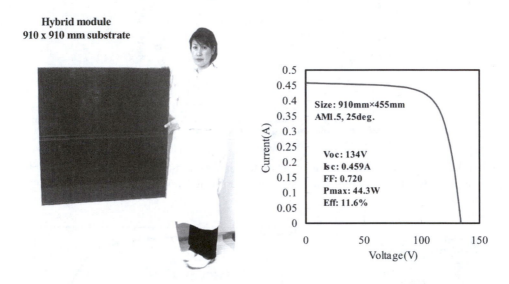

Figure 14. Photograph of 910×910 mm Hybrid Submodule, and Typical I-V Performance of Hybrid module Manufactured by Kaneka Solar Tech. Inc. (KST).

Cell size (cm^2)	801.6
Eff. (%)	13.37
Voc (V)	1.92
Jsc (mA/cm^2)	9.87
Fill Factor	0.705

Figure 15. Large area a-Si//nc-Si//nc-Si triple-stacked thin film solar cell module and its performance presented by Canon Inc.

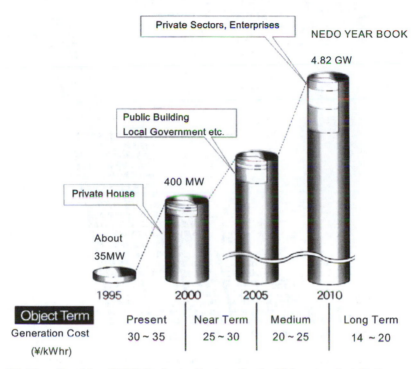

Figure 16. New Sunshine (NSS) Project milestone for the PV-system installation up to 2010.

In December 1994, the Japanese government identified an integral part of their national renewable energy policy to be "Fundamental Principles to Promote New Energy Developments and Utilization." Related policy actions such as tax reduction and government subsidy were approved by the Japanese Congress on April 10, 1997. The strategies are applied not only to whole ministries and government offices but also to local government authorities and private enterprises.

With this new government policy, development and promotion of PV technologies has shown great promise in Japan. An integrated installation of 400MWp PV modules by Fiscal Year (FY) 2000 and 4.82 GWp by FY 2010 for Japanese domestic use are scheduled as a milestone in the program as shown in Figure 16. Some government incentives include a special tax reduction for investments in renewable energy plants, financial support of one half subsidy on PV systems for public facilities (provided by the PV Field Test Experiments incentive program), and one third subsidy for private solar houses field testing (provided by the PV House Planning incentive program). Figure 17 shows the numbers of accepted PV houses built under this government subsidy in the past six years. As seen in the figure the number of government subsidy-accepted PV houses has doubled each year. On the other hand, in the PV Field Test Experiment, a total of 751 sites with 18.6 MW have been installed in the five years since 1998. A noticeable result of this project is an accelerated nationwide promotion of PV in practical usage. In fact, since 2002, 218 sites with 5.28 MW have been

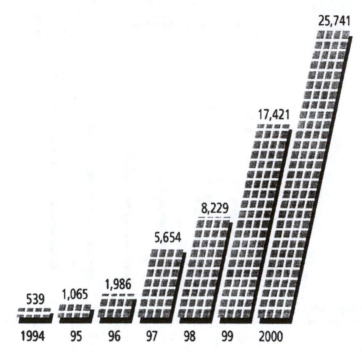

Figure 17. Number of accepted PV house monitor planning by the government subsidy.

installed. As the result of this accelerated promotion strategy, the sales price for a private home's 3kW solar photovoltaic system has decreased very sharply from 6M¥ (U.S. $50,000) in 1994 to 2M¥ (U.S. $16,667) in 2002.

Figure 18 shows recent transitions of the solar cell module annual production in Japan since 1993 as surveyed by the Optoelectronic Industry Technology and Development Association (OITDA, 2003b). As can be seen from the figure, a remarkable increase of the annual production has been seen since starting of the New Sunshine Project in 1993. The same kind of survey on the worldwide scale has also been carried out as shown in Figure 19 (Tomita, 2003). Before closing this section, as an example of the demonstration PV system, a picture of 630kWp Solar Ark is shown in Figure 20. The size of the Solar Ark is 315 m long by 37.1 m tall; it is installed at the front of Sanyo Electric Co. Ltd., Gifu Plant, Gifu, Japan.

5. Future Prospects and Roadmap for Solar Photovoltaics

As was mentioned in Figure 16, a target milestone set by the 1994 renewable energy initiative for volume PV installations in Japan by the year 2010 was set at 4.82GWp. There are many discussions concerning the target volume of nearly 5 GW in 2010. For example, if 200,000 solar PV houses were built annually in Japan (which corresponds to nearly 10% of all private houses), it will only take five years with 5 kW

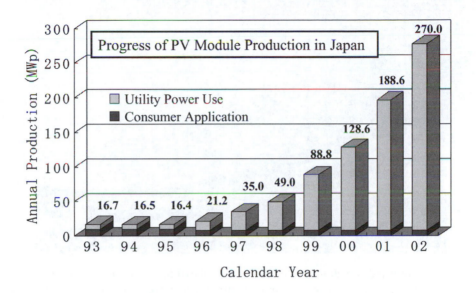

Figure 18. Recent growth of PV solar module shipment in Japan.

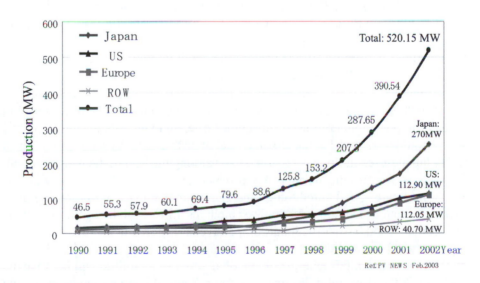

Figure 19. Recent transitions of World Cell/Module Productions in the world.

Figure 20. An outward appearance of the Solar Ark, one of the largest demonstration plants of 630 kWpeak having annual energy output of 530 MWhrs which corresponds to 95 tons of CO_2 reduction in carbon equivalent grams unit (presented by Sanyo Electric Co. Ltd.).

PV per home to achieve the target set for 2010. Of course there are also much more pessimistic opinions which take into account the recent downturn in the Japanese economy. Technological milestones for annual production, PV modules and system cost targets for 2005, 2010, 2020, and 2030 were investigated at the PV Technology Association (PVTEC) Future Vision Specialist Committee Meeting. The results are summarized in Table 3(a-b) (Izumina, 2003).

As described in Table 3-a, once the target module cost of ¥100/W has been achieved by 2010, the PV electricity generation cost could be levelized to that of hydroelectric power generation as shown in Table 3-b. By the year 2020, a levelized cost of 8.7-9 ¥/kWh would be achieved which would be competitive with combustion turbine electric power generation. If this comes true, an explosive expansion of the PV market could be realized at the targeted goal of 2015-2020. As seen from Figures 10 and 11, Japanese technology is regarded to be among the best in the world for bulk Si solar cells as the present mainstay PV product, however, this status may not be assured in the future. The United States and European countries are preparing multiple menus for future technologies of bulk Si modules. In view of the fulfillment of the domestic demands and the international competitiveness, more intensive technological development will be required.

On the other hand, a fabrication volume of an order of magnitude higher than the present level of thin-film modules made of amorphous silicon and nano-crystalline silicon could be expected using a continuous PECVD (Plasma Enhanced Chemical Vapor Deposition) system. In view of the technological potential for mass production,

Table 3-a. Milestones of the PV module Cost and Efficiency (PVTEC, Future Vision Committee Report).

Mass Production		Present	Medium Term	Long Term	
Phases		2000	~2010	2015~20	2020~30
Module Cost		¥300/W	¥100/W	¥75/W	¥50~60/W
Module Production		—	> 100MW/year	~250MW/year	~500MW/year
Module Efficiency	Bulk Sl	Cast Si 13~14%	Ultra-thin slice Si: 17% / Sliceless Si: 15%	Ultra-thin Si: ~20%	
	Film Si	—	> 12%	> 13%	~16%
	Film CIS	—	> 13%	~16%	~18%

Table 3-b. Targets and Implementation Schedule of PV module cost in Japanese Yen per Watt.

Expected Implementation[*1]		Present	Short-Term	Medium-Term	Long-Term	
		2000	around 2010	around 2010	~2020	~2030
Module Output		-	-	100 MW/year	250 MW/year	500 MW/year
Module Cost		¥300/W	¥200/W	¥100/W	¥75/W	¥50~60/W
Housing PV	PV System[*2]	¥873/W	¥530/W	¥300/W	¥230/W	-
	Module	¥590/W	¥340/W	¥170/W	¥130/W	-
	Inverter	¥100/W	¥50/W	¥40/W	¥30/W	-
	Auxiliary	¥67/W	¥40/W	¥30/W	¥20/W	-
	Installation	¥117/W	¥100/W	¥60/W	¥50/W	-
	Payback Yrs[*3]	-	36.3	15.7	11.4	-
	Power Price[*4]	¥55.8/kWh	¥33.9/kWh	¥19.28/kWh	¥14.7/kWh	-
	Added Value[*5]	-	- ¥2.7/kWh	- ¥2.7/kWh	- ¥4.9/kWh	-
	AV deducted	¥55.8/kWh	¥31.2/kWh	¥16.5/kWh	¥9.8/kWh	-
Industrial PV	PV System[*2]	¥1,033/W	-	¥243/W	¥178/W	(¥120/W)
	Module	¥502/W	-	¥145/W	¥110/W	-
	Inverter	¥170/W	-	¥28/W	¥15/W	-
	Auxiliary	¥128/W	-	¥26/W	¥14/W	-
	Installation	¥94/W	-	¥210/W	¥18/W	-
	Other Works	¥80/W	-	¥13/W	¥11/W	-
	Administration	¥59/W	-	¥10/W	¥10/W	-
	Power Cost[*4]	¥66.1/kWh	-	¥15.5/kWh	¥11.4/kWh	¥7.7/W
	Added Value[*5]	-	-	- ¥2.7/kWh	- ¥2.7/kWh	- ¥2.7/kWh
	AV deducted	¥66.1/kWh	-	¥12.8/kWh	¥8.7/kWh	¥5.0/W

NOTES:

1. This table shows the time of implementation, and the technological development is to have been finished a few years prior to this time.
2. The size of PV system is set to 4kW for the housing, and 100kW for the industrial. The price in parentheses is set for a plant of 100MW size.
3. The payback years for the housing PV installation is estimated on the assumption that the mean power price is ¥23/kWh and the interest rate 3%.
4. The estimation of unit power cost is based on the New Energy Subcommittee Report of the Comprehensive Energy Investigation Committee (June 2001) with the assumption that the interest rate is 3% and the payback 20 years (published by Energy Conservation and Renewable Energy Department, Agency for Natural Resources and Energy (ANRE), MITI, Japan).
5. The component of added value consists of reduction of CO_2 emission (¥2.7/kWh) and roof construction (¥2.2/kWh, for housing PV only).

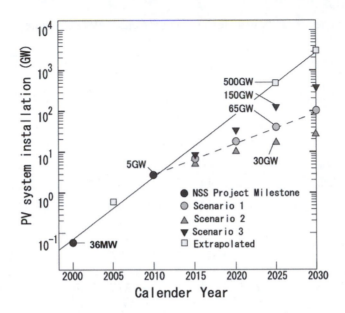

Figure 21. A scenario of the accumulated PV system installation in Japan.

a dramatic increase in market expansion would be induced, allowing the PV industry to greet an independence day as a clean energy enterprise.

Figure 21 presents a scenario of accumulated PV system installations in Japan up to 2030. Assuming the scenario 2, on estimated installation volume of 65 GW in 2025 might cover more than 40% of the peak savings of electrical energy even taking into account 45% of the peak saving use of PV system operation efficiency. With this scenario 1 in 2030, the estimated installation volume of 360GW is corresponding to 45 GW assuming 12.5% of PV system operation efficiency (360×0.125) of the real supplying power which can easily cover 70% of the electric power generated by fossil fuels in Japan.

Figure 22 shows simulation results of future world wide energy supply by (A) prospect of primary energy, and (B) the electricity generated by various energy resources. As can be seen in A, renewable energy including PV will become a major energy resource at about 2050. As for the electric power generation, 2040 will be a critical period when renewable energy becomes the majority in the 21st century. As described in the previous section, solar photovoltaic power generation is an almost maintenance-free clean energy technology. Therefore, the penetration of PV technology into utility power generation is of prime importance for market size expansion, using the economy of scale. With increasing feasibility of other power applications and full use of maintenance-free solar photovoltaics, many environment clean-up processes could be performed with solar photovoltaic power. Two examples include ashing of pollutant gas in the air by glow discharge decomposition and cleaning of water by electrochemical processing as shown in Table 4. Mass production of hydrogen energy

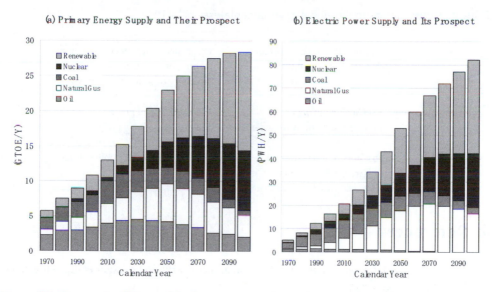

Figure 22. Prospects of the world primary energy demand (a) and that of electric power demand (b) for the 21st century calculated for the sustainable development scenario.

in the Sahara desert is planned using solar photovoltaics. The International Energy Association (IEA) Task IV has started a new project called VLS-PV (Very Large Scale Photovoltaic System) (Kurokawa, 1999). There exists a possibility of stopping desert expansion and "greening of the desert" by photovoltaic water pumping and planting of trees; the Gobi Project has been organized for this purpose (Nihon-Keizai-Shinbun, 1990). In 1992, a preliminary study team began a survey of natural conditions as a part of a Japan-Mongolia program of cooperation.

Results of these considerations indicate that an economically feasible age of photovoltaics will come earlier than expected in the near future, if efforts continue to be made to promote research and development efforts and market stimulation by means of government support and international cooperation. In any case, the most important emphasis should be placed on stopping the effects of contamination by fossil fuel burning with a worldwide energy policy and development of this novel clean energy technology for the future benefit of all mankind.

Let us embark on a new kind of energy revolution with clean energy photovoltaics. Might it be possible to accomplish the clean energy revolution within the coming 25 years? This might not be a fanciful dream — let us enjoy our new challenge!

Table 4. Contributions to the global environmental issues by PV.

Local	(1) Solar PV power generation	Clean sustainable energy resource
	(2) Cleaning of polluted air	Ashing of pollutant gases by glow discharge decomposition by PV
Environment	(3) Cleaning of water	Electrochemical processing by PV
	(4) Generation of hydrogen energy	Electrolysis of water of PV
Global	(5) Stop of desertification Greening of deserts	PV water pumping at plantations

6. References

Apfel, J.H., 1975. Optical coatings for collection and conservation of solar energy. Journal of Vacuum Science Technology. 12, p. 1016.

Arya, R., Bennett, H.W., Catalano, A., 1987. Effect of Super lattice Doped Layers on the Performance of a- Si p-i-n Solar Cells. In: K. Takahashi (Ed.), Technical Digest of PVSEC-3, Tokyo, Japan, pp.517-524.

Baraona, C.R., Brandhorst, H.W., 1975. V-Grooved Si Solar Cells. In: Proceedings of the 11th IEEE PVSC, Scottsdale, AZ, USA, p.44.

Boer, W. Den, Strijp, R.M. Van, 1982. Optical Optimization of Amorphous Silicon Solar Cells. In: Bloss, W.H., Grassi, G. (Eds.), Proceedings of the 4th E.C. Photovoltaic Solar Energy Conference, Stresa, Italy, pp. 764-768.

Brandhorst, H.W., 1976. Status of Silicon Solar Cell Technology. Japanese Journal of Applied Physics. 16 Suppl. 1, pp. 399-405.

Fan, John C.C., Palm, B.J., 1983. Optimal design of amorphous single-junction and tandem solar cells. Journal of Solar Cells, 10, 1, pp.81-98.

Fujimoto, K., Kawai, H., Okamoto, H., Hamakawa, Y., 1984. Improvement in the efficiency of amorphous silicon solar cells utilizing the optical confinement effect by means of a TiO2/Ag/SUS back-surface reflector. Journal of Solar Cells, 11, 4, pp. 357-366.

Green, M.A., Wenham, S.R., Blakers, A.W., 1987. Recent Advances in High Efficiency Silicon Solar Cells. In: Proceedings of the 19th IEEE PVSC, New Orleans, LA, USA (IEEE Publ. No. 87CH2400-0), pp. 6-12.

Hamakawa, Y., 2000. Recent Advances in Solar Photovoltaic Activities in Japan and New Energy Strategy Toward 21st Century. In: H. Scheer et al. (Eds.), Proceedings of the 16th EC-PVSC, Glasgow, Scotland, Joint Research Center, European Commission, PB1.2. pp. 2746-2751.

Hamakawa, Y., Takakura, H., 2000. Key issues for the efficiency improvement of silicon basis stacked solar cells. In: Proceedings of the 28th IEEE PVSC, Anchorage, Alaska, USA, pp. 766-771.

Hamakawa, Y., 2001. Recent Progress and Future Prospects of Solar PV Conversion. Chapter II of Japanese R&D Trend Analysis on Advanced Materials, Report No. 4-3, Kansai Research Institute, Kyoto, Japan, p. 31.

Hamakawa, Y., Tawada, Y., Nonomura, S., Okamoto, H., 1982. a-SiC/a-Si Heterojunction Solar Cells. In: Proceedings of the 16th IEEE PVSC, San Diego, CA, USA, pp.679-684.

Hamakawa, Y., Okamoto, H., Nitta, Y., 1979. A New Type of a-Si PV Cells Generating More than 2.0 V. Appl. Phys. Lett., 35, p.15. Japanese Patent Sho-54-32993 and U.S. Patent 4,271,328, March 20, 1979.

Hamakawa, Y., Matsumoto, Y., Xu-Zang-Yang, Okamoto, H., 1986. a-Si Basis Heterojunction Stacked Solar Cells. In: Y. Hamakawa et al. (Eds.), Materials Research Society, Symposia Proc. 70, in Palo Alto, Pittsburg PA, USA, pp. 481-486.

Hamakawa, Y., 1998. A Technological Evolution from Bulk Crystalline age to Multi-layers Thin-film age in Solar Photovoltaics. In: A.A.M. Sayigh (Ed.), Proceedings of WREC, Florence, Italy, Part I, pp.22-31.

Hata, R., 2002. Recent Trends of the Electric Power Demands, Topics Survey Committee Meeting of the Osaka Industrial Association, pp. 1-23, (Japanese).

Ito, K., 2000. Prospect of Fossil Fuels. In: Program Committee of the 3rd annual Meeting of Nuclear Fusion (Eds.), Energy for 21st Century, 3rd annual Meeting Report for Nuclear Fusion, pp. 12-13, (Japanese).

Izumina, M., 2003. Future Vision of PPV Technology. PVTEC NEWS, 41, pp. 3-10, (Japanese).

Kurokawa, K., 1999. A Preliminary Analysis of Very Large Scale Photovoltaic Power Generation System, VLS-PV System, IEA-Task IV Report, Published by IEA PVPS, pp. 1-148.

Ma, W., Horiuchi, T., Lim, C.C., Okamoto, H., Hamakawa, Y., 1993. 21% efficiency with a-Si// poly-Si stacked Solar Cells. In: Proceedings of the 23rd IEEE PVSC, p. 338.

Matsui, T., Yamazaki, T., Nagatani, A., Kino, K., Takakura, H., Hamakawa, Y., 2001. 2D-numerical analysis and optimum design of thin film silicon solar cells. Solar Energy Materials and Solar Cells, 65, pp. 87-93.

Miyazawa, K., Katoh, K., Kawamura, K., 1997. Organization and Structure of the New Sunshine Project. In: Report of Solar Energy Division Meeting, Technology and Industrial Council (MITI), pp.56-67, (http://www.miti.gov.my/) (Japanese).

Nakano, S., Takui, H., Haku, H., Nishiwaki, H., Tsuda, S., Kuwano, Y., 1987. High Quality a-Si Film and Super lattice structure p-Layer a-Si Solar Cells, In: Y. Hamakawa et al. (Eds.), MRS Symposia Proceedings Vol.70, 1986, pp.511-516.

Nihon Keizai Shinbun, 1990. Tree-Planting Greening of Gobi Desert by Photovoltaic Power Water-pumping (boxed news item), News Paper Release, Nihon Keizai Shinbun, May 11, p.35 (Japanese).

OITDA, 2000. News Letter No. 6, OITDA (Optoelectronic Industry and Technology Development Association), 2000, p. 1, (Japanese).

OITDA, 2003a. d-Blue Cell High efficiency Poly-Si Solar Cells. Column R&A (Research & Analysis) Optnews, 1, pp. 44-45, (Japanese).

OITDA, 2003b. Photon Energy Industry Survey Committee Annual Report. OITDA, pp. 1-5, (Japanese).

Saito, K., Ogawa, K., 2003. High Efficiency Large Area Solar Cells Using, ic-Si. In: M. Yamaguchi (Ed.), Technical Digest WCPEC-3, Osaka, Japan, S20-B9-04, (article in press).

SEIA, 1996. In: Directory of the US Photovoltaic Industry 1996 US Solar Energy Industries Association (SEIA), p.4.

Takakura, H., Hamakawa, Y., 1978. An Optimum Design of Hetero-face Solar Cells. Journal of Institute of Electrical Engineers of Japan, 47, 9, pp.872-877, (Japanese).

Takakura, H., 1992. Optimum Design of Thin-Film- Based Tandem-Type Solar Cells. Japanese Journal of Applied Physics, 31, Part 1, 8, pp. 2394-2399.

Tanaka, M., Okamoto, S., Kiyama, S., 2003. Development of HIT Solar Cell with more than 21% Conversion Efficiency. In: M. Yamaguchi (Ed.), Technical Digest WCPEC-3, Osaka, Japan, 4O-D10-01, (article in press).

Tawada, Y., 2003. Production of Amorphous Si//ic-Si Hybrid Modules. In: M. Yamaguchi (Ed.), Technical Digest WCPEC-3, Osaka, Japan, 5PL-D1-03, (article in press).

Tomita, T., 2003. Present Status of Photovoltaic Industry and Issues in the Future Energy. In: M. Yamaguchi (Ed.), Technical Digest WCPEC-3, Osaka, Japan, 6PL-D2-01, (article in press).

Wallace, W., Sabisky, E., Stafford, B., Luft, W., 1987. In: Proceecings of the 19th IEEE PVSC, New Orleans, LA, USA, pp. 593-598.

Yamamoto, K., 2003. Development of Nobel Hybrid Solar Cells. In: M. Yamaguchi (Ed.), Technical Digest WCPEC-3, Osaka, Japan, S20-B9-03, (article in press).

Zhao, J., Wang, A., Altermatt, P.P., Green, M.A., 2001. High Efficiency PERT Cells on High Quality N-Type CZ Si Substrates. In: Jinsoo Song (Ed.), Technical Digest of 12th PVSEC, Jeju, Korea, pp.19-22.

Chapter 4

III-V Compound Multi-junction and Concentrator Solar Cells

by

Masafumi Yamaguchi*,a, Tatsuya Takamotob and Kenji Arakia,c
aToyota Technological Institute, 2-12-1 Hisakata,
Tempaku, Nagoya 468-8511, Japan
bSharp Corporation, Hajikami, Shinjo, Nara 639-2198, Japan
cDaido Steel Corporation, Daido-cho, Minami, Nagoya 457-8545, Japan

Abstract

III-V compound semiconductor multi-junction (MJ) solar cells have the potential for achieving conversion efficiencies of over 50% and are promising for space and terrestrial applications. In this chapter, our R&D activities of high-efficiency III-V compound multi-junction, concentrator solar cells and modules have been presented and key issues for realizing super high-efficiency have also been discussed.

Conversion efficiencies of InGaP/InGaAs/Ge have been improved up to 31-32% (AM1.5) and 29-30% (AM0) by developing technologies such as double hetero wide band-gap tunnel junction, InGaP-Ge heteroface (Ge cell with an InGaP window layer) structure bottom cell, and precise lattice-matching of InGaAs middle cell to Ge substrate by adding indium into the conventional GaAs layer. For concentrator applications, grid structure has been designed in order to reduce the energy loss due to series resistance, and 37.4% (AM1.5G, 200-suns) efficiency has been demonstrated. In addition, we have realized high-efficiency and large-area (7,000 cm^2) concentrator InGaP/InGaAs/Ge 3-junction solar cell modules of an outdoor efficiency of 27% as a result of developing high-efficiency InGaP/InGaAs/Ge 3-junction cells, low optical loss Fresnel lens and homogenizers, and designing good thermal conductivity modules.

Future prospects are also presented. We have proposed concentrator III-V compound MJ solar cells as the 3rd generation solar cells in addition to 1st generation crystalline Si solar cells and 2nd generation thin-film solar cells.

---

* *Corresponding Author. E-mail: masafumi@toyota-ti.ac.jp*

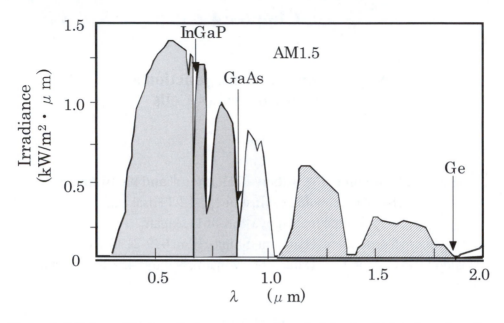

Figure 1. Efficient utilization of solar spectrum using principle of multi-junction solar cell.

1. Introduction

Multi-junction (MJ) tandem solar cells have the potential for achieving high conversion efficiencies of over 50% and are promising for space and terrestrial applications due to wide photo response as shown in Figure 1. One of the authors has been conducting his research on AlGaAs/GaAs 2-junction solar cells since 1982. His group has demonstrated 20.2% efficiency (Amano et al., 1987) by proposing double hetero-structure tunnel junction as a sub-cell interconnection in 1987 (Yamaguchi et al., 1987). A double-hetero (DH) structure tunnel junction was found to be useful for preventing impurity diffusion from the tunnel junction and improving the tunnel junction performance by the authors (Sugiura et al., 1988).

Based on such an activity, R&D project for "Super-high Efficiency MJ Solar Cells" has been conducted in Japan under support by NEDO (New Energy and Industrial Technology Development Organization) since FY (fiscal year) 1990 (Yamaguchi and Wakamatsu, 1996), in which challenges and efforts are made in the development of super-high efficiency solar cell technology, aiming at a dramatic increase in conversion efficiency of over 40% and developing innovational technologies as a long-term target for the early 21st century.

As a result of performance improvements in tunnel junction and top cell, over 30% efficiency has been obtained with InGaP/GaAs tandem cells by the authors (Takamoto et al., 1997a). Recently, InGaP/GaAs 2-junction solar cells have drawn increased attention for space applications because of the possibility of high conversion efficiency of over 30%. In fact, a commercial satellite (HS 601HP) was launched in 1997 with 2-junction GaInP/GaAs-on Ge solar arrays (Brown et al., 1997).

We have proposed AlInP-InGaP double hetero (DH) structure top cell, wide-bandgap InGaP DH structure tunnel junction for sub cell interconnection, and lattice-matched InGaAs middle cell. As a result of the above technology developments, the mechanically stacked InGaP/GaAs//InGaAs 3-junction cells (1cm^2) have reached the highest efficiency of 33.3% (1-sun world-record) at 1-sun AM1.5G following joint work by Japan Energy Co., Sumitomo Electric Co. and Toyota Tech. Inst (Takamoto et al., 1997b). Since FY2001, such an R&D project has been shifted to the project for "Concentrator MJ Solar Cells and Modules."

In this chapter, more recent results for high-efficiency III-V compound multi-junction (MJ) solar cells, concentrator MJ solar cells and modules conducted under the New Sunshine Project in Japan are presented. Those for space solar cells are also presented. Future prospects for super high-efficiency and low-cost MJ and concentrator cells and modules are also presented.

2. Historical Background

Tandem solar cells have been studied since 1960 by M. Wolf (Wolf, 1960). Table 1 shows progress of the III-V compound multi-junction solar cell technologies. J.C.C. Fan et al. (1982) encouraged R&D of tandem cells based on their computer analysis.

Although AlGaAs/GaAs tandem cells, including tunnel junctions and metal interconnectors, were developed in the early years, a high efficiency close to 20% was not obtained (Hutchby et al., 1985). This is because of difficulties in making high performance and stable tunnel junctions, and the defects related to the oxygen in the AlGaAs materials (Ando et al., 1987). A double hetero (DH) structure tunnel junction was found to be useful for preventing impurity diffusion from the tunnel junction and improving the tunnel junction performance by the authors (Sugiura et al., 1988). An InGaP material for the top cell was proposed by J.M. Olson et al. (1990). As results of performance improvements in tunnel junction and top cell, over 30% efficiency has been obtained with InGaP/GaAs tandem cells by Takamoto et al. (1997a). Recently, InGaP/GaAs 2-junction solar cells have drawn increased attention for space applications because of the possibility of high conversion efficiency of over 30%. In fact, the commercial satellite (HS 601HP) with 2-junction GaInP/GaAs-on Ge solar arrays was launched in 1997 (Brown et al., 1997).

Table 1. Progress of the III-V compound multi-junction solar cell technologies.

1960	Proposal of multi-junction solar cell	Wolf
1982	Efficiency calculation of tandem cells	Fan
1982	15.1% AlGaAs/GaAs 2-junction (2-J) cell	RTI
1987	Proposal of double-hetero structure tunnel junction for multi-junction interconnection	NTT
1987	20.2% AlGaAs/GaAs 2-J cell	NTT
1989	27.6% AlGaAs/GaAs 2-J cell (metal interconnection)	Varian
1989	32.6% GaAs/GaSb concentrator 2-J cell (mechanical-stacked, 100-suns concentration)	Boeing
1990	Proposal of InGaP as a top cell material	NREL
1990	27.3% InGaP/GaAs 2-J cell	NREL
1994	29.5% InGaP/GaAs 2-J cell	NREL
1996	30.3% InGaP/GaAs 2-J cell	Jpn. Energy
1996	Production of MJ cells	TECSTAR Spectrolab
1997	26.9% (AM0) InGaP/GaAs 2-J cell	Jpn. Energy & Toyota T.I.
1997	33.3% InGaP/GaAs//InGaAs 3-J cell (mechanical-stacked)	Jpn. Energy, Sumitomo & Toyota T.I.
1997	Commercial satellite with 2-J cells	Hughes
2000	31.7% InGaP/InGaAs/Ge 3-J cell	Jpn. Energy
2003	37.4% InGaP/InGaAs/Ge 3-J cell (200-suns concentration)	Sharp
2003	27% large-area (7,000cm^2) InGaP/InGaAs/Ge 3-J cell module (outdoor)	Daido Steel, Daido Metal, Sharp &Toyota T.I.

3. Key Technologies for High-efficiency Cells

Key issues for realizing high-efficiency tandem cells are discussed based on our results.

3.1 Selection of Top Cell Materials and Improving the Quality

Selection of top cell materials is also important for high-efficiency tandem cells. It has been found by the authors (Ando et al., 1987) that an oxygen-related defect in the AlGaAs top cell materials acts as the recombination center as shown in Figure 2. InGaP has some advantages as a top cell material latticed matched to GaAs or Ge substrates (Olson et al., 1990), such as lower interface recombination velocity, less oxygen problem and good window layer material compared to AlGaAs.

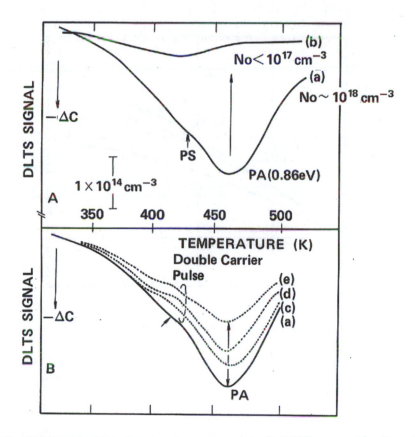

Figure 2. (A) Deep level transient spectroscopy spectra for MBE-grown $Al_{0.4}Ga_{0.6}As$ films with different oxygen concentrations, and (B) comparison of those under different carrier pulse conditions.

The top cell characteristics depend on the minority carrier lifetime in the top cell layers. Figure 3 shows changes in photoluminescence (PL) intensity of the solar cell active layer as a function of the minority carrier lifetime (τ) of the p-InGaP base layer grown by MOCVD and surface recombination velocity (S). The lowest S was obtained by introducing the AlInP window layer and the highest τ was obtained by introducing buffer layer and optimizing the growth temperature. The best conversion efficiency of the InGaP single junction cell was 18.5 % (Yang et al., 1997).

3.2. Low Loss Tunnel Junction for Intercell Connection and Preventing Impurity Diffusion from Tunnel Junction

Another important issue for realizing high-efficiency monolithic-cascade type tandem cells is the achievement of optically and electrically low-loss interconnection of two or more cells. A degenerately doped tunnel junction is attractive because it only involves one extra step in the growth process. To minimize optical absorption, forma-

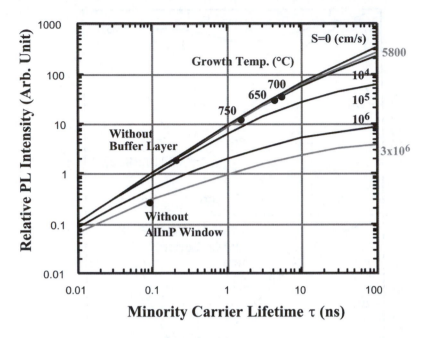

Figure 3. Changes in PL intensity of the solar cell active layer as a function of the minority carrier lifetime (τ) of the p-InGaP base layer and surface recombination velocity (S).

tion of thin and wide-bandgap tunnel junctions is necessary as shown in Figure 4. However, the formation of a wide-bandgap tunnel junction is very difficult, because the tunneling current decreases exponentially with increase in bandgap energy.

In addition, impurity diffusion from a highly doped tunnel junction during over-growth of the top cell increases the resistivity of the tunnel junction. As shown in Figure 5, a double hetero (DH) structure was found to be useful for preventing impurity diffusion by the authors (Sugiura et al., 1988). An InGaP tunnel junction has been tried for the first time for an InGaP/GaAs tandem cell in our recent work (Takamoto et al., 1997a). As p-type and n-type dopants, Zn and Si were used, respectively. Peak tunneling current of the InGaP tunnel junction is found to increase from 5mA/cm^2 up to 2A/cm^2 by making a DH structure with AlInP barriers. Therefore, the InGaP tunnel junction has been observed to be very effective for obtaining high tunneling current, and DH structure has also been confirmed to be useful for preventing impurity diffusion.

DH structure effect on suppression of impurity diffusion from the tunnel junction has been examined. Effective suppression of the Zn diffusion from tunnel junction by the InGaP tunnel junction with the AlInP-DH structure is attributed to the lower diffusion coefficient (Takamoto et al., 1999a) for Zn in the wider bandgap energy materials such as the AlInP barrier layer and InGaP tunnel junction layer.

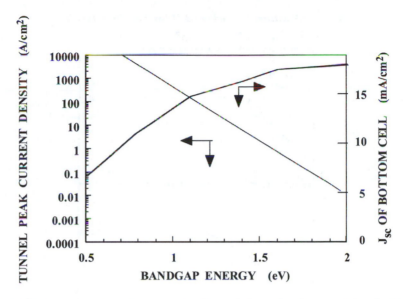

Figure 4. Calculated tunnel peak current density and short-circuit current density J_{sc} of GaAs bottom cell as a function of bandgap energy of tunnel junction.

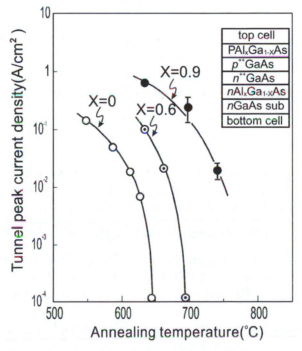

Figure 5. Annealing temperature dependence of tunnel peak current densities for double hetero structure tunnel diodes. X is the Al mole fraction in $Al_xGa_{1-x}As$ barrier layers.

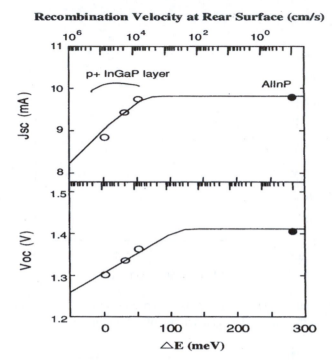

Figure 6. Changes in open-circuit voltage V_{oc} and short-circuit current density J_{sc} and of InGaP single-junction cells as a function of potential barrier ΔE.

3.3 Lattice Matching Between Cell Materials and Substrates

Reduction in the lattice constant mismatch between cell materials and substrate is also very important for obtaining high efficiency cells because misfit dislocations are generated in the upper cell layers which deteriorate the cell efficiency. Application of InGaAs middle cell (Takamoto, 1999b) lattice-matching to Ge substrates has demonstrated to increase the open-circuit voltage (V_{oc}) due to lattice-matching and short-circuit current density (J_{sc}) by decrease in bandgap energy of middle cell.

3.4 Effectiveness of Wide Bandgap Back Surface Field (BSF) Layer

Figure 6 shows changes in V_{oc} and J_{sc} of InGaP single-junction cells as a function of potential barrier ΔE. Wide bandgap BSF layer (Takamoto, 1999b) is found to more effective for confinement of minority carriers compared to highly impurity-doped BSF layers.

Table 2 summarizes the key issues for realizing super high-efficiency multi-junction solar cells.

Table 2. Key issues for realizing super high-efficiency multi-junction solar cells.

Key Issue	Past	Present	Future
Top cell materials	AlGaAs	InGaP	AlInGaP
3rd layer materials	None	Ge	InGaAsN etc.
Substrate	GaAs	Ge	Si
Tunnel junction	DH-structure GaAs tunnel J.	DH-structure InGaP tunnel J.	DH-structure InGaP or GaAs
Lattice matching	GaAs middle cell	InGaAs middle cell	(In)GaAs middle cell
Carrier confinement	InGaP-BSF	AlInP-BSF	Widegap-BSF
Photon confinement	None	None	Bragg reflector etc.

4. High-efficiency Multi-junction Cells

Japan Energy Co. has successfully fabricated monolithic cascade (MC) type 3-junction InGaP/InGaAs/Ge with efficiencies of 31.7% (1cm x 1cm) and 31.2% (5cm x 5cm) for 1 sun AM 1.5G. They achieved these high efficiencies by lattice-matching between middle cells and Ge substrates and introducing the C-doped AlGaAs/Si-doped InGaP hetero-structure tunnel junction with AlInP barriers (Takamoto et al., 2000). Figures 7 and 8 show a structure, I-V curve and spectral response of a high efficiency InGaP/GaAs/Ge 3-junction cell fabricated on a Ge substrate.

The mechanically stacked (MS) type 3-junction cells of monolithically grown InGaP/GaAs 2-junction cells and InGaAs bottom cells have reached the world-record efficiency of 33.3% (Takamoto et al., 1997b) at 1-sun AM1.5G following the joint work by Japan Energy Co., Sumitomo Electric Co. and Toyota Tech. Inst. as shown in Figure 9.

5. More Recent Approaches for High-efficiency MJ Solar Cells

Conversion efficiency of InGaP/GaAs based multijunction solar cells has been further improved by the following technologies. A schematic illustration of the InGaP/(In)GaAs/Ge 3-junction solar cell and key technologies for improving conversion efficiency are shown in Figure 10.

5.1 Wide Bandgap Tunnel Junction

A wide band-gap tunnel junction which consists of double-hetero structure p-AlInGaP/p-AlGaAs/ n-(Al)InGaP/n-AlInGaP increases the incident light into the (In)GaAs middle cell and produces effective potential barriers for both minority-carriers generated in the top and middle cells. The Voc and Isc of the cells are improved by the wide band-gap tunnel junction that reduces optical absorption in this layer. They

Structure of Triple-Junction (3J) Cell

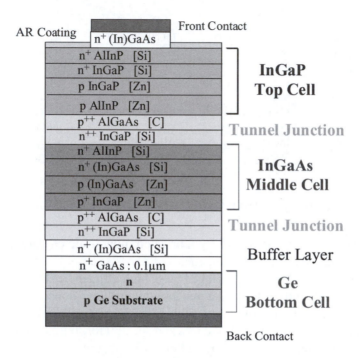

Figure 7. A structure of a high efficiency InGaP/GaAs/Ge 3-junction cell fabricated on a Ge substrate.

Characteristics of 3J Cell (x=0.01)

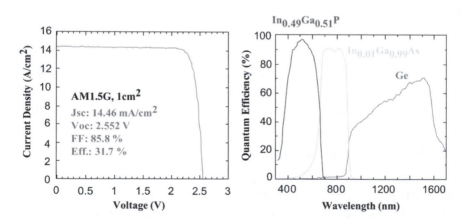

Figure 8. I-V curve and spectral response of a high efficiency InGaP/GaAs/Ge 3-junction cell fabricated on a Ge substrate.

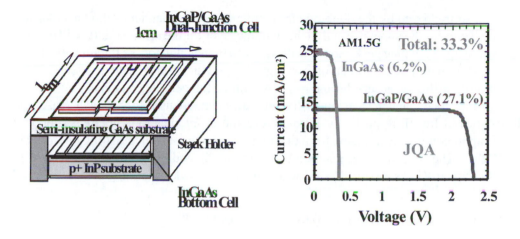

Figure 9. A structure and I-V curve of the world-record efficiency InGaP/GaAs//InGaAs mechanically stacked 3-junction solar cell.

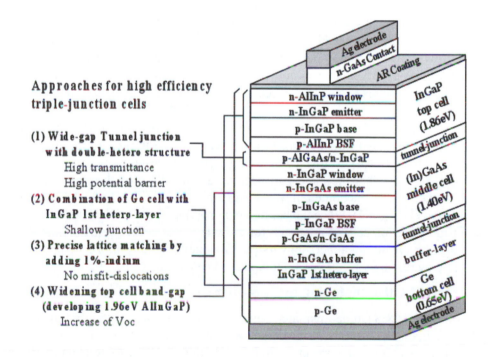

Approaches for high efficiency triple-junction cells

(1) **Wide-gap Tunnel junction with double-hetero structure**
 High transmittance
 High potential barrier

(2) **Combination of Ge cell with InGaP 1st hetero-layer**
 Shallow junction

(3) **Precise lattice matching by adding 1%-indium**
 No misfit-dislocations

(4) **Widening top cell band-gap (developing 1.96eV AlInGaP)**
 Increase of Voc

Figure 10. Schematic illustration of a triple-junction cell and approaches for improving efficiency of the cell.

are also improved by the double-hetero structure layers that act as the back-surface field layer for the upper cell and as the window layer for the lower cell. It is difficult to obtain high tunneling peak current with wide gap tunnel junction, so thinning the depletion layer width by the formation of a highly doped junction is quite necessary. Since impurity diffusion occurs during the growth of the top cell (Sugiura et al., 1988), carbon and silicon, which have low diffusion coefficient, are used for p-type AlGaAs and n-type (Al)InGaP, respectively. Furthermore, the double-hetero structure is supposed to suppress impurity diffusion from the highly doped tunnel junction (Takamoto et al., 1999a). The second tunnel junction between middle and bottom cells consists of p-InGaP/p-(In)GaAs/n-(In)GaAs/n-InGaP which have wider band-gap than middle cell materials.

5.2 InGaP/Ge Heteroface Structure Bottom Cell

InGaP/GaAs cell layers are grown on a p-type Ge substrate. The p-n junction is formed automatically during MOCVD growth by the diffusion of V-group atom from the first layer grown on the Ge substrate. So, the material of the first hetero layer is important for the performance of Ge bottom cell. An InGaP layer is thought to be suitable material for the first hetero layer, because phosphorus has a lower diffusion coefficient in Ge than arsenic, and indium has lower solubility in Ge than gallium. Figure 11 shows the change in spectral response of the triple-junction cell by changing the first hetero growth layer on Ge from GaAs to InGaP. Quantum efficiency of the Ge bottom cell was improved by the InGaP hetero-growth layer. In the case of GaAs hetero-growth layer, junction depth was measured to be around 1μm. On the other hand, thickness of n-type layer produced by phosphorus from the InGaP layer was 0.1μm. An increase in Ge quantum efficiency was confirmed to be due to a reduction in the junction depth.

It was found that the absorption edge of the InGaP top cell shifted to the longer wavelength region, by using the InGaP first hetero layer. Band-gap of the InGaP top cell reduced from 1.86 eV to 1.81 eV by changing the hetero-growth layer from GaAs to InGaP. The fact that the band-gap decreased with the growth temperature increased indicated this phenomenon was due to the ordering effect in the InGaP material (Gomyo et al., 1987). Since the band-gap narrowing of the top cell decreases Voc of the triple-junction cell, an approach for growth of less-ordering InGaP should be necessary. As a matter of a fact, conversion efficiency has been improved up to 30% (AM0) by increasing top cell band-gap up to 1.89 eV (King et al., 2002).

5.3 Precise Lattice-matching to Ge Substrate

Although 0.08% lattice-mismatch between GaAs and Ge was thought to be negligibly small, misfit-dislocations were generated in thick GaAs layers and deteriorated cell performance. By adding about 1% indium into the InGaP/GaAs cell layers, all cell layers are lattice-matched precisely to the Ge substrate. As a result, cross-hatch pattern caused by misfit-dislocations due to lattice-mismatch disappeared in the surface

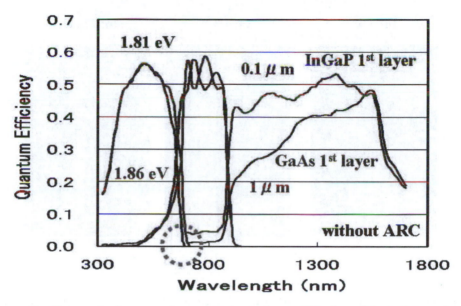

Figure 11. Changes in the spectral response due to the modification of the 1st hetero-layer from GaAs to InGaP (without anti-reflection coating (ARC)).

morphology of the cell with 1% indium, as shown in Figure 12. The misfit-dislocations were found to influence not the Isc but the Voc of the cell. Voc was improved by eliminating misfit-dislocations for the cell with 1% indium. In addition, wavelength of the absorption edge became longer and Isc of both top and middle cells increased, by adding 1% indium.

5.4 Widening of Top Cell Bandgap by AlInGaP

Now, we are developing AlInGaP top cells in order to improve Voc of the triple-junction cells. Current matching between top and middle cells should be done by controlling the top cell band-gap instead of thinning the top cell. In this case, Voc of the cell can be increased with keeping the maximum current. An AlInGaP cell with 1.96 eV band-gap and 2.5 μm thickness was found to attain high Voc of 1.5V with keeping the same Isc as the conventional InGaP top cells under AM1.5G condition. Figure 13 shows comparison of light I-V curve under AM1.5G condition. For AM0 condition, a further increase in the band-gap of up to about 2.0-2.03 eV is required for the AlInGaP cells, although that is dependant on the current matching requirement from the beginning of life (BOL) to the end of life (EOL).

The best data of the triple-junction cells in our laboratory are summarized in Table 3. Technologies described in sections 5.1-5.3 were applied to the fabrication of triple junction cells. Band gap of the InGaP top cell of about 1.82 eV is still low. By using an AlInGaP top cell with 1.96 eV, higher V_{oc} close to 2.72 V is predicted. Conversion

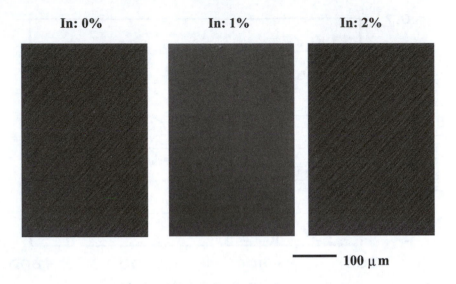

In: 0% **In: 1%** **In: 2%**

—— **100 μm**

Figure 12. Surface morphology of InGaAs with various indium composition grown on Ge.

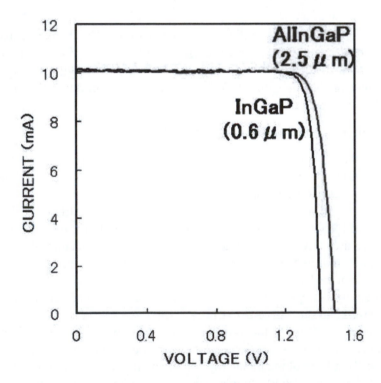

Figure 13. Light IV curves of the single junction AlInGaP and InGaP cells (AM1.5G, without ARC).

Table 3. Characteristics and predicted efficiencies of the triple-junction cells. The triple-junction cells using an InGaP top cell show the best data and those using an AlInGaP top cell show predicted properties.

	Voc (mV)	Jsc (mA/cm^2)	FF	Eff. (%)	Condition
InGaP	2567	14.1	0.87	31.5	AM1.5G, 25^{0}C
(Eg=1.82eV)	2568	17.9	0.86	29.2	AM0, 28^{0}C
AlInGaP	2720	14.1	0.87	33.3	AM1.5G, 25^{0}C
(Eg=1.96eV)	2721	17.9	0.86	31	AM0, 28^{0}C

efficiencies over 33% (AM1.5G) and close to 31% (AM0) are expected for the (Al)InGaP/InGaAs/Ge triple junction cells

For the next-generation multi-junction cells, one approach is an optimization of the band-gaps for utilizing solar energy with a wide energy range above the Ge band-gap. A lattice-mismatched AlInGaP(1.8eV)/InGaAs(1.2eV)/Ge (0.65eV) 3-junction cell and a lattice-matched AlGaInP (2.0eV)/GaAs(1.4eV)/GaInNAs(1eV)/Ge(0.65eV) 4-junction cell are the challenging structures (Kurtz et al., 1997). Another approach is to increase the number of junctions for utilizing the high-energy range. Many thin junctions with contiguous band-gaps can reduce the energy loss due to the difference between photon energy and band-gap energy. AlInGaP cells and InGaAsP cells lattice matched to GaAs provide various band-gaps and might establish the high efficiency cells.

6. High-efficiency Concentrator MJ Solar Cells

Based on the activities described above, a new R&D project for "Super-high Efficiency MJ Concentrator Solar Cell modules" is underway in Japan. The project is a collaborative effort between Sharp Co., Daido Steel Co., and Daido Metal Co. with support from NEDO. The target of this project by the end of March 2006 is 40% cell efficiency and 31% module efficiency under a 500-sun concentration.

In order to use a high efficiency multi-junction cell developed for 1 sun condition with a concentrator system operating under ~500 suns condition, reduction in energy loss due to series resistance is the most important issue. A cell size of 7mm x 7mm was chosen for the concentrator application considering the total current flow. Grid electrode pitching, height and width were designed to reduce series resistance. Figure 14 shows the fill factor (FF) of the cell with various grid pitching under 250 suns. Grid electrode with 5μm height and 5μm width was made of Ag. Grid pitch influences the

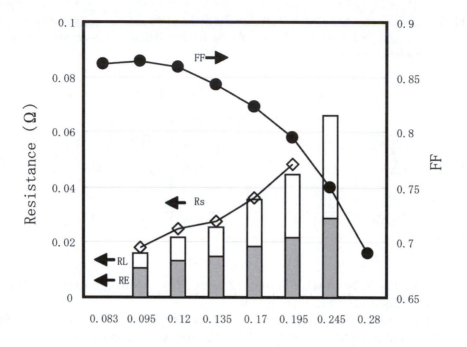

Figure 14. FF of the concentrator cells with various grid pitching under 250 suns light. Series resistance (Rs), lateral resistance (RL) and total electrode resistance (RE) are also shown.

lateral resistance between two grids (RL) and the total electrodes resistance (RE). Series resistance of the cell (RS), RE and RL are also shown in Figure 14. RE was measured directly after removing the electrode from the cell by chemical etching. RL was calculated by using the sheet resistance of the window and emitter layers. Based on the data in Figure 14, the grid pitch was determined to be 0.12 mm. To reduce the series resistance down to 0.01Ω and obtain high FF under 500 suns, the grid height should be doubled. High efficiency under <500 suns is thought to be obtained by the optimal grid design without any modification of the cell layer structure such as the emitter thickness and tunnel junction thickness from the cell developed for 1sun condition.

Figure 15 shows the most recent results of efficiency of a 3-junction InGaP/ InGaAs/Ge concentrator solar cell as a function of concentration ratio. The authors had earlier demonstrated a 36.5% efficiency of 3-junction solar cells (Takamoto et al., 2003). The concentration ratio is defined by the increase in Isc, Isc/Isc(1-sun) ratio. High conversion efficiency over 36% is measured under concentrated light with concentration ratio ranging from 30 suns to 200 suns. At a concentration ratio of 200-suns,

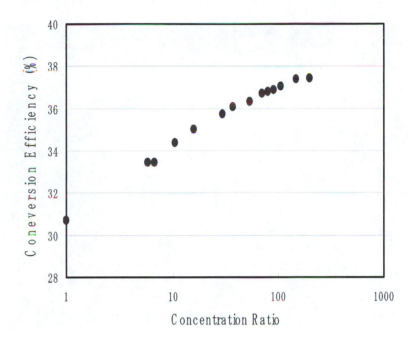

Figure 15. Efficiency of an InGaP/InGaAs/Ge concentrator 3-junction cell as a function of concentration ratio.

the conversion efficiency was measured to be 37.4%. This value is the highest ever reported for any solar cell.

7. High-efficiency and Low-cost Concentrator MJ Cell Modules

Research and development of III-V concentrator cells in Japan is now being conducted mainly with 3-junction monolithic modules. Besides research on cells, extensive studies are being conducted in (1) Non-imaging Fresnel lens concentrator, (2) Homogenizers, (3) New and simple module design (Araki et al., 2003).

A concentrator module of 7000 cm^2 area and 400 X concentration was fabricated with 36 pieces of the randomly-selected sub cell modules connected in series and the same number of the newly-developed dome-shape Fresnel lenses as shown in Figure 16. The module was designed and fabricated without any heat sinks or external cooling.

The efficiency on a hot summer day (35°C of ambient temperature) was 27.0 % as shown in Figure 17 (Araki et al., 2004). This value was well above the record efficiency of 22.7 % established by the Fraunhofer Institute Germany in October 2002

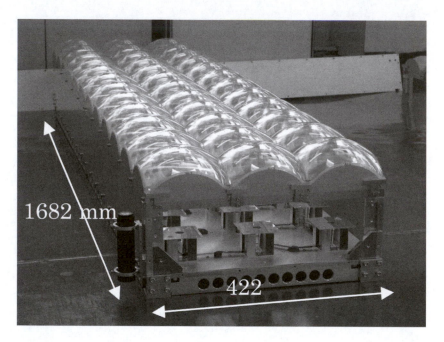

Figure 16. 7,000cm² and 400X concentrator module with the 36 receivers connected in series and dome-shaped Fresnel lenses made by injection mold.

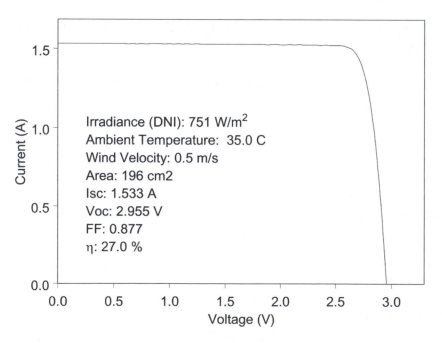

Irradiance (DNI): 751 W/m²
Ambient Temperature: 35.0 C
Wind Velocity: 0.5 m/s
Area: 196 cm2
Isc: 1.533 A
Voc: 2.955 V
FF: 0.877
η: 27.0 %

Figure 17. Outdoor evaluation of I-V with a plastic Fresnel lens (Including loss in concentrator optics and temperature rise).

with one-order less area (768 cm^2). This achievement resulted from several new technologies:

(1) High-pressure and vacuum-free lamination of the solar cell that suppresses the temperature rise to only 8 degrees under a 400 X geometrical concentration solar radiation.

(2) Direct and voids-free soldering technologies of fat metal ribbon to solar cells that suppresses hot spots and reduces resistance resulting in 300 times higher output current than normal non-concentration operation.

(3) A new encapsulating polymer that survives exposure of high concentration UV and heat cycles.

(4) Beam-shaping technologies that illuminate a square aperture of the solar cells from a round concentration spot.

(5) Homogenizer technologies giving uniform flux and preventing conversion loss due to chromatic aberrations and surface voltage variation.

(6) Assemble tolerance between optical lenses and solar cells as large as 1.75mm is allowed in this development. There is no need of special optical alignment. Even local mechanical industries can assemble the main body.

(7) Shaped non-imaging Fresnel lens made by low-cost injection molding that gives a variation in optical efficiency as low as 0.5% over the entire Fresnel lens.

The technological roadmap toward a 31% efficient module is shown in Table 4 (Araki et al., 2004). Since we have identified detailed technological problems and how to solve them, we expect to achieve the target well in advance.

8. High-efficiency and Radiation-resistant Space Solar Cells

Figure 18 shows effectiveness of radiation-resistance and high conversion efficiency of space cells from the point view of power density. Since radiation in space is severe, particularly in the Van Allen radiation belt, lattice defects are induced in semiconductors due to high-energy electron and proton irradiations, and the defects cause a decrease in the output power of solar cells. Further improvements in conversion efficiency and radiation-resistance of space cells are necessary for widespread applications of space missions.

Recently, InGaP/GaAs-based MJ solar cells have drawn increased attention because of the possibility of high conversion efficiency of over 40% and radiation-resistance. Figure 19 shows correlation between AM0 and AM1.5 efficiencies for various solar cells. Therefore, high-efficiency MJ solar cells developed for terrestrial use can be used as space solar cells as shown in Figure 19.

Table 4. Roadmap table of more than 31% efficiency module.

		Technologies in Nov. 2002	Technologies in Aug. 2003	Technologies in Mar. 2004	Technologies in Mar. 2005	Technologies in Mar. 2006
A	Cell Efficiency @ 1sun	30.1 %	30.3 %	32 %		
B	Cell Efficiency Concentration	34.4 %	35.3 %	37 %		40 %
C	Lens Efficiency	72.4 %	85.4 %	85.8 %		91 %
D	Homogenizer Efficiency	94.4 %	96.3 %	97.5 %	97.5 %	97.5 %
E	Ohmic Loss in Circuit	0.1 %	0.1 %	0.1 %	0.1 %	0.1 %
F	Spectrum Mismatching Loss	5.3 %	5.1 %	3 %		
G	Current Mismatching Loss	3.7 %	3.7 %	2 %	2 %	2 %
H	Loss by Temperature Rise	1.2 %	1.3 %	1.3 %		1 %
I	Total Efficiency	21.7 %	27.0 %	29 %	> 29 %	> 31 %

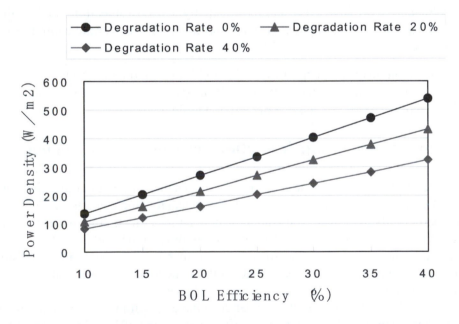

Figure 18. Effectiveness of radiation-resistance and high conversion efficiency of space cells from the point view of power density.

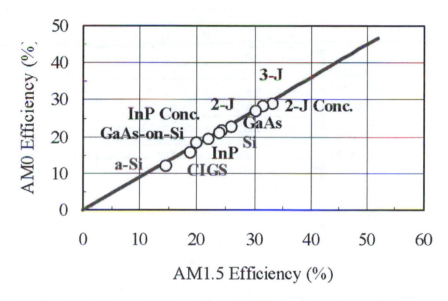

Figure 19. Correlation between AM0 and AM1.5 efficiencies for various solar cells.

Figure 20 shows a schematic diagram of a high-efficiency InGaP/GaAs 2-junction solar cell that was fabricated (Takamoto et al., 1999c). InGaP/GaAs cell layers were grown on a GaAs substrate by the metalorganic chemical vapor deposition (MOCVD) method. The minority carrier lifetime in the p-InGaP base layer and the recombination velocity at the interface between the AlInP window layer and the n-InGaP emitter were estimated to be around 10-50 ns and 5800 cm/s, respectively, using the time-resolved photoluminescence technique. The top and bottom cells were connected by a double-hetero structure InGaP tunnel junction. Previously, a world-record efficiency of 26.9% (Voc=2.451V, Isc=67.4mA, FF=88.1%) under AM0, 28°C conditions was measured at NASDA (National Space Development Agency of Japan) using a one light source simulator, for 4 cm^2 InGaP/GaAs 2-junctionm cells, as shown in Figure 21. More recently, an AM0 efficiency of 29.2% has been demonstrated for an InGaP/InGaAs/Ge 3-junction cell (4cm^2) (Takamoto et al., 2003) as shown in Figure 22.

Figure 23 shows the radiation resistance of InGaP/GaAs 2-junction cells against 1 MeV electron irradiation with fluences in the range from 3×10^{14} to 1×10^{16} cm^{-2}, along with those of InP, InGaP and GaAs-on-Ge solar cells (Yamaguchi et al., 2003). The radiation resistance of our tandem cell is similar to that of a GaAs-on-Ge cell. The remaining factor of Voc, Isc, FF and P$_{max}$ at EOL (1×10^{15} cm^{-2}) are 0.92, 0.83, 0.97 and 0.74, respectively. Degradation in our tandem cell performance is thought to be mainly attributed to large degradation in the GaAs bottom cell with a highly doped base layer.

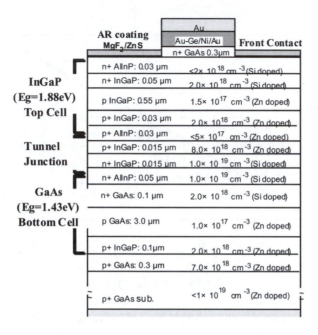

Figure 20. A schematic diagram of a high-efficiency InGaP/GaAs 2-junction solar cell fabricated.

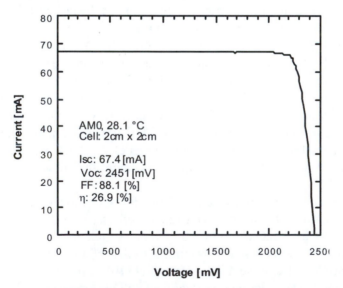

Figure 21. I-V curve of an InGaP/GaAs 2-junction cell with a conversion efficiency of 26.9% (AM0), which was measured by NASDA using a one light source simulator.

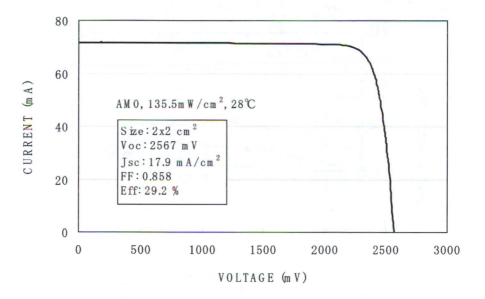

Figure 22. I-V curve of an InGaP/InGaAs/Ge 3-junction cell with a conversion efficiency of 29.2% at AM0.

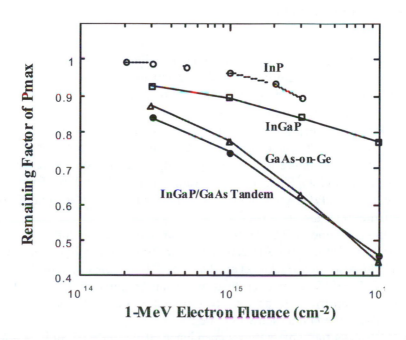

Figure 23. Radiation resistance of InGaP/GaAs 2-junction cells against 1 MeV electron (standard radiation to qualify devices for space use) irradiation, in comparison with that of InP, InGaP and GaAs-on-Ge solar cells.

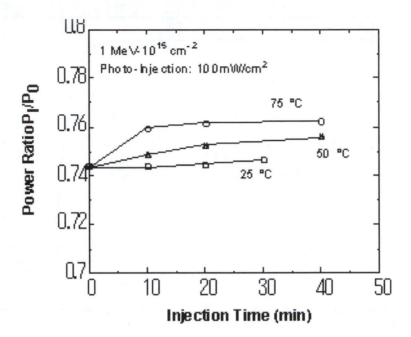

Figure 24. The maximum power recovery of the InGaP/GaAs tandem cell due to light illumination at various temperatures.

Figure 24 shows the maximum power recovery due to light illumination of 100 mW/cm^2 at various temperatures for 1-MeV electron-irradiated InGaP/GaAs tandem cells (Yamaguchi et al., 1997a). The ratios of maximum power after injection, P_I, to maximum power before irradiation, P_0, are shown as a function of injection time. Even at room temperature, photoinjection-enhanced annealing of radiation damage to InGaP/GaAs tandem cells was observed. The recovery ratio increases with an increase in ambient temperature within the operating range for space use. Such a recovery is also found in InGaP top cell layer (Yamaguchi et al., 1997b). Therefore, the results show that InGaP/GaAs tandem cells under device operation conditions have superior radiation-resistant properties.

Figure 25 shows the DLTS (Deep Level Transient Spectroscopy) spectrum of trap H2 (Ev+0.55eV) for various injection times at 25°C with an injection density of 100mA/cm^2. It is also found (Khan et al., 2000) by DLTS measurements that a major defect level H2 (Ev+0.55eV) recovers by forward bias or light illumination. Moreover, the H2 center is confirmed to act as a recombination center by using the double carrier pulse DLTS method as shown Figure 26.

Figure 27 shows the temperature dependence of the thermal and injection-enhanced annealing rates due to injection (forward bias), for radiation-induced defects in p-InGaP, determined from solar cell property (Jsc or L) and for the H2 trap observed by DLTS.

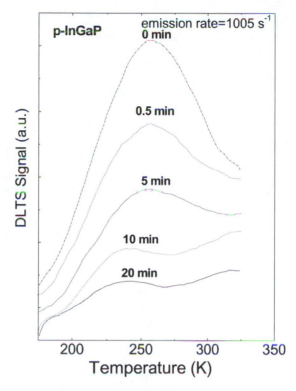

Figure 25. DLTS spectrum of trap H2 (Ev+0.55eV) with various time of injection at 25°C with an injection density of 100mA/cm².

As shown in Figure 27, under minority-carrier injection condition, the radiation-induced defects in InGaP show a large defect annealing enhancement factor of 6-7 orders of magnitude compared to the room-temperature annealing process. On the other hand, in the case of InGaP/GaAs tandem cells, the radiation-induced defects in InGaP top cell layer are thought to be annihilated under forward bias and light illumination conditions but those in GaAs bottom cell layer still remain. Therefore, the forward bias injection annealing rate for InGaP/GaAs tandem cells is lower by a factor of 2 than that for InGaP single-junction cells.

For InGaP/GaAs tandem cells, the photoinjection annealing rate is a factor of 3 lower than the forward bias injection annealing rate because the photoinjection (100mW/cm²) is thought to have lower injection current density (operating current density of about 15mA/cm²) compared to the conventional forward bias injection (100mA/cm²).

The enhancement of defect annealing in InGaP top cell layer under minority-carrier injection conditions is thought to occur as a result of the nonradiative electron-hole recombination process (Lang et al., 1976) whose energy E_R enhances the defect motion or the so-called Bourgoin mechanism (Bourgoin et al., 1975), i.e., the migration

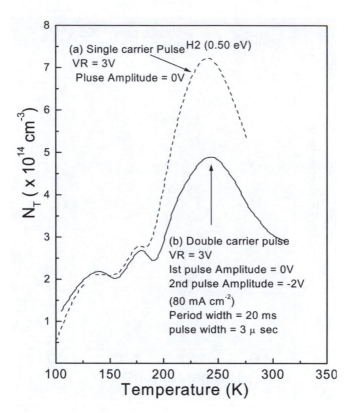

Figure 26. DLTS spectra showing the reduction in the peak height of H2 deep level on the minority carrier injection pulse voltage (a) 1msec single carrier pulse (hole) with a reverse bias of 3V and pulse height of 0V to fill the trap by hole and (b) 50μsec minority carrier injection pulse (-2V double carrier pulse) with a reverse bias of 3 V and current density of 80 mA/cm2.

induced by the alternative changes of the defect charge state. If the mechanism of the radiation damage recovery of InGaP top cell layer is due to the nonradiative electron-hole recombination process, the thermal activation energy E_A (1.1eV) of the defect is reduced to E_I (0.48~0.54eV) by an amount E_R (0.56~0.62eV). Thus electronic energy from a recombination event can be channeled into the lattice vibration mode which drives the defect motion: $E_I = E_A - E_R$.

However, further studies are necessary to clarify the mechanism of the minority-carrier injection-enhanced annealing phenomena and the origins of the radiation-induced defects in InGaP.

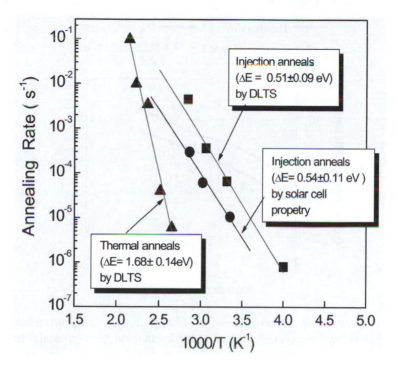

Figure 27. The temperature dependence of the thermal and injection-enhanced annealing rates due to injection (forward bias), for radiation-induced defects in p-InGaP, determined from solar cell property (J_{sc} or L) and for the H2 trap observed by DLTS.

9. Future Prospects

Multi-junction solar cells will be widely used in space because of their high conversion efficiency and better radiation-resistance. In order to apply super high-efficiency cells widely, it is necessary to improve their conversion efficiency and reduce their cost. Figure 28 shows theoretical conversion efficiencies of single-junction and multi-junction solar cells in comparison with experimentally realized efficiencies (Yamaguchi, 2002). Therefore, concentrator 3-junction and 4-junction solar cells have great potential for realizing super high-efficiency of over 40%. As a 3-junction combination, InGaP/InGaAs/Ge cell on a Ge substrate will be widely used because this system has already been developed. The 4-junction combination of an Eg=2.0eV top cell, a GaAs second-layer cell, a material third-layer cell with an Eg of 1.05eV, and a Ge bottom cell is lattice-matched to Ge substrates and has a theoretical efficiency of about 42% under 1-sun AM0. This system has a potential efficiency of over 47% under 500-suns AM1.5 condition.

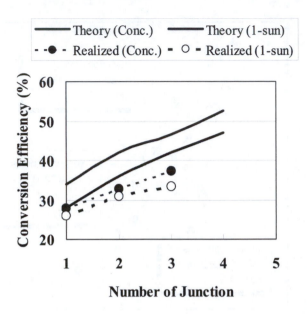

Figure 28. Theoretical conversion efficiencies of single-junction and multi-junction solar cells in comparison with experimentally realized efficiencies under 1-sun and concentration conditions.

We are now challenged to develop low-cost and high output power concentrator MJ solar cell modules with an output power of 400W/m^2 for terrestrial applications as shown in Figure 29.

Concentrator operation of the MJ cells is essential for their terrestrial applications. Since the concentrator PV systems have potential of cost reduction, R&D on concentrator technologies including MJ cells has started in Japan. Therefore, concentrator MJ and crystalline Si solar cells are expected to contribute to electricity cost reduction for widespread PV applications as shown in Figure 30 (Yamaguchi et al., 2004).

We are also now challenged to develop high-efficiency, light-weight and low-cost MJ solar cells for space applications.

10. Summary

Conversion efficiency of InGaP/InGaAs/Ge has been improved up to 31-32% (AM1.5) and 29-30% (AM0) as a result of technologies developed such as double hetero wide band-gap tunnel junction, InGaP-Ge heteroface structure bottom cell, and precise lattice-matching of InGaAs middle cell to Ge substrate by adding indium into the conventional GaAs layer. For concentrator applications, grid structure has been designed in order to reduce the energy loss due to series resistance, and a 37.4% (AM1.5G, 200-suns) efficiency has been demonstrated. In addition, we have realized

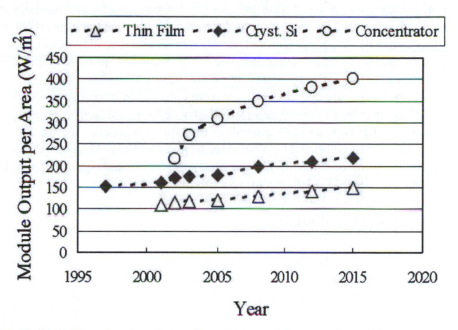

Figure 29. Projection of concentrator cell module output in comparison with output projection of crystalline Si and thin film solar cell modules.

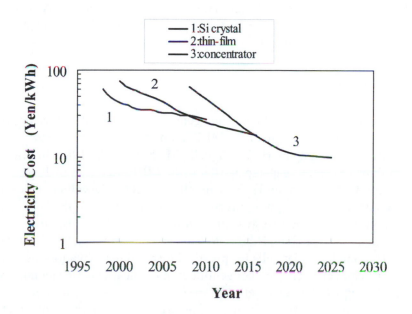

Figure 30. Scenario of electricity cost reduction by developing concentrator solar cells.

high-efficiency concentrator InGaP/InGaAs/Ge 3-junction solar cell modules of an outdoor efficiency of 27% as a result of developing high-efficiency InGaP/InGaAs/Ge 3-junction cells, designing grid contact with low series resistance, developing non-imaging Fresnel lens and 2nd optics (homogenizers) with low optical loss, and designing modules with low series resistance and better thermal conductivity.

Future prospects are also presented. We have proposed concentrator III-V compound MJ solar cells as the 3rd generation solar cells in addition to 1st generation crystalline Si solar cells and 2nd generation thin-film solar cells. We are now challenged to develop low-cost and high output power concentrator MJ solar cell modules with an output power of 400W/m2 for terrestrial applications and high-efficiency, lightweight and low-cost MJ solar cells for space applications.

11. Acknowledgements

This work is partially supported by the New Energy and Industrial Technology Development Organization as a part of the New Sunshine Program under the Ministry of Economy, International Trade and Industry, Japan. This work is also partially supported by the Japan Ministry of Education, Culture, Sports, Science and Technology as a part of the Private University Academic Frontier Center Program "Super-High Efficiency Photovoltaic Research Center." The authors would like to thank M-J. Yang, N.J. Ekins-Daukes, S. Lee, N. Kojima and Y. Ohshita of TTI; A. Yamamoto, H. Sugiura, K. Ando, C. Amano and S. Katsumoto of NTT Labs; M. Ohmori, H. Kurita and E. Ikeda of Japan Energy, T. Agui and M. Kaneiwa of Sharp; A. Kitamura of KEPCO; Y. Hayashi of ETL, and S. Wakamatsu of PVTEC for their fruitful discussion, kind cooperation and support.

12. References

Amano, C., Sugiura, H., Yamamoto, A., Yamaguchi, M., 1987. 20.2% efficiency AlGaAs/GaAs tandem solar cells grown by MBE, Appl. Phys. Lett. 51, 1998.

Ando, H., Amano, C., Sugiura, H., Yamaguchi, M., Salates, A., 1987. Nonradiative e-h recombination of mid-gap trap in AlGaAs by MBE, Jpn. J. Appl. Phys. 26, L266.

Araki, K., Kondo, M., Uozumi, H., Yamaguchi, M., 2003. Development of a robust and high efficiency concentrator receiver, In Proceedings of the 3rd World Conference on Photovoltaic Energy Conversion, pp.630, Osaka, Japan.

Araki, K., Kondo, M., Uozumi, H., Ekins-Daukes, N.J., Yamaguchi, M., 2004. Packaging III-V tandem solar cells for practical terrestrial applications achievable to 28% of module efficiency by conventional machine assemble technology. Solar Energy Materials and Solar Cells. In press.

Bourgoin, J.C., Corbett, J.W., 1975. Ionization effects on impurity and defect migration in semiconductors, In Lattice Defects in Semiconductors, Conf. Ser. No.23, ed. By Huntley F.A., pp. 149, (Inst. Phys., London and Bristol).

Brown, M.R., Goldhammer, L.J., Goodelle, G.S., Lortz, C.U., Perron, J.N., Powe, J.S., Schwartz, J.A., Cavicchi, B.T., Gillanders, M.S., Krut, D.D., 1997. Characterization testing of dual GaInP2/GaAs/Ge solar cell assemblies, In Proceedings of the 26th IEEE Photovoltaic Specialists Conference, pp. 805, (IEEE, New York), Anaheim, CA USA.

Fan, J.C.C., Tsaur, B-Y., Palm, B.J., 1982. Optimal design of high-efficiency tandem cells, In Proceedings of the 16th IEEE Photovoltaic Specialists Conference, pp.692, (IEEE, New York), San Diego, CA USA.

Gomyo, A., Suzuki, T., Kobayashi, K., Kawata, S., Hino, I., Yuasa, T., 1987. Evidence for the existence of an order state in $Ga_{0.5}In_{0.5}P$ grown by metalorganic vapor-phase epitaxy and its relation to band-gap energy,. Appl. Phys. Lett., 50, 673.

Hutchby, J.A., Markunas, Timmons M.L., Chiang P.K., R.J., Bedair, S.M., 1985. A review of multijunction concentrator solar cells, In Proceedings of the 18th IEEE Photovoltaic Specialists Conference, pp.20, (IEEE, New York), Las Vegas, USA.

Khan, A., Yamaguchi, M., Bourgoin, J.C., Takamoto, T., 2000. Room-temperature minority-carrier injection-enhanced recovery of radiation-induced defects in p-InGaP and solar cells, Appl. Phys. Lett. 76 (18), 2559.

King, R., Fetzer, C., Colter, P., Edmondson, K., Ermer, J., Cotal, H., Yoon, H., Stavrides, A., Kinsey, G., Kurzt, D., Karam, N., 2002. High-efficiency space and terrestrial multijunction solar cells through bandgap control in cell structures, In Proceedings of the 29th IEEE Photovoltaic Specialists Conference, pp. 776, (IEEE, New York), New Orleans, USA.

Kurtz, S.R., Myers, D., Olson, J.M., 1997. Projected performance of three- and four-junction devices using GaAs and GaInP, In Proceedings of the 26th IEEE Photovoltaic Specialists Conference, pp.875, (IEEE, New York), Anaheim, USA.

Lang, D.V., Kimerling, L.C., Leung, S.Y., 1976. Recombination-enhanced annealing of the E1 and E2 defect levels in 1-MeV-electron-irradiated n-GaAs, J. Appl. Phys. 47, 3587

Olson, J.M, Kurtz, S.R., Kibbler, K.E., 1990. A 27.3% efficient $Ga_{0.5}In_{0.5}P$/GaAs tandem solar cell, Appl. Phys. Lett. 56, 623.

Sugiura, H., Amano, C., Yamamoto, A., Yamaguchi, M., 1988. Double hetero-structure GaAs tunnel junction for an AlGaAs/GaAs tandem solar cell, Jpn. J. Appl. Phys. 27 , 269 .

Takamoto, T., Ikeda, E., Kurita, H., Ohmori, M., Yamaguchi, M., 1997a. Two-terminal monolithic $In_{0.5}Ga_{0.5}P$/GaAs tandem solar cells with a high conversion efficiency of over 30%, Jpn. J. Appl. Phys. 36, 6215.

Takamoto, T., Ikeda, E., Agui, T., Kurita, H., Tanabe, T., Tanaka, S., Matsubara, H., Mine, Y., Takagishi, S., Yamaguchi, M., 1997b. InGaP/GaAs and InGaAs mechanically-stacked triple-junction solar cells, In Proceedings of the 26th IEEE Photovoltaic Specialists Conference, pp. 1031, (IEEE, New York), Anaheim, USA.

Takamoto, T., Yamaguchi, M., Ikeda, E., Agui, T., Kurita, H., Al-Jassim, M., 1999a. Mechanism of Zn and Si diffusion from a highly doped tunnel junction for InGaP/GaAs tandem solar cells, J. Appl. Phys. 85, 1481.

Takamoto, T., 1999b. Studies on physics of high-efficiency InGaP/GaAs tandem solar cells, In PhD Thesis (Toyota Tech. Inst., Japan).

Takamoto, T., Yamaguchi, M., Taylor, S.J., Yang, M.-J., Ikeda, E., Kurita, H., 1999c. Radiation resistance of high-efficiency InGaP/GaAs tandem solar cells, Solar Energy Materials and Solar Cells 58, 265.

Takamoto, T., Agui, T., Ikeda, E., Kurita, H., 2000. High efficiency InGaP/InGaAs tandem solar cells on Ge substrates. In: Proceedings of the 28th IEEE Photovoltaic Specialists Conference, pp. 976, (IEEE, New York), Anchorage, USA.

Takamoto, T., Agui, T., Kamimura, K., Kaneiwa, M., Imaizumi, M., Matsuda, S., Yamaguchi, M., 2003. Multijunction solar cell technologies-high efficiency, radiation resistance, and concentrator applications, In Proceedings of the 3rd World Conference on Photovoltaic Energy Conversion, pp.581, Osaka, Japan.

Wolf, M., 1960. Limitations and possibilities for improvement of photovoltaic solar energy converters, Proc. Inst. Radio Engineers IRE 48, 1246.

Yamaguchi, M., Amano, C., Sugiura, H., Yamamoto, A., 1987. High efficiency AlGaAs-GaAs tandem solar cell with tunnel junction, In Proceedings of the 19th IEEE Photovoltaic Specialists Conference, pp. 1484, (IEEE, New York), New Orleas, USA.

Yamaguchi, M., Wakamatsu, S., 1996. Super-high efficiency solar cell R&D program in Japan, In Proceedings of the 25th IEEE Photovoltaic Specialists Conference, pp.9, (IEEE, New York), Washington DC., USA.

Yamaguchi, M., Okuda, T., Taylor, S.J., Takamoto, T., Ikeda, E., Kurita, H., 1997a. Superior radiation-resistant properties of InGaP/GaAs tandem solar cells, Appl. Phys. Lett. 70, 1566.

Yamaguchi, M., Okuda, T., Taylor, S.J., 1997b. Minority-carrier injection-enhanced annealing of radiation damage to InGaP solar cells, Appl. Phys. Lett. 70, 2180.

Yamaguchi, M., 2002. Present status of R&D for super-high-efficiency III-V compound solar cells in Japan, In Proceedings of the 17th European Photovoltaic Solar Energy Conference, pp. 2144, (WIP, Munich, Germany), Munich, Germany.

Yamaguchi, M., Khan, A., Dharmarasu, N., 2003. Analysis of superior radiation resistance of InP-based solar cells, Solar Energy Materials and Solar Cell, 75, 285.

Yamaguchi, M., Takamoto, T., Araki, K., 2004. Super high-efficiency multi-junction and concentrator solar cells, Solar Energy Materials and Solar Cells (to be published).

Yang, M-J., Yamaguchi, M., Takamoto, T., Ikeda, E., Kurita, H., Ohmori, M., 1997. Photoluminescence analysis of InGaP top cells for high-efficiency multi-junction solar cells, Solar Energy Materials and Solar Cell, 45, 331.

Chapter 5

Progress of Highly Reliable Crystalline Si Solar Devices and Materials

by

Tadashi Saitoh
Department of Electrical and Electronics Engineering
Tokyo University of Agriculture and Technology
Koganei, Tokyo 184-8588
Japan

Abstract

With the recent expansion of the photovoltaic market, the reliability of crystalline Si cells, modules and materials has become a great concern for customers. The reliability has been investigated worldwide since the Flat-Plate Solar Array Project of USA in early 1980's. In this paper, the issue is discussed from long- and short-term points of view. The long-term reliability includes more engineering matters, i.e. solarization of glass sheet materials, delamination, and discoloration of encapsulation materials. The short term relates to relative new phenomena regarding initial light-induced degradation of cell and module efficiencies due to minority-carrier lifetime degradation of crystalline Si materials. Forming a boron-oxygen complex or creating interstitial iron from iron-boron pairs is considered to generate the light-induced degradation of minority-carrier lifetimes. Two kinds of practical solutions to suppress light degradation are the usage of gallium dopant instead of boron and the reduction of interstitial oxygen and iron impurity.

Corresponding Author. E-mail: tadashi@cc.tuat.ac.jp

1. Introduction

In accordance with the recent photovoltaic (PV) market expansion, economic issues such as money and energy payback times have become a great concern for customers. As for the energy payback time, the reliability of PV modules has been investigated in real outdoor and indoor conditions (Christensen, 1985; Ross, Jr., 1981; Forman and Themelis, 1980; Atmaran et al., 1996; Kitamura and Matsuda, 2001; Hishikawa et al., 2002; Sakamoto and Oshiro, 2003; Osterwald et al., 2002; Rosenthal and Lane, 1991; Pern and Glick, 2000; Wohlgemuth and Peterson, 1992; Chianese et al., 2000; King et al., 1997; Shigekuni and Kumano, 1997). Current reliability issues should be discussed to achieve the reliability for more than 50 years as regards with encapsulation and back sheet materials, de-lamination and Si cell itself.

In the past, the reliability of the c-Si solar modules had been firstly investigated under the Flat-Plate Solar Array Project organized by Jet Propulsion Laboratory, USA, in early 1980's (Christensen, 1985). The reliability of the PV modules and arrays is second in importance only to cost reduction influencing the market acceptance of the PV systems. Because of their modular nature, the PV systems possess a higher than normal sensitivity to common mode failure. As part of the JPL FSA Project, a comprehensive engineering activity had been directed to understanding of the reliability of terrestrial PV arrays and deriving analysis and design tools useful for optimization and cost reduction of the PV arrays.

Based on the results with the subsequent ones, recent concern is to provide certifications the standard of safety for Photovoltaic Modules for module manufacturing companies, e.g. by Underwriters Laboratories in the USA and Japan Electrical Safety & Environment Technology Laboratories in Japan. However, cell researchers on the metastability study in silicon solar cells have been independent of module engineers on the reliability research of solar modules. At the 3rd World Conference on Photovoltaic Energy Conversion (WCPEC-3) held in Osaka in 2003, a special symposium was organized to understand the effect of the metastability in c-Si solar cells on the reliability of solar modules. It was pointed out that the reliability of solar modules should be discussed by dividing short-term and long-term issues. The short term includes more physics relating to the initial degradation of minority-carrier lifetimes and cell efficiencies. The long-term issue includes more engineering, i.e. delamination, yellowing, hot spot, etc. Even for the crystalline Si solar cells, we should use stabilized efficiencies, which means that efficiencies should be determined after a few hours of illumination and further degradation is marginal.

From the highly reliability point of view, crystalline Si (c-Si) solar cells and materials are not always perfect and need to investigate in more detail. A first paper on light-induced degradation of minority-carrier lifetimes was presented by Fischer and Pschunder (1973), who pointed out that solar cell efficiencies using low-resistivity, Czochralski-grown Si (Cz-Si) wafers decreased under AMO simulated sunlight. However, the light-induced degradation issue has not been paid much attention so far since high-efficiency c-Si solar cells were fabricated using medium-resistivity c-Si wafers without light-induced degradation. In recent years, light-induced degradation research has been extensively conducted for p-type Cz-Si crystals and solar cells in order to realize higher efficiencies (Glunz et al., 1998; Schmidt et al., 1997). The degree of the light-induced degradation depends on the metastable defects created under light irradiation. To eliminate the light-induced degradation, the reduction of oxygen content has been reported to be effective and in addition, Ga dopant instead of B was found to be effective to eliminate light-induced degradation of minority-carrier lifetimes (Hashigami et al., 1999; Saitoh et al., 1999; Saitoh et al., 2000). In the future, it is expected to fabricate more highly efficient solar cells and to provide more reliable silicon PV modules.

Multicrystalline silicon (mc-Si) crystals, which are primary materials in the current PV market, were also investigated under simulated sunlight irradiation (Nagel et al., 1997; Dhamrin et al., 2002; Takaki, 2001). The minority-carrier lifetimes of low-resistivity, mc-Si wafers also degraded, but not so much as compared with the Cz-Si wafers. This is due the fact that the oxygen content is lower than that of the Cz-Si wafers.

In this paper, research status and issues on the reliability of crystalline Si modules are described from the long-term reliability point of view including the solarization of glass sheet materials and yellowing of EVA encapsulation materials. The reliability of the Cz-Si and mc-Si solar cells is also discussed from the short-term point, i.e. initial cell efficiency reduction due to the light-induced degradation of minority-carrier lifetimes.

2. Reliability of Silicon Solar Modules During Exposure Testing

In the late 1970's to early 1980's, the reliability of Si solar modules had been systematically studied as part of Jet Propulsion Laboratory's Low-Cost Solar Array and afterwards Flat-Plate Solar Array Project (Christensen, 1981). The Low-Cost Solar Array Project (LSA) at JPL/DOE/NASA, initiated in 1975, was very effective to develop low-cost and long-life PV arrays and stimulation of commercial production. As indicated in Figure 1, the LSA project included R&D of low-cost solar array manufacturing and mass production technology transfer of the technology to industry and procurement of solar arrays.

Major accomplishments of the LSA Project (later the Flat-Plate Solar Array Project) were the establishments of a new low-cost, high-purity Si feedstock refinement process, significant advances in Si ribbon, higher efficiency cells and manufacturing processes, PV module design and evaluation and economic analysis capabilities. Several production process sequences from MG-Si to module assembly were compared to achieve $0.70/watt (in 1980 U.S. $). One of the candidates included refinement of MG-Si, sheet growth, ion-implantation, pulse anneal, metallization, antireflection film

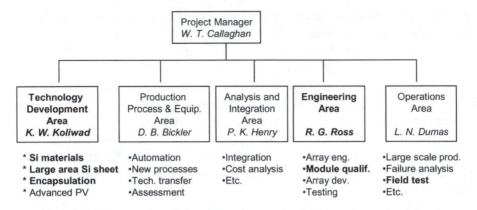

Figure 1. Low-Cost Solar Array Project organized by Jet Propulsion Laboratory from 1975 to 1980 (Christensen, 1985).

coating, encapsulation and so forth. As part of the LSA project, a comprehensive array engineering activity had been directed to understand the reliability attributes of terrestrial PV arrays and to define a rational approach to achieving high reliability at minimum cost (Ross, Jr., 1981). The reliability studies had been conducted in areas covering cell failure, interconnect fatigue, glass breakage and electrical insulation break-down. The reliability problem was also studied by taking accounts of electrically inter-connect literally thousands of nearly identical solar cells in series and in parallel to achieve the voltage and current requirements of the intended application. For ex-ample, a typical 250-volt residential array will require 500 to 600 series cells. This large number of series elements makes an array extremely sensitive to infrequent cell failures unless a high level of circuit redundancy is utilized. The reliability engineering issue is thus to achieve a high level of reliability at low cost by optimally trading off the available solutions. These include defining and achieving the appropriate piece part failure rates for the solar cells and interconnecting components, designing the appro-priate levels of fault tolerance into the array circuit, and selecting the optimal mainte-nance/replacement strategy.

Under the JPL Project, the cell failure rates as low as 0.0001 per year could be shown to significantly degrade the array performance if appropriate circuit redun-dancy is not applied. Several statistical design approaches had been noted which ad-dresses the areas of cell breakage, interconnect fatigue, glass breakage, and insulation breakdown. A quantitative design was conducted for a specified low level of compo-nent failures, and then to control the degrading effects of the remaining failures through the use of fault-tolerant circuitry and module replacement. With the component failure rates and with the use of multiple cell interconnects, series/paralleling and bypass diodes, it appeared possible to achieve high levels of array reliability with no module replacement for routine component failures.

Table 1. Module degradation at Nebraska cumulative field inspection results. Time period: 7/1977–10/1979 (Forman and Themelis, 1980).

Module inspected	2080 modules
Edge-seal delamination	1037 modules
Newly cracked cells	1044 modules
Delamination over cells	386 modules
Delamination over or around interconnect	65 modules

MIT Lincoln Laboratory had also conducted another outdoor testing of Si modules at a 25kW PV system located at Nebraska, USA (Forman and Themelis, 1980). A cumulative listing of physical defects observed from 1977 to 1979 is indicated in Table 1. Clearly, the most prevalent defect was encapsulant delamination around the edge seal, over cells and around interconnects.

As the Si solar module technology has become dominant for PV power generation, recent concerns have been raised about module lifetime and possible longer-term detrimental effects. In 1996, Florida Solar Energy Center reported no significant power loss over ten years of outdoor operation (Atmaran et al., 1996). However, the wet insulation resistance values of the majority of the modules were lower than the IEEE Standard 1262. The encapsulation discoloration did not appear to have any effect on the module power generation.

In Japan, two groups had conducted outdoor testing of PV modules at Rokko 500kW and Hamamatsu 100kW PV systems. The Rokko system consisted of 12,000 silicon solar modules installed for 10 to 15 years (Kitamura and H. Matsuda, 2001). About 3,000 modules were selected for visual inspection. Some of the modules showed the delamination between cells and EVA encapsulant, which indicated detrimental *I-V* curves, as shown in Figure 2. To investigate the degraded *I-V* curves, simulated *I-V* curves were calculated for Si solar modules with or without partial filter covering of cells. Almost all the degraded *I-V* curves could be understood except for some curious curves.

In 1990's, long-term stability of crystalline Si PV modules had also been systematically investigated in Japan by indoor and outdoor test conditions for ten years (Hishikawa et al., 2002; Sakamoto and Oshiro, 2003). About 2,300 modules of a 100 kWp PV system supplied from several manufacturers were located at Hamamatsu site of Japan Quality Assurance Organization. About 100 modules were also installed at 5 locations in Japan for data acquisition of *I-V* characteristics of the modules, irradiance and temperature. The modules exposed outdoors for 8 to 10 years and were occasionally removed for detailed indoor *I-V* measurements.

The system efficiency of the 100 kWp system showed monthly change due to the effects of temperature and solar spectra, whereas no distinct long-term change was

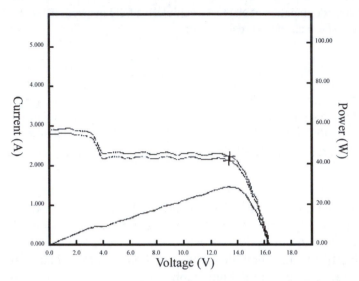

Figure 2. *I-V* curves for a degraded module at a Rokko test site near Kobe, Japan. The system had a 500 kWp capacity consisting of 12,000 crystalline Si modules (Kitamura and Matsuda, 2001).

observed after ten-year outdoor testing as shown in Figure 3. Certain types of the tested 2,300 modules showed a small degradation correlated to *Pmax* due to short-circuit current and fill factor changes. The reason of degraded fill factors was weakened adhesion strength of solder bonding and cell metallization on the cells. Another degradation was delamination of the encapsulant from the cell surfaces, which resulted in lowering short-circuit current density. No serious degradation problems such as failure of the modules were observed, which demonstrated that the reliability of the modules in the 1990's was improved than that of the modules manufactured in 1980's or before, suggesting the module reliability of 20 to 30 years.

Recently, interesting results were reported on a solar weathering of commercial PV modules including of single and multicrystalline Si modules (Osterwald et al., 2002). The weathering methods were real-time outdoors, accelerated one-sun exposures and 3-suns mirror enhancement (Outdoor Accelerated-weathering Test System, OATS), indoor under fluorescent light (UVA-340) and indoors under xenon arc (XR-260). Following the ASTM weathering standards, the exposure totals were quantified by the time integral of the UV irradiances. The weathering results showed a linear relationship between maximum power (*Pmax*) degradation and the total UV exposure dose for the four different types of commercial PV modules. The average degradation rate for the four modules types was 0.71% per year. The results showed little or no changes in fill factor and only slight drops in open-circuit voltage. Figure 4 shows one of the examples of the normalized *Pmax* and short-circuit current (*Isc*) as a function of the total UV exposure dose for a single crystalline Si module. As shown in this figure, losses of short-circuit current were responsible for the maximum power degradation.

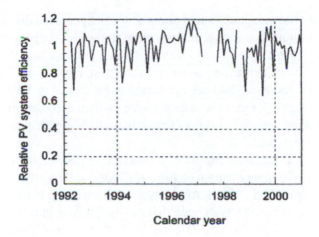

Figure 3. Typical long-term variation of relative PV system efficiency for a grid-connected 11 kWp Hamamatsu PV system consisting of 230 PV modules. No distinct long-term degradation from 1992 to 2001 was observed (Hishikawa et al., 2002).

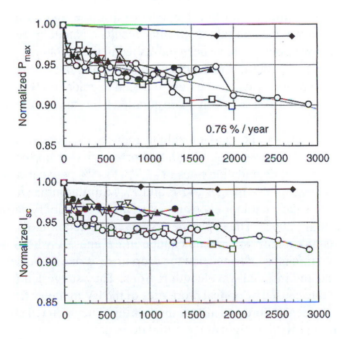

Figure 4. Normalized P_{max} and I_{sc} for a single crystalline Si module as a function of 300-400 UV dose. Exposure methods: control (♦), OATS 1x (●), OATS 3x (▲), real-time (▽), UVA (□) and XR-260 (○) (Osterwald et al., 2002).

The reason of the slow degradation of about 0.7% per year was discussed by four possible causes including obscuration or absorption in the glass superstrate, obscuration or absorption in the ethylene vinyl acetate (EVA) encapsulant, degradation of crystalline Si material or thermal degradation. In the past, degradation of P_{max} and I_{sc} in modules that show yellowing had been assumed to be caused by EVA degradation (Rosenthal and Lane, 1991), but without conclusive evidence to support this assumption. Because yellowing is such a striking change in module appearance, the assumption is perhaps a natural one.

In fact, a study of EVA discoloration as a function of accelerated exposure concluded that module performance cannot be fully attributed to the extent of encapsulant discoloration (Pern and Glick, 2000). An analysis of losses in one module from the Carrisa Plains system of the 1980s showed that only a 3% loss in I_{sc} could be attributed to EVA discoloration, and this module had severely browned areas directly over the solar cells (Wohlgemuth and Peterson, 1992). In addition, the analysis concluded that most of the degradation in the Carrisa system was caused by hot spots resulting from inadequate module bypass and blocking protection, not browning.

Judging from the appearance of the Si modules, it is doubtful that the short-circuit current degradation was caused by encapsulation yellowing or obscuration. Compared with the cell efficiency in a degraded module to one from an unexposed module, it appears that most of the degradation has occurred in the 800–1100 wavelength region, and not short wavelength region. Similar results of module weathering tests that show degradation of only I_{sc} have been published in the literatures (Hishikawa et al., 2002; Chianese et al., 2000). The degradation of Isc was not caused by encapsulation yellowing, but probably due to the light-induced degradation (LID) of B-doped and oxygen-contaminated c-Si solar cells. The detailed LID study will be described in a latter section.

Also visible in Figure 4 is the rapid initial light-induced P_{max} and I_{sc} degradation of about -2% to -4%. In Chianese et al., 2000, a shorter 1.5 year experiment was reported that showed initial degradation rates of -3.7% to -5% per year in six identical PV modules, again negligible changes to FF and V_{oc}. Similar initial I_{sc} degradation was reported in a few days outdoor exposure experiment of four crystalline Si modules (Hishikawa et al., 2002). The experiment showed reduction in I_{sc} of 1% to 5% within one year outdoor exposure and no change in FF and V_{oc} values. Spectral response of degraded modules indicated that the drop in J_{sc} is due to the decrease in the response of the red and infrared wavelength regions. They suggested that the initial I_{sc} change is probably caused by optical properties of the PV modules from the optical reflectance measurement before and after light soaking. They pointed out that further study is required to clarify the origin of the initial degradation.

3. Solarization of Front Glass Sheets

Low-iron glass containing cerium has been used in the manufacture of Si solar modules for its absorption of UV light below 350 nm. The UV absorption exhibited by Ce-doped glass has been shown to reduce yellowing of encapsulation EVA (ethylene vinyl acetate) caused by UV radiation. However, UV absorption of the Ce-doped glass introduces a decrease of transmission of the glass in the solar spectrum (solarization). The solarization results the reduction of module efficiency.

The NREL group investigated the solarization of float glass containing 0.2 wt.% Ce (King et al., 1997). The Ce element in the glass was considered to reduce ultraviolet radiation into the encapsulation materials and to suppress the yellowing of the materials. The glass samples were exposed with several different accelerated weathering systems and also outdoors at one terrestrial sun in Golden, Colorado. Glass samples were exposed indoors at one sun, and approximately two suns. Both of these weathering chambers provide exposure to filtered Xenon arc light and are operated at 60°C and 80% relative humidity. Under the accelerated weathering systems, the transmittance of the glass dropped about 2%. Most of the loss in transmittance occurs above 800 nm and has little effect on the visible wavelength range. Glass samples exposed outdoors only one month also showed a similar 2% drop in the solar-weighted transmission as shown in Figure 5. In addition, the solarization of the glass appears to be self-limiting to about 2% loss in the transmission.

The Ce-doped glass exhibits a loss in transmission from 800 nm and extending into near infrared region. A red shift in the high energy cutoff position is also observed as the Ce^{3+} initially present in the glass is oxidized to the Ce^{4+} state, resulting in an increase in absorption in the 330 nm to 400 nm region of the UV portion of the spectrum. This slight increase in optical density results in some additional UV screening potential of the glass and also gives the glass a very slight yellow color. It should be noted that this additional UV screening still does not result in the complete exclusion of damaging radiation, but simply attenuates an additional fraction of the light in this wavelength regime. This effect should provide more protection for UV-sensitive materials. The conversion of Ce^{3+} to Ce^{4+} is nearly complete after a photon dose equivalent to five years outside in Golden. The rate of change appears to slow as a function of exposure, resulting in what appears to be a self-limiting process. Thus, these preliminary ultra-accelerated tests suggest that the solarization process appears to be self-limiting, as there is little additional change in the transmittance spectra of the glass even after the equivalent of an 8-year photon dose.

4. Yellowing of Encapsulation Materials

Yellowing of encapsulated EVA materials under a long-time testing has been considered to be a primary reason for the stability of Si solar modules. NREL researchers have performed a systematic study to establish an adequate methodology for per-

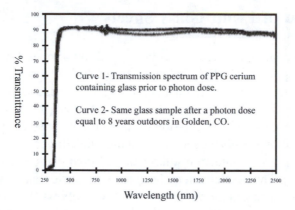

Figure 5. Comparison of the transmission spectra of Ce-doped glass sheet before and after photon exposure 8 years outdoors in Colorado (King et al., 1997).

forming accelerated exposure testing (Pern and Glick, 2000). When treated under about 6.5 UV suns condition, encapsulants of the three EVA formulations show transmission changes from -2.3% to -12.2 %, as indicated in Table 2. The UV exposure is equivalent approximately 3 years in the hot dry areas in Arizona. EVA A9918 exhibits the largest discoloration, followed by EVA 15295 and EVA V11.

The magnitude of cell efficiency loss under the acceleration test might not be proportional to the corresponding transmission loss. As seen in Table 2, the cell efficiency loss depends on short-circuit current loss except for sample C4. In the sample, fill factor loss is relatively large as compared to other samples. UV-induced discoloration of EVA materials was also investigated by fluorescence analysis. The EVA V11 in the C4 sample had not formed conjugated double bonds (polylenes). On the other hand, EVA 15295 and A9918 show a broad band absorption at around 525 nm relating to the yellowing or browning of the encapsulating materials.

To clarify the mechanism of the yellowing reaction in more detail, molecular substances in the EVA generated under accelerated weathering test were investigated by chemical analyses (Shigekuni and Kumano, 1997). The analyses showed that 2.6-di-t-butyl-4-methyl phenol of antioxidant and 2-hydroxy-4-octoxy-benzophenone of UV absorbent were consumed after the weathering test and 3,5-di-t-butyl-4-hydroxy-benzaldehyde having yellow color was newly produced.

5. Light-induced Degradation of Silicon Solar Cells

5.1 Past Works

As already mentioned, the light-induced degradation issue has not paid much attention so far since high-efficiency c-Si solar cells were fabricated using medium-range resistivity c-Si wafers without light-induced degradation. However, light-induced

Table 2. Measured changes of cell *I-V* parameters and percent transmission after accelerated testing (Pern and Glick, 2000).

Sample No.	EVA type	ΔJ_{sc} (%)	ΔFF (%)	ΔE_{ff} (%)	$\Delta \%T$ (%)
A4	15295	-5.3	0.02	-5.5	-4.8
C4	V11	-3.3	-5.5	-8.9	-2.3
D4	A9918	-5.2	-1.0	-7.1	-12.2

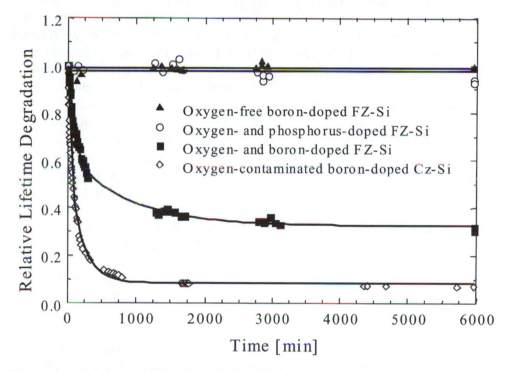

Figure 6. Light-induced lifetime degradation of oxygen-doped and oxygen-free silicon wafers (Glunz et al., 1998).

degradation has been extensively studied again to realize higher efficiencies using low-resistivity, p-type Cz-Si and mc-Si wafers. A typical example of the light-induced degradation for nitride-passivated float-zone and Cz-Si wafers is shown in Figure 6, indicating an initial rapid degradation of minority-carrier lifetimes in oxygen-contaminated and boron-doped wafers (Glunz et al., 1998). However, oxygen-free, boron-doped p-type and phosphorus-doped, n-type Cz-Si wafers show no light-induced degradation of minority-carrier lifetimes.

Therefore, the light-induced minority-carrier lifetime degradation in Cz-Si has been a significant matter since highly efficient solar cell performance was found to degrade. The degradation research on Cz-Si in the past decade has been making the identification of the responsible defect clearer to provide a practical solution for the defect suppression.

The fact that the defect correlates with substitutional boron, B_s, and interstitial oxygen, O_i, concentrations (Glunz et al., 1998) led ideas to suppress the minority-carrier lifetime degradation: usage of an alternative impurity such as gallium (Schmidt et al., 1997; Hashigami et al., 1999), and reduction or elimination of oxygen contamination by magnetic-field-applied Cz (MCz) or float-zone crystal growth (Glunz et al., 1998; Saitoh et al., 1999; Saitoh et al., 2000). For the conventional B-doped Cz-Si, several ways of optimized high temperature process, such as oxidation or phosphors diffusion, were found to reduce the degradation (Glunz et al., 1998) although the mechanism has not understood yet. Even at low temperature of 450°C, the defect is reducible by long time annealing for more than 30 hrs (Bothe et al., 2002). The cell performance losses by the degradation can be minimized in thin wafers for substrates (Damiani et al., 2002), in which even carriers with degraded diffusion length were effectively collected before being captured by the light-induced recombination center.

Data analysis with the Shockley-Read-Hall theory was applied to the carrier injection-dependent and temperature-dependent lifetime spectroscopy specified a probable energy range of the defect (Nagel et al., 1997). However, a very low defect concentration below the detection limit of deep level transient spectroscopy makes further identification difficult. That is the reason that most of evaluations for the B_s-O_i related defect were done only by the minority-carrier lifetime measurements.

Although a complete explanation for the degradation has not been realized, there are several defect models proposed, which are mostly based on B_s and O_i concentration dependence of the minority-carrier lifetime degradation. Bourgoin et al. (2000) reported O_i concentration dependence of the minority-carrier lifetime that follows a power factor of 3 and linear B_s dependence. They suggested a Jahn-Teller distortion effect on a configuration that consists of a B_s atom surrounded by three O_i atoms for the phenomena despite no experimental verification. The model assumes that injected minority carriers are trapped by the B_s-O_i complex to cause the bond distortion, which is supported by the fact that the application of forward voltage to a Cz-Si cell in the dark degrades the performance similarly to the light-induced degradation (Hashigami et al., 2002). On the other hand, Schmidt et al. (2002) proposed an oxygen dimmer model based on a result of quadratic increase of the minority-carrier lifetime degradation with increasing O_i concentration. Such a difference in O_i concentration dependence of the degradation is probably because of rather big scattering in the data plots due to the difficulty of satisfying a requirement that samples should have precisely the same resistivity and various O_i concentrations. It can be concluded that modeling for the Cz-Si-specific metastable defect requires more experimental support, confirmation for assumptions and further interpretation of the phenomena.

In addition, to such a microscopic interpretation of the degradation, the impact of the minority-carrier lifetime degradation on the solar cell performance degradation should be understood. There are several papers that predicted cell performance formation losses or the ideally factor degradation. However, the relation between the carrier lifetime and cell performance decay curves during the degradation has not yet been investigated.

In the next section, the relation between the carrier lifetime and the cell performance decay time for the Cz-Si solar cells is investigated using PC-1D cell simulation program. After that, the influence of illumination conditions on the cell performance decay time is investigated, including experiments using a spectroscopic illumination and a simulated sunlight illumination with various intensities. Finally, the responsible factor for the metastable defect activation is specified, and its role in the reaction is discussed.

5.2 Monocrystalline Silicon Cells

The solar cells used in our investigation were random-pyramid passivated emitter and rear cells fabricated from low-resistivity and oxygen-contaminated Cz-Si wafers (Hashigami et al., 2003). Before illumination testing, the cell samples were annealed at 200ºC for 20 min to deactivate the defects and then placed on a temperature-controlled chuck in a black box with optical filters. The sample was kept at an open-circuit condition during the illumination using a solar simulator. Open-circuit voltage and short-circuit current were simultaneously measured under a pulsed white light within three seconds at 25ºC.

The cell performance decay times were obtained by a simple exponential equation fitting. The original data after subtracting the final values were plotted on a semi-logarithmic scale and fitted to an exponential equation:

$$y(t) - y(t \to \infty) = \Delta y(t) = A \times exp(-t / \tau),$$ (Eq. 1)

where A is the amplitude, t is the illumination time and t is the decay time constant.

The cell samples were illuminated under white light with an intensity of 10 mW/cm^2, monitoring decay curves until the degradation saturated. Figure 7 shows the measured and simulated performance decay curves of the solar cell with base resistivity of 0.5 W·cm. The open symbols and the closed symbols represent measured values and simulation results, respectively. Minority-carrier lifetimes degraded extremely fast in the very beginning within 15 s followed by a second slower decay. The magnitude of the initial lifetime degradation was significant: 50 % of the initial value for the sample depending on the resistivity of the Cz-Si wafers. Such a multi-step decay was also observed in measured DV_{oc} and DJ_{sc} curves as well. DV_{oc} and DJ_{sc} curves were almost parallel, indicating that V_{oc} and J_{sc} degradation occurred simultaneously.

For a numerical comparison of the decay curves, all of the data plots were fitted to Eq. 1 modified to have double exponentials required for the best fittings. The fitting excludes the initial degradation. Resultant time constants are summarized in Table 3. The simulations indicated very similar curves to measured DV_{oc} and DJ_{sc} with almost the same decay time constants, explaining nicely the actual cell performance degradation, including the initial and double exponential degradation.

Table 3 . Decay time constants for solar cell performance and carrier lifetime degradation obtained from Figure 7 (Hashigami et al., 2003).

Resistivity		Decay time constants (min)					
		J_{sc}		V_{oc}		τ_{eff}	
Cell/wafer (Ω·cm)		τ_1	τ_2	τ_1	τ_2	τ_1	τ_2
	Measured	16.6	53.7	13.5	60.9	8.9	72.0
0.5/0.75	Calculated	19.0	86.2	14.8	73.2	---	---

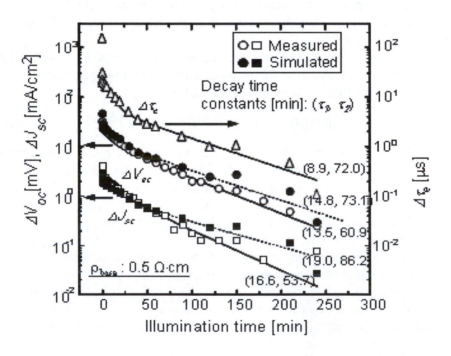

Figure 7. Measured and simulated decay curves of Si cell performances with base resistivity of 0.5 Ω·cm and effective minority-carrier lifetimes with resistivity of 0.75 Ω·cm (Hashigami et al., 2003).

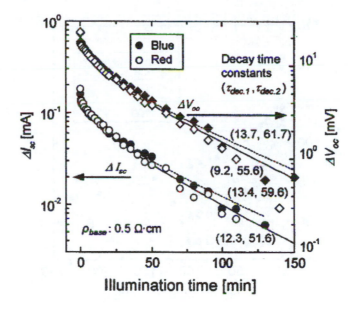

Figure 8. Effect of illumination spectra on *Voc* and *Isc* decay curves for a 0.5 Ω·cm solar cell (Hashigami et al., 2003).

Nevertheless, regarding the minority-carrier lifetime, the decay was distinctly steep in the beginning, for which the first time constant $t_{dec.1}$ was twice as fast as those for cell performance degradation. This is due the fact that the diffusion length after the very fast initial degradation was as long as the cell thickness. Concerning the multistep degradation, other reports can be found in literature as well. In our previous work, we found that the very fast initial degradation was due to metastable bulk defects that behave similarly to B_s-O_i related defects but that a distinct thermal property. This might be a suggestion that the light-induced defects specific to B-doped Cz-Si have some different configurations, although the reason for the double exponential second degradation is still to be found.

For a further interpretation of the phenomenon, the light wavelength dependence of the degradation was investigated. Blue and red light illuminations were carried out for a cell with a base resistivity of 0.5 W·cm. Blue and red light with an intensity of 1 mW/cm2 illuminated the samples for 270 min, when the degradation was almost saturated. Figure 8 shows DI_{sc} and DV_{oc} decay curves together with decay time constants. Very similar decay curves were obtained for I_{sc} and V_{oc} degradation regardless of the stress-light spectra. The decay time constants for I_{sc} were 13 and 60 min. for the blue light and 12 and 52 min. for the red light, respectively. A similar result was also obtained for V_{oc} degradation, although a small difference was observed in the time constants.

The results showed clearly that the degradation of the solar cell performance was not the function of the illumination spectra. In other words, light with energy higher

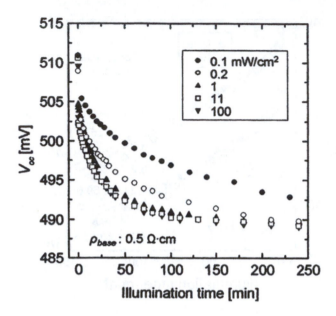

Figure 9. *Voc* decay curves of Si solar cells with a base resistivity of 0.5 Ω·cm under illumination with various light intensities (Hashigami et al., 2003).

than Si band gap activates the metastable defect effectively. This result is consistent with an experimental result reported that the activation energy of the defect is 0.4 eV (Schmidt et al., 2002). It is a rather small energy, so that the energy higher than the Si band gap would be sufficient to the defect activation.

In order to elucidate the effect of carrier injection level on the degradation, illumination intensity dependence of the degradation was investigated. Since excess carriers induce the degradation, the decay time constants would be expected to have a linear correlation with the injection level. However, even illumination with a low intensity was found to have a strong effect on the degradation. Figure 9 shows the decay curves of the *Voc* degradation under illumination with intensities from 0.1 to 100 mW/cm². Surprisingly, the decay curves were very similar in the range of the illumination intensities of 1 to 100 mW/cm². Under the illumination with an intensity of 0.2 mW/cm², the decay times were slowed but reached saturation after 240 min. Note that the magnitude of the degradation was uniformly about 20mV, in this case, even under weakest illumination.

The decay time constants of *Isc* and *Voc* obtained by the data fittings are plotted in Figure 10. Plots for *Isc* are represented with solid symbols and dotted lines while *Voc* with open symbols and solid lines. As predicted from very similar decay curves shown in Figure 10, the time constants showed almost flat characteristics in a wide range of the illumination intensities from 1 to 100 mW/cm². The results obtained good agreement with those of a carrier lifetime investigation. Under illumination intensities

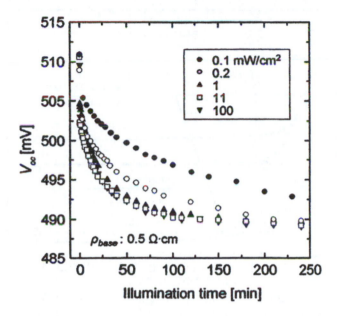

Figure 10. Illumination intensity dependence of decay constants of a Si solar cell with a base resistivity of 0.5 Ω·cm (Hashigami et al., 2003).

less than 1 mW/cm2, the time constants increased, which means that the decay curves slowed.

The decay time saturation implies that the defect activation in an electrical charging-related reaction via excess carriers. The metastable defect concentration is estimated be very low. Even in a high degradable material with a low resistivity less than 1 W·cm, the defect concentration was estimated at an order of approximately 10^{10}cm-3. Therefore, illumination with an intensity of 1 mW/cm2, corresponding to a photon flux of 10^{15}cm-2s-1, which is equivalent to the injected carrier concentration, may be sufficient to saturate the defect activation. According to this point of view, the independence of the decay time on carrier injection level is expected to pronounce more, and the decay time will become smaller in the samples with a higher resistivity that contains lower amount of the defect.

5.3 MulticrystallineSilicon Solar Cells

As mentioned above, many studies have been introduced to explain the light-induced lifetime degradation mechanism in boron-doped B-doped Cz-Si solar cells. However, more investigation is needed to understand the light-induced degradation phenomena in the lower quality multicrystalline Si (mc-Si) wafers due to the fact that the primary material in the current PV market is the mc-Si. The mc-Si wafers are fabricated by slicing Si cast ingots using the wire saw technology. Recently, the minority-carrier lifetimes of low-resistivity, mc-Si wafers was reported to degrade under

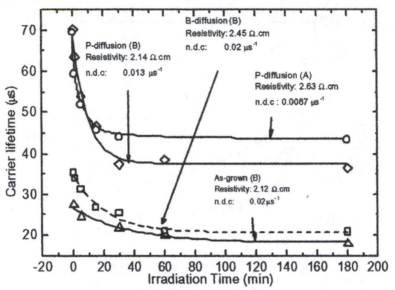

Figure 11. Light-induced lifetime degradation of multicrystalline cast Si wafers before and after P and B-diffusions and hydrogen passivation (Dhamrin et al., 2002).

illumination, but not so much as compared with the Cz-Si wafers (Boethe, Schmidt and Hezel, 2002; Damiani et al., 2002; Schmidt and Cuevas, 1999). This is due to the fact that the oxygen content is lower than that of the Cz-Si wafers.

Light-induced lifetime degradation of three kinds of mc-Si wafers and solar cells provided by different manufacturers has been investigated to study what we can call the through process disappearance of the light-induced lifetime degradation phenomena in mc-Si solar cells (Dhamrin et al. 2002). Firstly, the gradation mechanism in conventional mc-Si wafers was investigated. Figure 11 shows the degradation decay curves of minority-carrier lifetimes under 0.5 suns illumination up to 180 min. Carrier lifetimes of as-grown samples degraded to about 70% of its initial values, whereas B and P-gettered samples degraded to about 50% to 60% of its initial values. This does not indicate that as-grown samples are better because calculated normalized defect concentrations (n.d.c.) according to the equation 2 show different results.

$$v_{th}\sigma N_t(t) = \left(\frac{1}{\tau_a} - \frac{1}{\tau_b} \right)$$

(Eq. 2)

where $N_t(t)$ is the concentration of specific defects activated during the illumination, σ the defect capture cross section, v_{th} the thermal velocity and τ_a, τ_b the lifetimes before and after illumination.

Results show that n.d.c. of 0.02 ms^{-1} for as-grown decreased to about 0.013 ms^{-1} after P-diffusion and to 0.02 ms^{-1}, indicating the effectiveness of P-diffusion and

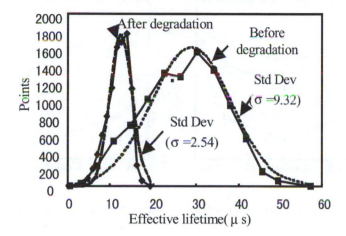

Figure 12. Distribution of carrier lifetimes for mc-Si wafers before and after light illumination (Takaki et al., 2001).

hydrogen passivation. A sample A was also treated by P-diffusion and hydrogen passivation leading to much smaller n.d.c. of about 0.0087 ms^{-1}.

It was also pointed out that the distribution of lifetimes indicated Gaussian and the mean value and standard deviation decreased after illumination as shown in Figure 12 (Takaki et al., 2001). The observation was consistent with the proposed model for the Cz-Si wafers although the interstitial oxygen concentration of the mc-Si wafers is considerably lower.

Screen-printed solar cells were fabricated using the cast mc-Si materials to investigate the light-induced degradation of solar cell characteristics. Before LID, the solar cells were annealed at 200°C for 20 minutse and then illuminated under one-sun solar simulator up to 21 hours. As shown Figure 13, a slight degradation existed in J_{sc} and efficiencies of three different Si solar cells. The degraded amounts were recovered after second annealing step at 200°C for 20 minutes. These results were similar for other mc-Si solar cells including those from electromagnetic Si materials.

There is uncertainty regarding with the measurement conditions of the back contacts before and after the second annealing step. The reasons that no big LID could be observed might be the low oxygen contents. This experiment raised a very interesting question why minority-carrier lifetimes degraded to almost 70% of its initial values whereas solar cell performance degraded very slightly.

To elucidate the LID of the mc-Si solar cells in more detail, the dependence of illumination spectra was also investigated (Kayamori et al., 2004). Figure 14 shows the relative J_{sc} decay curves under blue and red illuminations for a cell with a resistivity of 1.64 W·cm and oxygen concentration of 0.48 ppma. J_{sc} degraded rapidly within

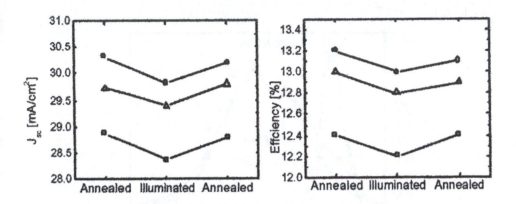

Figure 13. Influence of spectra on short-circuit current density, J_{sc}, and cell efficiency, E_{ff}, degradation and decay time constants under blue and red light illuminations (Dhamrin et al., 2002).

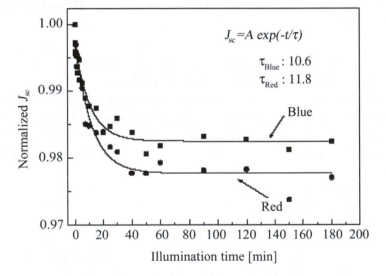

Figure 14. Influence of illumination spectra on Jsc degradation under blue and red lights for multicrystalline Si solar cells (Kayamori et al., 2004).

the first 40 minutes under both illuminations. The curves were fitted to a single exponential decay function 3.

$$J_{sc} = A \exp(-t/\tau) \tag{Eq. 3}$$

The two decay curves were similar to each other and resulted in decay time constants of 10.6 and 11.8 minutes for blue and red light, respectively. It means that wavelength does not affect the degradation of J_{sc}. This tendency was similar to the case of in B-doped Cz-Si solar cells reported by Hashigami et al. (2003). Relative J_{sc} measured at red light illumination degraded more than that at blue light illumination.

Then, the relative V_{oc} decay curves of the same cell were measured under blue and red illuminations. The relative V_{oc} degraded within 40 min illumination and both degraded less than J_{sc}. However, V_{oc} degraded under red light more rapidly than blue light and resulted in decay time constants of 14.8 and 6.43 min for blue and red light.

Degradation of relative J_{sc} in white, blue and red responses was also examined after blue light illumination for 3 hours. After 3 hours of illumination under blue light, red response of the solar cells degraded more as compared to white and blue responses. This means that the degradation generated by the blue light illumination existed deep in the bulk. The penetration depth of the blue light is less than $1 f \hat{E}m$, so that it seems to be difficult to say that the red response degradation was due to photon flux. It is possible that the carriers generated by the blue light diffused into the bulk and activate the LID metastable possible boron-oxygen defects.

6. Suppression of Lifetime Degradation

The proposed model of the boron-oxygen complex suggests that LID might annihilate by reducing the concentrations of either boron or oxygen. The expectation was confirmed through the international joint research, in which various kinds of Cz, MCz and FZ-Si wafers were investigated (Saitoh et al., 1999). The degree of efficiency degradation for solar cells is about 10% even if lifetime after illumination reduces to one tenth of the initial value as shown in Cz cells of #1 and #2 in Figure 15 (Glunz, 1999). MCz cells of #4 and #5 with a lower oxygen concentration showed no degradation although lifetimes of the MCz wafers decreased after illumination This suggests the boron-oxygen complex structure transforms at a high-temperature process. FZ-Si cells of # 6 and #7 and also Ga-doped Cz cells of #8 to #10 showed also no degradation irrespective of varying bulk resistivity.

During the international joint research, record cell efficiencies were fabricated using the high-quality Cz and FZ wafers. Fraunhofer Institute achieved record cell efficiencies of 22.7 and 22.5% (aperture area) using the B-doped MCz and Ga-doped Cz wafers. No degradation was observed for the B-doped MCz and Ga-doped Cz cells. In addition, a 10x10 cm²-area cell of 20.2 % was achieved using low-resistivity Ga-doped wafers. It is remarkable that the determined lifetimes are as high as 1 ms for 1 Ù·cm, Ga-doped Cz-Si wafers. In addition, Ga-doped Cz-Si wafers are superior to B-doped wafers in the whole resistivity region.

Lifetime stability of Ga-doped multicrystalline silicon wafers was also studied under illumination (Dhamrin et al., 2003). As shown in Figure 16, no degradation was observed in a Ga-doped sample with lifetime killer impurities reduced by phosphorus diffusion, so-called gettering, whereas a phosphorus-diffused, B-doped sample degraded to about 60% of its initial lifetime. The Ga-doped multicrystalline silicon wafers are a promising material for future photovoltaics (Dhamrin et al., 2004).

The LID of as-grown Ga-doped samples was also investigated using quasi steady-state photoconductance method (Dhamrin et al., 2003). There is no degradation in the

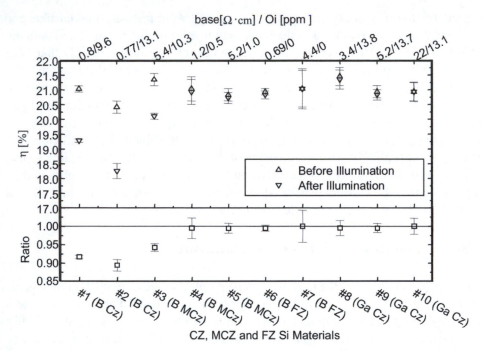

Figure 15. Cell efficiencies of RP-PERL (random pyramids, passivated emitter, real locally diffused) cells before and after illumination using 0.5 suns light. The lower graph gives the ratio between the values before and after illumination (Glunz et al., 1999).

injection level range from 10^{14} to 10^{16} cm^{-3} confirming no degradation in Ga-doped mc-Si wafers. The injection level dependence of the carrier lifetimes in p-type silicon wafers is due to the increase of SRH lifetime with increasing the excess carrier concentrations. The decrease of carrier lifetime with increasing the excess carrier concentration beyond 3×10^{15} cm^{-3} is due to the shallow-level recombination centers higher injection levels.

7. Conclusion

The reliability of silicon solar modules was investigated from long and short-term points of view. As for the long-term reliability, the solarization of glass sheet materials is a new matter, but the discoloration of encapsulation materials is not so detrimental. The short term includes light-induced degradation of minority-carrier lifetimes inducing initial degradation of cell and module efficiencies. The light-induced degradation of Cz-Si and multicrystalline Si cell performance was generated by forming boron-oxygen complex under illumination. Two kinds of practical solutions to suppress the light

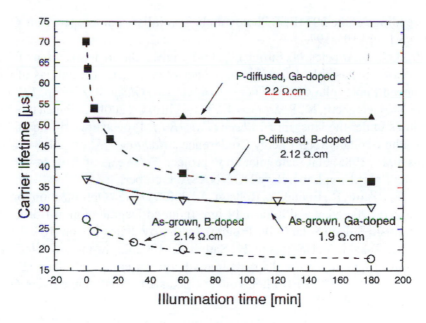

Figure 16. Effect of phosphorus gettering on light-induced lifetime degradation of boron and gallium-doped multicrystalline cast silicon wafers (Dhamrin et al., 2003).

degradation were introduced: the usage of gallium dopant instead of boron and the reduction of interstitial oxygen.

8. Acknowledgements

The author would like to appreciate the international joint research members for light-induced degradation, especially Dr. T. Abe of Shi-etsu Handotai and Dr. S. Glunz of FhG-ISE. He is indebted to Mrs. I. Yamaga and T. Hirasawa of Daiichi Kiden Co. for providing Ga-doped cast Si samples. Finally, he could not continue tenacious research without his devoted students at Tokyo University of Agriculture and Technology.

9. References

Atmaran, G.H., Ventre, G.G., Maytrott, C.W., Dunlop, J.P., Swamy, R., 1996. Long-term performance and reliability of crystalline silicon photovoltaic modules. In: *S. Bailey (Ed.),* Proceedings of the 25th IEEE Photovoltaic Specialists Conference, Washington DC, USA, pp.1279-1282.

Bothe, K., Schmidt, J., Hezel, R., 2002. Effective reduction of the metastable defects concentration in boron-doped Czochralski silicon for solar cells. In: *R. Arya (Ed.),*

Proceedings of the 29th IEEE Photovoltaic Specialists Conference, *New Orleans, La./USA,* pp. 194-197.

Bourgoin, J.C., de Angeles, N., Strobl, G., 2000. Light induced degradation of Si cells: Model of a metastable defect. In: *H. Scheer et al. (Ed.),* Proceedings of the 16th European Photovoltaic Solar Energy Conference, *Glasgow, UK,* pp. 1356-1358.

Chianese, D., Cereghetti, N., Rezzonico, S., Travaglini, G., 2000. Eighteen types of PV modules under the lens. In: *H. Sheer et al. (Ed.),* Proceedings of the 16th European Photovoltaic Solar Energy Conference, *Glasgow, UK,* pp. 2418-2421.

Christensen, E., 1985. Flat-plate solar array project: Ten Years of Progress. JPL-400-279, Jet Propulsion Laboratory/DOE/NASA, October.

Damiani, B., Ristow, R., Ebong, A., Rohatgi, A., 2002. Design Optimization for Higher Stabilized Efficiency and Reduced Light Induced Degradation in Boron Doped Czochralski Silicon Solar Cells. Progress in Photovoltaics, 10, pp. 185- 193.

Dhamrin, M., Takaki, A., Hashigami, H., Saitoh, T., 2002. Light-induced lifetime degradation of commercial multicrystalline silicon wafers. In: *R. Arya (Ed.),* Proceedings of the 29th IEEE Photovoltaic Specialists Conference, *New Orleans, La./USA,* pp. 395-398.

Dhamrin, M., Hashigami, H., Saitoh, T., Eguchi, T., Hirasawa, T., Yamaga, I., 2004. Extra-exceptional high carrier lifetimes in Ga-doped mc-Si wafers: Toward millisecond range. In: *H. Ossenbrink (Ed.), to be published in* Proceedings of the 19th European Photovoltaic Solar Energy Conference and Exhibition, *Paris, France.*

Dhamrin, M., Hashigami, H., Saitoh, T., 2003. Elimination of light-induced degradation with gallium-doped multicrystalline silicon wafers. Progress in Photovoltaics,11, pp. 231-236.

Fischer, H., Pschunder, W., 1973. Investigation of photon and thermal induced changes in silicon solar cells. In: *The IEEE (Ed.),* Proceedings of the 10th IEEE Photovoltaic Specialists Conference, *New York, USA,* pp. 404-411.

Forman, S.B., Themelis, M.P., 1980. Performance and reliability of photovoltaic modules at various MIT LL test sites. In: *The IEEE (Ed.),* Proceedings of the 14th IEEE Photovoltaic Specialists Conference, San Diego, CA/USA, pp.1284-1289.

Glunz, S.W., Rein, S., Warta, W., Knobloch, J., Wettling, W., 1998. On the degradation of Cz-silicon solar cells. In: *J. Schmid et al. (Ed.),* Proceedings of the 2nd World Conference on Photovoltaic Solar Energy Conversion, *Vienna, Austria,* pp. 1343-1346.

Glunz, S.W., Rein, S., Knobloch, J., Wettling, W., Abe, T., 1999. Comparison of boron- and gallium-doped p-type Czochralski silicon for photovoltaic application. Progress in Photovoltaics, 7, pp. 463-469.

Hashigami, H., Itakura Y., Saitoh, T., 2003. Effect of illumination conditions on Cz-Si solar cell performance degradation. J. Appl. Phys. 93, pp. 4240-424.

Hashigami, H., Itakura, Y., Saitoh, T., 2002. Performance degradation of Czochralski-grown silicon solar cells by means of current injection. Jpn. J. Appl. Phys. Part 2, 41, pp. L1191-L1193.

Hashigami, H., Wang, X., Abe, T., Saitoh, T., 1999. Comparison of dopant species on light degradation of carrier lifetimes for Cz-silicon wafers. In: *T. Fuyuki (Ed.)*, Technical Digest of the 11th International Photovoltaic Science and Engineering Conference, *Sapporo, Japan*, p. 978.

Hishikawa, Y., Morita, K., Sakamoto, S., Oshiro, T., 2002. Field test results on the stability of 2,400 photovoltaic modules manufactured in the1980's. In: *R. Arya (Ed.)*, Proceedings of the 29th IEEE Photovoltaic Specialists Conference, *New Orleans, La./USA*, pp.1687-1690.

Kayamori, Y., Hashigami, H., Dhamrin, M., Saitoh, T., 2004. Effect of illumination spectra on light-induced degradation in multicrystalline Si solar cells. In: *K. Kirtikara (Ed.)*, Proceedings of the 14th International Photovoltaic Science and Engineering Conference, *Bangkok, Thailand*, pp.303-304.

King, D.E., Pern, F.J., Pitte, J.R., Bingham, C.E., Czandema, A.W., 1997. Optical changes in cerium-containing glass as a result of accelerated exposure testing. In: *A. Rohatgi (Ed.)*, Proceedings of the 26th IEEE Photovoltaic Specialists Conference, *Anaheim, CA/USA*, pp.1117-1120.

Kitamura, A., Matsuda, H., 2001. Long-term degradation phenomena of crystalline Si solar modules. In: *J. Jang (Ed.)*, Technical Digest of the 12th International Photovoltaic Science and Engineering Conference, *Cheju, Korea*, pp. 757-758.

Nagel, H., Schmidt, J., Aberle, A.G., Hezel, R., 1997. Exceptionally high bulk minority-carrier lifetimes in block-cast multicrystalline silicon. In: *H. Ossenbrink (Ed.)*, Proceedings of the 14th European Photovoltaic Solar Energy Conference, *Barcelona, Spain*, pp. 762-765.

Osterwald, C.R., Anderberg, A., Rummel, S., Ottoson, L., 2002. Degradation analysis of weathered crystalline-silicon PV modules. In: *R. Arya (Ed.)*, Proceedings of the 29th IEEE Photovoltaic Specialists Conference, *New Orleans, La./USA*, pp.1392-1395.

Pern, F.J., Glick, S.H., 2000. Photothermal stability of encapsulated silicon solar cells and encapsulation materials upon accelerated exposures-II. In: *J. Benner (Ed.)*, Proceedings of the 28th IEEE Photovoltaic Specialists Conference, *Anchorage, Alas./USA*, pp.1491-1494.

Rosenthal, A.L., Lane, C.G., 1991. Field test results for the 6 MW Carrizo solar photovoltaic power plant. Solar Cells, pp. 563-571.

Ross Jr., R.G., 1981. Photovoltaic module and array reliability. In: *The IEEE (Ed.)*, Proceedings of the 15th IEEE Photovoltaic Specialists Conference, Orlando, FL/USA, pp. 1157-1163.

Saitoh, T., Hashigami, H., Rein, S., Glunz, S., 2000. Overview of light degradation research on crystalline silicon solar cells. Progress in Photovoltaics, 8, pp. 537-547.

Saitoh, T., Hashigami, H., Abe, T., Igarashi, T., Glunz, S., Rein, S., Wettling, W., Damiani, B.M., Rohatgi, A., Yamasaki, I., Sawai, H., Ohtuka, H., Warabisako, T., Zhao, J., Green, M.A., Schmidt, J., Cuevas, A., Metz., A., Hezel, R., 1999. Light degradation and control of low-resistivity Cz-Si solar cells", In: *T. Fuyuki (Ed.),* Technical Digest of the 11th International Photovoltaic Science and Engineering Conference, *Sapporo, Japan,* pp. 553-556.

Sakamoto, S., Oshiro, T., 2003. Field test results on the stability of crystalline silicon photovoltaic modules manufactured in the 1990's. In: *M. Yamaguchi (Ed.),* Proceedings of the 3rd World Conference on Photovoltaic Energy Conversion, *Osaka, Japan,* pp.1888-1891.

Schmidt, J., Aberle, A., Hezel, R., 1997. Investigation of carrier lifetime instabilities in Cz-grown silicon. In: *A. Rohatgi (Ed.),* Proceedings of the 26th IEEE Photovoltaic Specialists Conference, *Anaheim, CA/USA,* pp.13-18.

Schmidt, J., Bothe, K., Hezel, R., 2002. Formation and annihilation of the metastable defect in boron-doped Czochralski silicon. In: *R. Arya (Ed.),* Proceedings of the 29th IEEE Photovoltaic Specialists Conference, *New Orleans, La./USA,* pp. 178-181.

Schmidt, J., Cuevas, A., 1999. Electronic properties of light-induced recombination centers in boron-doped Czochralski silicon. J. Appl. Phys. 86, pp. 3175-3180.

Shigekuni, T., Kumano, M., 1997. Yellowing reaction in encapsulant of photovoltaic modules. In: *A. Rohatgi (Ed.),* Proceedings of the 26th IEEE Photovoltaic Specialists Conference, *Anaheim, CA/USA,* pp.1221-1223.

Takaki, A., Itakura, Y., Hashigami, H., Dhamrin, M., Glunz, S., Saitoh, T., 2001. Light-induced lifetime degradation in hydrogenated multicrystalline cast silicon substrates. In: *H. Ossenbrink (Ed.),* Proceedings of the 17th European Photovoltaic Solar Energy Conference, *Munich, Germany,* pp.1487-1490.

Wohlgemuth, J., Peterson, R., 1992. Reliability of EVA modules. Proceedings of the PV Performance and Reliability Workshop (SERI/CP-411-5184), National Renewable Energy Laboratory, Golden, CO, pp. 313-326.

Chapter 6

Recent Advances in Parabolic Trough Solar Power Plant Technology

David Kearney*
Kearney & Associates
PO Box 2568
Vashon, Washington, 98070 USA

Henry Price
National Renewable Energy Laboratory
1617 Cole Boulevard
Golden, Colorado, 80401 USA

Contributors

Nikolaus Benz, Daniel Blake, Avi Brenmiller, Douglas Brosseau, Robert Cable, Gilbert Cohen, Fritz-Dieter Doenitz, Robert Fimbres, Scott Frier, Randy Gee, Michael Geyer, Carin Gummo, Mary Jane Hale, Ulf Herrmann, Bruce Kelly, Cheryl Kennedy, Eckhard Lüpfert, Rod Mahoney, Eli Mandelberg, Kenneth May, Luc Moens, Paul Nava, Robert Pitz-Paal, Nicholas Potrovitza, Rainer Tamme, Eduardo Zarza

Abstract

Parabolic trough solar technology is the lowest cost large-scale solar power technology available today, primarily because of the development and experience associated with the nine large commercial-scale solar power plants that are operating in the California Mojave Desert. These plants, developed by LUZ International Limited and referred to as Solar Electric Generating Systems (SEGS), range in size from 14-80 MW and represent 354 MW of installed electric generating capacity. More than 2,000,000 m^2 of parabolic trough collector technology has been operating daily for up to 20 years at these plants, which as the year 2004 ended, had accumulated 154 years of operational experience. The SEGS parabolic trough collector technology has dem-

Corresponding Author. E-mail: dkearney@attglobal.net

onstrated its ability to operate in a commercial power plant environment like no other solar technology in the world. Although no new plants have been built since 1990, technology development has been active and several new 50 MWe parabolic trough projects are under development, with expected initiation of construction in 2005-2006 and plant start-ups in 2006-2007. Over the last 14 years significant advancements in collector and plant design have been made possible by the efforts of the SEGS plants operators, the parabolic trough industry, and solar research laboratories around the world. This chapter reviews the current state-of-the-art of parabolic trough solar power technology and describes the R&D efforts that are in progress to enhance this technology. The chapter also shows how the economics of future parabolic trough solar power plants are expected to improve.

1. Introduction

The SEGS parabolic trough power plants developed from the mid-1980's to the early 1990's have generated a great deal of interest in parabolic trough technology. Consisting of 9 plants from 14 MWe to 80 MWe in capacity with a cumulative total of 354 MWe, the SEGS facilities have continued to operate and to demonstrate reliable performance. In the last two years, development activity in commercial parabolic trough solar thermal power plants has accelerated, with developments in the United States and in other countries. Several industrial consortiums are actively developing new parabolic trough projects in sunbelt regions across the globe. This section provides an overview of parabolic trough power plant technology and describes the environment that existed during the development of the SEGS plants and describes some of the developments currently in progress in 2004 (Price et al., 2002).

1.1 Parabolic Trough Technology

Parabolic trough power plants[1] consist of large fields of parabolic trough collectors, a heat transfer fluid/steam generation system, a power system such as a Rankine steam turbine/generator cycle, and optional thermal storage and/or fossil-fired backup systems (Pilkington Solar International GmbH, 1996; Electric Power Research Institute, 1997). The solar field is made up of a large modular array of single-axis-tracking parabolic trough solar collectors, comprised of many parallel rows of solar collectors, usually aligned on a north-south horizontal axis. Each solar collector has a linear parabolic-shaped reflector that focuses the sun's direct beam radiation on a linear receiver located at the focal line of the parabola. The collectors track the sun from east to west during the day to ensure that the sun is continuously focused on the linear receiver. A heat transfer fluid (HTF) is heated up as high as 393°C as it circulates through the receiver and returns to a series of heat exchangers (HX) in the power block, where the fluid is used to generate high-pressure superheated steam (100 bar, 371°C). In the

---

[1] The plant parameters presented here are typical of current technology, such as the later SEGS plants.

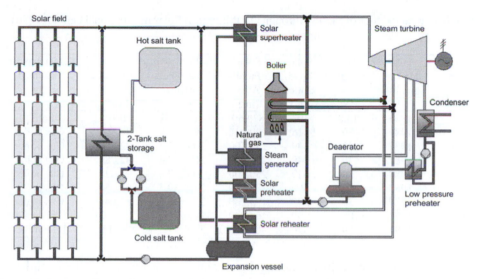

Figure 1. Process flow schematic of large-scale parabolic trough solar power plant (Flagsol GmbH).

Rankine cycle configuration, the superheated steam is then fed to a conventional reheat steam turbine/generator to produce electricity. The spent steam from the turbine is condensed in a standard condenser and returned to the heat exchangers via condensate and feed-water pumps to be transformed back into steam. Typically, mechanical-draft wet cooling towers supply cooling to the condenser, though dry cooling towers could be used as well, with some penalty in performance and cost. After passing through the HTF side of the solar heat exchangers, the cooled HTF is recirculated through the solar field. The plant parameters presented here are typical of current generation of SEGS plants.

The existing parabolic trough plants have been designed to use solar energy as the primary energy source to produce electricity. The plants can operate at full-rated power using solar energy alone at design-point radiation levels. During summer months, the plants typically operate for 10-12 hours a day on solar energy at full-rated electric output. To enable these plants to achieve rated electric output during overcast or nighttime periods, the plants also have been configured as hybrid solar/fossil plants; that is, a secondary backup fossil-fired capability can be used to supplement the solar output during periods of low solar radiation. As an alternative, thermal energy storage (TES) could be integrated into the plant design to allow solar energy to be stored and dispatched when power is required. Figure 1 shows a process flow schematic for a typical large-scale parabolic trough solar power plant.

1.2 SEGS Plant Development

Parabolic trough collectors capable of generating temperatures greater than 260ºC were initially developed for industrial process heat (IPH) applications. Several parabolic trough developers sold IPH systems in the 1970s and 1980s, but generally found

Table 1. Characteristics of SEGS I through IX (Pilkington Solar International GmbH, 1996).

SEGS Plant	First Year of Operation	Net Output (MW$_e$)	Solar Field Outlet (°C)	Solar Field Area (m^2)	Solar/Fossil Turbine Effic. (%)	Annual Output (MWh)	Dispatchability Provided By
I	1985	13.8	307	82,960	31.5/ NA	30,100	3-hrs TES
II	1986	30	316	190,338	29.4/37.3	80,500	Gas boiler
III	1987	30	349	230,300	30.6/37.4	92,780	Gas boiler
IV	1987	30	349	230,300	30.6/37.4	92,780	Gas boiler
V	1988	30	349	250,500	30.6/37.4	91,820	Gas boiler
VI	1989	30	390	188,000	37.5/39.5	90,850	Gas boiler
VII	1989	30	390	194,280	37.5/39.5	92,646	Gas boiler
VIII	1990	80	390	464,340	37.6/37.6	252,750	HTF heater
IX	1991	80	390	483,960	37.6/37.6	256,125	HTF heater

three barriers to successfully marketing their technologies. First, a relatively high marketing and engineering effort was required, even for small projects. Second, most potential industrial customers had cumbersome decision-making processes, which always took considerable effort but could result in a negative decision. Third, the rate of return for IPH projects did not always meet industry criteria. In addition, a highly reliable energy supply is much more important to industrial customers than the potential fuel savings from the IPH system.

In 1983, Southern California Edison (SCE) signed an agreement with LUZ International Limited to purchase power from the Solar Electric Generating System (SEGS) I and II plants. Later, with the advent of the California Standard Offer power purchase contracts for qualifying facilities under the U.S. Federal Public Utility Regulatory Policy Act (PURPA), LUZ was able to sign a number of standard offer contracts with SCE that led to the development of the SEGS III through SEGS IX projects. Initially, PURPA limited the plants to 30 MWe in size, though later this limit was raised to 80 MWe. In total, nine plants were built; representing 354 MWe of combined capacity. Table 1 shows the characteristics of the nine SEGS plants that LUZ built.

In 1991, LUZ filed for bankruptcy when it was unable to secure construction financing for its tenth plant (SEGS X). Although many factors contributed to the demise of LUZ, the basic problem was that the cost of the technology was too high to compete in the power market with declining energy costs and incentives. Lotker (1991) describes many of the events that enabled LUZ to successfully compete in the power market between 1984 and 1990 and many of the institutional barriers that contributed to its eventual downfall. However, the ownership of the SEGS plants was not affected by the status of LUZ because the plants had been developed as independent power projects, owned by investor groups, and continue to operate today in that form. Figure 2 shows the five 30-MW SEGS plants located at Kramer Junction, California. The large fields with rows of parabolic trough collectors are readily apparent. The five 30-MW power blocks can be observed near the center of each solar field. Details of the performance and operation of these plants are presented in Section 3.

Figure 2. SEGS III-SEGS VII solar plants at Kramer Junction, California. The large fields with rows of parabolic trough collectors are readily apparent. The five 30-MWe power plants can be observed near the center of each solar field.

1.3 Parabolic Trough Research and Development

Following the demise of LUZ, a number of events and R&D efforts continued the development of parabolic trough technology. In 1992, Solel Solar Systems Ltd. purchased the LUZ manufacturing assets, providing a source for the SEGS collector receiver technology. Solel has continued the development of the receiver and other parabolic trough technologies. In the same year, a five-year R&D program, designed to explore opportunities to reduce operations and maintenance (O&M) costs, was initiated between the operator of the SEGS III through SEGS VII plants (KJC Operating Company) and Sandia National Laboratories (SNL) (Cohen et al., 1999). This program resulted in a number of incremental advances in the technology that helped to significantly reduce O&M costs at existing plants. In 1996, the Direct Solar Steam (DISS) project (Zarza et al., 2004) was initiated at the Plataforma Solar de Almería (PSA) to test parabolic trough collectors that generate steam directly in the solar field. In 1998, a consortium of companies created the EuroTrough project to develop a next generation parabolic trough concentrator design. In 1999, an international workshop was held to develop a technology roadmap for parabolic trough technology. This roadmap has been used to develop an expanded R&D effort in parabolic trough technology by the U.S. Department of Energy. This effort has included materials R&D on receivers and reflective surfaces at NREL, evaluation and development of thermal energy storage options and subsystems, and development and field testing of an advanced collector design carried out by Solargenix. European work has also included concrete thermal storage development, a new commercial receiver design by Schott-Rohrglas, and EuroTrough prototype testing carried out in the U.S. and the Kramer Junction[2] site.

1.4 New Project Developments

1.4.1 GEF Projects

In 1996, the Global Environment Facility (GEF) approved $49 million (USD) grant for a parabolic trough project in Rajasthan, India. Subsequently, after an in-depth study to evaluate the future cost reduction potential of parabolic trough technology (Enermodal, May 1999), the GEF approved three additional $50 million grants for concentrating solar power technologies in Morocco, Egypt, and Mexico. However, the progress in developing these projects has been slow. This is in part due to a general restructuring of the power industry sector in these countries in an effort to move from government-owned power plants to private ownership. A number of other factors have also played a role in the slow implementation of these projects, including the lack of an established developer with a proven track record, the relatively high cost of near-term technology, and administrative procedures in the host countries.

1.4.2 Spanish Solar Tariff

During the late 1990s, interest in concentrating solar power plants continued in Europe because of rising fuel prices and CO_2 mitigation concerns. The southern European countries of Spain, Italy, and Greece have demonstrated the greatest interest in CSP technologies. In August 2002, a special solar premium tariff was extended to CSP plants in Spain. As a result, four 50-MWe parabolic trough plants are being developed in or planned in southern Spain.

1.4.3 Renewable Portfolio Standards

Energy shortages and price volatility in the western United States, a general increase in the public interest in green power technologies, and the implementation of renewable portfolio standards (RPS) in Arizona and Nevada have helped to boost interest in concentrating solar power (CSP) technologies. These renewable portfolio standards have resulted in a 1 MW trough plant being developed in Arizona and a 50-MWe trough project under development in Nevada. Formulation of an RPS in California has, similarly, turned the attention of both investor-owned utilities and municipal utilities to the potential of large solar plants based on trough technology.

1.4.4 CSP Southwest 1000 MW Initiative

A recent study by Sargent & Lundy (2003) evaluated the cost reduction potential of parabolic trough technology. The study confirmed a significant cost reduction potential and a resource potential in the southwestern United States sufficient to make

---

2 The SEGS plants are located at three different sites, with separate O&M companies: 44 MWe at Daggett, Calif., 150 MWe at Kramer Junction, and 160 MWe at Harper Lake.

the technology a viable and important power supply option. During the last two years, the CSP industry in the United States has worked with Congress and the Western Governors Association (WGA) to create a program to develop 1000 MW of CSP plants in the Southwest, taking advantage of the huge solar resource in that area. Interest by the four southwestern states of California, Nevada, Arizona, and New Mexico appears to be high. The WGA has identified CSP as a key large-scale solar technology, and is currently working to develop an implementation plan with aggressive goals for deployment.

As part of efforts to advance the President's National Energy Policy, the Department of the Interior's Bureau of Land Management (BLM) and the DOE National Renewable Energy Laboratory (NREL) identified and evaluated renewable energy resources, including CSP, on public lands (BLM, tribal, and Forest Service lands) (NREL, 2003). The report is being used to help federal land managers make decisions on prioritizing land-use activities to increase the development of renewable energy resources on public lands in the West.

1.4.5 Israel

Over the last few years, the Israeli Electric Corporation has been considering the development of up to five 100-MWe parabolic trough power plants in the Negev Desert. These plants would incorporate the latest developments in trough and solar power plant technology.

1.4.6 Global Market Initiative (GMI)

One factor in the emerging market interest in power plants utilizing CSP technology is the growth of an international initiative to accelerate and facilitate CSP plant development. International meetings to formulate and approve the initiative were held in Berlin (2003) and Palm Springs (2004) (Sandia National Laboratory's CSP Global Market Initiative website, 2004). The GMI was formally introduced at the International Conference on Renewable Energies 2004 in Bonn, Germany in June 2004.

2. Operating Experience of the SEGS Plants

The SEGS plants offer a unique opportunity to examine the operational track record of large parabolic trough plants, Even though the 9 plants in the Mojave Desert of California with a cumulative capacity of 354 MWe were the first such plants built, they all remain operational (in 2004) and the Kramer Junction site in particular provides excellent resource for performance and O&M data.

2.1 Operations and Maintenance (O&M) of Solar Power Plants

Parabolic trough solar power plants operate similar to other large Rankine steam power plants except that they harvest their thermal energy from a large array of solar collectors. The existing plants operate when the sun shines and shut down or run on fossil backup when the sun is not available. As a result the plants start-up and shut-down on a daily or even more frequent basis. Compared to a base load plant, this introduces more extensive service requirements for both equipment and O&M crews. The solar field is operated whenever sufficient direct normal solar radiation is available to collect net positive power. This varies due to weather, time of day, and seasonal effects due to the cosine angle effect on solar collector performance; generally, the practical lower limit of direct normal radiation in the plane of the collector for effective solar field operation is about 300 W/m². Since none of the plants currently have thermal storage[3], the power plant must be available and ready to operate when sufficient solar radiation exists. The operators have become very adept at keeping the plant on-line at minimum load through cloud transients to minimize turbine shut-downs, and at starting up the power plant efficiently from cold, warm or hot turbine status.

The O&M of a solar power plant is very similar to other steam power plants that cycle on a daily basis. The plants are staffed with operators 24 hours per day, using a minimal crew at night; and require typical staffing to maintain the power plant and the solar field. Although solar field maintenance requirements are unique in some respects, they utilize many of the same labor crafts as are typically present in conventional steam power plants (e.g., electricians, mechanics, welders). In addition, because the plants are off-line for a portion of each day, operations personnel can help support scheduled and preventive maintenance activities. A unique but straightforward aspect of maintaining solar power plants is the need for periodic cleaning of the solar field mirrors, at a frequency dictated by a tradeoff between performance gain and maintenance cost.

Early SEGS plants suffered from a large number of solar field component failures, power plant equipment not optimized for daily cyclic operation, and operation and maintenance crews inadequately trained for the unique O&M requirements of large solar power plants. Although the later plants and operating experience has resolved many of these issues, the O&M costs at the SEGS plants have been generally higher than LUZ expectations. At the Kramer Junction site, the KJC Operating Company's O&M cost reduction study (Cohen et al., 1999) addressed many of the problems that were causing high O&M costs.

Key accomplishments included:

- Solving HTF pump seal failures resulting from daily thermal and operational cycling of the HTF pumps;

- Reducing HCE failures through improved operational practices and installation procedures;

- Improved mirror wash methods and equipment designed to minimize labor and water requirements, and the development of improved reflectivity moni-

---

[3] The SEGS I plant initially had 3-hours of thermal energy storage, but the system was damaged in a fire in 1999.

toring tools and procedures, that allow performance based optimization of mirror wash crews; and

- Development of a replacement for flex hoses that uses hard piping and ball joints; resulting in lower replacement costs, improved reliability, and lower pumping parasitics.

Another significant focus of the study was the development of improved O&M practices and information systems for better optimization of O&M crews. In this area, important steps were:

- An update of the solar field supervisory control computer located in the control room that controls the collectors in the solar field to improve the functionality of the system for use by operations and maintenance crews;

- The implementation of off-the-shelf power plant computerized maintenance management software to track corrective, preventive, and predictive maintenance for the conventional power plant systems;

- The development of special solar field maintenance management software to handle the unique corrective, preventive, and predictive maintenance requirements of large fields of solar collectors;

- The development of special custom operator reporting software to allow improved tracking and reporting of plant operations and help optimize daily solar and fossil operation of the plants; and

- The development of detailed O&M procedures and training programs for unique solar field equipment and solar operations.

As a result of the KJC Operating Company O&M cost reduction study and other progress made at the SEGS plants, solar plant O&M practices have evolved steadily over the last decade. Cost effectiveness has been improved through better maintenance procedures and approaches, and costs have been reduced at the same time that performance has improved. O&M costs at the SEGS III-VII plants have reduced to about 25 USD/MWh. With larger plants and utilizing many of the lessons learned at the existing plants, expectations are that O&M costs can be reduced to below 10 USD/MWh at future plants.

2.2 Solar Plant Availability

Solar plants differ from conventional fossil and nuclear power plants in that they must harvest their fuel via the solar field. Thus the availability of the solar field becomes a key indicator of potential plant performance. Figure 3 shows the average solar field availability for the five 30MW SEGS plants at Kramer Junction from 1987 to 2003. Solar field availability refers to the percentage of the solar field that is available at any time to track the sun. The year 1987 was the first year of operation for the

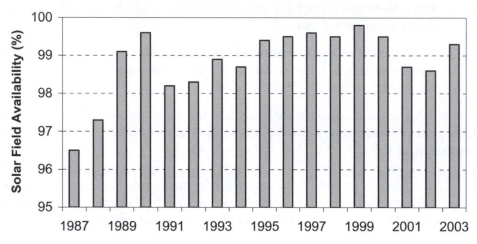

Figure 3. Average solar field availability for the five SEGS plants located at Kramer Junction, CA (KJC Operating Company).

first two plants at Kramer Junction. The data displays a steady trend of improved solar field availability through the life of the plants. The drop in availability in 1991 and 1992 was caused primarily by the bankruptcy of LUZ in the summer of 1991, which resulted in a change in plant ownerships and management of the O&M company that temporarily reduced or eliminated the supply of several key solar field spare parts. Now the availability is controlled by management decisions on the most cost-effective replacement strategy.

The SEGS power plants are conventional Rankine cycle steam power plants. For the most part, these plants have maintained good overall equipment availability. Although daily cycling of the plant results in a more severe service situation than base or intermittent load operation, daily nighttime outages allow some maintenance activities to be conducted while the plant is off-line, helping to maintain high availability during daytime solar hours. During normal day-to-day operation, it takes approximately 45 to 90 minutes to start up the plant, from initial tracking of the solar field to synchronization of the turbine generator. During the summer, the plant can be on-line in approximately 45 minutes. It takes up to twice as long in the winter because of the lower solar input to the plant. Once the plant is on line, the turbine can be ramped up to full load in a short time at a rate dictated by the thermal condition of the turbine (e.g., cold or warm). Because of their design warm-up characteristics, the natural-gas-fired boilers take longer to bring on line than the solar field and solar heat exchangers. The natural-gas-fired boilers must be warmed up more slowly to minimize thermal stresses on the boiler drum.

Since the total daily plant output varies significantly between the summer and winter seasons, the parabolic trough plants track the impact on availability as a function of lost solar generation. A full day outage in the winter may result in losing only 20% as much solar generation as would be lost by a full day outage in the middle of the

Table 2. Forced and scheduled outages for SEGS III-VII as a function of lost solar generation (KJC Operating Co., 2002).

	1997	1998	1999	2000	2001
Forced Outages	0.5%	1.0%	0.8%	3.7%	0.9%
Scheduled Maintenance	1.6	0.9	1.5	1.2	2.2
High Wind Outage	1.2	1.8	1.7	2.2	0.7
Force Majeure	0.0	0.1	0.0	0.0	0.0
Total	**3.3%**	**3.8%**	**4.0%**	**7.1%**	**3.8%**

summer. Thus, the plants schedule their annual maintenance outages during the November to February time frame when solar output is lowest. KJC Operating Company, the operator of the five 30MWe trough plants at Kramer Junction, has maintained detailed scheduled and forced outage data (KJC Operating Co., 2002). These plants typically schedule an 8-day outage each year and an extended outage (5-8 weeks) about every 10 years. They track availability as a function of the impact on solar generation. Table 2 shows forced and scheduled outages over a five-year period for these five plants as a function of lost solar generation. The high forced outage rate during 2000 was due to problems with tube leaks on the solar steam generators. These problems were resolved and the forced outage rate was reduced again in 2001. The period shown includes both routine annual and 10-year extended outages; specifically each year includes an 8-day outage at four of the plants and a 5-8 week outage at the fifth plant. High-wind outages occur when wind speeds exceed 35 mph and the solar field must be stowed to protect it from damage. Over this period the plants experienced a solar-output-weighted scheduled and forced outage rate of 4.4%. Without inclusion of the extended outages, the outage rate drops below 4%. This level of power plant availability is considered excellent for any power plant. The SEGS plants have a projected life of 30 years. The solar field and conventional steam cycle power equipment shows every sign of meeting and exceeding that lifetime.

2.3 Solar Electric Generation

The best performance indicator of the SEGS plants is the gross solar-to-electric performance.

Figure 4 shows the annual and cumulative gross solar electric generation for the nine SEGS plants through the end of 2003 (Frier, 2003). The increasing annual generation during the first 7 years shows the impact of new plants coming online. The dip in annual generation during 1992 was due to the Mount Pinatubo Volcano in the Philippines. The volcano erupted during the summer of 1991 and resulted in a noticeable reduction in direct normal solar radiation in the California Mojave Desert during 1992. Of significance is the sustained level of performance over the last 11 years. Cumulative solar generation from these plants should exceed 10 terawatt hours during 2004.

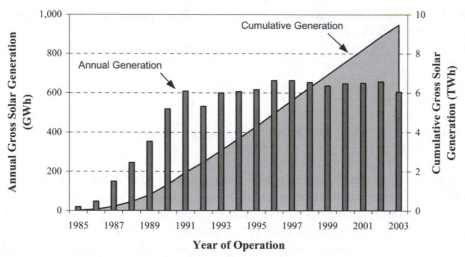

Figure 4. Annual gross solar generation for SEGS I-IX (Frier, 2003).

2.4 On-Peak Electric Generation

The SEGS plants sell power to the local utility, Southern California Edison (SCE). As part of their contract with SCE, the SEGS plants are required to generate power at a specified level during the utility's peak electric demand period.

Figure 5 is a graphical representation of the different SCE time-of-use (TOU) periods during the year. The summer on-peak period has the highest demand for power. The shaded region in the figure shows the time during the day when parabolic trough plants normally operate. In general, parabolic trough plants are well suited for generating power during the SCE summer on-peak TOU period.

To help ensure that the SEGS plants can operate at full rated output during the summer on-peak period, the SEGS plants have the capability to use a backup fossil energy for periods when solar energy is not available. Figure 6 shows gross electric output for three days in 1999 from one of the 30-MW SEGS plants at Kramer Junction. Day 172 is the summer solstice (June 21), which is the longest solar day of the year. On day 172, the plant operated from solar input only and the plant was able to operate during the summer on-peak period from 12 noon to 6 pm averaging above rated capacity for the period (30 MW net or approximately 33 MW gross). Days 260 and 262 represent 2 days near the fall equinox in September. Day 262 was a weekend day, so the plant operated on solar input only. The figure shows that the plant was not able to maintain full rated output on solar energy alone during the 12 noon to 6 pm time frame. Day 260 is a weekday with solar output in the morning similar to that of day 262, but the backup natural gas fired boiler was used in the afternoon to supplement the solar input to allow the plant to operate at rated capacity during the afternoon on-peak period from noon to 6 pm. This figure clearly illustrates how the hybrid SEGS plants have been able to operate and provide power to the utility when it is needed most.

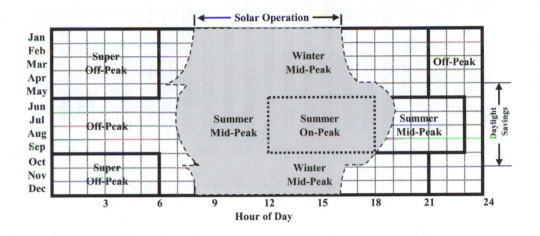

Figure 5. Power utility time-of-use and solar operation (Frier, 2003).

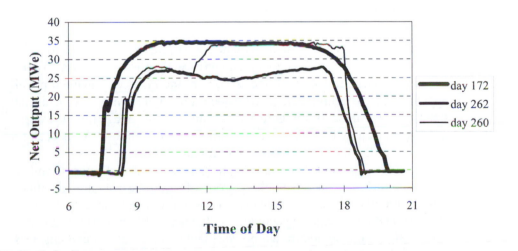

Figure 6. Electric output from 30-MW SEGS plant (KJC Operating Co., 2002).

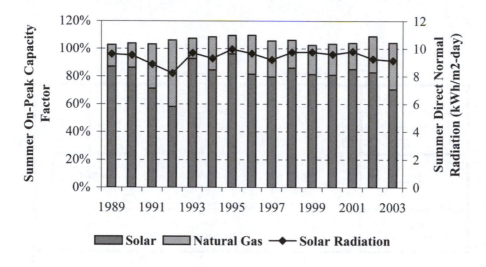

Figure 7. SEGS III-VII on-peak capacity for the last 15 years (KJC Operating Co., 2002).

Figure 7 shows the on-peak capacity factor for the five 30-MWe parabolic trough plants at Kramer Junction over the last 15 years. With the aid of the fossil backup, the plants have exceeded 100% of rated net capacity for every one of the last 15 years. The plants have averaged about 80% of rated capacity from solar energy alone with natural gas used to fill in to 100% capacity. Note that solar output was low in 1991 and 1992 as a result of the eruption of the Mount Pinatubo volcano.

3. Solar Collector Technology

This chapter deals specifically with linear parabolic trough collectors for concentrating sunlight. The trough-shaped parabola is covered with reflector material to concentrate the solar radiation in the focal line. To do so, the symmetry plane (optical axis) of the parabola has to be directed toward the incoming light from the sun, achieved by tracking the sun on a single axis. Figure 8 shows an example of a parabolic trough collector and illustrates how the direct beam component of sunlight reflects back to the receiver located at the focus of the parabolic mirrors.

The basic modular component of the solar field is the solar collector assembly (SCA). Each SCA is an independently tracking group of parabolic trough solar collectors made up of parabolic reflectors (typically mirrors); the metal support structure; the receiver tubes; and the tracking system that includes the drive, sensors, and controls. The solar field in a parabolic trough power plant is made up of hundreds, and potentially thousands, of SCAs. All these components are in continuous development, aiming at further cost reductions to enhance market opportunities.

Figure 8. Parabolic trough collector (Plataforma Solar de Almería (PSA), 2000).

3.1 Support Structure

All the collector components are mounted on the support structure, which maintains the optical alignment, withstands external forces (dominated by wind forces), and tracks the sun via the drive system. Until recently the LUZ designs used in the SEGS plants were the only commercial option for large electric systems, but presently new options are offered in the U.S. and Europe. Building on the experience and lessons learned by the SEGS plants, these options include several new parabolic trough collector designs under active development, as described below.

3.1.1 LUZ (SEGS plants)

The LUZ LS-3 collector was the final concentrator design used at the newest SEGS plants (SEGS VII-IX). A variation of the LS-3, which allows the collector to be tilted a few degrees, is used for the direct-steam generation test at the PSA. Although the operational experience of the LS-3 collector has been excellent (high tracking availability), the thermal performance and the maintainability (alignment) of the collector at the SEGS plants has not been equal to the earlier LS-2 design. LUZ changed from the LS-2 to the LS-3 design to reduce the collector cost for large field deployments. It is not clear if the expected capital cost benefit of the LS-3 design over the LS-2 was ever realized. Operational experience from the SEGS plants shows that performance and maintainability issues have likely offset any cost benefit that may have existed for the LS-3.

Solel Solar Systems Ltd. of Israel is currently offering improved versions of the SEGS LS-3 collector, building on the California experience and likely addressing both design and O&M issues noted from field practice.

3.1.2 EuroTrough

A European team has completed the development and testing on a next-generation parabolic trough concentrator, known as EuroTrough (Lüpfert et al., 2001). The EuroTrough concentrator is a torque box concentrator concept intended to improve on characteristics associated with the LS-2 and LS-3 collectors during fabrication and operation. The torque box design combines the torsional stiffness and alignment benefits of the LS-2 torque tube design (Figure 9a) with the reduced cost of an LS-3 like truss design (Figure 9b). Wind-load analysis and finite element modeling identified the design, which is composed of a rectangular torque box with mirror support arms, as the best concentrator design evaluated (Figure 9c). The rotational axis is at the center of gravity, a few millimeters above the torque box. The torque-box design has less deformation of the collector structure, which can result from dead weight and wind loading, than the LS-3 design. This reduces torsion and bending of the structure during operation and improves optical performance. The stiffer design allows the length of each collector to be increased from 100 meters to 150 meters. This decreases the total number of drives and flexhoses (or ball joint assemblies) required for a collector field, which reduces thermal losses and the total collector field cost.

The key feature of the EuroTrough collector is the 12-meter long steel space-frame structure with a square cross-section that holds the support arms for the parabolic mirror facets of 5.8 m aperture width. The box is constructed with only four different steel parts, which simplifies manufacturing processes and reduced costs for on-site assembly and erection. In addition, the EuroTrough design allows improved packing densities to help minimize freight costs. This design is expected to reduce glass breakage during high wind conditions. As a result of a new drive pylon design, the SCA can be mounted on an inclined site (3%), which can decrease site preparation costs.

Concentrator accuracy is achieved by combining prefabrication with on-site jig mounting. Most of the structural parts are produced with steel construction tolerances. One of the design objectives was to reduce the weight of the apparatus compared to that of the LS-3 collector structure. The steel structure now weighs approximately 26 kg/m^2, about 14% less than the available design of the LS-3 collector.

These improvements-reducing the variety of parts, lessening the weight of the structure, and using more compact transport-are assumed to result in cost reductions in on the order of another 10%. For the total collector installation, series production costs below 200 USD/m^2 of aperture area are anticipated.

Following initial mechanical and thermal testing at PSA, beginning in the year 2000 on a partial EuroTrough collector (Lüpfert, 2003), prototype testing of a full solar field "loop," designated SKAL-ET, was initiated in April 2003 within the existing SEGS V power plant at Kramer Junction, California (Herrmann, Graeter, and Nava, 2004).

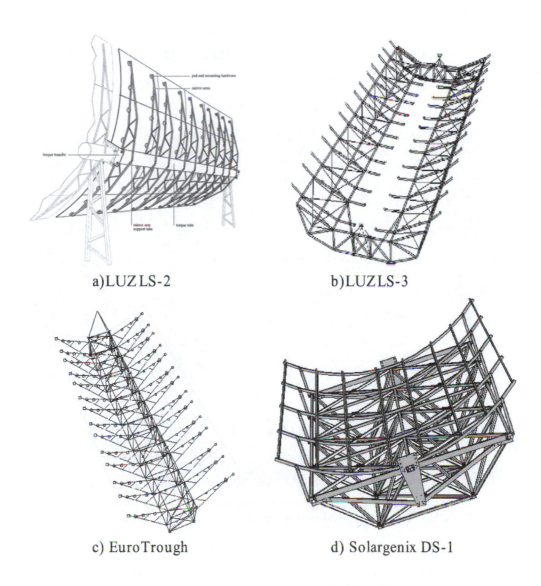

a)LUZ LS-2

b)LUZ LS-3

c) EuroTrough

d) Solargenix DS-1

Figure 9. Parabolic trough concentrator structure (a: KJCOC, b and c: EuroTrough; d. Solargenix).

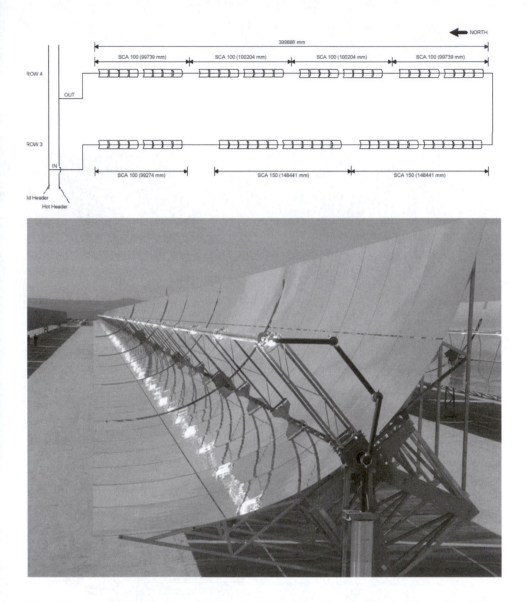

Figure 10. SKAL-ET demonstration loop in Kramer Junction, California (Flagsol GmbH).

The demonstration loop, shown in Figure 10, replaced eight standard LS-3 collectors. The 'cold' side of the loop consisted of a 100m SCA at the inlet of the loop and two adjacent long-version 150m SCAs (designed to reduce unit costs); the 'hot' side of the loop consisted of four 100 m SCAs. One complete row has a length of 400m, resulting in a collector aperture area of about 4,350 m² for the test loop.

Thermal test results showed that the 100m SCA using new Solel receivers obtained a thermal efficiency approximately 10% higher than the reference LS-3 SCA . Further, the 150 m SKAL-ET collector achieved an optical peak efficiency of 78%, matching the performance of the smaller 100 m collector. The high performance values have been achieved with Solel UVAC absorber tubes and the advanced manufacturing and assembly concepts in order to reach geometrical precision and radiation intercept at high quality level. The success of these efforts has also been shown by photogrammetric concentrator geometry analyses on large collector areas and flux and intercept measurements in the focal area (Lüpfert, 2004). The EuroTrough collector is likely to be implemented in several 50MWe solar power plant projects in Spain (approximately 500,000 square meters of solar field aperture area each) over the next few years.

3.1.3 Solargenix Energy

Solargenix Energy, formerly Duke Solar Energy, in Raleigh, North Carolina, has engineered an advanced-generation parabolic trough collector that also uses a space frame concentrator structure, a new two-speed hydraulic drive, and an advanced sun-tracking controller. This collector design is patterned after many of the characteristics of the LS-2 collector (Figure 9a), but is twice the length (i.e., ~100m long). The new design is thought to be superior to the LS-2 in terms of structural properties, weight, manufacturing simplicity, corrosion resistance, manufactured cost, and installation ease. Finite element models of the LS-2 and the new space frame design were developed to comprehensively assess both structures for comparison. The structural models show that the new space frame, though much lighter, achieves the high resistance to wind loads in a 100m length (i.e., high bending stiffness and torsional stiffness) that the LUZ LS-2 has demonstrated in a 50m length, which should yield excellent performance in the field.

The Solargenix design emphasizes simplicity of fabrication and field assembly of the space frame to reduce concentrator shipping costs from the factory. The space frame is composed of interconnected aluminum struts, arranged in a three-dimensional truss-like pattern (Figure 9d) and connected by a field-installed hub system. All the struts used in the space frame are 2 in. square extruded aluminum tubes, and the structure is easy to assemble. Drilled holes in each end of the struts are used to connect the struts to the hubs. In terms of weight, this space frame design has a significant advantage since it is about half the weight of the LS-2 structure. A lightweight structure is superior in terms of shipping, handling during manufacture, and field installation. The space frame also has greater corrosion resistance because it is made entirely of aluminum. The space frame is engineered to accept the standard silvered-glass mirrors that have demonstrated excellent reliability and lifetime in the operating LS-2 collector systems. Although the installed costs of the Solargenix parabolic trough are expected to be lower than those of the LS-2 collector, Solargenix anticipates a higher level of performance.

Sandia National Laboratories has completed thermal performance testing of a prototype Solargenix Energy parabolic trough collector module with Solel receivers. The Solargenix collector module, shown in Figure 11, has an aperture 5-meters wide and 7.84-meters long, yielding a total aperture area of 39.2 m². This size is identical to

Figure 11. Solargenix collector test at Sandia National Laboratory.

the LS-2 collector module, which is used in many of the large-scale SEGS systems in California. Since the LS-2 collector has also been thermally tested at Sandia, it provides a good point of comparison between the LS-2 and the Solargenix collector module test results.

The measured peak steady-state thermal conversion efficiency for the Solargenix prototype collector is presented graphically in Figure 12, which shows the measured peak efficiency (%) vs. fluid temperature above ambient temperature in degrees Celsius (ºC) at an average direct normal insolation (DNI) of 940 W/m². The peak steady-state thermal conversion efficiency at the "zero-heat-loss condition" (i.e., optical efficiency) of the Solargenix collector measured during testing is 78% (with an uncertainty interval reported by Sandia of approximately ± 2.5%). The increase in performance over the LS-2 collector is likely due to improved receivers over those used on the LS-2 collector.

The Solargenix collector will undergo further prototype testing during 2004-2005 at a facility adjacent to the planned 50-MWe Nevada power plant project, and will be implemented in 2005 at the Solargenix 1-MWe power plant project at an Arizona Public Service (APS) utility site north of Tucson, Arizona, USA.

3.1.4 Industrial Solar Technology (IST)

IST produces parabolic trough collectors that have been used primarily for lower temperature process heat applications. As part of NREL's USA Trough Program, IST

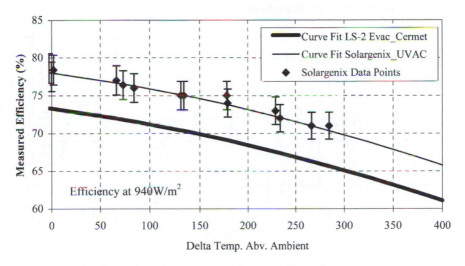

Figure 12. Sandia thermal performance test data on Solargenix prototype.

initiated a program to upgrade its collector to perform more efficiently at higher temperatures and to reduce the cost. IST converted its concentrator from aluminum to a galvanized steel structure; replaced the aluminized polymeric reflector with a thin, silvered-glass reflector; updated the collector's local and field computer controllers to use off-the-shelf hardware; and worked on developing a solar-selective absorber coating on the receiver to improve thermal performance and durability at higher temperatures. The new galvanized steel concentrator and control systems worked well in field tests, however, the thin glass used in the test did not appear to hold up in outdoor service.

Table 3 highlights the key elements of the new collector designs along with the original LUZ concentrator designs.

3.2 Reflector Development

3.2.1 Thick Glass Mirrors

The LUZ LS-3 parabolic trough concentrator uses a glass mirror reflector supported by the truss system that provides its structural integrity. The glass mirrors, supplied by Flagsol, Germany (formerly Pilkington Solar International), are made from a low-iron 4-mm float glass with a solar-weighted transmittance of 98%. The glass is heated on accurate parabolic molds in special ovens to obtain the parabolic shape. The mirrors are silvered on the back and then covered with several protective coatings. Ceramic pads used for mounting the mirrors to the collector structure are attached with a special adhesive. The mirrors have a solar-weighted specular reflectance of 93.5%. The accurate curvature allows more than 98.5% of the reflected rays to be incident on the linear receiver.

The operational experience with the mirrors has been very good. In general the mirrors are holding up well and maintaining high reflectivity. After more than 15 years

Table 3. Data on single-axis parabolic trough collectors.

Collector	Structure	Aperture width	Focal length	Length per element	Length per collector	Mirror Area per drive	Receiver Diameter	Geometric concentration	Mirror Type	Drive	Module Weight per m² [a]	peak optical efficiency
		m	m	m²	m	m²	m	sun			kg	%
LS-1	Torque tube	2.55	0.94	6.3	50.2	128	0.04	61:1	Silvered low-iron float glass	Gear	n/a	71
LS-2	Torque tube	5.00	1.49	8	49	235	0.07	71:1	Silvered low-iron float glass	Gear	29	73 [b]
LS-3	V-truss framework	5.76	1.71	12	99	545	0.07	82:1	Silvered low-iron float glass	Hydraulic	33	78 [c]
New IST	Space frame	2.30	0.76	6.1	49	424	0.04	50:1	Silvered thin glass	Jack screw	24	78
Euro-Trough	Square truss torque box	5.76	1.71	12	100-150	545-817	0.07	82:1	Silvered low-iron float glass	Hydraulic	26	78
Solargenix	Aluminium space frame	5.00	1.49	8	100	470	0.07	71:1	Silvered low-iron float glass	Hydraulic	24	78

Notes: a) Module weight is for the tracking parabolic concentrator unit and includes the structure, mirrors, receiver, and receiver supports. The pylons, drive system, and flexible interconnections are not accounted for in the module weight. b) Based on Sandia platform testing (Dudley, 1994). c) Luz assumed optical efficiency of 80% adjusted for bellows shadowing.

of service, the mirrors can still be cleaned to their as-new reflectivity. With the latest design, mirror failures have been infrequent. Still, failures have been experienced on the windward side of the field unless there is wind protection. In addition to presenting a safety hazard, mirror failures can cause damage to the receiver tube and can cause breakage of adjacent mirrors. Flagsol is working with the operator of the SEGS VIII and IX plants to test a stronger (5-mm) mirror for high wind perimeter locations. The company is also developing new mounting hardware to improve the transfer of wind loads to the steel structure (Lüpfert et al., 2001). New collector designs may move the pad-mounting locations for glass mirrors closer to the corners of the mirrors to further reduce loads on the glass. Due to environmental concerns, the protective layers on the mirror back surface no longer contain copper or lead. It is unclear at this time how these changes may affect mirror lifetimes.

3.2.2 Advanced Reflector Development

Alternatives to thick glass mirror reflectors have been under development for more than 20 years. Thin glass has been used and NREL has been working on polymeric reflectors since the 1980s. Polymeric reflectors are attractive because they are lightweight, easy to curve, and have the potential to be low cost. However, until recently none of these materials has demonstrated the cost, performance, and lifetime characteristics required for commercial trough development. Kennedy and Terwilliger provide updates on the status of the most promising alternative reflectors (Kennedy and Terwilliger, 2004).

- Thin glass mirrors: Thin glass mirrors (< 1mm thick) are relatively lightweight in comparison to thick glass and offer the promise of acceptable durability in the field. Nevertheless, thin glass mirrors are more fragile, which increases handling costs and breakage losses during manufacture. Thin glass mirrors can have initial solar-weighted reflectance of 93% to 96% depending on glass iron content, and cost in the range of $15 to $40/m^2. The solar experience with thin glass reflectors is mixed. Historically some good results have been experienced; however, more recently thin glass facets have not been durable in outdoor applications. NREL has conducted a number of tests on thin glass and believes that the protective backing on current generation of thin glass is not designed for outdoor use. This may in part be due to the elimination of lead from the paint used to protect the silver on the back of mirrors. Attempts to apply aftermarket products to improve outdoor durability have only succeeded in accelerating mirror degradation. If thin glass is to be used in the future, glass manufacturers will need to develop better protective backing for outdoor applications.

- Front surface mirrors: LUZ Industries Israel created a front surface mirror (FSM) that consists of a polymeric substrate with a metal or dielectric adhesion layer; a silver reflective layer; and a proprietary, dense, protective top hardcoat. The reflector has excellent initial reflectance. Durability testing of the LUZ prototype demonstrated outstanding durability with a solar-weighted specular reflectance >95% for more than 5 years of accelerated-exposure testing and >90% for more than 6 years. The accelerated-exposure testing operates continuously with light levels about equal to outdoor exposure and elevated temperatures of 60ºC and 60% relative humidity. A single day of testing (24 hours) is roughly equivalent to three times the outdoor exposure in terms of light intensity. Solel Solar Systems Ltd. has supplied new samples for evaluation, but they have not been under test long enough to validate the same performance as seen on the initial LUZ samples.

- Super Thin Glass: SAIC of McLean, Virginia, and NREL have been developing a front surface mirror concept with a hardcoat protective layer. The material uses an ion-beam-assisted deposition (IBAD) process to deposit the very hard (cleanable), dense (protective) alumina topcoat. The material can be produced on a roll-coater, with either a polymeric or a steel substrate. NREL has developed two additional hardcoats for use with front surface mirrors; they have demonstrated excellent optical characteristics, durability, and lifetimes similar to the LUZ FSM. Recent production cost forecasts indicates costs below $10/m^2 are possible at reasonable production volumes (Kennedy and Swisher, 2004).

- Aluminized reflector: Alanod, Germany has developed a front surface aluminized reflector that uses a polished aluminum substrate, an enhanced aluminum reflective layer, and a protective oxidized alumina topcoat. Although these reflectors have high initial solar weighted specular reflectance (~90%) they have experienced a rapid degradation in specular reflectance in field application. Alanod is working on abrasion resistive and corrosion protection coatings to resolve the problems. Alanod also has a new silver reflector with a solar weighted specular reflectance of 95%.

- Layered all-polymeric reflector: 3M has developed a nonmetallic, thin-film reflector for interior lighting that uses a multilayer "Radiant Film" technology. The technology employs alternating co-extruded polymer layers of differing refractive indices to create a reflector without the need for a metal reflective layer. The alternating polymer layers enable multiple Fresnel reflections at the interfaces of the respective layers, which results in a very high overall reflection over the visible wavelength bandwidth. This technology has the potential for very high reflectance (~99%) over more broadband wavelength regions without a metal reflective layer that has the potential to corrode. Spectral characteristics can be tailored to the particular application. Current samples under evaluation have exhibited high reflectance in a narrow band but have had a problem with ultraviolet (UV) durability. 3M plans to develop an improved solar reflector with improved UV screening layers and a top layer hardcoat to improve outdoor durability. However the UV screening layer will reduce the solar weighted reflectivity of the mirrors.

- Laminated silvered-polymer reflector: ReflecTech and NREL have developed a reflector constructed of multiple layers of polymer films with an inner layer of silver to provide for high specular reflectance. This special construction (patent pending) protects the silver layer from oxidation to achieve outdoor durability. The initial solar-weighted specular reflectance is ~93%, and the selling price is below $15/m^2 in volume. The reflective film possesses high mechanical stability, and is not subject to the "tunneling" problems (a delamination of the polymeric layer from the silver reflector layer) that have plagued other reflective film constructions. Accelerated-exposure test results indicate that a lifetime of at least 10 years is expected. The material would benefit from a hardcoat for improved washability. This material is now commercially available (www.reflectechsolar.com). The Solar Film is constructed with a removable pre-mask to protect the Solar Film until final usage. The Solar Film also has a pressure sensitive adhesive that allows application to smooth, nonporous surfaces. A peel-off liner protects the pressure sensitive adhesive.

Table 4 summarizes the characteristics of the reflector technology alternatives. At this point, thick glass will likely remain the preferred approach for large-scale parabolic trough plants, although alternative reflector technologies may be more important in the future as more advanced trough concentrator designs are developed. It should be noted that to use any of these advanced reflectors some form of structural

Table 4. Alternative reflector technologies (Kennedy, and Terwilliger, 2004).

	Solar Weighted Reflectivity (%)	Cost ($/m^2)	Durability	Abradable During Washing	Issues
Flabeg Thick Glass	94	40	Very good	Yes	Cost, breakage
Thin Glass	93-96	15-40	Poor	Yes	Protective backing for outdoor application
All-Polymeric	99	~10	Poor	No	UV protective coating needed with hard coat
ReflecTech Laminate	>93	~13	To be tested	No	Hard coat, scale-up production
Solel FSM	>95	NA	Good	Yes	Long-term durability unknown
SAIC Super Thin Glass	>95	~10	Good	Yes	No Manufacturer
Alanod	95	<20	Poor	No	Outdoor use

facet is also required. The discussion of structural facets is covered in the next section.

3.2.3 Structural Facets

Structural facets offer a potentially stronger mirror facet that can be integrated into the concentrator design and used as part of the concentrator structure. The goal is to create a stronger and less expensive reflector facet that can reduce the overall cost of the concentrator. The current focus is primarily on developing replacement facets for the existing SEGS plants as a way to validate designs with commercial plant field-testing.

- IST facet: Industrial Solar Technology developed a replacement facet for the LUZ LS-2 concentrator, and KJC Operating Company purchased several thousand for use in high wind locations. These facets used aluminum skins with a cardboard honeycomb core and 3M's EPC-305+ polymeric reflector. Initially these facets performed well, but later the water-soluble adhesive used to attach the skins to the honeycomb core reacted with the materials, causing corrosion of the aluminum skins and eventual blistering in the reflective material. The blistering significantly reduced the specular reflectance of the polymeric reflector. KJCOC also reported some change in the mirror curvature over time, thought to be due to moisture leakage into the cardboard honeycomb core.

- Paneltec facet: Paneltec Corporation developed a replacement facet for the LUZ LS-2 concentrator (Zhang et al., 1998). It uses steel skins with an aluminum honeycomb core material and thin glass for the reflector. The Paneltec facet used a vacuum-bagging manufacturing process that allowed a number of facets to be manufactured at the same time, all stacked on the same mandrel. Several hundred of the Paneltec facets were manufactured and are currently being field tested at the SEGS plants. Although they have only been in field service for a few years, they appear to be maintaining their optical accuracy and reflective quality. The primary problem with the Paneltec facet is its initial cost. The manufacturing process is labor intensive, largely because of the thin glass mirrors used for the reflective surface. The availability of an alternative reflector that would allow the manufacturing process to be simplified could dramatically improve the economics of the Paneltec facet.

- Solargenix facet: Solargenix is developing a fiberglass composite facet design (Wendelin and Gee, 2004). The facets are made of 3 to 4 mm fiberglass with a raised perimeter on the backside of the mirror. Metal strips are embedded in the perimeter of the facet to improve structural integrity and metal ribs have been added to help maintain the parabolic shape. Aluminum sheets with a thin film reflector are attached to the front side to provide the reflective surface. A number of prototype facets have been optically tested by NREL and show reasonable optical accuracy when new. However, a number of facets were put in an environmental chamber and exposed to ten temperature cycles between 10°F and 120°F over a five-day period. These facets showed signs of a reduction in optical accuracy. A number of facets are also being field tested by KJCOC, and appear to be performing satisfactorily.

3.3 Receiver Development

The parabolic trough linear receiver, also called a heat collection element (HCE), is one of the primary reasons for the high efficiency of the original LUZ parabolic trough collector design. The HCE consists of a 70-mm outside diameter (O.D.) stainless steel tube with a cermet solar-selective absorber surface, surrounded by an antireflective (AR) evacuated glass tube with a 115-mm O.D. The HCE incorporates conventional glass-to-metal seals and metal bellows to achieve the necessary vacuum-tight enclosure and to accommodate for differing thermal expansions between the steel tubing and the glass envelope. The vacuum enclosure serves primarily to significantly reduce heat losses at high operating temperatures and to protect the solar-selective absorber surface from oxidation. The vacuum in the HCE, which must be at or below the Knudsen gas conduction range to mitigate convection losses within the annulus, is typically maintained at about 0.0001 mm Hg (0.013 Pa; 10-4 torr). The graded-cermet coating is sputtered onto the steel tube to result in excellent selective optical properties with high solar absorptance of direct beam solar radiation and a low thermal emissivity at the operating temperature to reduce thermal re-radiation. The

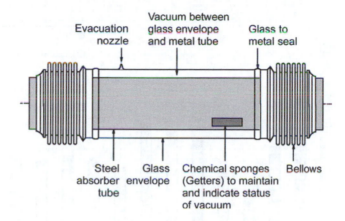

Figure 13. Heat collection element (HCE) (Flabeg Solar International, 1996).

outer glass cylinder has an AR coating on both surfaces to reduce reflective losses from the glass surfaces, thus maximizing the solar transmittance. Getters, which are metallic compounds designed to absorb gas molecules, are installed in the vacuum space to absorb hydrogen and other gases that permeate into the vacuum annulus over time. A schematic of an HCE is shown in Figure 13.

Although highly efficient, the original LUZ receiver tubes experienced high failure rates (approximately 4% to 5% per year). Failures included vacuum loss, breakage of the glass envelope, and degradation of the selective coating, which typically occurs with the presence of oxygen after the vacuum is lost or the glass envelope breaks. The degradation of the LUZ cermet coating is caused by oxidation of the cermet coating, which typically results in a progressively thicker white film forming on the inside of the glass envelope, eventually blocking most of the light from passing through the glass. Receivers with this failure take on the appearance of a fluorescent light bulb when the collectors track the sun, causing this failure to be referred to as a fluorescent tube failure. Although the collector can continue to operate with any of these failures, the collector thermal performance is significantly degraded (Dudley et al., 1994). While replacing damaged receiver tubes has a relative quick payback, it represents a significant maintenance cost.

In order to improve the reliability and lifetime of receivers it is necessary to have a clear understanding of the key causes of failure. KJCOC has developed a detailed database program that is used to track failures and repairs of solar field components (Price et al., 2004). The database includes information on virtually all repairs that have been made since initial operation of the plants, incorporating a system that uses a clear methodology for identifying key failure causes. In general, receiver failures result from: poor designs, inadequate manufacturing quality control, improper installation, operational excursions, improper solar field maintenance practices, and external events.

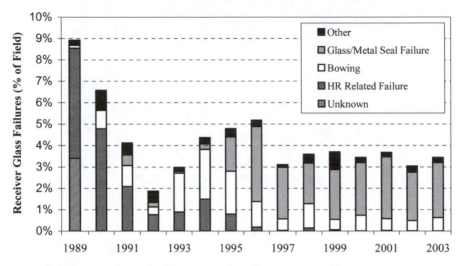

Figure 14. SEGS VI receiver glass failures by cause (Price et al., 2004).

Figure 14 shows the main causes of receiver glass envelope breakage at the SEGS VI plant using data from the KJCOC database.

There are 9600 receivers in the solar field at SEGS VI, thus each percentage point represents approximately 100 tubes. The data shows that there have been three primary causes of receiver glass tube failure at SEGS VI: hydrogen remover (HR) related failures in the early years, then bowing of the absorber tube, and finally glass-to-metal seal failures. All other failures combined (maintenance, wind, accidental, etc.) represent a small fraction of the failures in any year. The HR was a palladium membrane that looked like a small finger attached at one end of the receiver inserted into the vacuum annulus. It was designed to remove hydrogen gas that would build up in the vacuum annulus over time. The HR failures were the result of a design problem that caused the HR to press against the glass when the HR was exposed to concentrated sunlight. HR failures occurred until KJCOC rotated the tubes such that the HR was out of the direct focus or had installed better shielding to protect the HR. The increase in bowing was caused by initial receiver replacement procedures and early operational practices. Finally the glass-to-metal (G/M) seal failures occurred due to insufficient protection of the G/M seals. The low failures during 1992 are thought to be a result of the lower solar radiation that year resulting from the Mount Pinatubo volcanic eruption in the Philippines in mid 1991. The causes of many of the failures in 1989 are listed as unknown because these were replaced during the early post construction operation of the plant and the causes were not recorded in the database. Availability of the type of data shown in Figure 14 is helping the SEGS O&M companies and parabolic trough receiver vendors to improve O&M practices and the design of replacement receivers to reduce the failure rate of receivers for both existing and future parabolic trough plants. Failure rates below 1% per year are now believed possible based on new O&M procedures and new receiver designs.

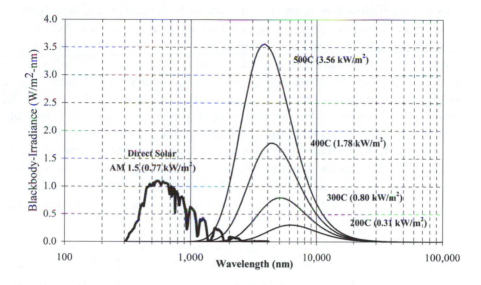

Figure 15. Solar spectra and blackbody irradiance (Kennedy, 2004).

The development of selective coatings has continued over the last 15 years. Improvements in the optical and thermal performance of new receiver designs are significantly increasing the overall solar-to-electric efficiency of parabolic trough plants. The next sections describe some of the new receiver technologies that are currently under development and the advanced receiver designs that are currently available.

3.3.1 Advanced Selective Coating Development

The performance of a solar absorber can be characterized by its solar absorptance (α) and thermal emittance (ε). Using Kirchoff's law, spectral absorptance can be expressed in terms of total reflectance $\rho(\lambda, \theta)$ for opaque materials, $\alpha(\lambda, \theta) = 1 - \rho(\lambda, \theta)$, and $\varepsilon(\lambda, T) = \alpha(\lambda, T)$, where $\rho(\lambda, \theta)$ is the sum of both collimated and diffuse reflectance, λ is the wavelength, θ is the incidence angle of light, and T is the given temperature. In addition, the thermal emittance is equal to the absorptance for any given wavelength and temperature. Figure 15 shows the wavelength dependent spectra for direct normal solar radiation with a standard air mass (AM) of 1.5 and the blackbody spectra for 200°C to 500°C. From the figure it can be seen that at low temperatures, 200°C and below, there is very little overlap between the solar and thermal spectra, with increasing overlap at higher temperatures. The figure also shows how the blackbody radiation losses rise as the working temperature is increased. For example, the black body radiation losses double by going from 400°C to 500°C. Ideal solar selective coatings would have a high absorptance or a low reflectance over the solar spectrum and a low thermal emittance or a high reflectance over the operating temperature blackbody spectrum. The optimal cutoff point between high absorptance and low emittance will depend on the operating temperature of the collector.

An extensive review was conducted of mid- to high-temperature selective absorber materials (Kennedy, 2002). Based on this work, selective absorber surface coatings can be categorized into a number of distinct types. Of these the metal-dielectric composite coating-or cermets-have proven to be the primary selective coating approach used for parabolic trough applications. LUZ initially used an electroplated black chrome selective coating on receivers at its first five SEGS plants. Later LUZ developed a graded-cermet coating which allowed higher operating temperatures and improved thermal and optical properties. In recent years a advancements have been made in selective coating technologies. These new coatings often combine multiple approaches for improving the selective nature of the coatings by using multi-layer cermets, the addition of anti-reflective surface texturing, or multi-layer antireflective coatings. In addition, coating development looks at ways to simplify the manufacturing process.

- Initially, LUZ Industries Israel manufactured the receivers for all the SEGS plant projects. In 1992, Solel Solar Systems acquired the LUZ receiver manufacturing line and began making spare parts for the SEGS facilities. Solel has continued to improve on the original LUZ graded-cermet coating. Solel improved the antireflective surface layer, changed materials to eliminate the fluorescing problem experienced with the LUZ cermet, and have further optimized (reduced) the thermal emittance of the coating.

- A new solar-selective absorber coating, known as Black Crystal, has been developed by Energy Laboratories Inc. and SNL (ELI, 2004), incorporating sol-gel overcoat(s) to mitigate oxidation at operating temperatures for an air-in-annulus receiver. On stainless steel substrates, the coating exhibits thermal stability at temperatures <375°C. Black Crystal is being used in some of the SEGS plants to refurbish damaged cermet HCEs that have lost vacuum or had their glass broken. It can be applied to new stainless steel tubing or to recycled stainless steel tubing (with degraded cermet). The recycled tubing can be straightened and must be prepared for the deposition of the Black Crystal absorber material. The coated steel tube can be reglazed with a conventional or AR-coated glass envelope. Given the temperature limitation of this coating, it is only appropriate for lower temperature parabolic trough power plant applications.

- Double-layer cermet coatings have been proposed to improve the thermo/optical properties of current receiver technology (Zhang et al., 1998). The double-layer cermet should be cheaper to produce than the current single-layer graded coatings. Centro de Investigaciones Energéticas, Medioambientales y Tecnológicas (CIEMAT) has developed a new solgel double layer selective coating with an AR coating which is stable in air at temperatures above 400°C (2003). Solgel is an inexpensive technique that can be used to produce coatings with special optical properties. The double cermet coating is comprised of alumina and platinum with a 40% metal content layer plus a 20% metal content layer. CIEMAT has also developed a sol-gel AR film to increase receiver glass transmittance. This AR film has a good mechanical durability and is suitable for the glass envelope of absorber pipes for parabolic troughs.

- In an effort to allow solar field operating temperatures above 400°C, NREL is working on developing a multi-layer cermet selective coating that has improved high temperature stability and lower emittance at higher temperatures (Kennedy, 2004).

- Figure 16 shows an example of specular reflectance of several selective coatings: black chrome (Lunde, 1980), the LUZ cermet coating (Dudley, Kolb, and Mahoney, 1994), and a high temperature coating under development by NREL. All the coatings have low reflectance (high absorptance) over the solar spectrum and high reflectance (low emittance) over the thermal black body spectrum. Table 5 shows the solar absorptance and the thermal emittance for these coatings and the other selective coatings mentioned above. The table provides two sources for the Solel UVAC receivers based on tests of two different samples. Figure 17 shows how the radiant thermal losses for each selective coating vary at different receiver surface temperatures. From the figure it is clear that the radiant thermal losses from the receiver increase significantly above 400°C. The new selective coatings can significantly reduce thermal losses at higher temperatures.

3.3.2 Solel Universal Vacuum (UVAC) Receiver

The Solel UVAC HCE is a derivative of the final LUZ HCE design used at the last SEGS plants and continues to be manufactured on the same assembly line. Solel has further developed and improved the receiver selective coating with the goal of improving receiver tube reliability and performance. Solel's most recent model is called the Universal Vacuum Collector (UVAC) HCE. UVAC HCE was designed to provide improved thermal performance, extended selective coating durability when exposed to air at operating temperatures, and better protection of the glass-to-metal seal. The enhanced thermal performance is obtained through improved glass envelope transmittance, increased solar absorptance, and decreased thermal emittance. The durability of the cermet coating at high temperature in air was improved by eliminating molybdenum in the original cermet coating that was causing the fluorescent problem. To improve the durability of the glass-to-metal seal, Solel has changed the manufacturing of the glass-to-metal seal and developed new external and internal protective shielding designs to protect both the inside and outside of the glass-to-metal seal.

NREL conducted a three-year testing program of the new Solel UVAC HCE in partnership with KJCOC to evaluate the reliability and performance (Price et al., 2004). The test included measurements of the thermo/optic properties, thermal per-

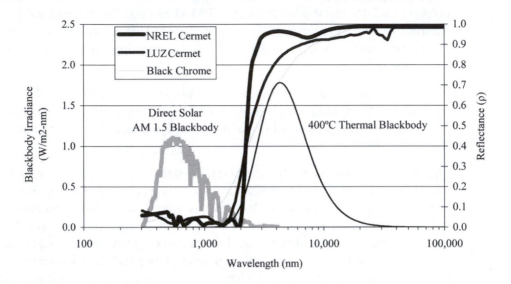

Figure 16. Spectral performance of an ideal selective solar absorber.

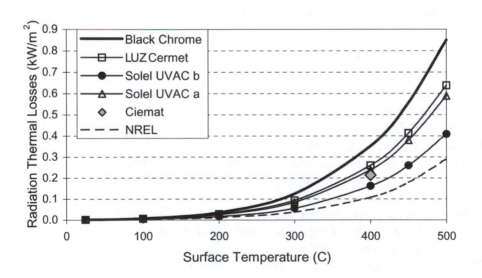

Figure 17. Radiation thermal losses.

Table 5. Selective coating thermo/optic properties.

Selective Coating	Black Chrome	Luz Cermet	Solel UVAC		Ciemat	NREL
Data Source	(Lunde, 1980)	(Dudley et al., 1994)	a (Price et al., 2004)	b (Solel Solar Systems, 2001)	(Ciemat, 2003)	(Kennedy, 2004)
Solar Absorptance	0.918	0. 938	0.954	0.944	0.940	0. 959
Thermal Emittance @						
25°C	0.047	0. 061	0.052	0.038		0.027
100°C	0.079	0. 077	0.067	0.045		0.033
200°C	0.117	0. 095	0.085	0.055		0.040
300°C	0.156	0. 118	0.107	0.070		0.048
400°C	0.197	0. 146	0.134	0.091	0.120	0.061
450°C	0.218	0. 162	0.149	0.102		0.070
500°C	0.239	0. 179	0.165	0.115		0.082

formance testing, and an evaluation of tube reliability. Measurements of the thermo/optic properties of the LUZ cermet and the Solel UVAC selective coatings are shown in Table 5. The UVAC selective coating is a significant improvement over the LUZ cermet. The table shows the solar absorptance and thermal emittance of both the UVAC tubes tested at KJCOC (UVAC a) (Price et al., 2004) and a later production batch (UVAC b) (Solel Solar Systems, 2001). KJCOC conducted thermal performance testing of 192 UVAC receivers on a loop of collectors at the SEGS VI plant.

A comparison of the UVAC receivers to a reference loop is shown in Figure 18 and Table 6. The UVAC loop clearly outperforms the non-UVAC loop in the field tests. The 192 UVAC receivers at SEGS VI and 72 UVAC receivers installed on LS-3 collectors at SEGS VII were monitored over a 3-year test period to observe failure rates. The field testing quickly identified that the initial internal glass-to-metal seal shield did not provide the expected protection. The redesigned internal and external shielding appears to adequately protect the glass-to-metal seal. Since proper shielding has been installed UVAC glass-to-metal failures have been significantly reduced. In addition, the composition of the new UVAC selective coating has eliminated the fluorescent tube coating failure experienced with the LUZ cermet. The selective coating does appear to degrade when exposed to air, but does not cause the white film on the inside of the receiver glass. Based on the field test experience, KJCOC believes the new UVAC design will have significantly lower failure rates and substantially higher thermal performance than previous receiver designs.

3.3.3 Schott Receiver

SCHOTT is developing a new parabolic trough receiver design (see Figure 19) (Price et al., 2004). The receiver will be interchangeable with the existing LUZ and Solel receivers and will also be vacuum insulated and have a highly selective cermet absorber coating. Details of the final coating have not yet been publicly disclosed, but the coating is expected to have thermal and optical properties competitive with other available receivers. The most significant innovations of the SCHOTT receiver are the

Table 6. Relative performance of SEGS VI UVAC test loop (Price et al., 2004).

Year	Summer	Winter	Comments
2000	101%	104%	Prior to UVAC Replacement
2001	121%	133%	After UVAC Replacement
2002	133%	135%	" "
2003	119%	117%	" "

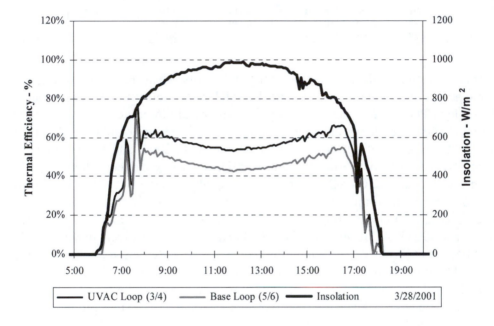

Figure 18. Solel UVAC test at SEGS VI-test loop thermal output for 3-28-01 (Price et al., 2004).

design of the bellows (see Figure 19), the glass-to-metal seal and the anti-reflective coating.

To reduce shading of the absorber tube by the bellows, a new configuration for the glass-to-metal seal and the bellows was developed. Compared to the existing design, where bellows and glass-to-metal seal are placed one after another, the Schott design places the components on top of each other. The bellows is embedded between the glass tube and the absorber tube, leading to a reduction in the optical blockage by the bellows. An advantage of the design is that the bellows is compressed at high temperatures when the absorber tube expands. This leads to a reduction in shading of the absorber by the bellows. The active length of the receiver is increased by about 2% over the LUZ design.

Figure 19. Schott receiver bellows design (Schott-Rohrglas GmbH).

As noted earlier, the glass-to-metal seal has proven to be the weakest link in the LUZ receiver design. When concentrated light hits the glass-to-metal seal it causes very high thermal gradients in the seal due to different coefficients of thermal expansion between the glass and metal. This causes stresses in the contact area, which can lead to breakage. The Solel receiver protects the G/M seal by mechanically shielding it from concentrated sunlight. The Schott design has addressed this problem by using a material combination in the glass-to-metal seal with similar coefficients of thermal expansion for both materials. The materials used are a standard nickel alloy and a newly developed borosilicate glass with high optical transmittance. The stress in the glass-to-metal seal is small even under highly concentrated solar flux.

Schott has also developed a new antireflective (AR) coating for the glass envelope with a transmittance of more than 96%. The SCHOTT AR-coating uses a sol-gel dipping technique to produce the coating. The weak point of competing AR-coatings is their low adhesion to the borosilicate glass substrate. They typically can be removed with a handkerchief. To overcome this weakness SCHOTT has developed a new composition of the AR-layer based on silicate, which improves its long-term stability against abrasion. The abrasion resistance of the AR-layer has been measured using a standard method developed by SCHOTT. A cylindrical standard eraser with a cross-section of 5 mm is moved under pressure ($0.5N/mm^2$ = 72 PSI) and the number of strokes needed to remove the layer is counted. The SCHOTT AR-coating tolerates at least 50 strokes whereas standard AR-coatings are removed after 5 strokes.

KJCOC and PSA are currently testing a number of prototype Schott receiver tubes to evaluate the mechanical integrity of the new design. These tubes were produced prior to completion of the final selective coating process and use a lower temperature selective coating. After a year in service none of these tubes at Kramer Junction have experienced a failure in the glass-to-metal seal or experienced a loss of vacuum. A number of tubes with the final cermet selective coating have been installed on the EuroTrough SKAL-ET collector test loop at SEGS V and are also being tested on an LS-2 collector mounted on the rotating platform at Sandia. Initial results indicate performance similar to the Solel UVAC receiver.

3.3.4 Low Cost Receiver Designs

The Solel UVAC and Schott receivers are obvious choices for new plants, but for replacement parts at existing plants, a lower cost and lower performance option may be preferable to the high-performance designs. A number of low-cost retrofit designs have been developed for use at the SEGS plants.

- Sunray Energy, the operator of the SEGS I and II plants (which operate at lower temperatures than the later SEGS plants) has developed retrofit receiver designs with support from SNL (San Vicente, Morales, and Gutiérrez, 2001). These designs allow receivers to be fabricated using recycled stainless steel tubing and also to be repaired in place in the field. Both receiver designs utilize a thin painted layer of Pyromark™ Series 2500 black paint for the absorber coating and on-site manufacturing processes for either full-length fused glass envelopes or full-length split glass envelopes ($\alpha=0.95$, $\varepsilon\sim0.85$ @ 400°C). The field repair returns approximately 80% of the performance of a new UVAC receiver at about 20% of the cost.

- Florida Power and Light (FPL) Energy, the owner and operator of SEGS VIII and IX, is implementing another low-cost retrofit design. For these plants, which operate at higher temperatures, a receiver retrofit program rehabilitates receiver tubes that have the glass broken off but still have a good cermet solar-selective coating. These receivers are refurbished using a special sol-gel overcoat (developed by SNL and Energy Laboratories, Inc.), which provides an oxidation barrier for the cermet that would normally degrade in air at operating temperatures. These tubes are then reglazed and reinstalled in the field. These refurbished HCEs return approximately 90% of the performance of a new UVAC receiver at about 30% of the cost (San Vicente et al., 2001).

3.3.5 Receiver Secondary Reflectors

A recent study was conducted to evaluate the potential benefits of non-imaging secondary reflectors for an LS-2 collector (Gee and Winston, 2001). The investigation included a parametric analysis to gain a better understanding of the potential performance advantages, including a small improvement in the optical intercept of a parabolic trough receiver (about 1%), and reduced receiver thermal loss (about 4%). Overall, the net performance advantage of the secondary reflector was calculated to be about 2%; that is, the entire trough collector field would have a 2% greater annual thermal energy output. The effect of rim angle of the primary concentrator was also investigated and the optical advantage was found to be virtually the same (from 70 to 80 degrees, with a slightly smaller advantage for a 90-degree rim angle). Finally, a method of manufacturing the secondary reflector was formulated, and cost analysis of the reflector was completed. The cost estimates indicate that the cost of a secondary reflector can add less than $60 to the cost of a 4m long evacuated receiver. At this

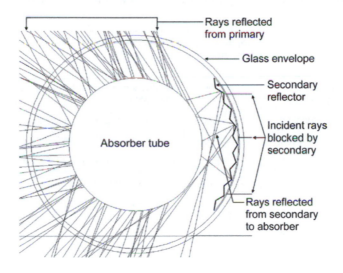

Figure 20. Trough receiver with secondary reflector (Solargenix Energy).

price, the addition of a secondary reflector offers only a modest performance enhancement to parabolic trough collectors. However, the design does achieve other indirect benefits, such as somewhat better flux uniformity around the absorber tube and an increased tolerance of the parabolic trough collectors to optical errors. For parabolic trough designs that can benefit from these other attributes, using a secondary reflector can be valuable. Figure 20 shows the output of ray tracing software modeling a parabolic trough receiver with a secondary concentrator.

For next-generation parabolic trough plants, the Solel UVAC or Schott receivers will be the receiver designs of choice. However, the design and coating developments currently under way are likely to result in further improvements in trough system cost and performance.

4. Heat Transfer Fluids and Thermal Storage

Parabolic trough solar collectors utilize an HTF that flows through the receiver to collect the solar thermal energy and transport it to the power block. The type of HTF used, and other factors, determine the operational temperature range of the solar field and thus the maximum power cycle efficiency that can be obtained. One of the potential advantages of parabolic trough technologies is the ability to store solar thermal energy for use during non-solar periods. Thermal storage also allows the solar field to be oversized to increase the plant annual capacity factor. In good solar climates, trough plants without thermal storage can produce an annual capacity factor of up to 30%.

Table 7. Heat transfer fluids with application in solar parabolic trough fields.

Fluid	Application temperature (°C)	Properties
Synthetic oil, e.g., VP-1 Biphenyl/diphenyl oxide	13 – 395	Relatively high application temperature, flammable
Mineral oil, e.g., Caloria	-10 – 300	Relatively inexpensive, flammable
Water, pressurized, + glycol	-25 – >100	Only low-temperature IPH applications
Water/steam	0 – >500	High receiver pressure required, thick-wall tubing
Silicon oil	-40 – 400	Odorless, nontoxic, expensive, flammable
Nitrate salt, e.g., HITEC XL	220 – 500	High freezing temperature, high thermal stability, corrosive
Ionic liquids, e.g., C_8mimPF_6	-75 – 416	Organic methyl-imidazole salts, good thermal properties, very costly, no mass product
Air	-183 – >500	Low energy density, only special IPH applications

Adding thermal storage allows the plant capacity factor to be increased to 50% or more.

4.1 Heat Transfer Fluid

The selection of the type of HTF that circulates through the solar field and steam generation system, will also affect the type of thermal storage technologies that can be used in the plant. Table 7 shows the available HTF options. The choice of the fluid is directly linked to the required application temperature and other plant features such as thermal energy storage.

Biphenyl/diphenyl-oxide, known by trade names Therminol VP-1 (Solutia, 1999) and Dowtherm A (Dow, 2001), in use at the latest SEGS plants, has shown excellent stability. Although it is flammable, safety and environmental protection requirements can be satisfied with reasonable effort. The primary limitations are the temperature range, the cost for the oil itself, and the need for heat exchange equipment to transfer thermal energy to the power cycle. In addition, because the fluid has a high vapor pressure, it cannot be easily used to store thermal energy for later dispatch.

4.1.1 Advanced HTF Options

During 2002, as part of the USA Trough Initiative, an industry/lab team led by Kearney and Associates evaluated the option of using inorganic molten-salts as the HTF and storage media in a trough solar power plant (Kearney, 2004). This direct approach allows the same fluid to be used in both the solar field and the thermal storage system, eliminating the need for the expensive heat exchangers between the solar field and storage system. In addition, the solar field can be operated to higher outlet temperatures, increasing the power cycle efficiency (from about 37% to over

40%) and further reducing the cost of thermal storage. The HTF temperature rise between the cold and hot tanks can potentially increase up to a factor of 2.5 (ΔT_{STOR} = 500-290 = 210°C verses ΔT_{STOR} = 385-300 = 85°C for the indirect TES system), reducing the physical size of the thermal storage system for a given capacity. Moreover, molten salt is cheaper and more environmentally benign than the present HTF. The primary challenge is that the lowest temperature molten salt available at a reasonable cost is HitecXL, which freezes at ~120°C (248°F). Because of this, considerable care must be taken to make sure that the salt HTF does not freeze in the solar field. The higher outlet temperature has some negative impacts as well, including higher heat losses from the solar field, concerns about the durability of the selective coating on the trough receivers, and the need for more expensive piping and materials to withstand the increased operating temperatures. However, considering all factors in an analysis of performance and cost, the study projected an attractive reduction in levelized electricity costs for this configuration. Updated results are presented later in this chapter.

Organic molten salts (termed ionic liquids in the chemical engineering and synthetic chemistry community) have been identified as a potential new type of heat transfer fluid for use with parabolic trough solar technology (Blake et al., 2003). Organic salts have many of the same attributes that make inorganic salts good heat transfer fluids, such as good thermal and heat transfer characteristics, good material compatibility, low flammability, low toxicity, and low vapor pressures. The primary advantage of organic salts over inorganic salts is that many remain liquid at room temperature. The other advantage to organic salts is that there is a wide range of cation and anion constituents that can be combined to create a customized fluid with the required properties for a specific application. The primary disadvantage is that organic salt chemistry is a relatively young field and only limited information on salts and the properties of those salts exists. In addition, most organic salts are produced in small quantities relative to the volumes required for a trough plant with thermal storage. Thus only limited information exists on the potential cost of such fluids. The primary challenge is to find an organic salt that can be used at high temperatures and can be manufactured at a reasonable cost.

In 2002-2003 NREL conducted an assessment of ionic fluids for thermal storage in a trough plant, such as heat transfer, materials compatibility, and factors influencing the production cost of representative ionic liquids. The work also included a compilation of a database of known materials that have potential to fit the application, utilization of existing capability to synthesize new examples of ionic liquids, and establishment of the capability for measuring thermal stability, and determining the mechanisms of decomposition. All these factors led the development team to focus on imidazolium salts. The next step focused on the identification of specific imidazolium salts that would be stable at higher temperatures. Methods were developed that gave a much better estimate of potential advanced fluid costs. However, it was concluded that no presently known imidazolium salts can meet the high temperature goals at or above 400°C (752°F) for an advanced fluid in a large parabolic trough power plant.

Rather, the maximum operating temperature appears to be in the 200-250°C range depending on the residence time and temperature distribution in the system.

It was recommended that the search for advanced fluids for the solar parabolic trough application should move to other classes of organic fluids or salts which were tentatively identified. One approach that is envisioned is a chemical modification of existing commercially available heat transfer fluids, while another is the study of functional fluids such as plasticizers and lubricants that are known to perform under extreme temperature fluctuations and that are well known to the chemical industry. The challenge will be to modify already known fluids such that they can satisfy the specifications required for use in a solar trough system, i.e., high thermal stability combined with very low vapor pressure and low melting point. Work on these new approaches is currently underway at NREL. Further, a model for the residence time/temperature distribution of the organic fluid should be developed in order to more accurately determine effective lifetime.

4.2 Thermal Storage

The potential for low-cost and efficient thermal storage is one of the claimed attributes of large-scale CSP technologies (Kearney and Price, 2004; Geyer, 1991; Pilkington Solar International GmbH, 1999). Thermal storage allows solar electricity to be dispatched to the times when it is needed most, and allows solar plants to achieve higher annual capacity factors.

The first SEGS plant used mineral oil HTF and included three hours of thermal storage, comprised of a two-tank system; one tank held the cold oil and a separate tank held the hot oil once it had been heated. This helped the plant dispatch its electric generation to meet the utility peak loads during the summer afternoons and winter evenings. The system worked well until 1999 when it was destroyed by a fire of the very flammable mineral oil resulting from a failure in its tank blanketing system. A mineral oil thermal storage system was also used at the Solar One steam central receiver demonstration power plant (Geyer, 1991). This system used a single-tank thermocline storage system with rock/sand filler. The storage system at Solar One worked well, although thermodynamically it was not well suited for integration with the central receiver steam conditions used at Solar One. The storage system also experienced fires related to the use of the mineral oil storage fluid. For operational reasons, the mineral oil HTF was replaced by a synthetic hydrocarbon fluid at the later, more efficient SEGS plants that operate at higher solar field temperatures. For these plants the two-tank storage system integrated directly with the solar field as used at SEGS I is not economically feasible because of the high cost of the synthetic HTF. In addition, the high vapor pressure of biphenyl/diphenyl-oxide would require pressurized storage vessels.

During the 2001-2002 period, a significant effort was undertaken to identify thermal storage technologies for parabolic trough solar power plants. A number of promising thermal storage options that could be used for higher temperature parabolic trough plants were identified (Pilkington Solar International GmbH, 1999). The initial goal was to develop a viable near-term thermal storage option that would be low-risk and could be used with confidence in near-term projects. An indirect 2-tank thermal storage system that uses the current HTF in the solar field and molten-salt in the storage system was proposed to meet the low-risk near-term thermal storage objectives (Kelly, 2001). While a viable and low-risk near-term approach, this system does not meet the longer-term cost goals for thermal storage. Several additional efforts have focused on opportunities for reducing the cost of thermal storage in future plants. These include the use of a single tank thermocline storage system, using molten-salt as both the heat transfer fluid in the solar field and the storage media in a single fluid loop (i.e., no heat exchanger), and the development of a new indirect molten-salt thermal storage system as part of the 2001 USA Trough Initiative.

The Kelly study (2001) considered a thermal storage design that would provide two hours of full-load energy to the turbine of an 80-MW SEGS plant (Figure 21). In this design, there is an oil-to-salt heat exchanger for the transfer of energy between the solar field HTF and the molten salt storage system. Although solar salt has a relatively high freezing point (~220°C (428°F)), the salt is kept in a low heat loss compact volume and is easily protected by heat tracing and systems that drain back to the storage tanks when not in use. By examining the experience at Solar Two, Kelly concluded that this thermal storage concept has low technical and cost risk. The study also found that the system had a specific cost of $40/kWht. Storage systems with more hours of storage relative to the turbine capacity would have lower specific costs, because the cost of the heat exchanger dominates the cost of the system. For larger storage systems lower specific costs on the order of $30/kWht are predicted.

The two parabolic trough 50MWe power plant projects are under development in Spain each include a 6-hour indirect thermal energy storage system of this type. In general, the TES capacity and solar field size in such systems can be optimized to maximize performance and minimize cost.

4.2.3 Thermocline Storage

One option for reducing the cost of thermal storage is to use a thermocline storage system. The thermocline uses a single tank that is only marginally larger than one of the tanks in the two-tank system. A low-cost filler material, used to pack the single storage tank, acts as the primary thermal storage medium. The filler displaces the majority of the salt in the two-tank system. With the hot and cold fluid in a single tank, the thermocline storage system relies on thermal buoyancy to maintain thermal stratification. The thermocline is the region of the tank between the two temperature resources.

During 2001-2002 a prototype test was conducted at the National Solar Thermal Test Facility at Sandia National Laboratory (SNL) of a molten-salt thermocline storage system (Pacheco et al., 2001). The test used quartzite rock and silica sand for the filler material, which replaced approximately two-thirds of the salt that would be needed in a two-tank system. In the SNL test, with a 60°C (140°F) temperature difference between the hot and cold fluids, the thermocline occupied between 1–2m of the 14m

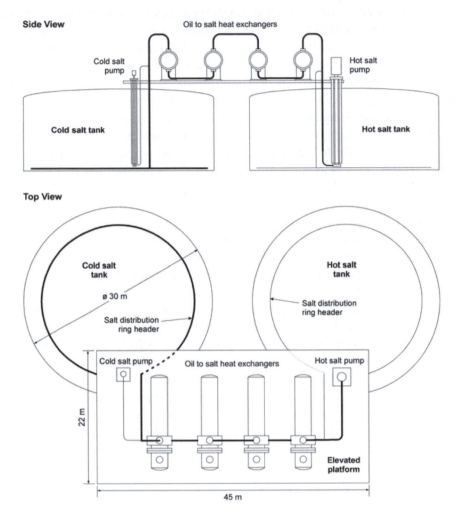

Figure 21. Two-tank indirect trough thermal storage design (Kelly, 2001).

tank height. For this reason, the thermocline storage system was thought to be best suited for applications with a relatively small temperature difference between the hot and cold fluids. The SNL testing showed that the thermocline maintained its integrity over a 3-day no-operation period. A cost comparison of two-tank and thermocline indirect molten-salt thermal storage systems with 3 hours of thermal storage for an 80MW plant was conducted and found that the thermocline system would be 35% cheaper than the two-tank storage system. However, no assessment was made on the potential impact that a thermocline storage system might have on the annual performance of the plant.

Subsequent analysis (Kelly, 2002) has shown that the size of the thermocline also depends on the method of operation of the thermocline system. While it is true that higher temperature differences will tend to cause a larger thermocline due to conduction and convection between the hot and cold temperature zones, the key factor appears to be the way in which the system is charged and discharged. It is crucial that the system is operated so that the thermocline is pushed partially out of the tank during charging or discharging of the tank. To accomplish this, the power plant must be designed to allow decay of the hot fluid from the storage system, and the solar field must be designed to allow an increase in the cold return temperature from the storage system to the solar field. This helps collapse the volume in the tank consumed by the thermocline gradient. If the thermocline is not collapsed during charge and discharge, it will eventually consume the entire tank. Collapsing the thermocline also helps to improve the economics of the TES system by increasing the storage utilization, defined as the fraction of the storage volume that can be used to store energy.

In 2003 SNL conducted additional tests to evaluate the durability of thermocline storage system filler materials (Brosseau et al., 2004). The primary objective of the tests was to determine if the selected filler materials could withstand thermal cycling at elevated temperatures in the molten salt HitecXL without decomposition over expected lifetimes and conditions of power plants with thermal energy storage. The thermal cycling test ran for 10,000 cycles in the 290ºC to 450ºC (842ºF) temperature range to simulate a 30-year plant life. Isothermal tests were also conducted at 450ºC (842ºF) and 500ºC (932ºF) for 1-year total immersion of the quartzite rock and silica sand filler materials in a ternary nitrate salt mixture (equivalent to HitecXL).

Evaluation of the filler materials at the conclusion of the tests indicate that the quartzite rock and silica sand appear able to withstand the molten salt environment quite well. No significant deterioration that would impact the performance or operability of a thermocline thermal energy storage system was evident. However, a number of recommendations for follow-on testing were given. There is need for additional testing of ternary molten-salt mixtures to understand their physical properties and freeze/thaw behavior. Better understandings of salt/metal interactions are desired, particularly at temperatures approaching 500ºC. A noticeable calcite formation on the filler material was detected during the testing. A better understanding of the long-term implications of continued buildup would be important for using Hitec XL in a thermocline application.

4.2.4 Molten-Salt HTF

Evaluation of this approach (Kearney et al., 2004) focused attention on the integration of thermocline storage with a molten salt HTF as a particularly important path for cost reduction. Design challenges exist, however. In a solar field with a molten salt HTF, one need is the simplification and cost reduction of the heat tracing and sealing of ball joints. Further, selective surface development was identified as a necessary step to achieve durability and good performance at the desired higher temperature levels made possible with the use of a molten salt. Clearly, prototype testing at small commercial-level capacities will be required for validation of both thermocline storage and a molten salt HTF solar field loop.

A key issue is selection of the appropriate salt. Nitrate salts were selected for the Solar Two project (Reilly and Kolb, 2001) because of their favorable properties compared with other candidates. In particular, these nitrate salts have low corrosion rates with common piping materials, are thermally stable in the upper temperature range required by steam Rankine cycles, have very low vapor pressures, are widely available, and are relatively inexpensive. Solar Salt was selected as the most practical salt for molten-salt power tower applications because the upper operating temperature limit (600ºC) allows the technology to be used with the most advanced Rankine cycle turbines. In addition, it is one of the lowest cost nitrate salts. However, a major disadvantage with Solar Salt is its relatively high freezing point of 220ºC. Hitec salt offers a lower freezing point of about 140ºC at a higher cost.

The freezing point is of major importance in a trough solar field because of the likely difficulties and cost associated with freeze protection, due to the need for extensive heat tracing equipment on piping and collector receivers. Primarily for this reason, a calcium nitrate salt mixture (basis of the commercial product HitecXL), with a lower freezing point of about 120ºC, was chosen for the evaluation. Other characteristics, like fluid cost, are important, but in the final analysis were deemed secondary to the risks associated with freezing.

The density, viscosity and heat capacity properties are generally similar for the nitrate salts, as shown in Table 8. Calcium nitrate salt has an upper operating temperature limit approaching 500ºC, but it is expected that the chemical stability of the receiver selective surface, not the salt, will be the limiting operational factor on the maximum operating temperature. The vapor pressures at these temperatures are very low, typically a fraction of a Pascal. Chemical reactivity and environmental issues are similar for the nitrate salts and are acceptable for this application.

Because thermal storage is an important issue for a trough system, the cost effectiveness of nitrate salts in a trough solar field were initially evaluated in terms of cost per unit thermal energy stored. That is, the costs were analyzed taking into account not only the raw costs of the salt constituents, but also the effective heat capacities of the salt solutions. Raw costs were based on dry industrial grade costs of the appropriate constituents or costs of commercial pre-mixed products. The temperature rise in the solar field was varied from 100ºC to 200ºC. The cost of the SEGS HTF (Therminol VP-1) was used for comparison at the 100 C point. The cost comparison is shown in Table 9. Thermal storage equipment is not included in this comparison.

The calcium nitrate salt (HitecXL composition) is significantly less expensive in terms of energy capacity than Therminol VP-1 at the same solar field temperature rise, and over 70% lower if used at a 200ºC rise. Although solar salt shows an even further cost reduction, the high freezing point poses difficult problems. Hitec has a lower cost and higher freezing point than the calcium nitrate salt, and remains an option. However, it does require an N_2 cover gas in the thermal storage tanks at atmospheric pressure to prevent the nitrite from converting to nitrate, thus raising its freezing point.

Table 8. Characteristics of the nitrate salts and therminol VP-1 (Kearney, 2004).

Property	Solar Salt	Hitec	Hitec XL (Calcium Nitrate Salt)	Therminol VP-1
Composition, %				74% diphenyl oxide, 26% biphenyl
NaNO$_3$	60	7	7	n/a
KNO$_3$	40	53	45	n/a
NaNO$_2$	-	40	-	n/a
Ca(NO$_3$)$_2$	-	-	48	n/a
Freezing Point, C	220	142	120	13
Upper Temperature, C	600	535	500	400
Density @ 300C, kg/m^3	1899	1640	1992	815
Viscosity @ 300C, cp	3.26	3.16	6.37	0.2
Heat capacity @ 300C, J/kg-K	1495	1560	1447	2319

Table 9. Effective storage fluid cost (Kearney, 2004).

Salt	Temperature Rise	Unit Cost	Storage Cost
	°C	$/kg	$/kWh$_t$
Hitec	200	0.93	10.7
Solar Salt	200	0.49	5.8
Hitec XL	200	1.19	15.2
(Calcium Nitrate)	150	1.19	20.1
	100	1.19	30.0
Therminol VP-1	100	2.2	57.5

4.2.5 Economic Evaluation of Storage Systems

Selected candidate TES systems were compared in (Kearney et al., 2004) on the basis of the levelized electricity cost to take both investment cost and performance into account. The systems that are either available now or appeared likely to be commercial in parabolic trough plants by approximately 2010 have been chosen for analysis. The systems are differentiated by the following characteristics:

- Type of storage system - indirect or direct;

- Configuration - two-tank or thermocline;

- Solar field HTF media - Therminol VP-1 or molten salt;

- Storage system media - molten salt type or molten salt/filler material type;

- Maximum working temperature of the storage system; and

- Storage system design capacity in terms of full-load electrical generation hours.

Design and capital cost parameters for the thermal storage systems evaluated are given in Table 10. They include variations in system type, configuration, HTF and storage media, and working temperature. Later analysis on these systems is also extended to additional consideration of receiver emissivity, storage capacity and higher operating temperature. A 3,600 MWh$_t$ capacity is equivalent to 12 hours storage capacity for a 100 MWe solar trough power plant, or 6 hours for a 200 MWe plant. An advanced organic HTF fluid system was not included because costs and technical data, while improved, are not known with confidence at this stage of development. It is clear from Table 10 that the investment cost projection for direct storage in a thermocline configuration offers an attractively low investment cost, reaching a level of specific cost that is expected to be of strong commercial interest. The levelized electricity costs (LEC) associated with the capital investments are also determined by the performance of the systems, largely influenced by the maximum temperature level of the HTF and storage media. These maximum temperatures establish turbine steam conditions, the parasitic power requirements in the solar steam system, and system heat losses. The performance model takes all these factors into account. The projections of levelized electrixity cost for trough plants with thermal storage are discussed in Section 6.

The direct systems analyzed above utilize the molten salt HitecXL, with a freezing temperature of 120ºC (248ºF). As discussed earlier, other salts with higher freezing temperatures but lower media costs could also be used. Solar salt (freezing point 220ºC (428ºF)) and Hitec (freezing point 142ºC (288ºF)) are candidates. Cases for these three molten salts were run for direct thermocline systems operating at 500ºC (932ºF) and an emissivity of 0.07 at a power plant capacity of 100MWe. The LEC results varied only slightly, less than ±0.10 cents/kWh, indicating that the higher costs for freeze protection heating offset the gains in lower salt investment cost.

5. Process Design Developments

All the SEGS plants have utilized a heat transfer fluid in the solar field to collect thermal energy, subsequently passing through a train of heat exchangers to generate steam for a conventional Rankine cycle power plant. Alternative fluids and power process concepts are currently under development to reduce cost, improve siting flexibility, or address specific market niches or opportunities.

5.1 Integrated Solar Combined-Cycle System (ISCCS)

The ISCCS integrates solar steam into the Rankine steam bottoming cycle of a combined-cycle power plant. The general concept is to oversize the steam turbine to handle the increased steam capacity. At the high end, steam turbine capacity can be approximately doubled, with solar heat used for steam generation, and gas turbine waste heat used for preheating and superheating steam., When solar energy is not available, the steam turbine must run at part load, which reduces efficiency. Doubling

Table 10. Comparision of thermal storage characteristics for 3600 MWh$_t$ capacity two-tank and thermocline/indirect and direct systems (approximately 12 hours of TES for a 100MWe plant) (Kearney et al., 2004).

Component	Indirect Storage System		Direct Storage Systems	
	Two-Tank	**Concrete**	**Two-Tank**	**Thermocline**
Solar Field HTF, type	Therminol	Therminol	HitecXL	HitecXL
Outlet Temperature	391°C (736°F)	391°C (736°F)	450°C (842°F)	450°C (842°F)
Storage Fluid, type	Solar Salt	n/a	HitecXL	HitecXL
Fluid mass (metric tons)	107,000		62,300	18,000
Fluid cost, (k USD)	51,200		71,200	26,000
Filler material, type	NA	High temp. conc.	NA	Quartzite
Filler mass (metric tons)		238,000		89,500
Filler cost, (k USD)		23,000		8,700
Tank(s), number	3 Hot, 3 Cold		2 Hot, 2 Cold	2 Thermocline
Tank cost, (k USD)	23,400	n/a	18,200	12,100
Tank dimensions	Each 43.7 m d x 12.5 m h	144,000m³	Each 42.0 m d x 12.5 m h	Each 40.5 m d x 14.0 m h
Salt-to-oil heat exchanger, (k USD)	9,000	n/a	0	0
Piping/Solar Field Heat Tracing (k USD)	0	0	10,600	10,600
Total, (k USD)	91,900	55,800	108,900	62,000
Specific cost, (USD/kWh$_t$)	26	16	33	19
	70		83	48
	66		113	100
(USD/kWh$_e$) Unit capacity, (kWht/m₃)				
Development status	early commercial	early prototype testing	pre-feasibility study	pre-feasibility study

the steam turbine capacity would result in a 25% design point solar contribution. Because solar energy is available only about 25% of the time, the annual solar contribution for trough plants without thermal storage would be only about 10% for a baseload combined-cycle plant. Adding thermal storage could double the solar contribution. Studies show (Kelly et al., 2001; Dersch et al., 2002), that the optimum solar contribution is typically less than the maximum; the more the steam turbine is oversized, the greater the off-design impact on the fossil plant when solar is not available. The ISCCS configuration is currently being considered for a number of GEF trough projects. The ISCCS improves the economics of trough solar technology because the incremental cost for increasing the steam turbine size on a combined-cycle plant is substantially lower than the cost of a stand-alone Rankine cycle power plant. In addition, the solar steam energy may, in some cases, be converted to electricity at a higher efficiency. An international team has conducted a study to evaluate ISCCS configurations at several locations for comparison with more conventional SEGS Rankine-cycle plants (Dersch et al., 2004). Although the study found that the ISCCS configurations generally offered the lower incremental cost for building parabolic trough power plants,

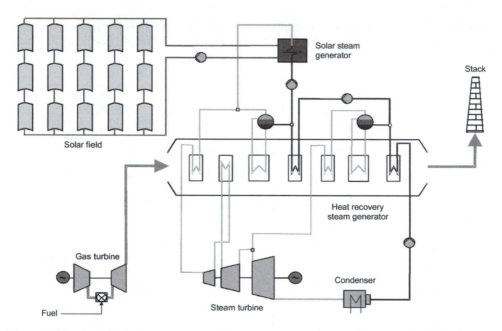

Figure 22. Scheme of an ISCCS power plant with a dual-pressure-reheat steam cycle using solar energy to replace latent heat of evaporation in the high-pressure part (Dersch et al., 2002).

it is not clear if the economic incentive will be sufficient to overcome the potential risk for developers to invest in these plants. Figure 22 shows the process flow diagram for an ISCCS plant.

5.2 Direct Steam Generation (DSG)

DSG refers to the generation of steam in the collector field, which eliminates the need for an intermediate HTF like Therminol VP-1. Although DSG increases the cost of the solar field piping by increasing the solar field fluid (steam) working pressure to above 100 bar, and may introduce two-phase flow in parts of the solar field, DSG has the potential to reduce the overall plant investment cost because it eliminates the HTF steam generation heat exchangers and all the elements associated with the HTF circuit (i.e., fire extinguishing system, oil expansion tank, oil tank blanketing system, etc.). Efficiency is increased by eliminating the heat exchange process between HTF and steam, reducing heat losses through improved heat transfer in the collector, increasing power cycle efficiency through higher operating temperatures and pressures, and through a reduction of pumping parasitics. One study indicates a 7% increase in annual performance and a 9% reduction in the solar system costs, resulting in an approximate 10% reduction in the solar portion of the levelized cost of energy (Svoboda et al., 1997). The study was performed for a small trough field in an ISCCS plant (contributing about 10-MWe to the system). The advantages may be greater for larger plants.

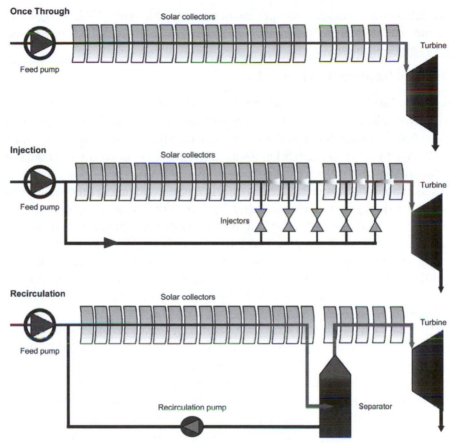

Figure 23. The three basic DSG processes: once-through (top), injection (center), and recirculation (bottom) (Zarza and Hennecke, 2000).

A prototype trough DSG row was successfully tested at the Plataforma Solar de Almería (PSA), in Spain (Zarza and Hennecke, 2000). Although it was initially assumed that the solar collectors would need to be tilted at 8º above horizontal to maintain the appropriate two-phase flow patterns in the receiver tube, DSG in the receivers of horizontal LS-3 collectors operated satisfactorily at the PSA. The DSG tests conducted at PSA have also examined once-through and recirculation modes of DSG collector field operation (Figure 23).

DSG technology may be best applied when used only to generate steam, as the advantage would be less for plants where solar energy is also used to preheat and superheat the steam. It is too early to fully assess the pros and cons of DSG in trough plants. The SEGS O&M companies have serious safety and maintenance concerns about having large solar fields of high-pressure steam, but DSG tests performed so far at the PSA are encouraging. Potential two-phase flow instabilities in long parallel rows or loops of collectors in a solar field are still to be fully evaluated in a prototype test, but are not considered to be a barrier.

Current thermal storage concepts are not compatible with DSG, though phase-change thermal storage (as yet undeveloped for trough solar fields) may be better suited for this application.

5.3 Organic Rankine Cycle (ORC)

Several solar and geothermal companies have investigated the integration of geo-thermal power plant technology with parabolic trough solar technology (Reflective Energies, 2000; Barber-Nichols, 2000). Systems under consideration range in size from 100 kWe to 10 MWe. ORCs have some advantages over steam-Rankine power cycles for small plants. ORCs can be much simpler because the working fluid can be condensed at above atmospheric pressures, and a non-condensing regenerator can be used in place of regenerative feed-water heaters. ORC systems operate at lower pressures, reducing the capital cost of components and operational pumping parasitics. Design studies indicate that optimized ORC systems could be more efficient than more complex steam cycles operating at the same solar field outlet temperature, especially at the sizes under consideration. One additional advantage of ORC systems is that they are more adaptable to dry-cooling than conventional SEGS wet-cooled steam cycle plants.

NREL analyzed a 1-MW ORC trough plant configuration (Figure 24) (Price and Hassani, 2002). The general concept is to create a small modular trough plant design that is highly packaged. The ORC technology reduces the need for on-site operations personnel, which helps to reduce the overall cost of electricity from these plants. Small geothermal plants have successfully operated as unattended power plants, and IST has demonstrated reliable unattended operation of trough solar fields. Modular plant designs that can be produced in quantities of 10-20 systems are expected to reduce the ORC power plant cost to about \$1/W. With current ORC cycles, electricity costs of about 20¢/kWh appear possible. An ORC optimized for a 300ºC operating temperature from a trough solar field should allow a significant increase in the ORC efficiency. In addition, at these temperatures, thermal storage is economically feasible, allowing solar capacity factors of 50% or higher to be achieved.

Arizona Public Service (APS) is currently developing a 1MWe parabolic trough ORC power plant. The plant will be located next to the existing APS Saguaro power plant, about 30 miles northwest of Tucson, Arizona. The parabolic trough solar field will consist of 10,340 m² of the new Solargenix collector technology described in section 3.1.3. The ORC system will be provided by Ormat and will use a recuperative cycle with wet cooling. The solar field will use a mineral oil type HTF and operate between 120ºC and 300ºC. The ORC power cycle will have a gross cycle efficiency of approximately 21%. Construction is expected to begin during the summer of 2004 and the plant should be in operation before the end of 2005. This plant will likely be the first parabolic trough power plant built since the SEGS developments ended in 1990.

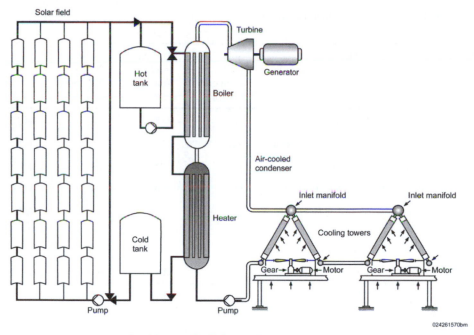

Figure 24. Basic organic Rankine cycle (Price and Hassani, 2002).

5.4 Power Cycle Overview

Table 11 presents a summary overview of the most common power cycles under consideration for use with parabolic trough solar technology, with typical design point process conditions and efficiencies.

6. Evaluating Trough Power Plant Economics

The SEGS plants were built between 1985 and 1990. During this period the right mix of economic incentives existed to make the projects economically attractive to investors, allowing the development of the SEGS I through IX plants. These included high value power purchase agreements, federal and state investment tax credits, solar property tax exemptions, and regulatory support. Subsequent to 1990, dropping energy prices, reduced state and federal incentives, and the onset of deregulation of the US power markets resulted in the halting of further development of new parabolic trough projects. Note that there was also a significant drop in the development of conventional power plants during this period in the United States due to increasing market uncertainty. In recent years, however, rising energy prices, a growing desire for green power sources, and green house gas mitigation concerns are increasing interest in large-scale solar power technologies. The primary issue for solar power technologies is whether they can vie with conventional power sources. In most cases, it is clear that

Table 11. Trough power cycle alternatives.

Plant/Cycle	Solar Field/ Turbine Working Fluid	Solar Field Outlet Temperatu re (°C)	Turbine Inlet Temperatu re (°C)	Solar Mode Efficiency (%)	Ref.
SEGS I	Caloria/steam	307	418	32[i]	[1]
SEGS III-V	Therminol VP-1/steam	349	327	31	[1]
SEGS VIII/IX	Therminol VP-1/steam	390	371	38	[1]
SEGS Salt HTF	Hitec XL/steam	450	430	40	[33]
Direct Steam Generation	Steam/steam	550	550	42	[6]
ISCCS	Therminol VP-1/steam	390	565	45[ii]	[35]
ORC - APS	Caloria/pentane	300	204	21	[42]

Notes: [i] Steam superheated by a natural gas fired superheater, [ii] Effective solar power cycle efficiency based on increase in electric output resulting from solar thermal input.

today's capital-intensive solar technologies cannot compete directly with fossil power plant technologies on the basis of levelized cost of energy alone. The next sections look at the potential for parabolic trough technology to achieve direct competiveness with fossil power.

This assessment focuses on power markets in the southwestern United States. This region combines some of the best solar resource sites in the world, proximity to population load centers with growing power demands, an extensive power transmission system, increasing conventional energy prices and growing support for renewable electric power sources.

6.1 Levelized Cost of Electricity (LCOE)

A financial figure of merit is needed for making comparisons between different technology options. In the case of electric technologies a levelized cost of electricity (LCOE) is typically used. This allows clear comparisons of current and future technologies and technologies that might not have the same lifetimes. One of the benefits of this metric is that it accounts for the financing structure and cost. The real LCOE is calculated in constant year dollars (dollars of a specific year). For most of this chapter 2004 US dollars are used. One problem with using a real LCOE is that it can be very different from the required tariff that a utility will need to pay for the power produced from a specific project. For additional clarity, a second financial metric is also used: the nominal LCOE. The nominal LEC is the price of electricity that would be required assuming that the same price is paid each year. Thus the price of power, in real dollars, decreases over time due to inflation. Both real and nominal LCOEs are used when appropriate.

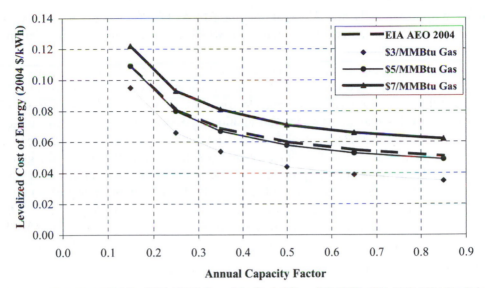

Assumes: Heat rate: 6350 Btu/kWh HHV, Overnight Capital Cost: $615/kWe, EIA AEO 2004 Gas Price:
Electric Generator (Pacific Region), IPP Financing, Equity: 15-year project life @ 12% IRR, Debt: 12-
year term @ 6.5%, Minimum DSCR: 1.3.

Figure 25. Cost of electricity from advanced combined cycle power plant.

6.2 Power Market

The primary application for parabolic trough technology is large-scale bulk power.
Because trough plants can be hybridized or include thermal energy storage, they can
provide firm capacity to the utility. Given typical capacity factors from 25% (solar
only) up to about 60% (enlarged solar field with thermal storage), trough solar power
would be able to provide firm peak to intermediate load capacity. Clearly a solar-only
plant would not be able to provide peak power 100% of the time because of periods of
extended cloud cover. However, in regions like the U.S. southwest, data from the
SEGS plants indicate they can generally meet the 80% on-peak capacity factor requirement that utilities like Southern California Edison typically require (Figure 7). In
the US southwest, parabolic trough plants are likely to be compared on a cost-of-
energy basis with natural gas fired combined-cycle power plants. The U.S. DOE
Energy Information Administration (EIA) forecasts energy prices in its Annual Energy Outlook (AEO). Figure 25 shows the cost of energy for power from a combined
cycle plant based on the plants annual operating capacity factor and the cost of natural
gas. The capital cost, O&M cost, and heat rate (efficiency) for the combined cycle
are based on the 2004 AEO assumptions (Energy Information Administration, 2004).
The cost of energy is calculated using financing assumptions from Platts Research
and Consulting (Leitner and Owens, 2003). The EIA's natural gas price forecast for
electric generators in the U.S. southwest is slightly above $5/MMBtu. Figure 25 shows
that a dispatchable solar plant with a 25% capacity factor would need to generate

power at approximately 8¢/kWh assuming the EIA AEO 2004 gas price forecast. A 50% capacity factor dispatchable solar plant would need to generate power at approximately 6¢/kWh.

In addition to the energy and capacity value of electricity produced from a dispatchable solar plant there are other economic values that can be introduced. Platts looked at the value of providing a stable electric source as a hedge against natural gas price volatility, finding. that this added an additional 0.5¢/kWh value to solar electricity (Leitner and Owens, 2003). In many regions there is also a green power premium that governments, utilities, or customers are either willing or required to pay for green power. While the value of green premiums varies significantly, green premiums in the U.S. of 0.5¢ to 2¢/kWh are considered feasible. In the near-term solar plants in the US could be competitive with conventional power sources at a cost of power of 8¢ to 10¢/kWh. In the longer-term the cost of solar power should ideally approach 6¢/kWh for direct competition with other technologies.

6.3 Parabolic Trough Performance and Economics Model

Solar plants rely on an intermittent fuel supply-the sun. It is necessary to model a plant's performance on at least an hourly time increment to accurately project the annual electrical output. This section looks at the performance and economics model used in this chapter to evaluate parabolic trough solar technologies (Price, 2003). The model performs a time-step performance simulation based on plant design and the desired operating strategy. The model is capable of modeling a Rankine cycle parabolic trough plant, with or without thermal storage, and with or without fossil-fuel backup. This tool has been used to predict capital and O&M costs, the annual performance of the trough plant configurations, and to conduct design optimization and economic analysis.

The model has been validated against the existing SEGS plants by comparing predicted with actual output from the plants, and is able to reproduce the actual gross solar output from the SEGS plants within a few percentage points on an annual basis. The calculated levelized cost of energy or LCOE (Owens, 2002) is based on current financial assumptions assumed to be available to a large-scale trough plant built in California and is stated in constant or real 2004 U.S. dollars. Unless otherwise noted, the analysis uses the 1999 insolation data from Kramer Junction, California (2,940 kWh/m²-yr).

6.3.1 Power Plant Financing

The manner of payment for a solar plant significantly affects the eventual cost of electricity from the plant. The money required to build a project is referred to as the capital investment. This is the complete cost including equipment, construction, and project development. There are a number of approaches that can be used to finance a power plant. The specific one used depends primarily on who will own the plant. Typically the owner is an investor owned utility, a government entity (a municipality) or a private owner known as an independent power producer (IPP). In the current power market, the IPP seems to be the preferred type of ownership for new power projects. Notably, this also is the most expensive type of ownership.

There are two major types of capital investments in a project: equity and debt. The parties that will own the plant make the equity investment. The equity investments in typical independent power producer (IPP) projects usually require a 12% to 18% internal rate of return (IRR) after taxes on established technologies. The debt investment is similar to a mortgage on a house. IPP projects typically use non-recourse debt, which simply means that the loan is secured by the cash flow of energy sales from the project and the debt investors cannot go after the owners if the project cannot make the loan payments. Because this increases the risk of payment, the debt interest rates on IPP projects are typically higher than would be necessary on other types of financing.

A primary difference between solar and fossil plants is that the solar plant has a large solar field that is equivalent to a 30-year fuel supply at the fossil plant, requiring a large front-end capital investment. Even if the capital cost of the solar field is the same as the fuel cost at the fossil plant, the cost of power from the solar plant will end up being more expensive primarily because of two factors. First, any capital investment must be paid back to investors at a high rate of return. Second, tax policy typically treats capital investment less favorably than expense-type investments such as fuel. Access to low-cost capital can significantly reduce the cost of solar power.

6.3.2 Financial Methodology

In order to determine the cost of electricity from a solar power plant, a 30-year cash flow analysis is performed. The analysis accounts for all the costs in the project, including such factors as the initial capital cost, O&M costs, fuel costs, insurance, taxes, financing and management fees, loan repayments, and return on investment to the owners. Key to getting realistic results is the use of appropriate assumptions in the analysis. Power industry financial experts and parabolic trough developers were contacted to determine appropriate financial parameters for the analysis discussed here. Table 12 shows the baseline financial assumptions used for a trough plant built in California. A more detailed discussion of project finance for parabolic trough plants is presented in (Kistner and Price, 1999). It should be noted that the price of individual projects discussed here can vary by 20-30% depending on the development costs for the specific location, configuration of the plant, exchange rate of the U.S. dollar with other currencies, current interest rates, and necessary financial risk premiums.

6.4 Trough Technology Assumptions

6.4.1 Baseline Trough System

The cost reduction of parabolic trough plant technology is discussed from a reference point of the SEGS plants. The efficiency of existing parabolic trough plants has been well characterized and provides a good basis for evaluating the potential perfor-

Table 12. Solar plant financing assumptions.

Project financial life	30-year
Equity internal rate of return	17%
Debt interest rate	6.5%
Debt term	20-year
Debt service coverage ratio	1.35
Construction loan	7%
Construction period	2-years
Annual insurance cost	0.5% of capital cost
Accelerated depreciation	5-year MACRS
Federal Investment Tax Credit	10%
Property taxes	California solar property exemption
Inflation	2.5%

mance improvements of future parabolic trough plants. The 30-MWe SEGS VI was selected for this analysis because it was the last SEGS plants to use the LS-2 collector for the full solar field. The LS-2 collector has demonstrated the best overall O&M characteristics of the three collector designs used at the SEGS plants. SEGS VI operates at the same process conditions likely to be used in initial new plants (steam conditions of 100 bar and 371°C). The SEGS VI plant is a hybrid plant and can produce electricity from both solar energy and natural gas. Federal law allows the SEGS plants to use up to 25% fossil fuel heat input into the steam on an annual basis. Table 13 lists the primary design characteristics, capital and O&M costs, annual performance, and nominal and real LCOE of the baseline system.

6.4.2 Near-Term Trough Technology

The near-term technology represents the technology of a plant that would be built today. For purpose of this analysis a collector recently tested at Sandia National Laboratories in Albuquerque was selected (section 3.1.3). This collector used the new SolarGenix structure with Solel UVAC receivers and thick glass FlagSol mirrors. The indirect 2-tank molten salt thermal energy storage technology is taken to be the near-term TES technology, based on its planned use in two Spanish 50-MWe parabolic trough plants. The near-term baseline plant size for this assessment is assumed to be 100 MWe. Although the three large trough plants currently under development are 50 MWe in size, the 50 MWe plant size is due to market constraints and not necessarily a reflection of an optimal design. There is a significant economy of scale with larger power plant capacities both in investment cost and O&M cost reductions. Table 13 lists the design characteristics, capital and O&M costs, annual performance, and nominal and real LCOE of the baseline SEGS VI and near-term parabolic trough system. Note that the baseline SEGS VI plant shows figures for both the solar only and the solar/gas hybrid plant.

Table 13. Baseline and near-term trough plant characteristics.

Site: Kramer Junction, CA Solar Resource: 8.05 kWh/m2-day	SEGS VI Solar/Hybrid 30 MWe	Near-Term Solar Only 100 MWe
In Service	1989	~2006
Solar Field		
Solar Multiple	1.1	1.5
Solar Field Size (km^2)	0.19	0.69
Land Area (km^2)	0.65	2.31
Heat Transfer Fluid	VP-1	VP-1
Solar Field Temp. (C)	293-391	293-391
Collector	LS-2	Solargenix
Aperture, m	5	5
Length, m	50	100
Receiver	Luz Cermet	Solel UVAC
Absorptance	0.93	0.96
Emittance @ 400C	0.15	0.14
Envelope Transmittance	0.935	0.96
Plant Performance		
Net Power (MWe)	30	100
Annual Capacity (%) [a]	22/34	29
On-peak Capacity (%)	85/ 100+	100+
Solar Mode Efficiency	10.6%	12.6%
Total Capital Cost (k$)	106,428	273,615
$/kW (w/o Land)	3,537	2,725
Operating & Maintenance Costs		
Non-Fuel Variable O&M ($/MWh)	4.94	1.98
Fixed O&M ($/kW-yr)	80.19	36.70
Financial Model Results ($/kWh)		
Nominal Levelized Cost of Energy	0.236/0.206	0.133
Real Levelized Cost of Energy 2004$	**0.168/0.146**	**0.094**

Notes: a) SEGS VI capacity factors are shown for both a solar only plant and a hybrid plant (combined 75% solar and 25% gas operation).

6.5 Design Optimization

One of the unique attributes of a solar thermal electric power plant is that the size of the solar field can be optimized for a specific power plant site. The optimum varies depending on various factors, such as location, solar resource at that location, and the valuation of specific power attributes from the plant. The optimum configuration is likely to change depending on which metric is used to optimize the plant design. Common metrics include: the lowest cost electricity, on-peak capacity factor, or highest net value (benefit - cost). In addition, if thermal energy storage is included, the optimum must consider both solar field size and thermal energy storage capacity.

6.5.1 Solar Multiple and Design Point

The solar multiple is the ratio of the solar energy collected at the design point to the amount of solar energy required to generate the rated turbine gross power. A solar

Table 14. Design point conditions for parabolic trough.

Metric	Value
Cos θ	0
Amb. Temp.	25 C
DNI	1000 W/m^2
Wind	5 m/s

Table 15. Trough solar field configurations.

Solar Multiple	Area m^2
1.0	455,224
1.2	549,279
1.5	684,717
1.8	820,156
2.1	959,356
2.5	1,139,941
3.0	1,365,672
3.5	1,595,165

multiple of 1.0 means that the solar field delivers exactly the amount of energy required to run the plant at its design output at the design point solar conditions. A larger solar multiple indicates a larger solar system. The design point is the reference set of conditions selected for designing the system. Solar multiples have commonly been used when examining proposed power tower systems, but were less relevant when designing the SEGS LUZ trough plants because, with the exception of SEGS I, the plants did not have thermal storage. However, the solar multiple is a useful metric for evaluating the performance and economics of plants over a range of solar field sizes. The design point conditions used for parabolic trough systems are listed in Table 14. They were chosen to represent a high, but not the peak, value of solar collection during a year. The design point is calculated for an incidence angle of zero degrees, which means that the sun is normal to the collector aperture. The wind speed of 5 m/s is typical of normal daytime conditions in the Mojave desert. For purposes of this assessment a range of solar multiples from 1.0-3.5 were used in a parametric analysis to find the optimum plant configurations.

Based on the design point conditions as described above, solar fields with solar multiples of 1.0, 1.2, 1.5, 1.8, 2.1, 2.5, 3.0, and 3.5 were evaluated. Table 15 shows the solar field size for each solar multiple. As a point of reference, the SEGS plants have solar multiples of about 1.1 to 1.2 based on the existing collector performance.

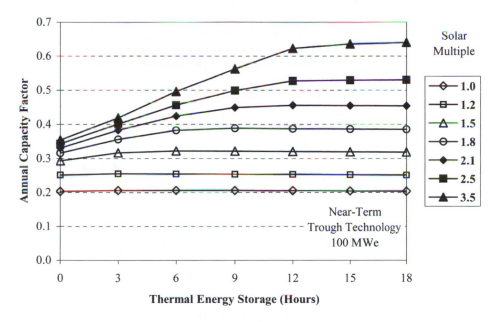

Figure 26. Annual capacity factor for near-term parabolic trough plan as a function of solar field size and size of thermal energy storage.

6.5.2 Plant Optimization

The following figures show the annual capacity factor, the on-peak capacity factor, and the levelized cost of energy for parabolic trough plants with different solar multiples (solar field sizes) and quantities of thermal storage. From Figure 26 it can be seen that parabolic trough plants can be designed to have annual capacity factors from 20% to over 60% in good solar resource regions. Note that thermal storage is only effective in this regard when the solar multiple is simultaneously increased.

Figure 27 shows how parabolic trough plants with various solar field sizes and amounts of thermal storage perform during the SCE summer on-peak time-of-use period. Trough plants without thermal energy storage and a solar multiple of 1.1 would be expected to achieve about 80% on-peak capacity from solar energy. This is what the SEGS plants report in Figure 7. However, by adding thermal storage or increasing the solar field size, the annual on-peak capacity factor can be increased to 100% from solar energy alone. It should be noted that the 100% on-peak capacity is achieved by operating up to 15% over rated capacity during some hours to compensate for other hours below 100% capacity.

6.6 Cost of Energy from Near-Term Plant

Figure 28 shows the real LCOE in 2004 dollars for 100 MWe parabolic trough plants with different sizes of solar field and amounts of thermal storage. From the figure the minimum cost of energy from a plant with no thermal energy storage occurs

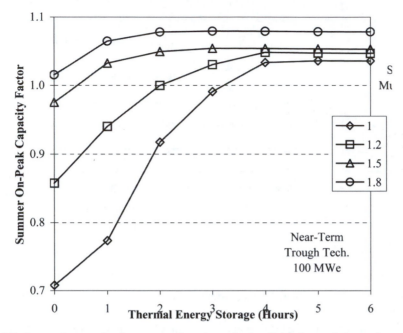

Figure 27. On-peak capacity factor as a function of solar field size and thermal storage capacity.

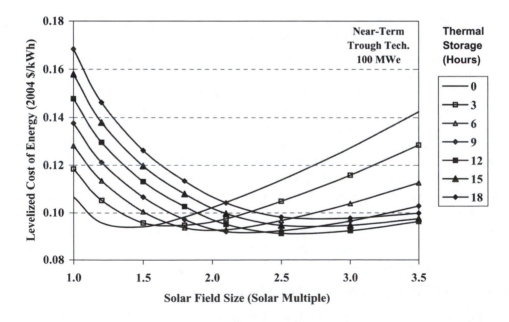

Figure 28. Levelized cost of energy as a function of solar field size and thermal storage capacity.

Table 16. Effect of solar resource on the cost of energy.

Site	DNI Resource kWh/m^{2}yr	LCOE 2004 $ $/kWh	Source
Kramer Junction, Calif.	2,940	0.094	a
Daggett, Calif.	2,792	0.099	b
Tucson, Arizona	2,636	0.103	b
Las Vegas, Nevada	2,606	0.108	b
Phoenix, Arizona	2,519	0.107	b
Alamosa, Colorado	2,530	0.121	b
El Paso, Texas	2,488	0.109	b
Cedar City, Utah	2,340	0.128	b
Reno, Nevada	2,333	0.127	b
Kahului, Hawaii	2,178	0.120	b

Source: a – KJC Operating Company, 1999 DNI data
b – NREL TMY 2 Data, http://rredc.nrel.gov/

with a solar multiple of about 1.5. The lowest cost of energy for a plant with thermal storage occurs with 12 hours of thermal energy storage and a solar multiple of about 2.5. However, the minimum cost of energy does not vary much for plants with 6 to 12 hours of thermal energy storage. Because the storage technology is relatively un-tested, a system with 6 hours of thermal energy storage and a solar multiple of 2.0 was selected for the near-term reference plant with thermal energy storage. Future plants with thermal storage are assumed to have 12 hours of storage and a solar multiple of 2.5.

6.7 Solar Resource

The direct normal solar resource has a significant impact on project economics. For the most part, the analysis presented in this paper uses the 1999 solar resource data from Kramer Junction, California. Table 16 shows the effect of site solar re-source on the cost of energy for the baseline 50 MWe near-term trough plant for other locations. This highlights the need for finding sites with good solar resources. This also shows the importance of having good solar resource data for the specific plant site.

6.8 Solar Versus Conventional Power

Parabolic trough plants can be designed in a number of configurations to provide different energy services to the grid. However, even in the best solar resource loca-tions, the cost of solar power from near-term parabolic trough plants is significantly higher than the cost of power from conventional power technologies, 6¢-8¢/kWh for gas fired advanced combined cycle power plants (with gas at $5/kJ), compared with ~9¢/kWh for a parabolic trough plant (in real 2004 US dollars). Although not currently directly economically competitive with fossil technology, rising fossil fuel prices are helping to close the gap between parabolic trough technology and conventional gas fired technology. The next sections look at the potential for further cost reduction in parabolic trough technology.

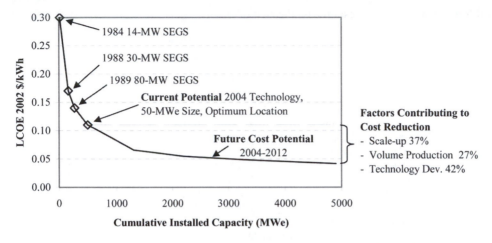

Figure 29. Parabolic trough cost reduction potential (Sargent and Lundy, 2003).

7. Cost Reduction of Future Parabolic Trough Electricity

Sargent & Lundy recently completed a study that looked at the future cost reduction potential of parabolic trough technology (Sargent & Lundy, 2003). The study found that parabolic trough technology had the potential to reach between 4¢ and 5¢/kWh (in real 2002 US dollars) based on plausible technology advancements, scale-up of plant sizes, and cost reduction due to future deployments. Figure 29 shows how the levelized cost of electricity is expected to come down with increased cumulative capacity and technology development. The sections that follow look at some of the factors that are expected to drive the technology down the cost reduction curve.

7.1 Advanced Solar Technology

Improvements in parabolic trough solar technology will help reduce the future cost of power. Table 17 shows how receiver and concentrator technology, in particular, affect the cost of energy for a 100 MWe trough plant. The first case shows the cost of energy using the original LUZ collector and receiver technology. The subsequent cases look at the impact of current receiver technology, larger concentrator sizes and advanced receiver technology. Case 3 is the assumed near-term technology. It should be noted that the calculations are based on the authors' estimates of cost and performance of each technology. These are first order estimates and do not reflect actual commercial offers by suppliers. However, it is clear that scale-up of the concentrator size and improvements in receiver technology offer significant opportunities for future cost reduction.

7.2 Advanced Thermal Energy Storage Technologies

Figure 30 compares three of the thermal energy storage configurations described in section 4.2 with the reference no-storage case. All storage cases offer an opportu-

Table 17. The influence of collector size and receiver technology on cost of energy.

Site: Kramer Junction, CA 100 MWe Trough Plant	Collector Size	Receiver	LCOE 2002 $/kWh	% of Case 1
Case 1 – Luz Technology	5 m x 50 m	Luz Cermet	0.1098	100%
Case 2 – Solel Receiver	5 m x 50 m	Solel UVAC	0.0985	90%
Case 3 – Solargenix Size	5 m x 100 m	Solel UVAC	0.0942	86%
Case 4 - EuroTrough Size	5.75 m x 150 m	Solel UVAC	0.0927	84%
Case 5 – Advanced Receiver	5.75 m x 150 m	Advanced $\varepsilon=0.10$	0.0862	79%
Case 6 – Advanced Receiver	5.75 m x 150 m	Advanced $\varepsilon=0.07$	0.0832	76%

nity for reducing the cost of energy from the plant. The direct systems assume Hitec XL molten-salt is used for the HTF/storage media, which allows higher solar field operating temperatures of 450ºC (842ºF). The higher temperature only makes economic sense if it is used with a thermocline storage system. Molten-salt storage systems with operating temperatures as high as 500ºC were also evaluated. However, these only showed marginal improvements when combined with advanced receiver technology, and higher cost when combined with current receiver technology.

7.3 Plant Scale-up

One of the primary opportunities for reducing cost is to increase the size of the power plant. In general, power plant equipment costs ($/kWe) decrease with increasing plant size (Sargent & Lundy, 2003). O&M costs also reduce with larger plant capacity because it typically takes a power plant O&M crew of about the same size to operate and maintain a 30-MWe steam plant as it would to operate and maintain a 200-MWe steam power plant. The solar field maintenance crew would be larger for the larger plant. The largest plant built by LUZ was limited to 80 MWe by then-current FERC rules for plants to qualify as renewal energy plants under applicable federal laws. LUZ considered larger plant sizes in the 150 to 200 MWe range (Lotker, 1991). The upper limit is defined by a tradeoff between economies of scale and the parasitics involved with the pumping of heat-transfer fluid through the solar field. With the replacement of flexible hoses with ball-joint assemblies, sizes of potentially 400 MWe or larger appear feasible due to lower pumping parasitics. Figure 31 shows the impact on the cost of energy for different size power plants. A 400-MW solar-only trough plant with current solar technology has the potential to produce power at less than 8¢/kWh compared to over 20¢/kWh for a 10 MWe plant. For small size plants, O&M costs represent an increasing portion of the total cost of energy.

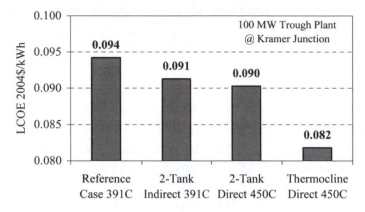

Figure 30. LEC comparisons of storage systems for 12 hours.

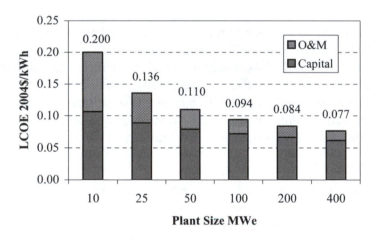

Figure 31. Impact of plant size on cost of energy.

7.4 Solar Power Parks

Many of the cost advantages achieved in scaling up a plant can also be achieved by siting multiple plants together in a power park. Comparisons of gas-turbine combined-cycle power plants show that the capital cost of plants with multiple gas turbines is similar to the cost of plants with a single gas turbine of similar capacity. Clearly gas turbines are a highly packaged technology, but a solar power park offers its own advantages for building multiple systems in a single facility. In the case of solar power plants, the mobilization costs are significant because a large portion of the total cost is related to on-site construction and assembly. A substantial amount of learning occurs during construction of the facility that can be applied to a next facility. Building multiple plants together as a single project can reduce engineering, development and financing costs, and offers the opportunity for improved competitive bidding of components. One site for multiple plants allows for shared common facilities and more effi-

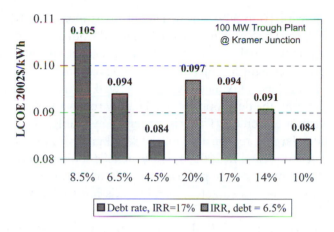

Figure 32. Effect of the cost of capital on the cost of energy.

cient O&M staffing and spare part inventories. It is difficult to quantify the impact of building plants in a solar power park. For the assessment presented later in this section, the following is assumed for a power park of four similar plants built over a five-year period: a 15% reduction in capital cost, a 25% reduction in project development expenses, and a 10% reduction in O&M costs.

7.5 Financing Cost of Capital

Figure 32 shows the impact on the cost of energy for different debt interest rates and equity IRRs when the other is held constant. The availability of low cost sources of debt and equity capital can significantly reduce the cost of energy from capital-intensive solar plants.

7.6 Tax Incentives

As previously mentioned, taxation policy tends to penalize capital-intensive solar projects. Without special property tax exemptions, solar power plants are required to pay property tax on the solar field land and equipment. Because the solar field represents a major portion of the total capital cost of the plant, property tax on this equipment represents a significant cost penalty for solar technologies. In contrast, fossil plants also do not pay sales tax on their fuel. To help achieve tax neutrality with fossil technology, solar plants should be exempted from paying sales tax on solar equipment. Because of the greater amount of capital investment for solar plants, state and federal governments collect more taxes on the income received by debt and equity investors. Thus, the state and federal governments can offer special incentives to help encourage investment in capital-intensive renewable technologies and still remain whole through increased tax revenues.

Historically, several types of incentives have been offered to renewable energy technologies. The SEGS plants benefited from federal and state investment tax credits (ITC) ranging from 10 to 50%[4] of the capital investment. A 10% federal ITC is currently still in place. The SEGS plants also benefited from a property tax exemption on all solar equipment, which is currently still in existence in California. The ITCs proved to be very successful for encouraging the development of the SEGS plants. Currently, production-based tax incentives are viewed by many as the preferred approach for encouraging the development of a healthy renewables industry. A 10-year 1.8¢/kWh production tax credit (PTC) is currently available to wind and biomass technologies and is largely responsible for the rapid growth in wind capacity in the United States in recent years. In 2005, a five-year 1.8¢/kWh production tax credit is available for solar technologies in place of the 10% investment tax credit. Unfortunately, the 5-year PTC is less attractive than the existing 10% ITC.

Figure 33 shows the affect of different tax incentives on the cost of power. The first bar represents the cost of a 100 MWe trough plant without any special tax incentives. The next bars show the impact of 10% and 30% ITCs, a 1.8¢/kWh PTC, and a property tax exemption. The 1.8¢/kWh PTC is essentially the same as the current 10% ITC for the 100 MWe plant. The next bar (CA) shows the reference plant case for incentives available to a plant built in California (10% ITC and property tax exclusion). The next bar shows the impact of adding the 1.8¢/kWh PTC to the existing California incentives. The last bar shows the impact of increasing the ITC to 30% and including the 1.8¢/kWh PTC. These incentives reduce the cost of power to about 7¢/kWh for the near-term solar-only 100-MWe trough plant.

7.7 Parabolic Trough Case Studies for Advanced Plant

A case study was analyzed to evaluate and compare the cost of near-term and future parabolic trough power plants. For this study, two time-frames are considered:

1) designs that might be used immediately for new parabolic trough power plants, and

2) those that might be built in a five to ten-year timeframe, based on technologies that are currently under development.

The technologies and designs assumed in the near term must have previously been demonstrated or otherwise considered ready for commercial application. These plants could be constructed and on line by 2006. Three potential near-term parabolic trough plant configurations are considered for the case study: solar only, hybrid, and thermal storage. The hybrid plant uses the AEO 2004 gas assumptions (Energy Information Adminstration, 2004).

The future plants will use technologies that are currently in development, that is, similar to today's systems but with design improvements to reduce costs and improve performance. They are the plant configurations that would be built in the 5-10 year

---

[4.] During 1984-86 the SEGS projects benefited from a 10% Federal Investment Tax Credit, a 15% Federal Energy Tax Credit, and a 25% California Solar Energy Tax Credit (Lokter, 1991).

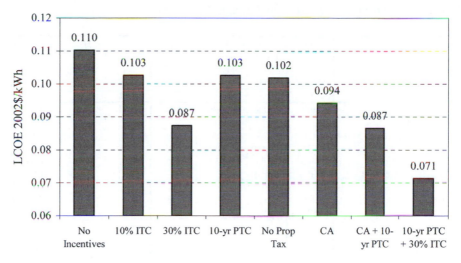

Figure 33. Effects of tax incentives on the cost of energy.

timeframe. These future power plant configurations are assumed to be a solar only or solar with thermal storage, with capacities from 100 to 200 MWe in size (although larger sizes may be possible). Advanced thermal energy storage technologies are assumed.

Table 18 shows the key design parameters for the baseline configuration, and the near-term and future parabolic trough power plant configurations.

Table 19 lists the performance of current, near-term, and future parabolic trough plants. Initial performance improvements are based on those that have already been demonstrated in field-testing. These include the use of the new Solel receiver and the replacement of flex hoses with ball joints, which will significantly reduce HTF pumping parasitics. Future efficiency gains are assumed to come from further improvements in the receiver selective coating and through increased solar field operating temperatures, which in turn lead to improved power cycle efficiency and further reductions in HTF pumping parasitics. The addition of thermal storage means that the solar field size can be increased, which results in increased annual capacity factors. Thermal storage also allows the power plant to operate closer to its design efficiency for longer periods, results in better utilization of the energy collected by the solar field, and requires fewer power plant start-ups per unit of generation.

Table 20 gives the capital cost, O&M cost, and levelized cost of energy for each trough plant configuration. Near-term costs based on cost models developed in collaboration with the parabolic trough industry are reasonably consistent with the prices currently being listed by industry. In the future scenario two capital cost cases are evaluated. The first considers the cost reduction due entirely due to technology advances. The second case also includes additional cost reduction achieved through competition and experience. In the cost reduction case solar field and power plant

Table 18. Trough plant configuration and design assumptions.

Case	Ref.	Next Plant Technology			Future Technology		
Project	SEGS VI Hybrid 30	Trough Solar 100	Trough Hybrid 100	Trough w/TES 100	Trough w/o TES 200	Trough w/TES 100	Trough w/TES 200
In Service	1989	2006	2006	2006	2011	2011	2011
Solar Field							
Solar Multiple	1.1	1.5	1.5	2.0	1.5	2.5	2.5
Solar Field Size (km^2)	0.19	0.69	0.69	0.91	1.28	1.04	1.01
Land Area (km^2)	0.63	2.31	2.31	3.07	4.30	3.51	3.40
Heat Transfer Fluid	VP-1	VP-1	VP-1	VP-1	VP-1	Hitec XL	Hitec XL
Solar Field Temp. (C)	293-391	293-391	293-391	293-391	293-391	293-450	293-500
Collector							
Aperture, m	5	5	5	5	5.75	5.75	5.75
Length, m	50	100	100	100	150	150	150
Receiver							
Absorptance	0.93	0.96	0.96	0.96	0.96	0.96	0.96
Emittance @ 400C	0.15	0.14	0.14	0.14	0.10	0.10	0.07
Envelope Transmittance	0.935	0.965	0.965	0.965	0.97	0.97	0.97
Bellows Shadowing	0.971	0.971	0.971	0.971	0.99	0.99	0.99
Thermal Energy Storage							
Storage Capacity (hrs)	0	0	0	6	0	12	12
Thermal Storage Media	NA	NA	NA	Solar Salt	NA	Hitec XL	Hitec XL

Table 19. Trough annual performance summary.

Kramer Junction 1999 Solar Resource 2940 kWh/m^2-year	Ref.	Next Plant Technology			Future Technology		
	SEGS VI Hybrid 30[a]	Trough Solar 100	Trough Hybrid 100	Trough w/TES 100	Trough w/oTES 200	Trough w/TES 100	Trough w/TES 200
In Service	1989	2006	2006	2006	2011	2011	2011
Plant Performance							
Net Power (MWe)	30	100	100	100	200	100	200
Annual Capacity (%)	22/34	29	40	41	30	56	57
On-peak Capacity (%)	88/100+	100+	100+	100+	100+	100+	100+
Solar Mode Efficiency							
Optical Efficiency	0.500	0.602	0.602	0.602	0.627	0.627	0.627
Receiver Thermal Loss	0.741	0.801	0.801	0.801	0.855	0.829	0.829
Piping Thermal Loss	0.964	0.968	0.968	0.968	0.968	0.972	0.958
Storage Thermal Loss	--	--	--	0.994	--	0.997	0.997
Dumped Energy	--	0.911	0.911	0.953	0.912	0.962	0.962
Power Plant Efficiency	0.349	0.364	0.364	0.368	0.365	0.392	0.403
Electric Parasitic Load	0.823	0.876	0.889	0.882	0.881	0.922	0.925
Power Plant Avail.	0.98	0.94	0.94	0.94	0.94	0.94	0.94
Solar-to-Electric Effic.	**10.6%**	**12.8%**	**13.0%**	**13.5%**	**14.3%**	**16.6%**	**16.9%**

Notes:
a) Based on actual operating data for 1999. No scheduled outage was taken at SEGS VI during 1999.

Table 20. Trough base case costs and financial results.

At Kramer Junction Radiation 8.05 kWh/m2/day	Next Plant Technology			Future Technology		
	Trough Solar 100	Trough Hybrid 100	Trough w/TES 100	Trough w/o TES 200	Trough w/TES 100	Trough w/TES 200
In Service	2006	2006	2006	2011	2011	2011
Plant Performance						
Annual Capacity Factor	29.3%	39.8%	41.2%	30.6%	57.9%	56.0%
Natural Gas Use (Btu/kWh)		3.04				
Plant Capital Cost						
Capital Cost ($/kWe)						
Base Case Capital Cost	2,736	2,957	3,888	2,262	4,156	3,783
Reduced Cost Case				1,727	3,330	2,978
Operating Costs						
Non-Fuel Variable O&M ($/MWh)	1.98	1.48	1.74	1.36	1.36	1.34
Fixed O&M ($/kW-yr)	36.70	37.59	43.90	23.61	45.54	43.88
Insurance (k$/yr)	1,362	1,473	1,936	2,251	2,069	1,883
Technology Case ($/kWh)						
Nominal Levelized Cost of Energy	0.133	0.135	0.130	0.101	0.099	0.092
Real LCOE 2004$	**0.094**	**0.096**	**0.092**	**0.072**	**0.070**	**0.065**
Reduced Cost Case ($/kWh)						
Nominal Levelized Cost of Energy				0.080	0.081	0.075
Real LCOE 2004$				**0.057**	**0.058**	**0.053**
All cases assume 30 year economic life, Equity IRR=17%, 6.5% Debt interest rate, California tax incentives.						

costs have been reduced by 20% in the future, with the following exceptions: mirror costs have been reduced to $25/m^2 (~58% of current price), and receivers have been reduced to $600/unit (~70% of the current price). These reductions are feasible, according to discussions with manufacturers, assuming high-volume production. The O&M cost model was developed in collaboration with KJC Operating Company and seems to be consistent with their O&M strategy. For purposes of comparing technologies both nominal and real levelized cost of energy prices are shown.

It is believed that the scenarios shown here display a realistic projection of parabolic trough plant electricity costs based on cases and assumptions that can be achieved with appropriate technology development, plant deployment levels, and regulatory incentives.

Figure 34 compares the cost of power from an advanced combined cycle plant with those of near-term and future parabolic trough technology. The future cost of power from parabolic trough plants appears to be competitive with power from the combined cycle plant. The near-term trough technology is not currently competitive with the combined cycle plant at current AEO fuel cost assumptions (approximately $5/kJ). Alternative approaches must be used to reduce the cost of near-term plants. The next section looks at financing and taxation incentives that might be used to reduce the cost of solar power.

7.8 Alternative Financing and Tax Incentive Schemes

Four alternative financing scenarios are considered for this analysis. The first assumes the current financial incentives for IPP power projects located in California. The second assumes a 10-year 1.8¢/kWh PTC in place of the current 10% ITC. The third assumes the 10% ITC and the 10-year 1.8¢/kWh PTC. The final case assumes that the project is purchased by a municipal utility. Municipal utilities have access to low cost financing with interest rates below 6%. The analysis looks at the cost of building a single 100-MWe plant and the cost of building a power park with four 100-MWe plants at a single site as sequential units one year apart. The cost reduction assumptions for the power park case are shown in section 7.4.

Figure 35 shows the results of the financial analysis for single parabolic trough plant and solar power park for each financing case. The cost of power for the single baseline 100-MWe trough plant is currently about 2¢/kWh higher that that of the advanced combined cycle plant with AEO 2004 natural gas price assumptions. Of the financing options and tax incentives considered, only low cost financing will reduce the cost of solar power to the competitive range with the advanced combined cycle plant. However, the power park development appears to offer another approach for achieving competitiveness with fossil technologies with standard IPP type financing. In the case of the 100 MWe reference plant, the ITC and the PTC offer a similar benefit; however, the PTC becomes much more attractive in the future when the capital cost, and thus the value from the ITC, is reduced.

Many factors have an effect on the cost of power: plant configuration, size, financing structure, and tax incentives. Even for current technology there is a wide range in costs for potential near-term plants. The costs presented here are for plants located in the Mojave Desert. The cost of power would be higher for locations with a lower solar resource.

7.9 Financing Summary

There is significant opportunity for reducing the cost of power from parabolic trough power plants. Under various realistic scenarios, future plants appear to have the potential to directly compete with fossil power. While increasing plant size offers the easiest opportunity for reducing the cost of power, a number of technology advances have been identified that can also significantly reduce costs. These include increasing the collector size, improvements in receiver selective coatings, and development of advanced thermal storage technologies.

Financial incentives, market incentives such as renewable portfolio standards, and other approaches such as hybridization or integration into combined cycle power plants may be necessary to encourage near-term projects to be realized and set the stage for accelerated growth of this attractive large-scale solar technology.

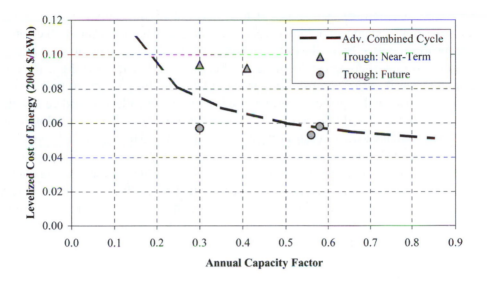

Figure 34. Comparison of advanced combined cycle and parabolic trough technology.

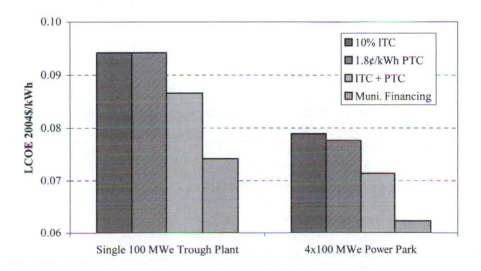

Figure 35. The cost of energy for near-term and future parabolic trough power plants with different financing assumptions.

8. Conclusion

The operating performance of the existing SEGS plants has shown that parabolic trough solar power plant technology is robust and an excellent performer in the commercial power industry. Since the last commercial parabolic trough plant was built in 1990, substantial technological progress has been realized. Together, these factors mean that the next generation parabolic trough plants are fully expected to be even more competitive, with enhanced features such as improved collectors and economical thermal storage. In addition, worldwide R&D efforts and commercial deployment are likely to continue to drive costs down and improve the performance and capabilities of this renewable energy option. Parabolic trough solar power technology appears to be capable of competing directly with conventional fossil-fuel power plants in mainstream markets in the relatively near-term. Given that parabolic trough technology utilizes standard industrial manufacturing processes, materials, and power cycle equipment, the technology is poised for rapid deployment as an important contributor to the continued growth of renewable energy power sources.

9. Acknowledgements

The authors would like to acknowledge the efforts and contributions of the many individuals, corporations, universities, laboratories, or others who have directly or indirectly contributed to this article, and in particular the Contributors noted at the beginning of this chapter. Specific thanks to the SEGS O&M companies, the European Commission, the German Ministry for the Environment (BMU), and the U.S. Department of Energy.

10. Nomenclature

AR	Antireflective
CIEMAT	Centro de Investigaciones Energéticas, Medioambientales y Tecnológicas, Almería, Spain
DISS	Direct Solar Steam
DOE	U.S. Department of Energy
DSG	Direct (solar) steam generation
ELI	Energy Laboratories, Inc.—Jacksonville, FL
FEA	Finite element analysis
FPL	Florida Power and Light —Harper Lake, CA
FSI	Flabeg Solar International, Köln, Germany
FSM	Front surface Mirror
GEF	Global Environment Facility of the World Bank
HCE	Heat collection element (receiver tube)
HTF	Heat transfer fluid

HX	Heat exchangers
IBAD	Ion-beam-assisted deposition
IPH	Industrial process heat
ISCCS	Integrated solar combined-cycle system
IST	Industrial Solar Technology
LEC	Levelized cost of energy
LS-3	LUZ System Three Parabolic Trough Collector
NREL	National Renewable Energy Laboratory
O.D.	Outside diameter
O&M	Operations and maintenance
ORC	Organic Rankine cycle
PSA	Plataforma Solar de Almería, Spain
PURPA	U.S. Federal Public Utility Regulatory Policy Act
SCA	Solar collector assembly
SCE	Southern California Edison Electric Utility
SEGS	Solar Electric Generating System
SNL	Sandia National Laboratories
UV	Ultraviolet
UVAC	Universal Vacuum (SOLEL HCE Receiver—most recent version)
ZSW	Center for Solar Energy and Hydrogen Research, Stuttgart, Germany

11. References

Alanod, 2004. MIRO-Silver Reflector (Website: www.alanod.com/).

Barber-Nichols, 2000. NREL n-Pentane Solar Rankine Cycle Analysis and Review. NREL/TP-550-31240, Golden, CO: National Renewable Energy Laboratory.

Moens, L., Blake, D.M., Rudnicki, D., Hale, M.J., 2003. Advanced Thermal Storage Fluids for Solar Parabolic Trough Systems. Transactions of the ASME, Vol. 125: 112-116.

Brosseau, D., Edgar, M., Kelton, J., Chisman, K., Ray, D., Emms, B., 2004. Testing of Thermocline Filler Materials and Molten-Salt Heat Transfer Fluids for TES Systems In Parabolic Trough Power Plants. Presented in Proceedings of 2004 International Solar Energy Conference, July 11-14, 2004, Portland, OR (ISEC2004-65144).

CIEMAT (Centro de Investigaciones Energéticas, Medioambientales y Technológicas), 2003. Parabolic Trough Collector Technology, Annual Report 2003 for Plataforma Solar de Almería (PSA), Tabernas, Spain, pp. 47-49.

Cohen, G., Kearney, D., Kolb, G., 1999. Final Report on the Operation and Maintenance Improvement Program for CSP Plants. Report No. SAND99-1290, Albuquerque, NM: Sandia National Laboratory.

CSP Global Market Initiative, 2004. (Website: www.energylan.sandia.gov/sunlab/documents.htm).

Dersch, J., Geyer, M., Herrmann, U., Jones, S. A., Kelly, B., Kistner, R., Ortmanns, W., Pitz-Paal, R., Price, H., 2004. Trough Integration into Power Plants: A Study on the Performance and Economy of Integrated Solar Combined Cycle Systems. Energy, Vol. 29, Issues 5-6, pp. 947-959, April-May 2004 (Website: www.elsevier.com/locate/energy).

Dersch, J., Geyer, M., Hermann, U., Jones, S.A., Kelly, B., Kistner, R., Ortmanns, W., Pitz-Paal, R., Price, H., 2002. Solar Trough Integration into Combined Cycle Systems. Campbell-Howe, R. (Ed.). Proceedings of Solar 2002: Sunrise on the Reliable Energy Economy, June 15-20, Reno, NV.

Dow Chemical, 2001. DowTherm: A Synthetic Organic Heat Transfer Fluid. Product Information Report No. 176-01463-1101-AMS (Website: www.dowtherm.com).

Dudley, V.E., Kolb, G.J., Mahoney, A.R., 1994. Test Results: SEGS LS-2 Solar Collector. Report No. SAND94-1884, Albuquerque, NM: Sandia National Laboratory.

Electric Power Research Institute (EPRI), 1997. Renewable Energy Technology Characterizations. EPRI Topical Report No. TR-109496, Palo Alto, CA: Electric Power Research Institute.

Energy Information Administration, 2004. Annual Energy Outlook 2004 with Projections to 2025. Washington, D.C. (Website: www.eia.doe.gov).

Energy Laboratories, Inc., 2004. High Expectations in Many Industries for Innovative New Selective-Surface Coating: Black Crystal (Website: www.solgels.com/BCinfo.htm).

Enermodal, 1999. Cost Reduction Study for Solar Thermal Power Plants: Final Report prepared for The World Bank by Enermodal Engineering Limited, Kitchener, Ontario, Canada, May 5.

Frier, S., 2003. SEGS Overview: Presentation to Global Market Initiative, Palm Springs, CA (unpublished presentation).

Gee, R., Winston, R., 2001. A Non-Imaging Secondary Reflector for Parabolic Trough Concentrators. Unpublished Report (not public) to National Renewable Energy Laboratory, by Duke Solar Energy, Raleigh, NC.

Geyer, M., 1991. Thermal Storage for Solar Power Plants. Pp. 199-214 In: Winter, C., Rizmann, R., Van-Hull, L. (Eds.). Springer-Verlag, Berlin, Germany, Ch. 6 Solar Power Plants.

Herrmann, H., Graeter, F., Nava, P., 2004. Performance of the SKAL-ET collector loop at KJC Operating Company. Presented at the 12th SolarPACES International Symposium, Solar Power and Chemical Energy System, Oaxaca, Oax., Mexico, October 6-8.

Jenkins, A., Reilly, H., 1995. Tax barriers to solar central receiver generation technology. Pp 505-509 In: Stine, W.B., Tanaka, T., Claridge, D.E. (Eds.). Solar Engineering 1995, Vol. 1. Presented at the ASME/JSME/JESE International Solar Energy Conference, Maui, HI, March 19-24.

Kearney, D., Price, H., 2004. Assessment of Thermal Energy Storage for Parabolic Trough Plants. Presented in the Proceedings of the 2004 International Solar Energy Conference, July 11-14, Portland OR (ISEC2004-64144).

Kearney, D., Kelly, B., Herrmann, U., Cable, R., Pacheco, J., Mahoney, R., Price, H., Blake, D., Nava, P., Potrovitza, N., 2004. Engineering Aspects of a Molten Salt Heat Transfer Fluid in a Trough Solar Field. Energy, Vol. 29, Issues 5-6, pp. 861-870.

Kelly, B., Herrmann, U., Hale, M.J., 2001. Optimization Studies for Integrated Solar Combined Cycle Systems. Pp. 393-398 In: Proceedings of Solar Forum 2001, Solar Energy: The Power to Choose, April 21-25, 2001, Washington, D.C.

Kelly, B., 2001. Thermal Storage Oil-to-Salt Heat Exchange Design and Safety Analysis. USA Trough Initiative, Task Order Authorization Number KAF-9-29765-09 to National Renewable Energy Laboratory, by Nexant, Inc., A Bechtel Technology & Consulting Company, San Francisco, CA , November 30.

Kelly, B., 2002. Thermocline Storage Performance Studies, Storage Performance Studies. USA Trough Initiative, Thermal Storage Analysis, prepared for the National Renewable Energy Laboratory, by Nexant, Inc., a Bechtel-Affiliated Company, San Francisco, CA, September 30.

Kennedy, C., 2002. Review of Mid- to High-Temperature Solar Selective Absorber Materials. NREL/TP-520-31267, Golden, CO: National Renewable Energy Laboratory.

Kennedy, C., 2004. Selective Coating Development. U.S. DOE Solar Program - Parabolic Trough Research and Development, FY04 2nd Quarter Activity Status Report, NREL/Sandia Trough Tasks FY04 for U.S. Department of Energy submitted by SunLab (Doug Brosseau), Albuquerque, NM.

Kennedy, C., Swisher, R., 2004. Cost Analysis of SolarReflective Hard-Coat Materials Deposited by Ion-Beam-Assisted Deposition. Presented in the Proceedings of the 2004 International Solar Energy Conference, July 11-14. Portland, OR (ISEC2004-65112).

Kennedy, C., Terwilliger, K., 2004. Optical Durability of Candidate Solar Reflectors. Presented in the Proceedings of the 2004 International Solar Energy Conference, July 11-14. Portland, OR (ISEC2004-65111).

Price, H., Kistner, R., 1999. Financing Solar Thermal Power Plants. NREL/CP-550-25901, Golden, CO: National Renewable Energy Laboratory

KJC Operating Company, 2002. SEGS Acquaintance and Data Package, Boron, CA.

Laing, D., Steinmann, W., Tamme, R., Richter, C., 2004. Solid Media Thermal Storage for Parabolic Trough Power Plants. Presented at the 12th SolarPACES International Symposium, Solar Power and Chemical Energy System, Oaxaca, Oax., Mexico, October 6-8.

Leitner, A., Owens, B., 2003. Brighter than a Hundred Suns: Solar Power for the Southwest. NREL/sr-550-33222, National Renewable Energy Laboratory, Golden, CO. Report by Platts Research and Consulting, Boulder, CO., Subcontractor.

Lotker, M., 1991. Barriers to Commercialization of Large-Scale Solar Electricity: Lessons Learned from the LUZ Experience. Report SAND91-7014, Albuquerque, NM: Sandia National Laboratory.

Lunde, P., 1980. Solar Thermal Engineering: Space Heating and Hot Water Systems. John Wiley & Sons, New York, NY, p. 136.

Lüpfert, E., Neumann, A., Riffelmann, K.J., Ulmer, S., 2004. Comparative Flux Measurement and Raytracing for the Characterization of the Focal Region of Solar Parabolic Trough Collectors. Presented in the Proceedings of the 2004 International Solar Energy Conference, July 11-14. Portland, OR (ISEC2004-65157).

Lüpfert, E., Geyer, M., Schiel, W., Esteban, A., Osuna, R., Zarza, E., Nava, P., 2001. EuroTrough design issues and prototype testing at PSA. In the Proceedings of the ASME International Solar Energy Conference Forum 2001: Solar Energy: The Power to Choose, Washington, D.C., April 21-25, 2001, pp. 387-391.

Lüpfert, E., Zarza, E., Geyer, M., Nava, P., Langenkamp, J., Schiel, W., Esteban, A., Osuna, R., Mandelberg, E., 2003. EuroTrough Collector Qualification Complete Performance Test Results from PSA. Pg 1/8-1/8. Proceedings of the ISES Solar World Congress, Göteborg, Sweden.

National Renewable Energy Laboratory, 2003. Assessing the Potential for Renewable Energy on Public Lands. Report DOE/GO-102003-1704, Golden, CO: National Renewable Energy Laboratory.

Owens, B., 2002. Economic Valuation of a Geothermal Production Tax Credit. Report NREL/TP-620-31969, Golden, CO: National Renewable Energy Laboratory.

Pacheco, J., Showalter, S., Kolb, W., 2001. Development of a Molten-salt Thermocline Thermal Energy Storage for Parabolic Trough Plants. In: Kleis, S., Bingham, C. (Eds.), Solar Engineering 2001, Proceedings of the International Solar Energy Conference. Washington, D.C., April 21-25.

Paneltec Corporation, 2000. Structural Facet Fabrication, Phase II Manufacturability Study, by Paneltec Corporation, 1218 Commerce Court, Lafayette, CO 80026. Report Number BD-4142, Albuquerque, NM: Sandia National Laboratories.

Pilkington Solar International GmbH, 1996. Status Report on Solar Thermal Power Plants. ISBN 3-9804901-0-6, Köln, Germany.

Pilkington Solar International GmbH, 1999. Survey of Thermal Storage for Parabolic Trough Power Plants. Report NREL/SR-550-27925, Golden, CO: National Renewable Energy Laboratory.

Price, H., 2003. A Parabolic Trough Solar Power Plant Simulation Model. In: Proceedings of ISES 2003: International Solar Energy Conference, Hawaii Island, Hawaii.

Price, H., Hassani, V., 2002. Modular Trough Power Plant Cycle and System Analysis. Report NREL/TP-550-31240, Golden, CO: National Renewable Energy Laboratory.

Price, H., Hale, M.J., Mahoney, A.R., Gummo, C., Fimbres, R., Cipriani, R., 2004. Developments in High-Temperature Parabolic Trough Receiver Technology. Presented in the Proceedings of the 2004 International Solar Energy Conference, July 11-14, 2004, Portland, OR (ISEC2004-65178).

Price, H., Lüpfert, E., Kearney, D., Zarza, E., Cohen, G., Gee, R., Mahoney, R., 2002. Advances in Parabolic Trough Solar Power Technology. Journal of Solar Energy Engineering, Vol. 124, no. 2, pp. 109-125.

Reflective Energies, 2000. The Solar Trough Organic Rankine Electricity System (STORES). NREL Subcontract No. AAR-9-29442-04, Final Report, November 2000, by E. Prabhu, Mission Viejo, CA: Reflective Energies.

Reilly, H.R., Kolb, G., 2001. An Evaluation of Molten-Salt Power Towers Including Results of the Solar Two Project, SAND2001-3674, Unlimited Release, Printed November 2001, Albuquerque, NM: Sandia National Laboratory.

San Vicente, G., Morales, A., Gutiérrez, M.T., 2001. Preparation and Characterization of Sol-Gel TiO2 Antireflective Coatings for Silicon. Thin Solid Films, Vol. 391, number 1, pp. 133-137.

Sargent & Lundy, L.L.C., Consulting Group, 2003. Assessment of Parabolic Trough and Power Tower Solar Technology Cost and Performance Forecasts, prepared for Department of Energy and National Renewable Energy Laboratory, Final Report SL-5641, May 2003, Chicago, IL: U.S. Department of Energy.

Solel Solar Systems, 2001. Solartechnik Prufung Forschung: October 29, 2001—Solel Absorber Sample (SOL10111000Z), Beit Shemesh, Israel.

Solutia, 1999. Therminol® VP-1 Heat Transfer Fluid. Technical Bulletin 7239115B (Website: http://www.therminol.com).

Svoboda, P., Dagan, E., Kenan, G., 1997. Comparison of Direct Steam Generation vs. HTF Technology for Parabolic Trough Solar Power Plants-Performance and Cost. In: Claridge, D.E., Pacheco, J.E. (Eds.), Presented at the 1997 International Solar Energy Conference, April 27-30, 1997, Washington, D.C., pp. 381-388, sponsored by The Solar Energy Division, ASME.

Wendelin, T., Gee, R., 2004. Optical Evaluation of Composite-Based Reflector Facets for Parabolic Trough Concentrators. Presented in the Proceedings of the 2004 International Solar Energy Conference, July 11-14, 2004, Portland, OR (ISEC2004-65151).

Workshop on Thermal Energy Storage for Trough Power Systems, 2003. NREL (Website: http://www.eere.energy.gov/troughnet/documents/thermal_energy _storage.html)

Zarza, E., Hennecke, K., 2000. Direct Solar Steam Generation in Parabolic Troughs (DISS): The First Year of Operation of the DISS Test Facility at the Plataforma Solar de Almería. In: Kreetz, H., Lovegrove, K., Meike, W., (Eds.), Proceedings of the 10th Solar PACES International Symposium on Solar Thermal Concentrating Technologies, Solar Thermal 2000, Pp 65-70. Sydney, Australia, 8-10 March.

Zarza, E., Valenzuela, L., León, J., Hennecke, K., Eck, M., Weyers, H.D., Eickhoff, M., 2004. Direct Solar Steam Generation in Parabolic Troughs (DISS) Final Results and Conclusions of the DISS Project. Energy, Vol. 29, Issues 5-6, pp. 633-959, April-May 2004, (Website: www.elsevier.com/locate/energy).

Zhang, Q., Zhao, K., Zang, B., Wang, L., Shen, A., Zhou, Z., Lu, D., Xie, D., Li, B., 1998. New Cermet Solar Coatings for Solar Thermal Electricity Applications. Solar Energy, Vol. 64, Nos. 1-3, pp. 109-114.

Chapter 7

Solar Pond Technologies: A Review and Future Directions

by

Aliakbar Akbarzadeh*, John Andrews
Energy Conservation and Renewable Energy Group
School of Aerospace, Mechanical and Manufacturing Engineering
RMIT University, Bundoora East Campus
PO Box 71, Bundoora, Victoria, Australia

Peter Golding
Department of Metallurgical & Materials Engineering
University of Texas at El Paso
500 W University Avenue
El Paso, Texas, 79968 USA

Abstract

Solar ponds, specifically salinity gradient solar ponds, present a number of advantages compared with other solar thermal technologies. They are relatively simple in concept, use abundant natural resources, and can provide heat upon demand since they include innate storage. They utilize established chemical, civil, materials, mechanical and industrial engineering practices, adapted for practical use in harnessing renewable energy. Solar pond technology has been significantly advanced during the last 15 years with three international forums focusing on their science and development. This chapter reviews the advances that have occurred, especially during the past 20 years. Pond hydrodynamics are increasingly well understood, at least for ponds using common salt, and operational stability has been maintained in field-operated ponds over annual cycles. Demonstrations of a broad range of new technologies, systems and components have provided valuable operational test results and a firm basis for predicting the performance of larger-scale commercial installations. A variety of useful applications have been tested and their feasibility demonstrated, including

---

* *Corresponding Author. Email: aliakbar.akbarzadeh@rmit.edu.au*

industrial process heating, electric power generation, mariculture, biogas production, desalination, and chemical (in particular salt) production. Key research projects at the Royal Melbourne Institute of Technology (RMIT) University in Australia and the University of Texas at El Paso, USA, that have contributed to these advances are described. Solar pond developments in other institutions and countries are reviewed. Priority areas identified as requiring more research and development to realize more economically viable systems include heat engines for electrical power generation from solar ponds, in particular trilateral and Kalina cycle engines; lining techniques; the use of alternative salts to sodium chloride; salt recycling systems; solar pond–multistage and/or multiple-effect flash desalting systems; and integration of solar ponds into salinity mitigation and arable land reclamation projects, mariculture and biotechnology applications.

1. Introduction

Solar ponds occur naturally: they were discovered, not invented. Naturally occurring solar ponds–called heliothermal lakes in the limnological literature since solar radiation heats their saline lower layers to well above ambient temperatures–are found in a wide variety of locations, including Rumania, Venezuela and even Antarctica (Hammer, 1986). The solar pond concept first appeared in the scientific literature at the turn of the last century, as an explanation for the unexpected increase in temperature with depth observed in several natural salt lakes in Transylvania, a region of Rumania (Ziegler, 1898). It was realized that the density gradient in a shallow body of saline water, created by layers with increasing salinity with depth, can suppress convection and hence trap solar radiation in the bottom region of the pond. The temperature in this region thus rises and it acts as a heat store. Kalecsinsky (1902) was the first to suggest that artificial solar ponds could be harnessed as an inexpensive source of heat in a variety of applications while the sun is shining or not, in winter as well as summer.

Development of artificial solar ponds was initiated by Bloch in Israel in 1958 and continued by a group led by Tabor through 1966 (Tabor, 1981). Solar pond development in Australia commenced in 1964, and in the US in 1973 (Golding, 1981; Rabl and Nielsen, 1975). Since that time many research groups around the world have worked on solar pond science, performance modeling, monitoring the parameters of experimental ponds, and applying solar ponds for the production of heat in industry, electricity, and fresh water by desalination, and in salinity mitigation schemes. The largest solar ponds ever constructed–with a total area of 250,000 m^2 in the northern part of the Dead Sea–were successfully used to generate up to 5 MW of electrical power in the mid 1980s.

At near the half-century mark after the first engineering interest in solar ponds, and more than ten years on from three international symposiums that focused upon their science and development (Huacuz, Dominguez and Becerra, 1987; Folchitto and

Principi, 1990; Golding, Sandoval and York, 1993), it is timely that progress is re-viewed, and some suggestions proffered for future directions in this ever fruitful area of scientific and technological endeavor. In the present review we document advances, to encourage ongoing interest in solar pond science and development of the technol-ogy for practical applications. The review is offered in the spirit of encouraging debate about 'where to from here for solar ponds?'.

We first describe the work in our own countries, Australia (section 2) and the USA (section 3), with particular emphasis on that in our own academic institutions, RMIT University and University of Texas at El Paso. Researchers at both institutions have operated key solar pond research and development programs during the last 20 years. We then present a brief review of work on solar ponds in other countries (section 4), based upon contributions received from researchers involved. In conclu-sion (section 5) we draw on the foregoing material to present an overview of the current status of solar pond work, from research to commercialization, and offer some suggestions for future directions for this fascinating technology.

2. Solar Pond Advances in Australia

2.1 Solar Ponds in Australia

Research on solar ponds in Australia dates back to 1964, following a visit to the Commonwealth Scientific and Industrial Research Organization (CSIRO) by Harry Tabor from Israel. A small experimental pond was constructed at Highett, Victoria, and provided the first experience of establishment and operating a solar pond. In 1979 research into solar pond use for electricity supply to isolated communities in outback Australia began at the University of Melbourne, when Aliakbar Akbarzadeh left Iran to work with the university's solar energy research group led by William (Bill) Char-ters. Two small experimental rooftop ponds were constructed at the Parkville cam-pus. Based on the successful operation of these facilities Akbarzadeh and Charters received support from the National Energy Research, Development and Demonstra-tion Council (NERDDC) to establish two 900 m^2 solar ponds utilizing bittern (satu-rated salt brines) at the Laverton salt works of the Cheetham Salt company. These ponds—one lined and the other unlined—were operated for three years (Akbarzadeh, MacDonald and Wang, 1983; Magasanik and Golding, 1985). In 1983, a 1600 m^2 solar pond was constructed and coupled to a 20 kW$_e$ organic Rankine cycle engine at Alice Springs in the Northern Territory by Australian Solar ponds (Collins, 1985) with the support of a grant from the National Energy Research, Development and Demonstra-tion Council.

2.2 RMIT University Solar Pond R&D Program

R&D work on solar ponds at RMIT University, Melbourne, started in 1986 by setting up small solar ponds using above-ground swimming pools. Extensive R&D

work was conducted to investigate the effect of sloping walls on the stability of solar ponds (Akbarzadeh and Manins, 1988; Akbarzadeh, 1989). These investigations were supported by laboratory experiments where salinity and temperature conditions similar to those in solar ponds were created in small tanks. One of the main findings was that convective layers generated by side-wall heating mainly have a local effect only and cannot be a source of extensive instability in large-scale salt gradient solar ponds.

In 2000 a 50 m^2 experimental solar pond was constructed at the Bundoora Campus of RMIT. This pond was circular of diameter 8 m and depth 2.5 (Figure 1). The pond, partly above-ground, was equipped with an observation window through which water clarity could be observed. The gradient in the pond was maintained by feeding salt (NaCl) through a cylindrical salt charger to the bottom at a height of 80 cm from the pond floor (Jaefarzadeh and Akbarzadeh, 2002). Continuous surface washing using mains water supply maintained the salinity of the top convective layer at a low level. Floating rings were used to limit surface wave action due to wind, and hence reduce the thickness of the upper convective zone.

Attempts have been made to maintain clarity in this experimental solar pond using brine shrimps, *Artemia salina* and *Artemia fransiscana* (Jaefarzadeh and Akbarzadeh, 2002). The shrimps swim in the pond feeding on algal populations and detritus, which are the main sources of turbidity. *Artemia fransiscana* brine shrimps in particular have been found to improve the clarity so much that the bottom of the pond could easily be seen. Measurements indicated turbidity in the upper convective and gradient zones was generally less than 1 NTU. However, the population of brine shrimps decreased gradually to extinction. A possible explanation of this loss was lack of dissolved oxygen in the water. A further introduction of brine shrimps is under way under different conditions as they provide a very convenient and effective means of maintaining clarity provided their numbers can be sustained.

Research at RMIT is also examining further the suitability of bittern for use in salinity-gradient solar ponds. Bittern, a saturated salt solution incorporating $MgCl_2$, is a by-product of commercial salt production often present as a waste product at evaporative salt works. A 15 m^2 circular experimental bittern solar pond with a depth of 2 m has been constructed at Bundoora (Figure 2) (Chinn, Akbarzadeh and Dixon, 2003). The vertical pond wall is insulated using spray foam. The bittern was obtained from a salt works in Pyramid Hill, located 250 km north of Melbourne. The density gradient was established using Zangrando's (1980) technique and pond operation commenced in August 2002. The density and temperature profile development is presented in Figures 3 and 4 respectively. It can be seen that the bittern pond had a very thin upper convective zone (less than 20 cm), attributed to its small surface area and strong salinity gradient. A maximum temperature of 75°C has been found to be relatively easy to establish and maintain. No chemical or biological treatments have been performed on the pond to improve water transparency.

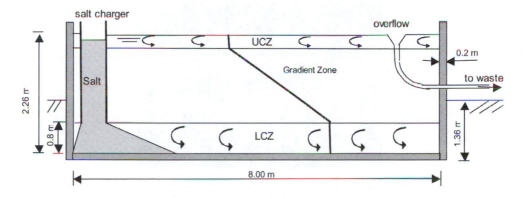

Fig. 1a.

Fig. 1b.

Figure 1. (a) Schematic view of the RMIT experimental solar pond with three zones, salt charger and surface washing system. (b) RMIT's experimental 50 m^2 solar pond at Bundoora, Melbourne. The west-facing window facilitates the visual observation of pond phenomena including water clarity at different depths. The floating rings and salt charger are clearly shown.

2.3 Demonstration of a Solar Pond Heating System for Salt Production

In early 2000, RMIT University, in partnership with two private Australian companies, Pyramid Salt Pty Ltd and Geo-Eng Australia Pty Ltd, began the 'Solar Pond Project' to demonstrate and commercialize solar pond systems for heating, electricity generation and combined heat and power. Stage 1 of the project has focused on solar ponds for industrial process heating, in particular the drying process in commercial salt production. The project partners received a $A550,000 grant over two years (Febru-

Figure 2. RMIT MgCl$_2$ Solar Pond. It has a total are of 15m^2 and is 2m deep. The walls of the pond were insulated using spray foam.

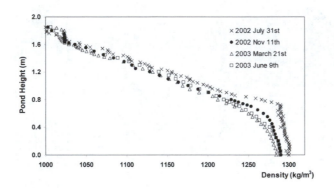

Figure 3. Development of the density profile in the RMIT bittern solar pond since the start of operation in 2002.

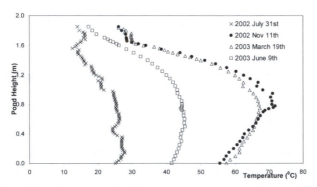

Figure 4. Development of the temperature profiles in the RMIT bittern solar pond since the start of operation in 2002.

ary 2000 to January 2002) through the Australian Greenhouse Office's Renewable Energy Commercialization Program (RECP) for this first stage of the project, which has aimed to make solar pond technology available in Australia and overseas as an economically attractive alternative for industrial process heating and other heating applications in regional areas remote from the natural gas distribution system.

It is planned that stage 2 of the Solar Pond Project, which has yet to commence, will extend the commercialization of solar pond technologies to electricity generation and the provision of combined heat and power.

The planned outcomes of stage 1 were as follows:

1. A demonstration solar pond supplying process heat to industrial processes at Pyramid Salt's site in Pyramid Hill, northern Victoria;

2. An evaluation of the technical performance and economic viability of solar ponds for process heating;

3. An assessment of the potential for solar pond process heating to reduce greenhouse gas emissions in Australia;

4. A heightened industry and public awareness about the commercial potential of solar ponds for supplying process heat, and reducing greenhouse gas emissions; and,

5. To be in a position to begin commercialization of solar pond design, construction and commissioning first in Australia and subsequently overseas.

In fulfillment of outcome 1, a demonstration solar pond supplying process heat for commercial salt production at Pyramid Salt's site in Pyramid Hill, northern Victoria, has been constructed and is now operating successfully (Figure 5). Its performance is being continuously monitored. The demonstration solar pond facility was officially opened by The Hon. Dr. Sharman Stone, MP, Parliamentary Secretary to the Federal Minister for Environment and Heritage, on Tuesday 14 August 2001. Gwen Andrews, Chief Executive, Australian Greenhouse Office, also spoke at the opening (Figure 6).

The demonstration solar pond has a surface area of around 3,000 m^2 and is just over 2 m deep. The pond is lined with Millenium liner supplied by Nylex and is insulated beneath the liner to reduce heat loss to the ground. The design of the pond was undertaken by RMIT University, in close collaboration with Geo-Eng using its civil and environmental engineering expertise.

The preparation of brines of various concentrations required in the construction of the salinity profile was carried out in Pyramid Salt's existing evaporation ponds. Pyramid Salt is an already established commercial producer of salt from saline groundwater (3% salinity) that is pumped to the surface as part of a salinity mitigation scheme. To maintain the salinity profile, concentrated brine is periodically added to the lower layer of the pond, and the surface is continuously flushed with bore water (3-4% salinity).

Figure 5. The 3,000 m² demonstration solar pond constructed at Pyramid Hill.

A grid of plastic rings floating on the surface of the pond is used to suppress wave action that could cause mixing and disturb the salinity gradient (Figure 5). Water clarity is maintained using brine shrimp that feed on the algae (section 2.2).

The heating system supplies hot air for use in the final crystallization phase in the commercial production of high-purity 'flake salt'. Heat is extracted from the pond by circulating fresh water through a heat exchanger located in the hot bottom layer, and then passing this heated water through a second heat exchanger 200 m away to deliver heat to the application (Figure 7). The heat extraction, transfer and delivery system was designed using a computer model developed by the RMIT Energy CARE group. Pyramid Salt installed the required ducting and heat exchanger in its salt production facility.

A monitoring system has been designed and installed by Geo-Eng Australia to measure and record the following main variables:

- Temperature at various depths in and beneath the pond;

- Heat flux under the pond;

- Groundwater pressure under the pond and adjacent to the pond area;

- Meteorological parameters such as dry bulb temperature, relative humidity, wind speed, solar radiation and rainfall; and,

- Instantaneous thermal power and cumulative energy delivered to the application.

Operation commenced in early 2001 when the first layers of saline solution were added. Process heat delivery commenced in June 2001. The maximum pond tempera-

Figure 6. The Hon. Sharman Stone, Member of Parliament, turns on the heating system at the opening of the Pyramid Hill solar pond in August 2001. Main group (from left to right): Sharman Stone, Peter Wood (Managing Director, Geo-Eng Australia), Professor Aliakbar Akbarzadeh (RMIT Energy CARE Group), Gwen Andrews (Chief Executive, Australian Greenhouse Office), Dr. Fouad Abo (Technical Director, Geo-Eng), John Ross (front, Director, Pyramid Salt), Dr. John Andrews (Manager, Solar Pond Project).

ture reached 55 ºC in February 2002, despite additions of cool saline solution infused to adjust the density gradient, and the fact that heat had been extracted regularly at rates up to 50 kW over a period of the proceeding seven months.

The average rate of heat delivery to the application was just over 40 kW during February 2002, with the maximum being 50 kW in mid December 2001. It is confidently expected that the design output of 60 kW (annual average) will be attained.

The analysis conducted in this demonstration project indicates that a commercial 1 hectare solar pond heating system at an appropriate location is likely to be economically competitive over 10 years compared to LPG as well as electricity from coal-fired ppower stations in Australia. The unit cost of heat delivered by such a system is estimated to be about \$A23/GJ.* This unit cost is under half that typical for electric heating, and just below that for LPG at \$A24/GJ. The corresponding simple payback periods are just over three years against electricity and just over seven years against LPG. Further cost reductions for heat supplied by solar ponds are expected as commercialisation proceeds. The prices of electricity, LPG or other fuels against which a solar pond heating system competes will of course vary from location to location, and will almost certainly increase over time.

---

*\$A = Australian dollar.

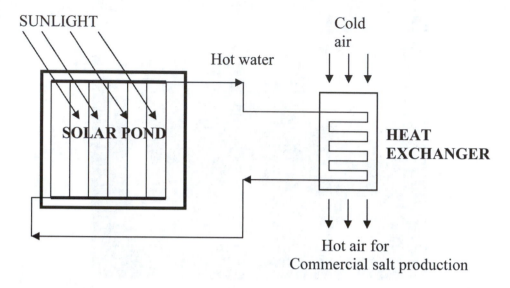

Figure 7. Schematic of a solar pond supplying heat for salt production, as at the Pyramid Hill solar pond.

The costs of constructing a solar pond heating system will also vary from site to site due to factors such as differences in terrain, soil type, the availability of brine/salt, the cost of flushing water from bores, and the distance between pond and heating application. Insolation will also vary. Therefore it will generally be necessary to conduct a site-secific economic evaluation of solar pond heating systems to make a realistic appraisal of economic viability.

The 0.3 hectare demonstration solar pond heating system at Pyramid Hill will reduce greenhouse gas emissions from electricity generation by almost 900 tons per year, and almost 18,000 tons over a 20-year lifetime. A one hectare pond will reduce emissions by some 3 760 tons/year. On a projection that over a five-year period from 2003 the annual rate of installation of solar pond heating systems rises linearly to 1 km^2/y in 2008, the annual greenhouse savings in 2008 would be around 874,000 tons per year, and the cumulative savings over the five-year period would be some 2 Mtons. On a more conservative assumption that there was a linear increase to constructing just 0.1 km^2/y of solar ponds, the annual greenhouse savings in 2008 would be around 87,000 tons per year, and the cumulative savings over the five-year period would be some 0.2 Mtons.

Work is continuing on the evaluation of the market potential and hence the greenhouse gas reduction potential of solar pond heating systems in Australia. Geo-Eng Australia Pty Ltd (MPW Development Pty Ltd) and Pyramid Salt Pty Ltd, in conjunction with RMIT University, are now in a position to offer a range of solar pond technologies and systems, and associated services, to prospective clients on a commercial basis.

2.4 Development of More Economically Competitive Heat Engines for Generating Electricity from Solar Ponds and Other Low Temperature Heat Sources

Solar ponds, along with a number of other renewable energy sources such as solar thermal collectors and geothermal sources, have the potential to supply low-temperature heat at competitive costs (Andrews and Akbarzadeh, 2002). However, to date there has been very little commercial exploitation of such heat sources for generating electricity, despite their abundance and the reduction in greenhouse gas emissions from burning primary fossil fuels that their utilization would yield.

Organic Rankine Cycle (ORC) engines have been developed specifically to produce electric power from lower-temperature heat sources (80-150°C) of the kinds just outlined. Thermodynamic performance is limited by the Rankine cycle, so designs have low net thermal-to-electric energy conversion efficiencies (~7 %), which adversely affects their economic viability.

In a solar pond–ORC engine system for combined heat and power, the solar pond supplies the heat for the ORC unit, and the hot water issuing from the evaporator of the ORC engine is still hot enough to be used for industrial process heating before it returns to be reheated by the pond. ORC engines have been employed to generate electricity from solar ponds in the Dead Sea, Israel (150-kW$_e$ and 5-MW$_e$, 1983–1990), at Alice Springs, Australia (nominal 15-kW$_e$, from 1984 to 1986), and El Paso, University of Texas (70-kW$_e$, 1986–present) as discussed in Section 3. Solar pond–ORC engine systems for power generation or combined heat and power would be especially suited for use in salinity mitigation schemes in regions adversely affected by high salinity. Power generated from solar ponds can not only be used locally for groundwater pumping required in salinity mitigation schemes, but also be supplied to the main grid and hence earn revenue that will enhance the economic viability of such schemes. The critical absent technology for solar pond power generation is an economical heat engine.

ORC engines have also been used to generate electricity from relatively low-temperature geothermal heat sources in Australia. In 1985, a 15-kW$_e$ ORC engine was installed at Mulka Station, South Australia, to generate power from hot water at 83°C; and since 1987 a 150-kW$_e$ geothermal-power station using an ORC engine has been run by CEB/ERGON Energy at Birdsville, Queensland, using geothermal hot water at 98°C. Enreco Pty Ltd, an Australian company designed and installed both these systems. Importantly ORC (or modified) heat engines have the further potential to generate power from lower-temperature heat currently wasted in power stations and industrial processes.

There is increasing research interest internationally in improving the net energy conversion efficiency and hence lowering the unit costs of power produced from heat engines using heat sources in the 60–100°C temperature range (Yamamoto et al., 2001; Smith, 1999; AGAM Energy Systems, 2001). Essentially three avenues of inquiry are being actively pursued.

The first is design changes in conventional ORC engines that reduce losses and increase the net power generated (Hung, 2001; Yamamoto et al., 2001). Such design changes include:

- Increasing the effectiveness of the evaporator and condenser, and use of counterflow preheating;

- Use of refrigeration and hydrocarbon working fluids (such as isobutane, n-butane and isopentane) including nonazeotropic mixtures of these;

- Improving the adiabatic efficiency of screw expanders and reducing thermal losses;

- Investigating optimum rotor tip speeds and oil injection/separation for screw expanders;

- Use of higher efficiency electric pumps;

- Minimization of losses through viscosity and turbulence in fluid flow; and,

- Reduction or elimination of parasitic energy inputs, by use of gravity-feed and lock-tank configurations.

In the latter case, a thermosyphon Rankine engine has been developed in which gravity is used to return the condensate back to the evaporator, thus eliminating the pump used in conventional Rankine cycle engines (Nguyen et al., 1994; Akbarzadeh et al., 2001). Several prototypes in the 100 W_e range using water as the working fluid have been constructed to demonstrate the concept, and a 2 kW_e is under test at RMIT University. Research is also under way to combine the use of a gravity-assisted thermosyphon Rankine engine with in-pond heat exchangers for the evaporator and condenser sections of the engine. In this passive configuration, heat is extracted from the lower convective zone and rejected to the upper convective zone. With water as the working fluid in the in-pond ORC engine, the height and thus pressure difference between condenser and evaporator provide sufficient head to allow the condensate to return to the evaporator without the need for an auxiliary pump.

The second avenue of inquiry is the use of different thermodynamic cycles such as the recently proposed Kalina cycle (Smith, R.W., 1996), the Goswami combined power/cooling cycle (Xu et al., 2000; Hasan and Goswami, 2002; Hasan and Goswami, 2003; Lu and Goswami, 2003; Tamm and Goswami, 2003; Tamm et al., 2003; Vijayaraghavan and Goswami 2003; Goswami et al., 2004; Tamm et al., 2004) and 'trilateral' cycles (Taniguchi et al., 1988; Smith, 1993; Smith and Pitanga Marques da Silva, 1994; Smith, Stosic and Aldis, 1996; and Smith, 1999) that promise to be more cost-effective than the existing ORC engine configurations.

In the Kalina cycle, the working fluid is a mixture of water and ammonia (Kalina, 1984). By varying the ammonia fraction of the mixture throughout the cycle, more efficient internal heat exchange is achieved through better temperature matching between the working fluid and hot and cold sinks (Dejfors, Thorin and Svedberg, 1998; Heppenstall, 1998). Smith (R.W.) (1996) has claimed that a Kalina cycle can achieve a net plant efficiency of nearly 59% – several percent greater than the best achieved so far by any Rankine cycle engine. In the trilateral cycle, the working fluid becomes a two-phase mixture of gas and liquid as it expands in a specially-designed two-phase expander, thus potentially increasing the adiabatic efficiency of the expander (see, for example, Smith, 1993; Smith, Stosic and Aldis, 1996; and Smith, 1999). Hence the net energy conversion efficiency of the overall heat engine is increased, with corresponding potential reductions in the unit cost of power generated. The Goswami combined power/cooling cycle uses ammonia/water mixture as the working fluid and absorption condensation which allows the expansion of working fluid in the turbine to temperatures much below ambient. This cycle gives a second low efficiency of over 60% for a heat source temperature of 360K, while producing power and refrigeration in the same cycle (Goswami, et al., 2004).

The third avenue of inquiry is employing direct flash evaporation of liquid heat sources (hot water or brines), together with novel condensing systems, such as the AGAM 'green engine' employing flash evaporation of hot brine and a brine-air condenser (AGAM Energy Systems, 2001).

Such systems employing flash evaporation of the liquid heat source (usually hot water or brine) and introduction of the consequent vapor (or two-phase liquid-vapor mixture) directly into the expander have the potential to reduce significantly the losses due to the temperature difference between the heat source and vaporised working fluid in a conventional evaporator. The Israeli company, AGAM Energy Systems (2001) have proposed systems of this kind, but working prototypes have not yet been constructed.

RMIT University is currently pursuing all three of these research avenues for developing more cost-effective heat engines capable of generating electricity from heat sources in the 60 to 100°C range, with solar ponds as a prime application. Anticipated outcomes include a small-scale prototype of an improved heat engine for economical power production from solar, geothermal, and waste heat sources that would be suitable for further development and commercialization. The target is to reduce the capital cost per kW of power generated from its present range of $A8,000 -10,000 to under $A6,000.

The successful development of a cost-effective heat engine for generating power from lower-temperature heat sources would yield benefits such as:

- more sustainable energy supply systems with reduced greenhouse gas emissions;

- support for regional development;

- improved economics of salinity mitigation; and,

- manufacturing and power industry development and employment.

2.5 Integration of Solar Ponds in Salinity Mitigation Schemes

Following European settlement of Australia, agricultural and other development has led to a major change in the vegetation of large parts of the land surface. Over large areas, native deep-rooted trees have been replaced by shallow-rooted pastures and legume crops. These shallow-rooted plants use less water than the trees, and thus allow more water to flow down through the soil into the groundwater system. In irrigation areas, water is diverted from rivers onto the land, thus increasing the volume of water reaching the groundwater system. Consequently land uses in Australia over the last two centuries have significantly changed the groundwater movement and storage patterns. Over time the groundwater level has risen, bringing salts previously stored deep in the soil profile to the surface. This process has resulted in salts accumulating in surface soils, inhibiting or preventing plant growth (Figure 8). The flow of saline groundwater directly into streams renders the stream water unfit for use.

More than $A130 million of agricultural production is lost annually from salinity in Australia, and dryland salinity is ruining 2.5 million hectares of land (National Action Plan for Salinity and Water Quality, 2003). Groundwater levels are still rising over large areas, and the area of land affected by salt continues to increase. The Commonwealth and State Governments in Australia are implementing a National Action Plan for Salinity and Water Quality, involving a commitment of $A1.4 billion over seven years for applying regional solutions to salinity and water quality problems. The aim is for all levels of government, community groups, individual land managers and local businesses to work together in tackling salinity and improving water quality.

A number of salt interception schemes involve the use of evaporation basins, into which saline groundwater is pumped. Solar radiation evaporates the water, leaving the salt behind. There are currently around a hundred evaporation basins in use in the Murray Darling Basin region in the state of Victoria, ranging in size from 3 to 2,500 hectares with depths ranging from 0.3 to 5 m.

Solar ponds could be incorporated in such evaporation basins to produce heat and/ or electricity from otherwise unproductive land (Akbarzadeh, Earl and Golding, 1987). If evaporation ponds are established in a chain, the first few ponds in the chain provide ideal opportunities for creating salt-gradient solar ponds. These first ponds are of low salinity, with a flow- through of water to maintain the required low surface salinity of a solar pond. The final pond in the chain can provide the source of highly saline brine to maintain the salinity of the bottom of the solar pond. This proposal is shown in Figure 9. It would thus be possible to establish solar ponds in up to half of the total evaporation basin without any loss of evaporative surface area. While the surface of the solar pond is acting as an evaporation surface, heat can be withdrawn from the bottom of the pond for use in power production and any other applications.

As evaporation basins are being built and used for salinity control purposes, the cost of heat and possibly electricity generated by a solar pond is reduced. The incremental basin construction costs for solar ponds would be small and salt is provided to the site at no cost. Burston (1996) performed a comprehensive study into the integra-

Figure 8. Agricultural land affected by salinity as evidenced by dead trees.

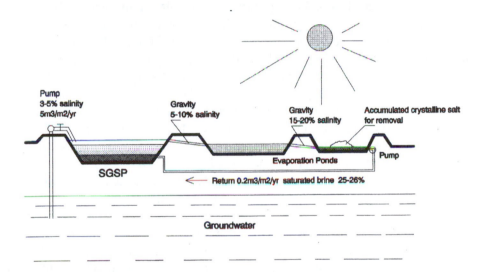

Figure 9. Integration of a salt-gradient solar pond into a salinity control scheme (Source: Burston, 1996).

tion of solar ponds in salinity mitigation schemes in Victoria, proposing a methodology for such integration, conducting case studies of pilot solar ponds at existing evaporation basins, and evaluating the economics of the concept.

Following on from the success of the Solar Pond Project at Pyramid Hill, which itself is located in a salt-affected area in northern Victoria a proposal had been developed for incorporation of solar ponds in future larger scale salinity mitigation schemes being planned in that area. It is estimated conservatively that 100 hectares of solar pond coupled to a 4 MW$_e$ generator could supply nearly 17 GWh of electricity annually. This is far in excess of the electricity requirements—mainly for pumping—of most salinity mitigation schemes so that excess power generated would be available for other local uses or feeding back into the main grid. Other alternatives would be to use the heat collected and stored in the pond in commercial salt production or for desalination using a multistage flash process.

3. Solar Pond Advances in the USA

3.1 Solar Ponds in the USA

Solar pond research and development began in the USA 30 years ago, with Ari Rabl and Carl Nielsen of the Ohio State University investigating space heating and grain drying applications. Experimental and practical studies were conducted of how ponds functioned and what limits might be expected on their performance (Rabl and Nielsen, 1975; Nielsen, 1988). Subsequently, several solar ponds were built in Ohio. Nielsen and associates operated two ponds at the Ohio State University, in Columbus, Ohio, for over twenty-five years. One was located at the site of an annual Farm Science Review. It was built specifically to demonstrate grain-drying. A brine-to-air heat exchanger transferred heat drawn from the pond to dry corn stored in silos. Another 400-m^2 circular pond was built adjacent to the university dairy. Many important learning experiences were provided at these two facilities. The results achieved at these and other pond projects in the US have been detailed in a review of advances in solar pond technology (Nielsen, 1988), and a comprehensive treatise on salinity-gradient solar ponds (Nielsen, Hull and Golding, 1989).

During the early 1980s a number of solar ponds were constructed and operated in the US, including significant projects operated by federal and state government authorities. The Tennessee Valley Authority operated a 3,700 m^2 pond in Chattanooga, Tennessee, and the California Department of Water Resources operated twin solar ponds in Los Baños, California (Engdahl, 1987). The latter were designed to demonstrate solar pond use as part of a desalination system, and they included small-scale tests of electric power production in the US, using an ORC engine using a screw compressor, and bench-scale electric power driven vapor-compression desalination units. These large government-funded projects often yielded important operating experience and identified problems to avoid. Meanwhile, new advances in solar pond technology development had been occurring, including research at The University of New Mexico in Albuquerque, and The University of Texas at El Paso (Zangrando, 1979; Zangrando and Bertram, 1984; and Reid, McLean and Lai, 1985). In particular, the El Paso solar pond project, initiated in 1983, has since provided a research-bed for many important studies. Together, the Ohio State University solar ponds and the University of Texas at El Paso solar pond have provided the bulk of the advances in understanding and operating of solar ponds in the US to the present time.

3.2 The University of Texas at El Paso Solar Pond Project

3.2.1 Major Milestones

The University of Texas at El Paso (UTEP) Solar Pond Project began in 1983 with the US Bureau of Reclamation (USBR) agreeing to fund an unsolicited proposal from Robert Reid, then Chairman of Mechanical & Industrial Engineering Department at UTEP. The first solar pond with an area of 3,700 m^2 was established at a food processing facility owned by Bruce Foods, and located 20 km northwest of downtown El Paso, in a former fire-protection reservoir for an adjacent clothing factory (Reid et al., Swift, 1985). This solar pond project has now been conducted by UTEP for 20 years, with Bruce Foods (best known for their *Casa Fiesta* brand of Mexican foods), the State of Texas, and the USBR providing the key funding support.

Major rebuilds of the original *Hypalon*-lined pond occurred in 1984 (new *XR-5* membrane liner and adjustable steel pipe-constructed heat extraction equipment), 1991 (*in situ* cleaning of pond brine), 1992-95 (polypropylene sidewall lining and lower region geo-synthetic clay-based lining installed and tested, new heat extraction and monitoring systems), and 1996-97 (new polypropylene liner installed and new gradient-modification systems).

Major application milestones have been demonstration of:

- process heat delivery, in 1984;

- the first grid-connected electric power production from a solar pond in the US using an Ormat ORC engine-generator, in 1985;

- the first desalting of brackish water in the US using a *Spinflash* falling-film multi-effect multi-stage evaporation system, in 1987;

- year-round high-performance operation of the solar pond, in 1989;

- solar pond-coupled desalination using multi-stage flash evaporation in 1991;

- vapor-compression desalting powered by electricity from an ORC engine in 1992;

- solar pond driven biogas generation from food processing wastes in 1995; and,

- thermally-driven membrane distillation in 1991.

In 1993, connectivity for heat exchange and transfer to the Bruce Foods food processing plant was reinstated, to provide industrial process heat for cooking and canning operations, displacing natural gas that is the traditional source for generating industrial heat. A recent photograph of the El Paso solar pond is shown in Figure 10.

Figure 10. The 3,700 m² solar pond located at Bruce Foods in northeast El Paso, Texas, and operated by at the University of Texas at El Paso.

3.2.2 Technological Advances

Major technological advances achieved over the 20 years of the UTEP solar pond project include:

- Advancing solar pond project design, operation and management in all its various aspects;

- Inventing scanning injection technology and using it to successfully build, alter and maintain salinity gradient profiles, and clean a pond *in situ;*

- Creating new methods for regular monitoring, automated control and stable operation;

- Demonstrating the first solar pond to operate in a closed salt-cycle (zero discharge) system;

- Demonstrating leadership in the US in solar pond technology and practical applications for community-public use and for industrial, manufacturing and commercial enterprises; and,

- Providing reliable, long-term experience on solar pond operations, incorporating continuity of projects and personnel that compounds learning and knowledge.

In addition to these direct advances achieved in solar pond technologies, there have been several spin-off research and application achievements. Notable examples include the use of solar ponds as salt repositories in inland desalting applications; the use of solar ponds within the process for refining of chemicals produced through solution mining; and the potential use of so-called 'produced water' (oil well produced brines) in petroleum recovery.

3.2.3 Continuous High Performance Operation

By far, the greatest contribution of the UTEP project to advancing solar pond technology has been determining the equipment, procedures and practices needed for high-performance operational reliability. Previously, operational reliability at high pond temperatures (that is, above 75°C and below the boiling point of the storage-zone brine) had seldom been achieved continuously. Highest recorded temperatures in newly-installed facilities around the world had always climbed impressively initially, but then deteriorated, rapidly on occasion, within months of being achieved. Experienced operators in Israel (specifically at En Boqeq and Bet Ha Arava) and in Australia (specifically at Alice Springs) did achieve relatively high operational performance for periods of months, but could not sustain such performance on an annual basis. If they had done so, it would no doubt have been reported.

High-performance operation at the El Paso solar pond was achieved during 1989 and maintained for a period of three years, with storage zone temperatures being maintained above 80°C year-round. It ceased then only because of liner failure. During the summer of 1992, a maximum electric power output of 120 kW$_e$ was achieved using an Ormat ORC engine coupled to the Salinity gradient Solar Pond (SGSP). The unit was installed in 1985 and demonstrated grid-coupled electric power production for the first time in the US. The Ormat unit was a 350-kW$_e$ nominal unit downgraded for operation at solar pond temperatures. As such, it had large heat exchangers ideal for rapid removal of heat from brine.

The ORC unit was inadvertently pivotal in the success of maintaining performance, since it was able to remove heat at a rate far in excess of the total input power from incident sunlight (around a maximum of 840 kW reaching the storage zone on a summer day), and enabled storage zone temperatures to be readily controlled. The capacity to withdraw energy rapidly prevents the possibility of thermal runaway (boiling). There is a need to have back-up systems in place for all key functions – solar pond monitoring, salinity profile modification, recharge of storage zone salt, pumping in and out of the pond and related brine management (evaporation) ponds, and heat extraction capability – are all of paramount importance. These systems provide a level of safety in operation appropriate to well-engineered facilities. At El Paso, should the ORC engine have failed there were three other systems available for heat removal:

industrial process heat exchange, delivery of heat to desalting equipment, and finally, in case of emergency, evaporator-pond recycling.

Reliable high-performance operation requires reliable monitoring of the key solar pond system data, and instrumentation is a key part of the system. Advances in monitoring technology and methods at the El Paso solar pond enable accurate solar pond profiling to be readily achieved. Pond operating status is determined by carefully and accurately monitoring salinity and temperature profile data, and it is also important to monitor water clarity. A highly automated monitoring system is in place, but backed-up by semi-automated monitoring systems, and manual monitoring of key data is also possible.

Salinity monitoring is a perfect example. Samples extracted from the pond can be analyzed for salt content in three alternative ways: (i) fed into an industrial-style Dynatrol density meter for (automated) measurement of salt concentration (the mV data being fed into a data logger attached to a computer; (ii) fill a hand-held Anton Par density meter for immediate (semi-automated) measurement of salinity (compensated for temperature) via a digital display (3-and-1/2 digits), and (iii) float calibrated hydrometers [4 x 0.5000 specific gravity (SG) range across SG 1.000 to 1.2000] providing for accurate (manual) monitoring by hand and eye. The first step allows for regular automatic monitoring. It utilizes samples withdrawn through a 0.2-cm gapped cylindrical diffuser positioned by a vertical drum-scanning mechanism that enables needed spatial resolution (samples are routinely measured at 5 cm intervals and at 2.5 cm intervals if needed, or even 1 cm for fine detail studies). The second step can be readily used to test samples provided by the same mechanism. The observer easily records the output. The third requires a half-litre sample to be collected and stored for later use. It typically requires several hours of meticulous lab work to achieve excellent (5 cm profile) salt content results. The benefit of having multiple systems in place and standard methods for measurement, recording and analysis cannot be overemphasized.

Parameters regularly measured include salinity, temperature, turbidity and pH profiles, all obtained via automatic monitoring equipment assembled and tested at the El Paso solar pond. Lu (1994) has detailed instrumentation system research and development, and the practices used.

3.2.4 Analysis and Prediction of Internal Stability

The ability to analyze, predict, and when necessary adjust stability within the gradient region of an operating solar pond is a vital advance. The El Paso solar pond uses common salt, like the RMIT pond at Pyramid Hill. Xu, Golding and Nielsen (1987) introduced the concept of a stability margin (SM) number for calculating internal stability (maintaining non-convection) in a sodium-chloride salinity-gradient solar pond. The SM number is a ratio of locally existing salinity gradient strength and the salinity gradient strength required to satisfy the dynamic stability criterion, as established by

Weinberger and discussed further by Zangrando (1979), and Zangrando and Bertram (1984). This ratio can be mathematically expressed:

$$SM = [dS_{measured}/dz] / [dS_{required}/dz]$$

where $[dS_{measured}/dz]$ is the measured salinity gradient (in units of %-salt by weight/ m) as computed from measured adjacent values of salinity and $dS_{required}/dz$ is the indicated theoretical salinity gradient value (in units of %-salt by weight/m) required to satisfy the Weinberger dynamic stability criterion, for the existing temperature gradient in the locale at height z, within the stratified gradient-zone region under consideration.

In principle, the SM number must be >1 in order to sustain local stability in the gradient zone. However, Xu, Golding and Swift (1991) found that in practical operation gradient dissolution (commonly called 'breakdown') can be anticipated, and occurs when the calculated SM number declines to a value <1.6 at a given locale in the case of the El Paso pond. The variance between theory and practice may be attributable to wall effects, wind-driven locomotion or similar phenomena, or other subtle physical perturbation effects that may be present. Whatever the cause, the practical result is that the gradient becomes unstable and breaks down when the SM number falls below 1.6.

Given this behavior, the local SM number can be utilized to establish an operational safety limit. If the margin approaches a predetermined safety limit, corrective measures must be used to maintain the internal stability within the gradient zone. At the El Paso pond, an operational safety limiting value of 2.5 has been used very successfully. If the SM number approaches this limit, timely action (ideally within days rather than weeks) can be initiated (as described below) to adjust the local salinity (or temperature) profile to maintain internal stability within the gradient zone, and hence sustained operability.

The ability to collect credible data on salinity and temperature profiles, described above, provides the data needed for gradients to be accurately estimated.

An example of a typical SM number plot is shown in Figure 11 to illustrate the differences between the results obtained. Although errors resulting in scatter of the SM number are reduced by the curve fit method, care must be exercised in interpreting the data fits. Curve-fitting might hide actual (real) discontinuities in the SM number profile, smoothed over erroneously through application of the curve fit algorithm. For this reason, it is highly recommended that both the straight-line and curve-fit methods be used to analyse stability in critical situations. In the absence of gradient modifications or local heating, diffusion dominates as the physical salt transport and heat transfer mechanism, and stability margins are expected to be smooth. If this is the case the curve-fit method works well. However, during or just after gradient modification (using injection or extraction of brine), or when there is local heating or cooling (induced by reduced local clarity or heat removal), the straight line method can effectively provide useful information to analyse the local stability of a specific region.

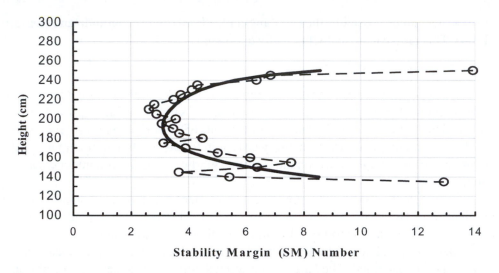

Figure 11. Stability Margin (SM) number comparison showing the scatter of computed data when using the straight line method for the El Paso solar pond. The minimum SM is located centrally in the gradient region. From the graph we can infer that the lower gradient-zone boundary is in the vicinity of 140-cm height, and the upper gradient-zone boundary is in the vicinity of 250 cm, on this occasion during summer 1999 when the total pond depth was typically ~300 cm.

3.2.5 Adjustment of Local Stability Using a Scanning Injection Technique

Careful attention to design and installation of the salinity profile, coupled with establishing and maintaining pond clarity through to the storage zone (enabling penetrating solar radiation to reach the storage-zone region), provide for optimum operating performance. Experience demonstrates, however, that operators need to be prepared for worst-case scenarios. Akin to installing sprinkler systems in buildings, one does not anticipate using them to put out a fire but they provide an accepted, necessary level of safety. With a solar pond, the SM number allows operators to see if there is cause for alarm. A localized SM number <2.5 warns the operator that the salinity gradient in that localized region is insufficient to inhibit convection (driven by the destabilizing temperature gradient) and the need for gradient modification is indicated. This may be achieved by salinity gradient strengthening (by scanning injection at the El Paso pond); by weakening the temperature gradient (by scanning injection or scanning withdrawal); or, in extreme cases, by removal of heat from the storage zone.

If the margin of dynamic stability (and by default, static stability) at some locale within the gradient is allowed to decline (say to SM<2), destabilization and 'instability' occurs, resulting in a tendency for internal convective layers to form. In extreme cases, when operating at high performance, the breakdown of the entire gradient can occur, usually beginning at sloping walls (Nielsen, 1980; Akbarzadeh, 1984; Akbarzadeh and Manins, 1988).

The ability to adjust the local stability by modifying the salinity gradient is another important advance achieved at the El Paso solar pond. It has been found from experience that salinity gradient modification via scanning injection (or extraction) is the most effective method. Scanning injection was invented in the UTEP solar pond project. Graduate student Robert McElroy, constructed the first scanners, and another graduate student, Venegas (1992), who made major contributions to the project after joining UTEP from the Division of Non-Conventional Energy at the Electric Research Institute of Mexico, reported the results of the method and application. The first simple scanner used a gear-reduced winch from a four-wheel drive vehicle, driven by a 12-V DC battery, and mechanically-tripped switches were used to change the direction of motion at operator-set extremities along the cable holding the diffuser.

The scanning injection technique offers several advantages over fixed-level injection. Primarily, in practice, it produces a smooth change to an existing profile. This occurs because the perturbation being introduced by injection is not limited to one (ideally horizontal) intrusion. Instead, the injected brine is assimilated across a vertical region. A second very significant advantage is that large injected-fluid volumetric flow rates can be achieved, so the change can be caused in a timely fashion within a large pond.

Prior to the introduction of this method, slow, meticulous and laborious injection methods were the only understood techniques for internal gradient adjustment (Nielsen, 1983). These methods correctly recognized the importance of the injection diffuser design, and fluid properties (and the change in fluid properties with depth in a solar pond) provided challenges for designers and operators (Zangrando and Johnstone, 1988; Liao, Swift and Reid, 1988). The Froude number is the critical thermo-fluid property to control when practicing injection in a double-diffusive system (Liao, 1987).

The scanning injection technique has the advantage of limiting sensitivity to Froude number. All that is required is that a Froude number of ≥ 18 be achieved to ensure adequate mixing of the injected fluid upon entry. This is accomplished by maintaining a high injection flow rate.

Diffuser design for scanning injection has also been advanced and designs simplified considerably. At the El Paso solar pond, early diffuser designs incorporated circular plates with lateral spacers to maintain even gap width. A design injection flow rate of 0.189 m^3/min (50-gpm) was first tested, this was successfully doubled to 0.378 m^3/min (100-gpm) to improve the time required to achieve a modification. Given available pumping capacity and the size (volume) of the EPSP, a design fluid flow of 0.567 m^3/min (150-gpm) is currently used. To distribute this flow, while maintaining a Froude number = 18, the latest diffuser design utilizes a 10-cm (nominal 4-inch diameter) PVC pipe in which is cut 14 slots, each 1.2-cm^2 in area. The diffuser, represented in Figure 12, is fed through a central tee. The design flow rate is 0.568 m^3/min (150-gpm), with the injection fluid velocity being ~ 2.8 m/s (8.5 ft/sec).

The latest diffuser is located under a pier jutting into the reconfigured 3000 m^2 pond, and is raised and lowered using a drum-cable system, powered by a DC motor. A precision potentiometer is coupled to the winch drum to provide vertical position

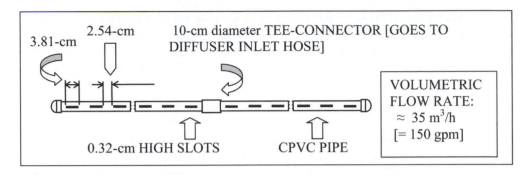

Figure 12. A schematic front-view of the features of the current plastic pipe injection diffuser design *circa* 2003 routinely used for gradient injections at the El Paso solar pond.

sensing. The calibrated position of the scanner, the range and direction of motion (upward and downward), are controlled via computer and associated relay circuitry. When the diffuser, moving in a given direction, reaches the preset limit points, the computer-controller reverses the direction of the motor automatically. The range of the injection scan is entered by an operator. The use of scanning injection techniques in two gradient failure and repair events at the El Paso pond is reported by Xu, Swift and Golding (1992). The same scanning injection diffuser system used for gradient modification purposes can also be used to establish the salinity gradient in the pond (Lu and Swift, 1996).

3.2.6 Control of Gradient Zone Boundaries

Controlling the position of zone boundaries is necessary to maintain operational performance. In the absence of monitoring and control, the gradient of a well maintained pond system slowly (of the order of cm/month) shrinks, as a result of erosion of the upper or lower boundaries, either individually or concurrently, in the absence of preventive or corrective measures. The optimum performance is achieved with the thinnest possible upper convective zone and a specified gradient thickness that depends on internal stability criteria, heat extraction and the application temperature.

Boundary interface motion has been carefully studied at the El Paso solar pond. The ratio of the molecular diffusivities of salt (κ_s) and heat (κ_t), is the Lewis number ($\tau = \kappa_s/\kappa_t$), and provides a suggested equilibrium density stability ratio (see Zangrando and Fernando, 1991). According to their model, boundary stability is provided for when the density ratio, Rρ equals $\tau^{-1/2}$. For the lower boundary of the main gradient of the El Paso solar pond, the practical operating equilibrium condition is found to be ≈ 6 ± 1. If Rρ is not in this range, corrective action such as heat extraction, injection or extraction of brine, prevents boundary migration. The upper boundary of the main gradient at the pond, if protected by a slab of fresh water, has been found to be in equilibrium when Rρ ≈ 16 ± 1. When there is surface layer protection the equilibrium condition occurred when Rρ ≈ 19.5 ± 1.5.

The equilibrium position is determined by the position of the heat extraction diffuser. The control of the lower boundary movement is normally accomplished by controlling the heat extraction rate. Optimum pond operation, in which the temperature is a maximum for the solar input and heat removal rate, is achieved with a gradient thickness where the temperature gradient approaches zero at the lower boundary. When this condition is met the boundary erosion driven by thermal convection is small and the gradient boundary adjusts to a level only slightly above the level of the heat extraction diffuser.

Upper-zone boundary control has been regularly achieved using two procedures. The climate at the El Paso location is characterized by a windy season during March through May each year. During that time protection of the upper boundary of the main gradient has been achieved using a small secondary surface gradient zone, added specifically for the purpose. The mini-gradient is established by floating a layer of fresh water on top of the pond before the windy season commences. This surface gradient zone gradually disappears during the windy season, driven by evaporation and wind-mixing. A second procedure used to control upper boundary level is surface zone redistribution. This procedure involves partially siphoning out the surface zone and rebuilding the desired depth by appropriate fresh water injection, using the scanning techniques described above.

Finally, it will be noted that the control of zone boundaries is more complicated at the El Paso solar pond where a closed salt cycle is maintained. Surface washing, possible at some sites, is not practical, where evaporation pond area is limited to about a half of the total pond surface area. This is not the case at the Pyramid Hill solar pond site, discussed in section 2.

3.2.7 Attaining and Maintaining Pond Clarity

At the El Paso solar pond the salt source was crystalline salt transported from the Carlsbad area of New Mexico. The salt was mixed with potable ground water, the resulting brine being of high transparency. This is advantageous, for the higher the transparency the better the performance. The maintenance of high transparency has been maintained through controlling biological growth. The control method consists of monitoring the pH, described earlier, and adding hydrochloric acid in regions of the profile where values approach 5.5 ± 0.5. Algal blooms have been observed to appear in the pond when $5.5 \leq pH \leq 8.0$. Acidification provides for low maintenance and is reliable for inhibiting biological growth blooms. Turbidity measurements provide the means for monitoring and controlling the quality of brine clarity, and typical values are ≤ 0.5 nepthalometric turbidity units (NTUs).

During the years 1984 through 1991, the heat extraction system at the El Paso solar pond was carbon steel-based. Diffusers were also constructed of metals. This situation was discontinued when the pond was dismantled after leakage was detected and subsequently led to relining of the pond (see below). In 1991, *in situ* clarification of the pond gradient was undertaken and was very successful, the stable and non-convecting gradient proving to be a bonus for flocculation. The ability to clean a pond without halting operation was first tested by Golding and Nielsen at the Farm Science Review pond in Columbus, Ohio and the application to a large-scale pond is another significant UTEP achievement. The excellent results achieved demonstrate a useful back-up method for retrieving clarity in the event that unexpected events cause deterioration to an inoperable situation.

Chemical application for maintaining clarity at the El Paso solar pond contrasts with the 'natural' approach used at the Pyramid Hill pond in Australia. As discussed earlier, the use of brine shrimp has proved valuable as a harvesting clarifier in that case (Jaefarzadeh and Akbarzadeh, 2002). This eliminates the need for adding chemicals, provided the pond conditions allow for maintenance of an active and prolific brine shrimp population.

3.2.8 Salt Inventory Analysis and Methods for Recycling Salt

At El Paso salt balances are regularly performed, as part of a process of maintaining salt inventory. The solar pond is located far from the closest commercial sources of salt and brine (~400 km), so salt or concentrated brine is a valuable commodity. The total inventory at the site is of the order of 2000 metric tons, half being used in the currently configured pond and the remainder in evaporation and concentration ponds located adjacent to the main site. Monitoring a salt balance is necessary for recycling planning, and also important in detecting leakage rates. Recycling methods involve batch removal of surface zone brine (two or three times a year) and re-concentrating the brine for re-injection into the storage zone, or for use in scanning re-injection, should it be deemed necessary (see Section 3.2.5).

Methods of concentrating brine have been tested, including a method developed by Ormat in Israel using spray evaporation, use of vertically-hung wetted netting and traditional pan evaporation. The latter is the simplest but requires a considerable investment in area and particularly lining cost, especially in the El Paso location where the sandy desert soil is permeable. Studies of reconcentration schemes have been conducted in Australia by Alagao, Akbarzadeh and Johnson (1994).

3.2.9 Containment, Lining Materials and Leak Detection

Operations at the El Paso solar pond have been interrupted throughout 20 years by the failure of lining systems. An original Hypalon (butyl rubber) liner pre-existed in the firewater pond that became the solar pond. It was replaced because of seam failure. When a liner fails, it is often at the seams or in places where the liner is creased. A woven synthetic liner with a Dacron polyester core, with the trade name XR5, was installed in the pond in 1994 and lasted through seven years of service. Failure was not caused by wicking but by mechanical failure. Slumping along the walls of the sloping pond embankments caused stresses and many of the failures were observed to occur in those regions. Leak detection methods, using techniques developed by Thomas Owen at the Southwest Research Institute in San Antonio, enabled mapping of electric and magnetic field strength patterns at the liner-fluid interface.

These showed that multiple leaks existed. Over 100 leaks were located in the pond, particularly in the region of slumps located at the top of the storage zone region. The failure of the liner (as low as 10 % of the initial tensile strength), which caused the loss of substantial salt inventory, ended the demonstration of high-performance year-round operation, and caused a shift of focus to lining technologies and their potential use in solar ponds. Incidentally, the apparent downward movement of the salinity profile (akin to a 'falling pond', see Nielsen 1975) was the key to realizing the pond was leaking, and regular maintenance of a salt balance was performed and confirmed leakage that led to pond shutdown. This reiterates the importance of keeping good records and analyzing data from an operation facility.

Interest in buried liners and combination plastic – earth (clay) lining systems by the US Bureau of Reclamation (USBR), followed the use of one such system in the largest artificial solar pond constructed to date: the Bet Ha Arava pond in Israel. A geo-synthetic clay lining system was installed in a reshaped containment in the EPSP and operations resumed in Spring 1995. Geosynthetic clay liners (or GCLs) are a new subset of so-called geosynthetic materials. They are manufactured in rolls from factory- fabricated thin layers of bentonite clay sandwiched between two geotextiles or bonded to a geomembrane. Structural integrity is maintained by needle punching, stitching or physical bonding. They are often used as a composite component beneath a geomembrane or by themselves as primary or secondary liners. The GCL offered several potential advantages over membrane liners. GCL is self-healing and puncture-proof with easy installation and predictable permeability. The El Paso solar pond was reconstructed with a GCL system during 1994-95. Before installing the new lining, the pond shape was modified by tamped fill to produce smooth walls transitioning to the floor, reducing the storage zone volume, and in so doing the surface area was reduced to 3,000 m^2.

A GCL provided by Gundle Lining Systems was installed in the lower half of the pond and a polypropylene side-wall lining was used. A drainage system was constructed beneath the pond to measure losses. A percolation rate of 2×10^{-6} cm/sec was determined. For comparison, a typical percolation rate for tamped earth is 10^{-8} m/sec ($\equiv$ to a brine loss of 0.4 m/year). This is six to eight times the transport rate by upward diffusion through the gradient zone and is unacceptable (Hull, Nielsen and Golding, 1989). It is equivalent to a loss of 0.8 m of gradient depth per year, with a lower zone salt concentration of 300 kg/m^3, and would carry away about 240 kg/m^2 of salt per year. Within two years the sidewall lining (0.1016 cm thick polypropylene) also failed on the banks above the water line. The entire pond was relined in 1997 using 0.1524 cm polypropylene. The lessons include the lining of solar ponds is never to be taken for granted, and leaks of some kind are expected. Designers need to take this into account when developing plans for future large pond systems.

3.3 Applications Specific To High-Performance Solar Ponds

3.3.1 Electric Power Production Using ORC Heat Engine

The production of electric power from solar ponds in the US has been demonstrated at Los Baños and at El Paso (Engdahl, 1987; Hull, Nielsen and Golding, 1989). An Ormat ORC engine has been operated at the El Paso solar pond for over 15 years. Apart from regular maintenance of seals and bearings, the only overhaul required in that time has been the dismantling and cleaning of the large tube-in-shell heat exchangers. Another issue, not apparent in 1985 when the engine was first operated, is its use of R-114, an ozone depleting substance, as the working fluid.

Performance has always been reliable and the engine is easy to start and stop, so that has been possible to use it for providing peaking power (nominally net 70 kW$_e$) to the adjacent Bruce Foods plant. Since the electric tariff is based upon peak power consumption, electric utility bills have been significantly reduced by running the ORC engine during the middle of the day when most power was being consumed. Peak-power operation was also practiced at the Bet Ha Arava pond in Israel, and is one economically feasible use of solar-pond generated electricity.

The ORC engine at El Paso also provided a ready means of extracting heat on demand, which is effective for control of the lower zone-gradient zone salinity boundary. On the basis of operational practice, the maximum output from the engine increases above 100 kW$_e$ when the available temperature difference (which determines thermodynamic efficiency) is at least 60°C. Continuous operation of the engine caused the lower zone temperature to be reduced, since the solar input is less than the extracted output. This design allowed the engine to be used as a control device during high-temperature storage zone testing (> 90°C) and avoid 'run away', the situation where excellent thermal performance leads to boiling in the storage zone of the pond.

3.3.2 Desalination using solar ponds

Desalting of brackish water (specifically groundwater sources in El Paso) is one of the major applications of interest in the El Paso solar pond project. The technology for separating water from dissolved solids is ancient (evaporation) but requires energy to drive the process. The use of solar energy to drive desalination process equipment enables a plant to be self-sufficient for energy. In the case of a solar pond it is also possible to utilize the by-product (usually concentrated salt solutions or brines) to construct additional solar pond capacity, thus supporting more energy for increased desalting (see section 2.5). Solar pond-powered desalination is one of the most promising renewable energy systems for producing large quantities of potable water.

In 1987, UTEP demonstrated groundbreaking solar pond-powered desalting in the US using Falling-Film, Multi-Stage (FFMS) Thermal Evaporator Technology. Advanced designs provided excellent results from two novel 24-stage falling-film evaporators featuring a rotating flasher stack. These units, known as Spinflash systems and designed and built by Ferris Stanniford under contract to the USBR, featured rotating flasher stacks (Swift et al., 1990). The thermodynamic performance of the units was excellent. Unfortunately the mechanical systems, specifically seals between stages, required substantial maintenance. These extremely compact systems produced 15,000 to 20,000 l/day output, using electricity provided by the ORC engine, and heat provided from the pond via heat exchange. Further, the surface of the pond provides an ideal cooling water heat sink. In March 1990, a second Spinflash desalting unit began producing pure water from brackish sources on a 24-hour basis.

The use of a more rugged multi-effect multi-stage (MEMS) flash evaporation unit (featuring 4 stages and 3 effects) began with installation and operational testing during 1990 through 1992. Unlike conventional evaporators that use vacuum pumps, the Licon Industries unit tested used jet pumps (eductors) to produce evacuation. The evaporator was constructed largely of fiberglass and CPVC, and the tube heat exchanger bundles of stainless steels and titanium. The unit was designed by Bill Williamson from Licon to produce an output of 700 l/h. The most important variables for the MEMS unit operation are flash range (ΔT between surface and storage zones), concentration level of the brine rejected after concentration, and the circulation rate of the first effect. Experience in operation showed the MEMS unit can operate effectively at a first-stage vapor temperature $\geq 60°C$ (ideal for heat delivery from a solar pond), and at a very high concentration ratio (with reject brine at near saturation conditions). Therefore the MEMS process has application for desalting saline waters of substantive concentration, well above those practical using, for example, reverse osmosis. A second, smaller Licon unit of around 200 l/h nominal capacity, was installed in 1999, and used to establish further relationships between distillate production rate and operating conditions (Lu, Walton and Swift, 2001) (Figure 13).

Since 1999, research at UTEP has focused upon a search and test of other low-temperature thermal desalination technologies; and development of a systems approach to solar pond engineering that integrates salinity-gradient solar ponds with multi-process desalination systems and brine concentration technologies.

Brine concentrate management or disposal is one of the major issues for application of desalination at inland sites. Inland desalination of brackish water requires consideration not only of the equipment and energy required to drive the process, but also management and/or disposal of brine concentrate in an environmentally-appropriate and cost-effective manner. Combining solar pond technology with desalination can lead to a 'zero-discharge' operational system. The need for disposal is removed by using the reject concentrate from the desalination processes in the solar pond (hence zero discharge). In order to study the technical and economic feasibility of this systems approach and to gather data and information for developing a 'zero discharge desalination plant', a brine concentrator and recovery system purchased by the USBR, was assembled at the El Paso solar pond in 1999, and tested during 2000. The research and results have been analyzed and reported by Becerra (2001).

Figure 13. A multi-effect multi-stage desalination system coupled to the El Paso solar pond.

3.4 Economics of Solar Pond Applications

With the exception of solution mining ponds, the largest salinity gradient solar ponds built in the US have been at Chattanooga, Tennessee (4,700 m²) (decommissioned in 1989), and at El Paso (3,700 m²). Because currently-operating ponds from which heat is extracted are smaller than expected commercial size, the associated economic projections of delivered energy costs should be treated as only rough estimates. With this proviso, a comprehensive study on the economics of solar ponds was conducted by Esquivel (1992), based partly upon the operation and performance of the El Paso facility. The economic data presented here for utilizing solar ponds for industrial process heat and generating base-load electric power were published by Esquivel et al. (1993).

The sizes of solar ponds in the US likely to find greatest application are in the range 10,000 to 100,000 m². Larger facilities than this would probably use units of up to 200,000 m² rather than one extremely large pond, for reasons of operational safety and reliability. The costs associated with excavation and berms (earthen dam walls), lining, and salt recyling (evaporation ponds needed, if desired), favor large sizes and make ponds of small areas less economically viable.

Esquivel examined the economic feasibility of solar ponds for industrial process heating as a function of salt costs, liner costs, and pond size, and determined that solar ponds can be economically feasible for those sites where land, water (can be brackish or sea water), and solar radiation are favorable. Such sites can be found in the majority of the central and southern US. For industrial process heat at medium temperatures (50 - 90°C), the levelized energy cost was found to range from $US(1992)6.60/GJ for a 1-hectare (Ha) pond to $US(1992)1.30/GJ for a 100-Ha pond at sites having similar climate conditions to El Paso, Texas. The levelized cost for heat delivered by solar ponds is compared with the delivered costs of heat from both natural gas and

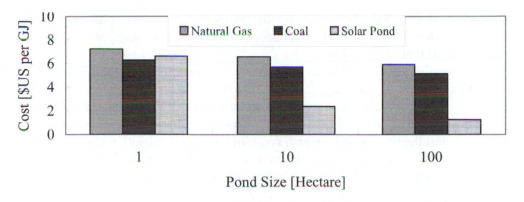

Figure 14. The estimated cost of industrial heat [$US(1992)/GJ] as a function of unit operating pond size in hectare. Adapted from Esquivel (1992).

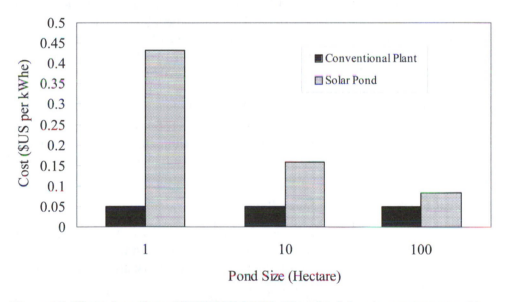

Figure 15. The estimated cost [$US(1992)/kWh] of base-load electric power generated by a solar pond in the US as a function of solar pond surface area (adapted from Esquivel, 1992).

coal in Figure 14. It can be seen that the unit cost of supplying industrial process heat from a solar pond is less than that for either natural gas or coal at even modest pond sizes.

In the case of electric power production from solar ponds, maximum revenue is generally obtained by maximizing operating hours of the engine or in a base load mode of operation (depending on peak power costs), because of the relatively high capital cost of the ORC engine. Figure 15 shows that the production of base load electricity using solar ponds is more expensive than current electrical base load generation technologies. However, it can become more cost competitive when the pond size is large, say greater than 100 Ha, and if the impact of environmental costs associated with burning fossil fuels are considered.

Desalination of brackish water appears to be an extremely useful application of solar pond technology (subsection 3.3.2). In addition to providing clean renewable energy to power the desalination processes, salinity gradient solar ponds can utilize the waste brine. A recent study at UTEP has shown the cost of water produced by a zero discharge desalination system – which combines a solar pond with membrane filtration, thermal desalination and brine concentrator – ranges from $1.06/m^3 for a 3 800 m^3 per day plant to $0.92/m^3 for a 75,700 m^3/day plant (Becerra, 2001).

3.5 Other Solar Pond Advances in the US

3.5.1 Sea Water Solar Ponds

A seawater solar pond is formed by floating fresh water over the surface of seawater, thus forming a narrow but stable density-gradient layer between fresh surface water and seawater that constitutes the storage zone of the pond. The seawater solar pond was conceived of and tested in a limited way by Preston Lowrey at San Diego State University. Lowrey demonstrated the principal features of seawater solar pond technology in small above-ground ponds (Lowrey et al., 1989) until 1995. The concept of the seawater solar pond was subsequently adopted by researchers in a Texas Solar Pond Consortium, led by the University of Texas, coupled with the University of Houston and Texas A&M University, and the initiative was experimentally tested in fish farming ponds located in the region of the Gulf of Mexico-coast region of Texas.

3.5.2 Fish Farming Utilizing Solar Ponds

It was a research group at the University of Houston, Texas, who further developed the sea water solar pond concept. The University of Houston innovation was to include a 'thermal refuge' area for (outdoor) marine culture (mariculture) ponds located on the Texas Gulf coast for raising fish.

Fish farming is a rapidly growing industry in the southern US. The rearing of fish is hampered, especially in winter months, by sudden cold spells that negatively impact toddler fish. The small fish have little resilience to reduction in farm-pond temperature. By placing fresh (underground-sourced, bore or so-called well) water on top of sea water in existing mariculture ponds, the heat losses through the surface of the pond can be greatly reduced and the storage zone water temperature kept above the lethal lower limit for extended periods. The use of seawater solar ponds for overwintering of fish in mariculture ponds along the Texas Gulf coast has been studied experimentally and has been shown to be both practical and effective. Covering the whole surface of the mariculture pond is not necessary. The University of Houston research project investigated several thermal refuge designs and testing of these was con-

Figure 16. East refuge area of a redfish mariculture pond with instrumented pier, at Palacios, Texas. Note the wind speed and direction tower and early curtain design at refuge opening to pond.

ducted at, a commercial fish farm, Redfish Unlimited, adjacent to Tres Palacios Bay, located at Palacios, Texas (Hart Shi, Kleis, and Banncrot, 1994).

Results were obtained in large ponds, averaging 26,300 m^2 in surface area (Figure 16). Each pond had a refuge at the corner. The refuge was constructed by extending a jetty (40 m long from the pond side, 20 m away from the north end). Thus the refuge area has dimensions of 20 m x 40 m, and is enclosed on three sides. Fresh bore (well) water was supplied to the surface of each refuge area at flow rates of up to 2,270 l/min. Two test ponds were nearly identical in construction and were instrumented to allow direct comparisons of refuge designs and modes of operation. Test data were taken over several years. Guo (1996) reports operating procedures which provided safe and reliable refuge operation during several cold spells.

4. Review of International Solar Pond Work

4.1 Introduction

In this section we review work on solar pond technologies in a number of other countries around the world. We have made every effort to be as inclusive as possible in this review by inviting leading researchers to contribute a short summary of relevant work in their respective countries. Inevitably, however, the breadth and depth of our international review has had to rely upon the material contributed. Nevertheless we believe it does provide a useful contemporary overview of international work on solar ponds.

Key references giving more detail on work in each of the countries covered are cited in the text. Contact names and addresses for more information are provided in an Appendix.

4.2 Israel

Early solar pond research took place in Israel in the 1950s and there was a resurgence of interest following the oil crisis in the 1970s. The successful pioneering of solar pond technology in Israel, including electricity generation on a relatively large scale, has played a major role in stimulating the increasing interest in solar ponds in the rest of the world. Solar pond studies were initiated in Israel in 1958 by Bloch and carried out by a group led by Tabor until 1966. This period resulted in published reports of operating experience from several ponds and several theoretical papers, including a hydraulic theory of heat extraction, and a theory of stability and energy balance (Weinberger, 1963; Tabor and Matz, 1965; Tabor, 1981).

Following the 1973 oil crisis, the Israeli solar pond research and development program restarted in 1974 with considerable government backing. Work has focused primarily on electricity generation using the solar pond to provide heat for an organic vapor Rankine-cycle turbine. The turbines used had been previously developed in Israel by Ormat Turbines Ltd. A solar pond at the Ormat plant in Yavne was used to operate a 6 kW_e turbo generator starting at the end of 1977.

The solar pond at Ein Boqeq was the first pond in the world to generate commercial electricity, supplying a peak output of 150 kW_e at a resort located on the shore of the Dead Sea from December 1979 (Tabor and Doron, 1986). Encouraged by the success of the Ein Boqeq demonstration, Ormat constructed a 5 MW_e solar pond power plant near Bet Ha Arava in the north of the Dead Sea (Figure 17). Two solar ponds, the first 50,000 m^2 in area and the second 200,000 m^2 were constructed and used to provide the thermal energy input to the power plant (Assaf, 1976; 1984). This solar pond power station was connected to the grid in 1984 and operated for one year. The ponds were located at the northern tip of the Dead Sea, south of Jericho and adjacent to Jordan. The geopolitical situation did not favor continuation of power plant operation at the remote site, and the plant was decommissioned in 1990.

After this intensive period of investigation during the 1970s and 1980s, interest in solar pond technology in Israel declined for some time, mainly due to the availability of low- priced fossil fuels once again. The renewed interest in solar ponds that has occurred in recent years both in Israel and internationally has stemmed mainly from their environmental benefits, such as low greenhouse gas emissions, rather than oil scarcity or security of supply. There has also been a shift in focus from electricity generation to direct use of heat from solar ponds for low temperature applications such as industrial process heating, space heating and desalination. Israeli experiences indicate that such applications are certainly feasible where solar ponds are used to replace heat from the more expensive fossil fuels or electricity.

Based on a US patent, Assaf from AGAM Energy Systems Ltd (2001) disclosed a new type of vapor pressure sink that is claimed to be extremely effective in dry and

Figure 17. The Bet Ha Arava 5 MWe solar pond power plant incorporated a 200,000 m² surface area pond having variable depth, and two 50,000 m² ponds. The control platform and instrumentation pier of the 200,000m² solar pond can be seen in the foreground.

hot climates. Incorporating such a sink in a steam turbine driven by the vapor produced by flashing the hot concentrated brine from the bottom of a solar pond, AGAM has proposed a solar power unit capable of delivering 1 MW$_e$ over a period of 5,000 h/year using the hot brine extracted from the bottom of a 16 hectare salinity-gradient solar pond.

In addition to the above work, researchers at Technion-Israel Institute of Technology have been working on the use of solar ponds to produce low-grade steam (Gommed and Grossman, 1988). By using an absorption heat transfer, these researchers have been able to produce low-grade steam from the ponds heat at a coefficient of performance of about 0.5.

A substantial amount of research has also been conducted in recent years in Israel on the physics of solar ponds. For example, Tanny and Gotlib (1995) at the Institute of Soil, Water and Environmental Sciences in Bet Dagan have been studying the double-diffusive system of solar ponds and its stability under different conditions. In another line of work, Tanny, Chai and Kit (1995) have investigated the turbulent mixing across density interfaces, which may be present in salinity-gradient solar ponds.

4.3 India

The Field Research Unit of Tata Energy Research Institute (TERI) at Pondicherry was among the pioneers to build and operate salt gradient solar pond (100 m²) in India in the 1970s. Thereafter TERI got actively involved in drawing up a National Solar Pond Plan (1983) and also in evaluating the feasibility of establishing a large-scale solar pond in India (1985). These efforts culminated in 1987 in the construction of a 6,000 m² solar pond project in Bhuj, in the state of Gujarat in western India (Motiani et

Figure 18. A panoramic view of the 6,000 m² solar pond at Bhuj, India, with heat exchanger in the foreground.

al., 1990, 1993; Kishore and Kumar, 1996; Kumar and Kishore, 1999a,b) (Figure 18). The Bhuj solar pond supplied heat to a milk-processing diary. The pond attained a maximum temperature of 99.8°C under stagnation in May 1991 but developed leakage soon after. A failure analysis subsequently indicated that the leakage was caused by the combination of high stagnation temperature and large air pockets below the liner. The lining scheme was re-designed and the pond re-established in June 1993. Hot water supply to the dairy started in September 1993. Over 20 million liters of hot water were supplied to the diary between September 1993 and March 1997. In the earth-quake of 2000, the dairy was totally devastated, but the solar pond is still there, albeit non-functional.

The major achievements and findings of this project can be summarized as fol-lows:

- Successful demonstration for the first time in India of the use of a solar pond for supplying process heat to an actual industrial user.

- Indigenous development of a cost-effective 'sandwich' type lining scheme, based on locally available materials.

- Successful maintenance of the stability of the pond.

- Achievement of the design capacity of supplying 80,000 liters of hot water per day to a nearby dairy at a delivery temperature of 70ºC or above.

Figure 19. The CSMCRI solar pond, with the rings acting as wave breakers.

- Savings to the dairy of more than 935 tons of lignite per year when the solar pond was working at full capacity. The simple payback period worked out to be less than 5 years without any subsidy or tax incentives.

The Central Salt and Marine Chemicals Research Institute (CSMCRI) in India has a solar pond R&D program currently focusing on the design and installation of a demonstration experimental solar pond (up to 10,000 m² area) in a coastal area to provide process heat to salt-based chemical industries (Major references describing CSMCRI solar pond activities are (Kishore and Joshi, 1984; Joshi and Kishore, 1985; Mehta et al., 1988). Recently, the Institute has submitted a proposal to set-up a 10,000 m² solar pond in Greater Rann of Kutch to the Ministry of Non-Conventional Energy Sources, Government of India, in New Delhi, in association with M/s Agrocel Industry, Bhuj, India. Thermal energy collected and stored in the pond will be used to operate a multistage flash desalination plant to produce potable water from saline water.

Major findings over the past ten years are:

- Experience on CSMCRI pond indicates that the thickness of the upper convective zone remains between 50-70 cm. It is difficult to control the thickness of this zone, especially in the pre-monsoon months.

- Radiation reaching in the lower convective zone is around 12-15% of the insolation incident on the surface of the pond. Clarity of pond brine does not remain uniform throughout the day and year.

A photograph of experimental facilities at CSMCRI is provided in Figure 19.

A substantial amount of theoretical and computer analysis has been performed at Indian Institute of Technology on thermal performance and stability of salt gradient solar ponds (e.g. Kaushika et al., 1996; Singh and Kaushika, 1994).

4.4 China

Research and development work on salinity-gradient solar ponds has been carried out in China since the late 1980s. Researchers at Capital Normal University in Beijing have studied the stability of solar ponds by using devices developed to measure salinity by the methods of refractive index and also the speed of light (Ye et al., 1987; Min et al., 1995; and Ma, 1994).

Work has also been conducted on the application of solar ponds to aquaculture (Song, Lu and Li, 1988; Li, Ge and Wang, 1993). This program was supported by the Chinese Ministry of Science and Technology from 1991 to 1995, and a demonstration was carried out in 1996. The main objective of the program was to use a solar pond to provide heat for overwintering of an aquatic product, the heat from the solar pond effectively increasing the production rates of anaerobic digesters.

The common process for overwintering aquatic products employed a boiler, but the difficulty in maintaining a uniform temperature distribution in the fishing ground reduced the survival rate of the products significantly. The temperature variation in the lower convective zone of solar ponds in China was just the 2 to 3°C daily temperature fluctuation. This small variation not only assisted in maintenance of the required temperature over winter, but also extended the growth period of aquatic products.

Two solar ponds of different sizes were built. One was rectangular with vertical walls of area just over 1000 m^2, and used bittern as a working medium. But, because of the breeding of algae (which could be cleared away by CuSO$_4$ solution) and other brine insects (which could not be eliminated completely by usual methods), the transparency of the pond to solar radiation decreased significantly. The maintenance of the temperature and salinity gradients was also difficult. For these reasons, this solar pond had to be shut down after three years operation. The second pond constructed was circular with tilted walls of area 20 m^2, and the working medium was MgCl$_2$ salt solution. After three years operation, successful results were obtained with this pond for overwintering two kinds of aquatic product.

From 1997 to 2002, several more solar ponds for overwintering aquatic products were constructed and operated, especially in coastal regions such as Shandong Province. The performance of these ponds was satisfactory.

Work has also been carried out on the thermal stability criterion of fully-saturated solar ponds (Li and Li, 1989; Li and Lu, 1991; Li, 1995), that is, ponds with solid salt lying on the bottom that is not dissolved in the layer of saturated brine. The idea of a fully-saturated solar pond was developed through both theoretical and experimental investigation of temperature and salinity stability of ponds. At first, the phenomenon of full saturation and the formation of a self-equilibrium system were observed and analyzed; then the general equations of a double-diffusive system with constant and varying temperature and salinity at the boundary of the non-convective zone and lower

Figure 20. An above-ground solar pond at Ferdowsi University of Mashad, Iran.

convective zone were introduced. In particular, the thermal stability criterion of a fully-saturated solar pond was proposed. The comparison of theoretical predictions with experimental data under natural conditions with industrial salt solutions proved satisfactory.

4.5 Iran

Research into solar ponds in Iran started in 1975 at Shiraz University, focusing mainly on mathematical modeling and performance prediction through computer simulation (Akbarzadeh and Ahmadi, 1980). At present R&D work on solar ponds is carried out at Ferdowsi University in Mashad in the north east of Iran. In 1998 an experimental, outdoor laboratory-scale salt gradient solar pond of area 4 m² and depth 1.08 m was built at Ferdowsi University. The pond was above ground with double-sided vertical walls, and its bottom and sides thermally insulated with 12 cm thick polystyrene sheets (Figure 20). Meteorological parameters such as global radiation, wind speed, relative humidity and ambient temperature as well as pond temperature at 15 points in depth were measured every minute by the corresponding sensors. Concentrated saline solution was injected once a week to the lower zone to compensate for the reduction of salinity caused by upward diffusion. The surface of the pond was washed once a week. The pond performance was monitored continuously for about four years.

The following conclusions have been drawn from the experiment (Jaefarzadeh, 2000):

- The surface of the pond was unprotected, so the thickness of the upper convective zone (UCZ) grew during summer to a maximum of 30 cm, while vanishing in the cold months when the surface was covered by a layer of ice.

- The lower convective zone (LCZ) was 40 cm thick with a sharp interface with the non-convective zone (NCZ) during the summer when circulatory currents were active, but became thin (~25 cm) and diffusive during the winter when currents were weak.

- The maximum temperature of the LCZ was 64°C and occurred in August, while the minimum temperature of the LCZ was 15°C in February. The temperature of the LCZ would have been more if there were a thicker NCZ and inclined walls to increase the solar penetration. While the temperature of the LCZ decreased monotonically from summer to winter, the temperature of UCZ fluctuated with ambient temperature.

4.6 Italy

Solar pond activities at the Ancona University, Italy, began in 1982. Since then, a variety of research activities have been undertaken to investigate the fundamental dynamical processes, influences of external factors, and operating practice. Many feasibility and some engineering design studies have been conducted for the use of solar pond to heat greenhouses and to produce fresh water. An experimental 625 m² tilted-wall solar pond was built in 1986 with a total depth of 3.5 m (Figure 21).

For ten years the heat extraction system from the thermal storage zone consisted of four separated cross-linked polyethylene pipes wound as a spiral and submerged below the interface between the convective and the salinity-gradient zones, each one extracting heat from a quarter of pond area.

Currently the foci in the Ancona solar pond R&D program are desalination for fresh water production and desalting agricultural water, and process heating applications. Heat is now extracted from the lower convective zone by pumping out brine and passing it through a titanium shell and tube heat exchanger of the desalination unit. The cooled brine is then returned to the pond at a lower level than it was removed. Experimental tests have been performed to verify the performance of components and equipment. From a technical viewpoint the solar pond desalinator may be judged as satisfactory and the performance of the plant has been experimentally confirmed.

As part of a scientific collaboration between Ancona University and the Ecole Nationale de Ingenieurs de Tunis (ENIT), a feasibility study of an integrated project for the sustainable development of the el Bibane-Jedaria regions located in the southeast of Tunisia along the Mediterranean sea has been conducted. Here the small population is distributed over a relatively large area of land bordering the sea with

Figure 21. The 625 m^2 solar pond at the Renewable Energy Laboratory, Department of Energy, University of Ancona, Italy (latitude 43°47' N).

scarce bio-resources, but ample seawater, solar energy and land. The possibilities of using a large solar pond coupled to a multistage flash system to desalinate seawater to produce fresh water for drinking and irrigation, as well as supply heat for greenhouses to cultivate early vegetables, have been investigated.

A major solar pond project was built and operated successfully in Margarita di Savoia on the southeast coast of Italy by Agip Petroli during 1987 through 1993 (Folchitto, 1993). It used bittern (SG 1.24) available on the site from a local sea salt works. The 25000 m^2 pond demonstrated the use of solar ponds for desalination using thermally driven vapor compression via a vapor ejector. An average production in excess of 300 l/h was achieved. There were no significant interruptions to operation, as the pond provided heat 24 hours per day, and the results of the program were reported by Folchitto (1993). The project required a large volume of bittern to be clarified to fill the pond. Innovations achieved in relation to this clarification included an external flocculation system that achieved clarification by flocculation, agglomerant additives and a carefully-dosed. hypochlorite sterilizer.

4.7 Tunisia

Since the mid 1980s work has been under way in Tunisia on various aspects of solar pond technology from science to applications. The R&D work includes the design, construction and operation of small (4 m^2) to medium size (1500 m^2) salt-gradient solar ponds (Figure 22).

Figure 22. 1,500 m² salt-gradient solar pond in Tunisia.

A number of applications for solar ponds have been considered, including greenhouse heating, space heating for offices, and desalination using Multistage Flash (MSF) Systems. Apart from the small 4 m² pond constructed using natural brine containing a mixture of salts, the other solar ponds were set up using NaCl. The performance of all these ponds has been monitored and temperatures up to 100°C have been achieved.

Work has also been conducted on stability of double diffusive systems using the shadowgraph method for visualization. In this study the stability of the lower interface has been investigated by studying the stability of a stratified medium heated from below.

Work has also been carried out on computer modeling of a 17300 m² closed-cycle salt-gradient solar pond providing heat to a desalination plant (Belghith, 1998; Ouni, 2003). The model takes into account heat and salt diffusion and also internal gradient stability. The model has been validated against data taken from the solar pond operated by University of Texas at El Paso.

4.8 Japan

R&D work has been performed at Oita University in Japan on salt-less solar ponds. In such a pond, a system of submerged semi-transparent multiple layers of insulation decreases heat loss by convection from a body of water exposed to solar radiation, so that it serves as a solar collector and heat store. Both mathematical modeling and computer analysis of the performance has been performed on small laboratory-scale salt-less solar ponds (3.13 m² area and 1 m deep) (Kamiuto, Nagumo, and Elbikai, 1990). Conclusions have been drawn on the optimum pond temperature to maximize electrical power production (Kamiuto, 1987). It has also been concluded that the maximum final conversion efficiency from solar radiation to electricity is less than 3% under Japanese meteorological conditions. No further work has been pursued since 1993.

4.9 Taiwan

Two experimental salinity-gradient solar ponds were built at the Energy & Resources Laboratories, Industrial Technology Research Institute, in Taiwan in 1984 to study the thermal performance of the solar ponds and their application to heat fish farming pools (Lin, 1986). Computer simulation and experimental measurements were carried out and compared to each other. Each pond had an area of 1,400 m^2 with 3 m depth. One pond had plastic liner on the side walls and bottom surfaces. The lower storage zone was 1.2 m thick, the gradient zone 1.1 m, and the surface zone about 0.2 to 0.4 m. These two ponds were located in Taiwan city, southern Taiwan, because of its relatively high solar radiation and yearly average ambient temperature, and the large area of salt farms available near the sea shore. Sea water was concentrated to construct the gradient layer in the solar ponds. The major problem of using sea water is the water quality control because of the growth of aquatic plants within the ponds, which will decrease the transmission of the solar flux.

The highest temperature obtained in the lower layer of the pond was 80°C in August 1985. The salinity in the upper surface zone changed from 2.6% to 6.8% and the zone thickened from 40 cm to 53 cm during a two month period when heat was extracted from the pond. The concentration of the other two layers did not change significantly. Most heat loss (about 50%) was to the soil underneath the pond in summer, while upward conduction loss to ambient air dominated in winter. Heat extracted from the solar pond was supplied to a fish farming pool. Based upon practical test conditions, the total heat that could be extracted annually was estimated at around 3×10^8 to 5.4×10^8 kCal by using PVC tube-bank heat exchangers with flow rates ranging from 3,600 kg/h to 12,000 kg/h.

Following construction, operation and performance evaluation of this kind of solar pond, the project was terminated in 1986.

4.10 Portugal

Between 1982 and 1991 a 1,000 m^2 salt gradient solar pond was operated at Pegoes in Portugal as a test device and a source of heating for a nearby greenhouse (Collares Pereira, Joyce and Valle, 1983; Collares Pereira and Joyce, 1985; Green, Joyce and Collares Pereira, 1987; Joyce, 1991). The inclination of the walls was 45º with a total depth of 3.5 m (Figure 23). Heat extracted from the pond via a submerged heat exchanger at the top of the lower convective zone was used to heat green houses with a total area of 7,209 m^2. The above research and development activity involved collaboration between the Economy Ministry of Portugal and the Technical University of Lisbon, as well as ENIT in Tunisia (see section 4.6 above).

Over its eight-year life the pond reached a maximum temperature of 67ºC and a minimum of 30ºC. The pond was fully instrumented and employed a wave suppression

Figure 23. 1,000 m² salt gradient solar pond at Pegoes, Portugal.

system mainly composed of 5 cm width polypropylene strips floating on the surface making squares of 1 m sides.

In addition to heating the greenhouse (by soil heating) the pond was used to investigate the dynamics of the double diffusive processes and also to verify the parametric models developed for the prediction of pond's performance (Joyce et al., 1992). Extensive work has also been done on the stability analysis of the gradient layer, which included field observations laboratory testing and computer modeling (Giestas, Pina and Joyce, 1996; Giestas, Joyce and Pina, 1997). A future potential for application of solar ponds as a renewable source of energy in marine salt works has been identified in both Portugal and Tunisia.

4.11 Mexico

A substantial amount of R&D has been performed at the National University of Mexico since the late 1980s on the suitability of clay as a liner for solar ponds. Almanza et al. (1988) and Almanza and Castaneda (1993) have done tests on the effects of salt solutions on the permeability of clay soils. Several laboratory test facilities have been designed and constructed simulating the temperature and hydraulic gradients, which may exist in the clay when used as a liner in solar ponds (Figure 24).

The experimental results show that lower permeability can be obtained by using a bentonite rather than a kaolinite clay. It has also been demonstrated that with clay as liner the permeability can be brought down to 10^{-8} cm/s. Since the cost of clay liner is expected to be two to three times less than conventional membrane liner, compacted clay is proposed as a more cost effective barrier in large-scale solar ponds.

Figure 24. Test facilities at the National University of Mexico for measuring permeability of clay liner under solar pond conditions.

Studies of solar pond systems and technologies have been undertaken since 1985 by the Electric Research Institute of Mexico, Department of Non-Conventional Energy, led by Jorge Huacuz Villimar. The Instituto De Investigaciones Electricas sponsored the first international conference on solar ponds, held in Cuernavaca in 1987 (Huacuz, Dominguez and Becerra, 1987). Researchers from the Department of Non-Conventional Energy built the first artificial solar pond in Mexico. It was located at Zacatapecas, and featured a buried plastic liner covered with local clay. Researchers from the Electric Research Institute participated in an international exchange program on solar ponds with the University of Texas at El Paso over the 1987 to 1995 period.

5. Current State of the Art and Future Directions

During the last decade key research projects around the world have substantially advanced salinity-gradient solar pond technologies. Landmark successes in operational practice and applications have been achieved. Multiple approaches to solar pond design have been explored to produce success in practice. A variety of applications have been demonstrated. The results are impressive. Reliable solar pond operations have been achieved during the last decade. The extent to which the future potential of solar ponds as a sustainable energy source is realized will nevertheless depend largely upon factors external to technology breakthroughs, in particular the degree of emphasis placed by governments on reducing greenhouse gas emissions in line with commitments given in the Kyoto protocol.

On the basis of the foregoing review, we have summarized in Table 1 the work being done on solar ponds around the world under a range of categories. It is evident from the table that there are many organizations around the world continuing work in the area of the science of solar ponds, and theoretical and computer modeling of pond characteristics and performance. The principal areas of solar pond science that are currently being actively investigated globally are:

- Maintenance of salinity profiles, including the behavior of upper and lower boundaries of the gradient layer, as well as clarity of the pond water.

- Investigation into different methods of heat extraction in order to increase the theoretical limits on the thermal efficiency of solar ponds.

- Computer modeling and performance prediction is being carried out by many institutions.

The most fruitful lines of inquiry into solar pond science over the coming few years are likely to be investigation of:

- Alternative salts and new lining techniques that may increase stability and reduce construction costs. Leak detection methods and salt recycling methods may also be further improved.

- Developing simple methods to limit climatic effects such as wind-driven mixing and the growth of the upper convective layer. Reducing the thickness of the upper convective layer can improve performance and efficiency of a solar pond.

- Chemical engineering of artificial solar pond liquids, leading to the development of a new genre of solar ponds.

- Ways by which the maximum achievable efficiency in solar ponds can be increased. At present the solar ponds efficiency of conversion from solar to thermal energy is low and is about one third to a half that of conventional flat plate solar collectors. Therefore to supply the same thermal load solar ponds require an area of two to three times that required by flat plate solar collectors. This is obviously a limitation, and efforts will be needed to increase the efficiency of solar ponds and thus reduce the required area to meet a given thermal load.

Table 1. Summary of some of the main areas of activity in relation to solar ponds in the countries reviewed in this paper.

COUNTRY/INSTITUTION	AREAS OF CURRENT INTEREST								
	Solar pond science, computer modelling	Industrial process heat	High-performance solar pond operation	Electricity generation	Salinity mitigation	Desalination	Aquaculture	Commercial Space heating	Waste brine management
Australia									
RMIT University	•	•							
USA									
University of Texas El Paso	•	•	•	•	•	•			•
University of Houston	•					•	•		
Israel									
ORMAT		•		•					
AGAM	•	•		•		•			
Universities	•	•							
India									
TERI, Pondicherry	•	•							
CSMCRI	•	•				•			
Indian Institute of Technology	•								
China									
Capital Normal University	•						•		•
Iran									
Ferdowsi University	•								
Italy									
Ancona University		•			•	•			
Tunisia									
Faculty of Sciences of Tunis		•				•			
Japan									
Oita University	•							•	
Taiwan									
ITRI	•						•		
Portugal (INETI-DER)	•	•							
Mexico (Ciudad Universitaria)	•								

The main facets of solar pond design, construction, operation and performance that require further research and development are:

- Lining. A major part of the cost of solar pond construction is still the lining. Work is required to develop construction techniques that allow solar ponds to operate without expensive lining, while minimizing percolation.

- The maintenance of high-performance (>80°C) operation of solar ponds. There is a need to demonstrate a feasible lifetime for high-performance solar ponds. Now that annual operation has been demonstrated, the question arises: what is the operational lifetime of a facility and what limits that useful lifetime? Economic models using, say, a 20-year lifetime may be grossly under- or overstating pond life.

- Reduction of operational maintenance costs. This reduction is necessary (especially for high performance ponds), since operator costs impact significantly on project viability.

- Testing of natural clarity control methods in high-performance ponds. Such testing can provide valuable information.

- Construction and operation of solar ponds using bittern. Bittern solar ponds are very suitable when constructed near salt works where bittern is a waste product from commercial (sodium chloride) salt production. Research needs to be carried out on the stability, maintenance and heat production of bittern solar ponds.

- Salt recycling systems for sites where continuous surface washing cannot be practiced. Such sites are effectively zero discharge locations.

Computer models based on theoretical and empirical analysis are now available that can estimate characteristic parameters of solar ponds such as efficiency and solar pond performance, such as temperature of the bottom layer and heat extraction rate given solar radiation, ambient temperature, wind speed and ground temperature. However, most such models are not user-friendly as well as not being generally available. Theoretical and computer modeling activity to overcome these limitations would be very useful.

In regard to applications of solar ponds, the present review has found (see Table 1) that industrial process heating and desalination are presently of greatest interest. In addition, there is continuing activity on the development and use of solar ponds for power generation, and increasing interest in new applications such as salinity mitigation, aquaculture, greenhouse heating, biogas production, and waste brine or other saline effluent management.

Important areas that require more research and development in order to realize more economically viable systems are:

a. Use of solar ponds with heat engines for electrical power generation, in particular engines that utilize novel cycles such as trilateral and Kalina;

b. Investigation and testing *in situ* heat engines with evaporator and condenser located entirely within a solar pond. This configuration can reduce power plant capital costs very substantially, since heat exchanger areas are large for low- temperature applications.

c. Desalination systems, such as solar pond – multistage and/or multiple-effect flash desalting systems, and compact solar pond desalting systems.

d. Systems engineering of hybrid solar pond stations (as in (a), (b) and (c) but combining heat and power, such as an electric power station coupled with a desalting plant.

e. Application of the above systems in salinity mitigation schemes and arable land reclamation projects.

f. Solar pond integration in mariculture and biotechnology applications.

Following this R&D there will be a need for field trials and demonstration of the most promising technological options identified by theoretical and experimental work.

There is a growing consensus that solar ponds can play a valuable role as integral components of salinity mitigation schemes. Solar ponds can provide heat for desalination, commercial salt production, other chemical manufacturing, brine concentration, and possibly for power generation as well. The design, development, and demonstration on a large-scale of such integrated systems using industrial ecology principles (Allenby and Richards, 1994) is in our view another priority area for the solar pond field. In Australia, for example, the deployment of solar ponds within planned salinity interception and mitigation schemes in the Murray-Darling river basin is a major opportunity. The same is true in the US with respect to salinity control projects such as the in the Salton Sea region, the largest inland body of water in the state of California, and in areas of the agriculturally-essential San Joaquin Valley in California (Morrison and Cohen, 1999; Grimes, 2001). According to a multi-party, inter-agency technical committee evaluating drainage water treatment programs in California, the use of salinity gradient solar ponds may play an important long-term role in drainage treatment for regions of the San Joaquin Valley (Alemi, 1999). Countries in the Middle East, and neighboring countries in the Mediterranean and beyond such as Turkey, also offer multiple sites for promising applications both in relation to salinity projects, industrial process heating and other applications.

While natural salt lakes are interesting in their own right, they also have historical importance for solar ponds because many of the advances in the understanding of double-diffusive convection were made by scientists who were studying these lakes and similar phenomena in the oceans (Nielsen, Hull and Golding, 1989). Although natural solar ponds are relatively rare, saline lakes are common landscape features on every continent. The estimated global volume of saline lake water (=100,000 km³) is almost as great as the volume of the world's fresh water (Hammer 1986). The eight largest saline lakes occupy an area of over 50,000 km². The potential for solar pond applications in the vicinity of salt lakes is enormous, and the potential benefits are huge. They include applications of increasing relevance to a world desperate for arable land and fresh water.

Solar ponds for industrial process heating in regions remote from natural gas distribution systems are ready for commercialization but have so far not been taken up by private firms and offered as attractive commercial solutions. It will be essential to establish that the key prerequisites for the consideration of solar ponds are satisfied in any particular application: that is, environmental acceptability, salt and land availability, suitable terrain, high insolation, and practical local application for the energy and/or fresh water produced. This is a commercialization enterprise that deserves more encouragement by governments, though solar ponds for heating applications are more likely to be a niche technology rather than being universally deployed.

Excellent sites for solar ponds are abundant throughout the world and many solar pond applications are poised for commercialization. Yet most importantly vision is needed to recognize that globally our future quality of life depends upon increasing integration of renewable energy technologies into energy supply networks. Solar ponds as a renewable energy source that can simultaneously yield environmental benefits deserve to play a role in this historic transition to a truly sustainable energy system.

6. Acknowledgements

We sincerely thank all those researchers engaged in solar pond work around the world for their generous assistance in the form of contributed summaries of the activities being conducted at their institutions. Of course, the responsibility for what is presented here remains with us.

7. References

AGAM Energy Systems, 2001. Heat pump/engine system and a method for utilizing same, US Patent No. US 6,266,975 BI, 31 July, Inventor Gad Assaf.

Akbarzadeh, A., 1984. Effect of sloping walls on salt concentration profile in a solar pond. Solar Energy, 33, 137.

Akbarzadeh, A., 1989. Convective layers generated by sidewalls in solar ponds: observations. Solar Energy, 43 (1), 17-23.

Akbarzadeh, A., Ahmadi, G., 1980. Computer simulation of the performance of a solar pond in the southern part of Iran. Solar Energy, 24 (2), 143.

Akbarzadeh, A., Manins, P., 1988. Convective layers generated by sidewalls in solar ponds. Solar Energy, 41, 521-9.

Akbarzadeh, A., Earl, G., 1992. Solar pond and salinity control in Victoria. In: Proceedings of Australian and New Zealand Solar Energy Society 'Solar 1992' Conference, Darwin, Australia, July.

Akbarzadeh, A., Johnson, P., Nguyen, T., Mochizuki, M., Mashiko, M., Sauciuc, I., Kusaba, S., Suzuki, H., 2001. Formulation and analysis of the heat pipe turbine for production of power from renewable sources. Applied Thermal Engineering, 21, 1551-1563.

Akbarzadeh, A., Mac Donald, R.W.G., Wang, Y.F., 1983. Reduction of surface mixing in solar ponds by floating rings. Solar Energy, 31 (4), 377-80.

Alagao, F.B., Akbarzadeh, A., Johnson, P.W., 1994. The design, construction and initial operation of a closed-cycle, salt-gradient solar pond. Solar Energy, 53 (4), 343-51.

Alemi, M., 1999. Drainage Water Treatment, Final Report Drainage Water Treatment Technical Committee, San Joaquin Calley Drainage Implementation Program and the University of California Salinity/Drainage Program, Department of Water Resources, Sacramento, California.

Allenby, B.R., Richards, D.J., 1994. The greening of industrial ecosystems. National Academy Press, Washington.

Almanza, R., Martinez, A., Segura, G., 1988. Study Kaolinite Clay as a Liner for Solar Ponds. Solar Energy, 42 (5), 395-403.

Almanza, R., Castaneda, R., 1993. What type of clay can be used as liner for NaC1 ponds? Solar Energy, 5 (4), 293-7.

Andrews, J., Akbarzadeh, A., 2002. *Solar Pond Project: Stage 1 – Solar Ponds for Industrial Process Heating*, End of project report for project funded under Renewable Energy Commercialization program, Australian Greenhouse Office (RMIT, Melbourne).

Assaf, G., 1984. Segregated solar pond, U.S Patent No. 4,475,535, Oct. 9.

Assaf, G., 1976. The Dead Sea: A scheme for a solar lake. Solar Energy, 18, 253.

Becerra, H.R., 2001. Performance Analysis of a Brine Concentrator and Water Recovery System. M. Sci. in Mech. Eng. Thesis, The University of Texas at El Paso.

Belghith, A., Guizani, A., Ouni, M., 1998. Simulation of the transient behavior of a salt gradient solar pond in Tunisia. Renewable Energy, 14 (1), 69-76.

Burston, I., 1996. Application of salinity-gradient solar ponds in a salt-affected area of Victoria. M. Eng. Thesis, Department of Mechanical Engineering, RMIT University, Melbourne, Australia.

Chai, A., Tanny, J., 1995. Hydrogen bubble visualization of a sheared density interface. Experiments in Fluids, 10 (5), 296.

Chinn, A., Akbarzadeh, A., Dixon, C., 2003. The design, operation and maintenance of an experimental magnesium choloride pond. In: Proceedings of Australia Solar Energy Society Conference, Melbourne Australia, November.

Collares Pereira, M., Joyce, A., 1985. A salt gradient solar pond for green house heating application. Conferencia International da ISES, Montreal, Canada.

Collares Pereira, M., Joyce, A., Valle, L., 1983. A salt gradient solar pond for greenhouse application: performance during the first operating months. Congresso da ISES, Perth, Australia.

Collins, R., 1985. The advanced Alice Springs solar pond. In: Proc. International Solar Energy Society, Montreal, 1479.

Dejfors, C., Thorin, E., Svedberg, G., 1998. Ammonia–water power cycles for direct-fired cogeneration applications. Energy Conversion Management, 39 (16-18), 1675-81.

Engdahl, D.D., 1987. Technical Information Record on the Salinity-Gradient Solar Pond System at the Los Banos Demonstration Desalting Facility. California Department of Water Resources, San Joaquin District.

Esquivel, P., 1992. Economic feasibility of utilizing solar pond technology to produce industrial process heat, base load electricity, and desalted brackish water. M. Sci. Thesis, The University of Texas at El Paso.

Esquivel, P., Swift, A.H.P., McLean, T.J., Golding, P., 1993. Solar pond economics for industrial process heat, baseload electricity and water desalting. In: Proceedings of the Third International Conference on Solar ponds, El Paso, Texas, May, 360-79.

Folchitto, S., 1993. A successful experience of coastal solar pond water clarity and establishing and maintenance. In: Proceedings of the Third International Conference on Solar Ponds, El Paso, Texas, May, 258-90.

Folchitto, S., Principi, P., (Eds.), 1990. Proceedings of the 2nd International Conference Progress in Solar Ponds, Rome, Italy.

Giestas, M., Joyce, A., Pina, H., 1992. The dynamics of free convection in a solar pond. World Renewable Energy Congress, Reading, UK, September.

Giestas, M., Joyce, A., Pina, H., 1997. The influence of non-constant diffusivities on solar ponds stability. International Journal of Heat and Mass Transfer, 40 (18), 4379-91.

Giestas, M., Pina, H., Joyce, A., 1996. The influence of radiation absorption on solar ponds stability. International Journal of Heat and Mass Transfer, 39 (18), 3873-85.

Golding, P., 1981. Meromictic Water Reservoir Solar Collectors. In: Proceedings of the International Solar Energy Society, Perth, Australia.

Golding, P., Sandoval, J., York, T., (Eds.), 1993. Proceedings of the Third International Conference Progress in Solar Ponds, El Paso, Texas.

Gommed, K., Grossman, G., 1988. Process steam generation by temperature boosting of heat from solar ponds. Solar Energy 41(1), 81-9.

Goswami, D.Y., Vijayaraghavan, S., Lu, S., Tamm, G., 2004. New and Emerging Developments in Solar Energy. Solar Energy, vol. 76:1-3, pp. 33-43, 2004. Also in Proceedings of the World Solar Forum, International Solar Energy Society (ISES), Adelaide Australia, November 2001.

Goswami, D.Y., Xu, F., 1999. Analysis of a New Thermodynamic Cycle for Combined Power and Cooling Using Low and Mid Temperature Solar Collectors. Journal of Solar Energy Engineering Vol. 121, No. 2, pp. 91-97.

Green, A.A., Joyce, A., Collares Pereira, M., 1987. The measurement of radiation transmission in a salt gradient solar pond. 10th Biennial Congress of ISES, Hamburg.

Grimes, P.M., 2001. Urbanization and Water in the Northern San Joaquin Valley. Master of Science in Civil Engineering Thesis, University of California at Davis.

Guo, Q., 1996. Thermal performance of the sea water solar pond. Master's Thesis, Department of Mechanical Engineering, University of Houston.

Hammer, U.T., 1986. Saline lake ecosystems of the world, Junk Publishers, Boston.

Hart, R.A., Shi, Z., Kleis, S.J., Bannerot, R.B., 1994. Sea water solar ponds applications for mariculture. In: Burley, S., et al., (Eds.), Solar '94, Proceedings of the 1994 ASES Annual Conference, San Jose, CA, 197-202.

Hasan, A.A., Goswami, D.Y., 2003. Exergy Analysis of a Combined Power and Refrigeration Thermo-dynamic Cycle Driven by a Solar Heat Source. ASME Journal of Solar Energy Engineering, Vol. 125, No. 1, pp. 55-60.

Hasan, A.A., Goswami, D.Y., 2002. First and Second Law Analysis of a New Power and Refrigeration Thermodynamic Cycle Using a Solar Heat Source. Solar Energy Journal, Vol. 73, No. 5, pp. 385-393.

Heppenstall, T., 1998. Advanced gas turbine cycles for power generation: a critical review. Applied Thermal Engineering, 18, 837-46.

Huacuz, J.M., Dominguez, M.A., Becerra, H.R., 1987. In: Proceedings of the Conference, International Progress in Solar Ponds, Cuernavaca, Mexico.

Hull, J.R., Nielsen, C.E, Golding, P., 1989. Salinity Gradient Solar Ponds. CRC Press, Boca Raton, Florida.

Hung, Tzu-Chen., 2001. Waste heat recovery of organic Rankine cycle using dry fluids. Energy Conversion and Management, 42, 539-53.

Idso, S.B., Idso, K.E., 1984. Conserving heat in a marine microcosm with a surface layer of fresh or brackish water: the 'semi-solar pond'. Solar Energy, 33 (2), 149-54.

Jaefarzadeh, M.R., 2000. On the performance of a salinity-gradient solar pond. Journal of Applied Thermal Engineering, 20, 243-52.

Jaefarzadeh, M.R., Akbarzadeh, A., 2002. Towards the design of low maintenance salinity gradient solar ponds. Solar Energy, 73 (5), 375-84.

Joshi, V., Kishore, V.V.N, 1985. A numerical study of the effects of solar attenuation modeling on the performance of solar ponds. Solar Energy, 35 (4), 377-80.

Joyce, A., 1991. Conductivity probe for measuring small variations in salinity gradient of a solar pond. Solar World Congress, Denver.

Kalecsinsky, A., 1902. Uber die ungarischen warmen und heissen Kochsalzseen als naturaliche Warmeaccululatoren. Ann. Physkik IV, 7, 408.

Kalina, A.I., 1984. Combined Cycle System with Novel Bottoming Cycle. J. Engineering for Gas Turbines and Power, 106, 737-742.

Kamiuto, K., 1987. Determination of the optimum pond temperature for maximizing power production of a convecting solar pond thermal energy conversion systems. Applied Energy, 28, 47-57.

Kamiuto, K., Nagumo, Y., Elbikai, I., 1990. Transient thermal characteristics of a small , saltless solar pond with one semi transparent air filled surface insulation layer. Solar Energy, 45 (4), 189-92.

Kaushika, N.D., Singh, T.P., 1994. On the stability theory of salt gradient. Energy Conversion Mgmt, 35, No. 2, pp. 117-119.

Kaushika, N.D., Avanti, P., Arulanantham, M., Singh, T.P., 1996. Double diffusive zone–boundary migration in solar ponds. International Journal of Energy Research, 20, 1027-35.

Kishore, V.V.M., Joshi, V., 1984. A practical collector efficiency equation for non-convective solar ponds. Solar Energy, 33 (5), 391-5.

Kishore, V.V.N., Kumar, A., 1996. Solar pond: An exercise in development of indigenous technology at Kutch, India. Energy for Sustainable Development, 3 (1).

Kleis, S.J., Sanchez, L.A., 1991. Dependence of sound velocity on salinity and temperature in saline solutions. Solar Energy, 46 (6), 371-6.

Kumar, A., Kishore, V.V.N., 1999. Construction and operational experience of a 6000 m^2 solar pond at Kutch, India. Solar Energy, 65 (4), 237-49.

Kishore, V.V.N., Kumar, A., 1999. Development of a large solar pond for a dairy plant in India. SunWorld, 23(1), 23-4.

Li, S.S., Li, Q.W., 1989. Preliminary exploring for the fully-saturated solar pond. In: Proc. ISES, Kobe, Japan, 1428.

Li, S.S., Lu, Z.A., 1991. Experimental study of mini-saturated solar pond under natural conditions. In: Proc. ISES, Denver, Colorado, 2297.

Li, S.S., 1995. Thermal stability criterion of fully-saturated solar pond. Acta Energiae Solaris Sinica, 16(4): 333-9.

Li, S.S., Ge, H.C., Wang, S.P., 1993. A preliminary exploration for division into districts of solar ponds in China. In: Proc, of the 3rd international Conference on Progress in Solar Ponds. May 23-27, El Paso, Texas.

Liao, Y., 1987. Gradient stability and injection analysis for the el paso solar pond. Master's Thesis, University of Texas at El Paso.

Liao, Y., Swift, A.H.P., Reid, R.L., 1988. Determination of the critical Froude number for gradient establishment in a solar pond. In: Proceedings of the ASME Solar Energy Division Conference, Denver.

Lin, J., 1986. The operation and heat extraction experience of EMRO's solar pond. In: Proceedings 21st IECEC, San Diego, CA, USA, August, 1184-9.

Lowrey, D.P., Ford, D.P., Collado, F., Morgan, J., Frusti, E., 1989. Combining mariculture and sea water-based solar ponds. Solar Engineering–1989, Proceedings of the 1989 ASME-SED International Solar Energy Conference, San Diego, April, 425-432; also Trans. ASME J. Solar Energy Engineering, 112 (2), 90-7, May 1990.

Lu, H., 1994. Monitoring and data analysis for solar pond operation. M. Sci. in Mechanical Engineering Thesis, The University of Texas at El Paso.

Lu, H., and Swift, A.H.P., 1996. Reconstruction and operation of the el paso solar pond with a geosynthetic clay liner system. In: Proceedings of the Joint International Solar Energy-ASME Solar Energy Division Conference, San Antonio, Texas.

Lu, H., Walton, J.C., Swift, A.H.P., 2000. Zero discharge desalination. Desalination & Water Reuse, 10 (3), 35-43.

Lu, H., Walton, J.C., Swift, A.H.P., 2001. Desalination coupled with salinity-gradient solar ponds. Desalination, 136, 13-23.

Lu, Huanmin., 2003. Personal communication.

Lu, S., Goswami, D.Y., 2003. Optimization of a Novel Combined Power/Refrigeration Thermodynamic Cycle. ASME *Journal of Solar Energy Engineering,* Vol. 125, No. 2, May, pp. 212-217.

Ma, W.Q., 1994. A magnetic levitated density meter and its calibration. Acta Energiae Solaris Sinica, 15(1), 83 –7.

Magasanik, D., Golding, P., 1985. Solar Pond Technology and Application, Consultancy, University of Texas at El Paso.

Matz, R., Tabor, H., 1965. A status report on a solar pond project. Solar Energy, 9, 177-182.

Mehta, A.S., Pathak, N., Shah, B.M., Gomkale, S.D. 1988. Performance analysis of bittern based solar pond. Solar Energy, 40(5) pp 469-75.

Min, Z.J., Li, J.P., Jiang, S.H., Song, A.G., Li, S.S., 1995. Measurement of salt salinity in solar pond by supersonic method. Acta Energiae Solaris Sinica, 16 (2), 224-8.

Morrison, J., Cohen, M., 1999. Restoring California's Salton Sea. Borderlines 52, 7 (1), January.

Motiani, M., Kumar, A., Kishore, V.V.N., Rao, K.S., 1990. Constructional details of a 6,000 sq.m. solar pond at Kutch Dairy, Bhuj. 2nd International Conference on Progress in Solar Ponds, Rome, Italy.

Motiani, M., Kumar, A., Kishore, V.V.N., Rao, K.S., 1993. One year performance of 6,000 sq.m. solar pond at Bhuj. 3rd International Conference on Progress in Solar Ponds, El Paso, Texas, USA.

National Action Plan for Salinity and Water Quality, 2003. Factsheets, "Frequently Asked Questions" and "Australia's Salinity Problem", from http://www.napswq.gov.au/publications/factsheets.html.

Nguyen, T., Johnson, P., Akbarzadeh, A., Gibson, K., Mochizuki, M., 1994. Journal of Heat Recovery Systems and CHP; 4, 333.

Nielsen, C.E., 1975. Control of gradient zone boundaries. In: Proceedings of the International Solar Energy Society, Atlanta, 1010.

Nielsen, C.E., 1980. Nonconvective salinity gradient solar ponds. In: Dickinson, W.C., Cheremisinoff, P.N. (Eds.) Solar Energy Technology Handbook, Marcel Dekker, New York.

Nielsen, C.E., 1983. Experience with Heat extraction and zone boundary motion. In: Proceedings of the American Solar Energy Society, Minneapolis, 405.

Nielsen, C.E., 1988. Salinity gradient solar ponds. *In:* Böer, K.W., (Ed.), Advances in Solar Energy, vol. 4, American Solar Energy Society and Plenum Press, pp. 445-498.

Ouni, M., Belghith, A., Lu, H., Guizani, A., 2003. Simulation of the control of a salt gradient solar pond in the South of Tunisia. Solar Energy, 75 (2), 95-101.

Ouni, M., Guizani, A., Belghith, A., 1998. Simulation of the transient behaviour of a salt gradient solar pond in Tunisia. Renewable Energy, 14(1-4), 69-76.

Rabl, A., Nielsen, C.E., 1975. Solar ponds for space heating. Solar Energy, 17, 1.

Reid, R.L., McLean, T.J., Lai, C.H., 1985. Feasibility study of a solar pond/ industrial process heat/ electrical supply for a food canning plant. ASME Solar Energy Division, Knoxville, 263.

Reid, R.L., Swift, A.H.P., Boegli, W.J., Castaneda, B.A., Kane, V.R., 1986. Design, construction and initial operation of the El Paso solar pond. In: Proc. ASME Solar Energy Division Conference, Anaheim, CA, 304.

Smith, I.K., 1993. Development of the trilateral flash cycle system. Part 1: fundamental considerations. In: Proceedings of the Institution of Mechanical Engineers, Part A, 1993, 207 (A3), 179-94.

Smith, I.K., 1999. Review of the development of two-phase screw expanders. City University, London (available from http://www.staff.city.ac.uk?~sj376/smith99.htm).

Smith, R.W., 1996. Kalina combined cycle performance and operability. Joint Power Generation Conference, Vol.2, ASME.

Smith, I.K., Pitanga Marques da Silva, R., 1994. Development of the trilateral flash cycle system. Part 2: increasing power output with working fluid mixtures. In: Proceedings of the Institution of Mechanical Engineers, Part A, 1994, 208 (A2), 135-44.

Smith, I.K., Stosic, N., Aldis, C.A., 1996. Development of the trilateral flash cycle system. Part 3: the design of high efficiency two-phase screw expanders. In: Proceedings of the Institution of Mechanical Engineers, Part A, 1996, 210 (A2), 75-93.

Song, K.H., Lu, H.M., Li, S.S., 1988. Experimental study on heating anaerobic digesters by solar pond. Acta Energiae Solaris Sinica, 9 (1): 41-8

Swift, A.H.P., Golding, P., Grant, K., Boegli, W.J., 1990. Demonstration of solar pond coupled desalting using a multistage, falling film, flash evaporator. In: Folchitto, S., Principi, P. (Eds.), Progress in Solar Ponds, 2nd International Conference, Rome, Italy, March 1990, pp. 399-416.

Tabor, H., 1981. Solar Ponds. Solar Energy, 27,181.

Tabor, H., Matz, R., 1965. Solar pond status report. Solar Energy, 9, 177-92.

Tamm, G., Goswami, D.Y., Lu, S., Hasan, A.A., 2004. Theoretical and Experimental Investigation of an Ammonia-Water Power and Refrigeration Thermodynamic Cycle. In: Proceedings of the World Solar Forum, International Solar Energy Society (ISES), Adelaide Australia, November 2001. Also in Solar Energy Journal, Vol. 76, pp. 217-228.

Tamm, G., Goswami, D.Y., 2003. A Novel Combined Power and Cooling Thermodynamic Cycle for Low Temperature Heat Sources–Part II: Experimental Investigation. ASME Journal of Solar Energy Engineering, Vol. 125, No. 2, May, pp. 223-229.

Tamm, G., Goswami, D.Y., Lu, S., Hasan, A., 2003. A Novel Combined Power and Cooling Thermodynamic Cycle for Low Temperature Heat Sources—Part I: Theoretical Investigation. ASME Journal of Solar Energy Engineering, Vol. 125, No. 2, May, pp. 218-222.

Taniguchi, H., Kudo, K., Giedt, W.H., Park, I., Kumazawa, S., 1988. Analytical and experimental investigation of two-phase flow screw expanders for power generation. Journal of Engineering for Gas Turbines and Power: Transactions of the ASME, 110, 628-35.

Tanny, J., Chai, A., Kit, E., 1995. On the law of turbulent enhancement across a density interface. Fluid Dynamics Research, 15 (1), 69-74.

Tanny, J., Gotlib, V.A., 1995. Linear stability of a double diffusive Layer with variable fluid properties. International Journal of Heat and Mass Transfer, 38, 9, pp. 1683-1691.

Tarbor , H., Doron, B., 1986. Solar Ponds-Lesson learned from the 150 kW(e) power plant at Ein Boqek. Proc ASME Solar Energy Div., Anaheim, California, 344.

Venegas, C., 1992. Development of a simple and practical method to modify the structure of the salinity gradients of a solar pond. Master's Thesis, University of Texas at El Paso.

Vijayaraghavan, S., Goswami, D.Y., 2003. On Evaluating Efficiency of a Combined Power and Cooling Cycle. In: Proceedings of the ASME Symposium on Thermodynamics and Design, Analysis and Improvement of Energy Systems, ASME International Mechanical Engineering Congress and Exposition, November 2002. New Orleans, LA, ASME Book AES Vol. 42. Also published in the ASME Journal of Energy Resources Technologies, Vol. 125, No. 3, pp.221-227.

Walton, J.C., Lu, H., Solis, S.S., 2000. Air gap membrane distillation by salinity gradient solar pond. Proceedings of ASME Solar Energy Conference, Madison, Wisconsin.

Weinberger, H., 1963. The physics of the solar pond. Solar Energy, 8, 2.

Xu, F., Goswami, D.Y., Bhagwat, S.S., 2000. A Combined Power/Cooling Cycle. Energy: The International Journal Vol. 25, No. 3, February, pp. 233-264.

Xu, H., Sandoval, J.A., Lu, H., Y'barra, A., Golding, P., Swift, A., 1993. Operating Experience with the El Paso Solar Pond. In: Golding, P., Sandoval, J., York, T., (Eds.), Proceedings of the 3rd International Conference on Progress in Solar Ponds, El Paso, TX, 69-84.

Xu, H., Golding, P., Nielsen, C.E., 1987. Prediction of Internal Stability in a Salt Gradient Solar Pond. Proceedings of the Conference International Progress in Solar Ponds, Electric Research Institute of Mexico, Cuernavaca, March 29-April 3, 107-120.

Xu, H., Golding, P., Swift, A.H.P., 1991. Method for Monitoring the Stability Within a Solar Pond Salinity Gradient. In: Mancini, T.R., Watanabe, K., Klett, D.E., (Eds),

Proceedings of the ASME Solar energy Division Conference–"Solar Engineering–1991," ASME Press, New York, 75.

Xu, H., Swift, A.H.P., Golding, P., 1992. Two gradient failure and repair events as experienced at the El Paso solar pond. In: Stine, W., Kreider, J., Watanabe, K., (Eds.), Trans. Am. Soc. Mech. Eng. Solar Energy Div.: 'Solar Engineering-1992,' ASME/Japan Solar Energy Soc./Korean Solar Energy Soc., 673 - 80.

Yamamoto, T., Furuhata, T., Arai, N., Mori, K., 2001. Design and testing of the Organic Rankine cycle. Energy, 26, 239-51.

Ye, X.M., Song, A.G., Li, S.S., 1987. Refractive index of salt solution in solar pond. Acta Energiae Solaris Sinica, 8 (1x), 95-7.

Zangrando, F., Bertram, L.A., 1984. The effect of variable stratification on linear double-diffusive stability. J. Fluid Mech., 125, 55.

Zangrando, F., 1979. Observation and analysis of a full-scale experimental solar pond. Ph.D. thesis, University of New Mexico.

Zangrando, F., 1980. A simple method to establish salt gradient solar ponds. Solar Energy, 25, 467.

Zangrando, F., Fernando, H.J.S., 1991. A predictive model for migration of double-diffusive interfaces. Solar Energy, 113, 59-65.

Zangrando, F., Johnstone, H.W., 1988. Effect of injection parameters on gradient formation. Proceedings of the ASME Solar Energy Division Conference, Denver.

Ziegler, G., 1898. An den Herausgeber des Prometheus: Absonderliche Temperaturehaltnisse in einem Solbenhalter. Prometheus, 9: 79 and discussion 9, 325.

7. **Appendix: Contact List**

Australia

Aliakbar Akbarzadeh
Energy Conservation and Renewable Energy Group,
School of Aerospace, Mechanical and Manufacturing Engineering,
RMIT University, Bundoora East Campus,
PO Box 71, Bundoora, Victoria,
Australia 3083
Email:aliakbar.akbarzadeh@rmit.edu.au

John Andrews
Research Fellow,
Energy Conservation and Renewable Energy Group,
School of Aerospace, Mechanical and Manufacturing Engineering,
PO Box 71, Bundoora, Victoria,
Australia 3083

RMIT University, Australia.
Email: andrews.john@rmit.edu.au

China

Li Shen Sheng
Department of Physics
Capital Normal University
Beijing 100037
People's Republic of China
Email: lijingm@agri.gov.cn

India

Amit Kumar
Fellow
Renewable Energy Technology Applications
Tata Energy Research Institute
New Delhi – 110 003
India
Email: akumar@teri.res.in

A.S. Mehta
Head
Business Development Discipline
Central Salt and Marine Chemicals Research Institute
Bhavnagar 364002
India
Email: salt@csir.res.in

N.D. Kaushika
Centre for Energy Studies
Indian Institute of Technology
Hauz Khas,
New Delhi 110016
India
Email: ndkaushika@yahoo.com

Iran

M.R Jaefarzadeh,
Department of Civil Engineering
Ferdowsi University
Mashad

Iran
Email: Jafarzadeh@hotmail.com

Israel

Gershon Grossman
Faculty of Mechanical Engineering
Technion, Israel Institute of Technology,
Hafa 3200, Israel
Email: merg0@techunix.technion.ac.il

Danny Rosing
Email: drosing@netvision.net.il

Joseph Tanny
Institute of Soil, Water and Environmental Sciences
Agricultural Research Organization
The Volcani Center
POB 6, Bet Dagan 50250
Israel
Email: tanai@volcani.agri.gov.il

Italy

Paolo Principi
Renewable Energy Laboratory
Dipartimento di Energetica
Universita degli Studi di Ancona
Via Brecce Bianche
I-60100 Ancona, Italy
Email: p.principi@unian.it

Japan

K.Kamiuto
High Temperature Heat Transfer Laboratory
Department of Production Systems Engineering
Oita University
Dannoharu 700, Oita 870-1192, Japan
Email: kamiuto@cc.oita-u.ac.jp

Mexico

Rafael Almanza
Institute de Ingenieria

Universidad Nacional Autonoma de Mexico
Ciudad Universitaria, 04510, Mexico, D.F. Mexico
ras@pumas.lingen.unam.mx

Portugal

Manuel Collares Pereira, and
Margarida Giestas Lima
INETI-DER, Edificio G
Estrada do Paco do Lumiar, 22
1649-038 Lisboa, Portugal
Email: margarida.giestas@mail.ineti.pt

Taiwan

Jian-Yuan Lin,
Researcher & Manager
Refrigeration Systems Laboratory
Energy & Resources Laboratories,
ITRI
Hsichu, Taiwan
Email: jyuanlin@itri.org.tw

Tunisia

M.Ouni
Department of Physics
Faculty of Sciences of Tunis
Tunis 1060, Tunisia
email: mohamed.ounit@fst.rnu.tn

United States of America

Peter Golding
Department of Metallurgical & Materials Engineering,
University of Texas at El Paso,
500 W University Avenue, El Paso, TX 79968, USA
Email: pgolding@utep.edu

Huanmin Lu
University of Texas at El Paso
500 W University Avenue, El Paso, TX 79968, USA
Email: huanmin@utep.edu

Stan Kleis
Department of Mechanical Engineering
Engineering Building One,
University of Houston, Houston, TX 77204-4006, USA
Email: kleis@uh.edu

Chapter 8

Passive Cooling of Buildings

by

Matheos Santamouris
Group Building Environmental Studies
Physics Department
University of Athens
Panepistimiopolis, Zografou, 157 84
Athens, Greece

Abstract

This chapter presents the state of the art on passive cooling technologies for buildings. The penetration of air conditioning is increasing rapidly all around the world. This has an important energy and environmental impact. During recent years, passive cooling techniques have received great attention which has resulted in important developments. The first part of the chapter discusses the actual penetration of air conditioning systems and problems resulting from using mechanical cooling. The second part classifies passive cooling systems while the third and fourth parts deal with the urban microclimate and possible improvements to urban thermal conditions. Finally, the fifth part discusses the recent progress on heat and solar control, thermal comfort and heat inertia techniques as well as heat dissipation methods.

Corresponding Author. E-mail: msantam@cc.uoa.gr

1. Introduction

Construction is one of the most important economic sectors worldwide. The total world's annual output of construction is close to $3 trillion and constitutes almost one-tenth of the global economy (CICA, 2002). About 30% of the business is in Europe, 22% in the United States, 21% in Japan, 23% in developing countries and 4% in the rest of the developed countries.

Buildings use almost 40% of the world's energy, 16% of the fresh water and 25% of the forest timber (UNCHS, 1993), while is responsible for almost 70% of emitted sulphur oxides and 50% of the CO_2 (Der Petrocian, 2001).

Energy consumption of the building sector is high. Although figures differ from country to country, buildings are responsible for about 30-40% of the total energy demand. Application of intensive energy conservation measures has stabilized energy consumption for heating in developed countries. However, energy needs for cooling increases in a dramatic way. The increase of family income in developed countries has made the use of air conditioning systems highly popular. In Europe the main commercial market for cooling and air conditioning systems totals 8 billion Euros. Almost 6% of office, commercial and industry buildings are cooled, making a total volume of about 20 million cubic meters (Adnot, 1999). The volume of air-conditioned buildings in Europe is expected to increase four times by the year 2010.

In the United States, the penetration of air conditioning is extremely high. More than 3.5 billion m^2 of commercial buildings are cooled. The total cooling energy consumption for the commercial sector is close to 250 Twh/y, while the necessary peak power demand for summer cooling of the commercial buildings is close to 109 GW.

The impact of air conditioner usage on electricity demand is an important problem as peak electricity load increases continuously, forcing utilities to build additional plants. In parallel, serious environmental problems are associated with the use of air conditioning.

Passive and hybrid cooling techniques involving microclimate improvements, heat and solar protection, and heat modulation and dissipation methods and systems can greatly contribute to buildings' cooling load reduction and increase thermal comfort during the summer.

Results of the European Research Project PASCOOL (Santamouris and Argiriou, 1997) showed improved knowledge on this specific topic and develop design tools, advanced techniques to better implement natural cooling techniques and new techniques to characterize the performance of passive cooling components have been developed as an aid to designers (Santamouris and Argiriou, 1997).

This paper aims to present the actual state of the art in the field of passive cooling of buildings. The first part of the paper discusses the actual penetration of air conditioning systems and problems resulting from using mechanical cooling. The second part classifies passive cooling systems while the third and fourth parts deal with the urban microclimate and possible improvements to urban thermal conditions. Following that we will discuss recent progress on heat and solar control, thermal comfort and heat inertia techniques as well as heat dissipation methods.

2.   On the Cooling Needs of Buildings

Increased living standards in the developed world using non-climatically responsive architectural standards have made air conditioning quite popular. Importantly, this has increased energy consumption in the building sector. Actually there are more than 240 million air conditioning units and 110 heat pumps installed worldwide according to the International Institute of Refrigeration (IIR) (IIR, 2002). IIR's study shows that the refrigeration and air conditioning sectors consume about 15% of all electricity consumed worldwide (IIR, 2002). In Europe it is estimated that air conditioning increases the total energy consumption of commercial buildings on average to about 40 $kWh/m^2/year$ (Burton, 2001).

It is evident that the total energy consumption of buildings for cooling purposes varies as a function of the quality of design and climatic conditions. In hot climates, as in the Mediterranean, commercial buildings with appropriate heat and solar protection and careful management of internal loads may reduce their cooling load down to 5 $kWh/m^2/year$, (Santamouris and Daskalaki, 1998), while buildings of low quality environmental design may present loads up to 450 $kWh/m^2/y$ (Santamouris, 1997). Under the same climatic conditions and when internal gains are not important, such as in residential buildings, the use of air conditioning may be completely avoided when efficient solar and heat protection as well as heat modulation techniques are used.

2.1 Recent Penetration of A/C Systems

Referring to the IIR data there are almost 79 million room air conditioners, 89 million duct-free and split systems, 55 million ducted split systems, 16 million unitary commercial systems and almost 856,000 water-based air conditioners.

Annual sales of air conditioning equipment approach a level of $60 billion, of which $20.9 billion are spent for room air units, $15.7 billion for packaged systems, $6.5 billion for rooftop units, and $12.3 billion for residential heat pumps (IIR, 2002). This is almost equivalent to 10% of the automobile industry on a worldwide basis.

The air conditioning market is expanding continuously. According to the Japan Air Conditioning and Refrigeration News (JARN) and Japan Refrigeration and Air Conditioning Industry Association (JRAIA) in 1998 the total annual number of sales was close to 35,188,000 units; by 2000 it had increased to 41,874,000 units and to 44,614,000 units by 2002, with a predicted level of 52,287,000 units in 2006 (JARN and JRAIA, 2002).

Most of the units are installed in North America, where the sales are not expected to increase further. In Europe, an increase of 22.3% is expected between 2002 and 2006, while the corresponding increases are expected to be 39.2% for the remainder of Asia, 23.2% for Oceania, 13.6% for Africa, 13.3% for South America and 10.5% for Middle East (Table 1).

The predicted 2002-2006 progression is more marked for room air conditioners, (+20%) as shown in Table 2, than for packaged air conditioners as shown in Table 3, (JARN and JRAIA, 2002).

Air conditioning penetration in Europe is much lower than in Japan and the USA. According to the "Energy Efficiency of Room Air-Conditioners" (EERAC) study, the 1997 penetration rate of room air conditioners was less than 5% in the residential sector and less than 27% in the tertiary sector (Adnot, 1999). The corresponding penetration rate in the tertiary sector is almost 100% in Japan and 80% in the USA, while almost 85% and 65% respectively of the residential buildings in Japan and USA have at least one air conditioner (see Table 4).

As shown in Table 5 almost 74% of the total stock of air conditioners in Europe are installed in Southern countries (Adnot, 1999). Italy has the highest number of installed units (29%), while split systems are the most common units and represent more than 60% of the appliances (Adnot, 1999).

In the United States, the number of households having a central air conditioning has increased from 17.6 million in 1978 to 47.8 in 1997. In the same period the number of households with room air conditioners increased from 25.1 to 25.8 millions (see Table 6) (EIA, 1997). The electricity consumption for air conditioning during the same period has increased from 0.31 to 0.42 Quadrillion of Btu (kWh or Twh) The main reason that electricity consumption has not followed the rate of penetration of air conditioners is the increased efficiency of air conditioners.

The US commercial sector's total energy cooling consumption is close to 250 Twh/y, while the necessary peak power demand for summer cooling is close to 109 GW. As shown in Table 7 in 1997 American households spent almost $140 per year for air conditioning, while almost 40% used their air conditioner all summer (EIA, 1997).

Most office buildings in the US have at least one air conditioning system (Table 8). Packaged air conditioners, central systems and heat pumps are the more popular systems (EIA, 1995).

2.2 Main Problems of Air Conditioning

There are different problems associated with the use of air conditioning. Apart from the serious increase of the absolute energy consumption of buildings, other important impacts include:

- The increase of the peak electricity load;

- Environmental problems associated with the ozone depletion and global warming;

- Indoor air quality problems.

Table 1. Actual and forecasted air conditioning sales in the world (JARN and JRAIA, 2002).

	1998 Actual	1999 Actual	2000 Actual	2001 Actual	2002 Projected	2003 Forecast	2004 Forecast	2005 Forecast	2006 Forecast
World total	35,188	38,500	41,874	44,834	44,614	46,243	47,975	50,111	52,287
Japan	7,270	7,121	7,791	8,367	7,546	7,479	7,344	7,459	7,450
Asia (excl. Japan)	11,392	11,873	13,897	16,637	16,313	17,705	19,227	20,890	22,705
Middle East	1,720	1,804	1,870	1,915	1,960	2,010	2,060	2,112	2,166
Europe	1,731	2,472	2,709	2,734	3,002	3,157	3,318	3,489	3,670
North America	10,437	12,408	12,322	11,894	12,521	12,522	12,524	12,525	12,525
Central & South America	1,588	1,665	2,109	1,939	1,866	1,906	1,973	2,043	2,114
Africa	511	670	664	758	781	806	833	861	887
Oceania	539	487	512	593	625	659	693	731	770

in thousands of units

Table 2. Actual and forecasted room air conditioning sales worldwide (JARN and JRAIA, 2002).

	1998 Actual	1999 Actual	2000 Actual	2001 Actual	2002 Projected	2003 Forecast	2004 Forecast	2005 Forecast	2006 Forecast
World total	35,188	38,500	41,874	44,834	44,614	46,243	47,975	50,111	52,287
Japan	7,270	7,121	7,791	8,367	7,546	7,479	7,344	7,459	7,450
Asia (excl. Japan)	11,392	11,873	13,897	16,637	16,313	17,705	19,227	20,890	22,705
Middle East	1,720	1,804	1,870	1,915	1,960	2,010	2,060	2,112	2,166
Europe	1,731	2,472	2,709	2,734	3,002	3,157	3,318	3,489	3,670
North America	10,437	12,408	12,322	11,894	12,521	12,522	12,524	12,525	12,525
Central & South America	1,588	1,665	2,109	1,939	1,866	1,906	1,973	2,043	2,114
Africa	511	670	664	758	781	806	833	861	887
Oceania	539	487	512	593	625	659	693	731	770

in thousands of units

Table 3. Actual and forecasted packaged air conditioning sales worldwide (JARN and JRAIA, 2002).

	1998 Actual	1999 Actual	2000 Actual	2001 Actual	2002 Projected	2003 Forecast	2004 Forecast	2005 Forecast	2006 Forecast
World total	35,188	38,500	41,874	44,834	44,614	46,243	47,975	50,111	52,287
Japan	7,270	7,121	7,791	8,367	7,546	7,479	7,344	7,459	7,450
Asia (excl. Japan)	11,392	11,873	13,897	16,637	16,313	17,705	19,227	20,890	22,705
Middle East	1,720	1,804	1,870	1,915	1,960	2,010	2,060	2,112	2,166
Europe	1,731	2,472	2,709	2,734	3,002	3,157	3,318	3,489	3,670
North America	10,437	12,408	12,322	11,894	12,521	12,522	12,524	12,525	12,525
Central & South America	1,588	1,665	2,109	1,939	1,866	1,906	1,973	2,043	2,114
Africa	511	670	664	758	781	806	833	861	887
Oceania	539	487	512	593	625	659	693	731	770

in thousands of units

Table 4. Penetration of room air conditioners in the tertiary and residential sector in US, Japan and Europe for 1997 (Adnot, 1999).

COUNTRY	TERTIARY	RESIDENTIAL
Japan	100%	85%
USA	80%	65%
Europe	<27%	<5%

Table 5. 1996 stock of different systems of room air conditioners in Europe (Adnot 1999).

	Split	Multi-split	Windows	Single-duct	Total
Austria	33 400	21 300	16 600	7 700	79 000
France	752 000	183 850	106 500	216 750	1 259 100
Germany	198 600	59 600	74 500	193 400	526 100
Greece	138 000	51 830	555 000		744 830
Italy	1 504 697	90 177	134 860	382 006	2 111 740
Spain	972 000		245 000	152 000	1 369 000
Portugal	267 157	30 143	17 720	7 800	322 820
UK	516 690		54 867	107 755	674 412
Others	119 160	31 100	44 700	116 040	315 660
Total EU	4 501 534	468 000	1 249 747	1 183 451	7 402 662
%	61%	6%	17%	16%	100%

Table 6. Consumption of electricity for air-conditioning and associated factors by survey year (EIA 1997).

Survey Year	Household Electricity Consumption for Air-Conditioning (TWh)	Number of Households with Central Air-Conditioning (million)	Number of Households with Room Air-Conditioning (million)	Average SEER of Central Air-Conditioning Units Sold During the Year
1978	1.06	17.6	25.1	7.34
1980	1.09	22.2	24.5	7.55
1981	1.12	22.4	26.0	7.78
1982	1.13	23.4	25.3	8.31
1984	1.09	25.7	25.8	8.66
1987	1.50	30.7	26.9	8.97
1990	1.64	36.6	27.1	9.31
1993	1.57	42.1	24.1	10.56
1997	1.43	47.8	25.8	10.66

Table 7. Electric air conditioning expenditures and usage – 1997 (EIA, 1997).

Average Annual Expenditures for Households Having Electric Air Conditioning	Of all Households, Percent Having Electric Air-Conditioners	Of all Households Having Electric Air Conditioners, Percent That Used Them All Summer	Of all Households, Percent That Have Electric Air Conditioners and Used Them All Summer
$140	72	40	30

Table 8. Number and percent of office buildings in each size category by cooling (EIA, 1995).

	Number of Buildings (thousand)			Percent of Buildings		
	Small	Medium	Large	Small	Medium	Large
All Office Buildings						
Residential-Type Central A/C	122	53	-	30	20	-
Heat Pumps	91	43	5	22	16	12
Individual A/C	-	11	-	-	4	-
District Chilled Water	-	-	3	-	-	7
Central Chillers	-	11	15	-	4	39
Packaged A/C Units	126	124	12	31	48	30

High peak electricity loads oblige utilities to build additional plants in order to satisfy the demand, but as these plants are used for short periods, the average cost of electricity increases considerably. Southern European countries face a very steep increase of their peak electricity load mainly because of the very rapid penetration of air conditioning. For example, Italy faced significant electricity problems during the summer of 2003 because of the high electricity demand of air conditioners. Actual load curves for 1995 and the expected future peak electricity load in Spain, are given in Figure 1 (Adnot, 1999). It is expected that the future increase of the peak load may necessitate doubling installed power by 2020.

Air conditioning's significant increase in peak electricity demand is partly blamed for the California energy system collapse. Besant-Jones and Tennebaum noted that in the summer of 2002 the demand for electricity was increased by air conditioning loads because of the highest recorded temperatures for 106 years (Besant-Jones and Tenenbaum, 2001). As a consequence the electricity demand increased and the supply started to fall below demand–thus dramatically increasing California's electricity prices. The market price in the day-ahead Cal PX which was between $25 and $50/MWh in the early months of 2000, had increased to $150/MWh during the summer months of the same year (Besant-Jones and Tenenbaum, 2001).

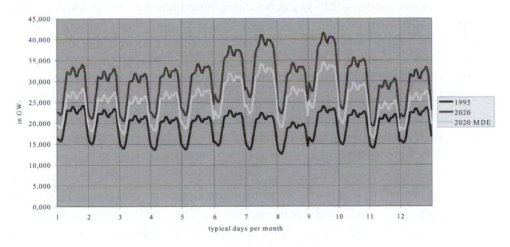

Figure 1. Load curves for 1995 and 2020 in Spain (Adnot, 1999).

The main environmental problems of air conditioning are associated with:

- Emissions from refrigerants used in air conditioning which adversely impact ozone levels and global climate. Refrigeration and air conditioning related emissions represent almost 64% of all CFC's and HCFC's produced (AFEAS, 2001).

- Cooling systems' energy consumption contribute to CO_2 emissions.

Refrigerant gases used in air conditioning are CFC's, HCFC's or HFC's. Chlorofluorocarbons have a very important impact on ozone depletion and they also exert global warming effects. According to the Montreal Protocol, CFC's production and use was banned by 1996 in the developed countries and must be stopped by 2010 in the developing countries. In Europe, "Regulation 2037/2000" totally bans their use for maintenance and servicing of equipment as of January 1, 2001.

Hydrochlorofluorocarbons, HCFC's have less of an impact on ozone depletion and a lower global warming potential than the CFC's. The Montreal Protocol has banned the use of HCFC's in developed countries by 2030 and by 2040 in developing countries. In Europe, HCFC's production will be banned as of 2025, and by 2010 the use of virgin HCFC's will be banned for maintenance and servicing of equipment.

Hydrofluorocarbons, HFC's, do not have an ozone depleting effect and their global warming impact is less than CFC's. An important market shift is being made to HFC's (R-407C and increasingly R-410A) in Japan—where more than 50% of room air conditioners produced are HFC models. Unlike the Japanese and European markets, the US industry keeps R22 (HCFC) equipment longer and aims to shift directly to R410A instead of going through R407C (HFCs).

Air conditioning systems may be an important source of indoor contamination. Cooling coils and condensate trays can become contaminated with organic dust. This may lead to microbial growth. The organic dust may also cause mold and fungal growth in fans and fan housings. Inefficient and dirty filters may also lead to unfiltered air in the building. Contaminated emissions from cooling towers may cause spread of diseases like *Legionelea* from poorly maintained systems. A very comprehensive analysis of all studies related to indoor air quality problems caused by HVAC systems is given by Limb (2000) and Lloyd (1992).

2.3 Recent Developments on the Field of Air Conditioning, High Efficiency A/C Systems

Air conditioning efficiency has increased substantially recently. According to the International Institute of Refrigeration (IIR), the air conditioning industry has improved equipment efficiency in line with recent regulations (IIR, 2003). As mentioned previously, Japan's target COP level for 2004 is almost twice as high as that in 1985. In parallel certain manufacturers in the US have begun marketing air conditioners achieving Seasonal Energy Efficiency Ratios (SEER) of 16 to 19.5.

In fact, a recent study investigating potential improvements of room air conditioners in Europe, shows that an increase of the COP up to 30% is quite feasible (Adnot, 1999). A list of possible improvements and the corresponding increases of the COP are given in Table 9. As a result of this study, the European Commission has published Directive 2002/31/EC dated March 22, 2002, related to "energy labeling of household air conditioners." According to the directive, household air conditioners will have to be labeled mentioning "the annual energy consumption (kWh) in cooling mode, the cooling output, and the energy efficiency ratio full load."

In parallel to room air conditioners there is an important improvement of chiller efficiency. According to Bivens, the chilller industry achieved a 33% reduction in energy consumption between 1978 and 1998 (Bivens, 1999). The efficiency of absorption chillers running with natural gas or waste heat has also improved tremendously (Bivens, 1999). While the COP of a typical absorption system is close to 0.7, the use of multi-effect technology may increase it up to 1.5 (IIR, 2002).

2.4 Alternative Techniques to Air Conditioning - Passive Cooling

Addressing successful solutions to reduce energy and environmental effects of air conditioning is a strong requirement for the future. Possible solutions include:
1. Adaptation of buildings to the specific environmental conditions of cities in order to efficiently incorporate energy efficient renewable technologies to address the radical changes and transformations of the radiative, thermal, moisture and aerodynamic characteristics of the urban environment. This involves the use of passive and hybrid cooling techniques to decrease cooling energy consumption and improve thermal comfort.

Table 9. Possible improvements of room air conditioners and the corresponding COP (Adnot, 1999).

No	Scenario	Efficiency/COP
0	Existing Situation	2.72
1a	Increase of frontal coil area (evaporator+condenser) by 15%	2.81
1b	Increase of frontal coil area (evaporator+condenser) by 30%	2.88
2a	Increase of coil depth (evaporator+condenser) by adding 1 row of tubes	2.97
2b	Increase of coil depth (evaporator+condenser) by adding 2 rows of tubes	3.09
3a	Increase of coil fin density (evaporator+condenser) by 10%	2.76
3b	Increase of coil fin density (evaporator+condenser) by 20%	2.80
4	Addition of subcooler	2.75
5	Improvement of fins	2.85
6	Improvement of tubes	2.87
7a	Improvement of fans using PSC motors	2.74
7b	Improvement of fans using ECM motors	2.75
8a	Improvement of compressor efficiency by 5%	2.79
8b	Improvement of compressor efficiency by 10%	2.87
8c	Improvement of compressor efficiency by 15%	2.94
9	Increase of heat transfer area in coils (combination of scenarios 1b, 2b and 3b)	3.22
10	Improvement of fins and tubes - increase of heat transfer coefficient (combination of scenarios 5 and 6)	3.14
11	Scenario 10 + Improvement of compressor efficiency by 15%	3.39
12	Scenario 9 + Improvement of compressor efficiency by 15%	3.48
13	Scenario 9 + Scenario 10	3.32
14	Scenario 9 + Scenario 10 + Improvement of compressor efficiency by 15%	3.58

2. Improvement of the urban microclimate to fight the effect of heat island and temperature increase and the corresponding increase of the cooling demand in buildings. This may involve the use of more appropriate materials, increased use of green areas, use of cool sinks for heat dissipation and appropriate layout of urban canopies.

Additionally, alternative strategies may be followed to decrease the impact of air conditioning, such as:

- Utilizing centralized or semi-centralized cooling production, and distribution networks based on renewable energies or waste heat (district cooling), together with demand-side management actions like local or remote cycling, (Papadopoulos et al., 2003).

- Using more efficient air conditioning equipment for individual buildings with optimized COP curves, using renewable sources or waste heat.

None of these can be seen as isolated areas of concern. The interrelated nature of the parameters defining performance efficiency of buildings during summer requires that practical actions be undertaken as an integrated approach.

3. Principles of Passive Cooling Techniques

Passive cooling techniques in buildings have proven to be extremely effective and can greatly contribute in decreasing the cooling load of buildings. Efficient passive systems and techniques have been designed and tested. Passive cooling has also proven to provide excellent thermal comfort and indoor air quality, together with a very low energy consumption.

When a building's internal and solar gains are sufficiently reduced, a *lean climatization concept* can be developed (Reinhart et al., 2001). The term *lean* signifies that the system is energy efficient so that only the amount of electricity needed to run fans and circulation pumps is required to maintain comfortable indoor temperatures year-round.

3.1 Classification of Passive Cooling Techniques

Passive cooling techniques can be classified in three main categories (Santamouris and Assimakopoulos, 1996):

a. **Solar and Heat Protection Techniques.** Protection from solar and heat gains may involve: Landscaping, and the use of outdoor and semi-outdoor spaces, building form, layout and external finishing, solar control and shading of building surfaces, thermal insulation, control of internal gains, etc.

b. **Heat Modulation Techniques.** Modulation of heat gain deals with the thermal storage capacity of the building structure. This strategy provides attenuation of peaks in cooling load and modulation of internal temperature with heat discharge at a later time. The larger the swings in outdoor temperature, the more important the effect of such storage capacity. The cycle of heat storage and discharge must be combined with means of heat dissipation, like night ventilation, so that the discharge phase does not add to overheating.

c. **Heat dissipation techniques.** These techniques deal with the potential for disposal of excess heat of the building to an environmental sink of lower temperature. Dissipation of the excess heat depends on two main conditions: 1) The availability of an appropriate environmental heat sink; and 2) The establishment of an appropriate thermal coupling between the building and the sink as well as sufficient temperature differences for the transfer of heat. The main processes of heat dissipation techniques are: ground cooling based on the use of the soil, and convective and evaporative cooling using the air as the sink, as well as water and radiative cooling using the sky as the heat sink. The potential of heat dissipation techniques strongly depends on climatic conditions. When heat transfer is assisted by mechanical devices, the techniques are known as hybrid cooling.

3.2 Microclimate Issues

Climate is the average of the atmospheric conditions over an extended period of time over a large region. Small-scale patterns of climate, resulting from the influence of topography, soil structure, and ground and urban forms, are known as Microclimates. Temperature, solar radiation, humidity and wind are the principal parameters that define the local climate. The energy balance of the 'earth surface-ambient air', system in an area is governed by the energy losses, gains, and the energy stored in the urban infrastructure and mainly in the opaque elements of buildings, such as Energy Gains = Energy Losses + Energy Storage. Thus:

Energy gains involve the sum of both solar and long wave radiation emitted by opaque elements (building, streets, etc.), as well as the anthropogenic heat, related to transportation systems, power generation and other heat sources.

Energy losses involve sensible or latent heat transfers resulting from evaporation, heat convection between the surfaces and the air, as well as heat transfer between the area and the surrounding environment.

Improvement of the local microclimate during the summer period, mainly decreases of the ambient temperature, necessitates reduction of solar and heat gains and an increase of thermal losses. Appropriate landscaping involving the use of reflective–cool–materials, vegetation and water sources can highly contribute to decrease ambient temperatures.

Given that the world is becoming more and more urbanized, specific attention must be given to the urban climate. The thermal balance in the urban environment differs substantially from that of rural areas. More thermal gains are added such as high anthropogenic heat released by cars and combustion systems, higher amounts of stored solar radiation, and blockage of the emitted infrared radiation by urban canyons. Thus, the global thermal balance becomes more positive and this contributes to the warming of the environment.

As a consequence of heat balance, air temperatures in densely built urban areas are higher than the temperatures of the surrounding rural country. This phenomenon is known as 'heat island', which has an adverse impact on the energy consumption of buildings for cooling. Also, wind speed between buildings, is seriously decreased compared to the undisturbed wind speed. This phenomenon, known as 'canyon effect,' is mainly due to the specific roughness of a city and channeling effects through canyons.

Improvement of the urban microclimate is a priority area of research and has received a great deal of attention during the last few years. Techniques to fight heat island and improve the thermal balance of cities have been proposed, developed and implemented successfully.

3.3 Solar Control and Heat Protection

'Solar control' deals with the permanent or temporal reduction of transmitted solar radiation through a transparent building component. In cooling-dominated climates solar control devices should reduce as much as possible the solar gains and still admit sufficient daylight and visual contact with the outside. This may be achieved either with the combination of a window with an external or internal shading device or by using variable transmittance glazing (switchable glazing). The overall solar control performance of a window is expressed in terms of the so-called g-value which is the sum of the transmitted solar radiation and of the heat gains resulting from the sunlight absorbed in the glazing unit.

Important research has been carried out to improve the global thermal and visual performance of solar control devices, which are commercially available (Wilson, 1999, 2000). Switchable glazing technology has been considerably improved and electrochromic glazing is commercially available. Multifunctional façade components able to provide ventilation, daylight, solar control and other energy benefits have been designed and tested and are expected to enter the market soon (Cromvall, 2001).

Heat protection of the building has to do mainly with the flow of heat between the ambient environment and the building. This is characterized by the U value of both the opaque and the transparent components of the building. During the last years tremendous improvements have been achieved regarding the thermal quality of transparent elements. Low –e coated glazing, noble gas filled windows, triple glazed windows and transparent insulation are already in the market. As shown in Figure 2, low –e windows in Germany now hold a market share of about 90% (Reinhart et al., 2001). According to the European Glass Industry, replacement of single glazed windows in Europe with double low –e glass, will save almost 75×10^{15}J, 10.4 million Euros, and 57.7 million tons of CO_2 (GEPVP, 2001). Similarly, replacement of simple double glazed windows with low –e glass will contribute to save almost 340×10^{15}J, 3.95 million Euros and 24.6 million tons of CO_2.

3.4 Heat Modulation

A very efficient way to reduce indoor air temperatures and cooling load peaks, is to store excess heat in the structural materials of the building, which is referred to as 'thermal mass.' Constructed of material with high heat capacity, such as poured concrete, bricks and tiles, it is typically contained in walls, partitions, ceilings and floors.

The rate of heat transfer through building materials and the effectiveness of thermal mass is determined by a number of parameters and conditions. Optimization of thermal mass levels depends on the properties of the building materials, building orientation, thermal insulation, ventilation, climatic conditions, use of auxiliary cooling systems, and occupancy patterns. For a wall material to store heat effectively, it must have high thermal capacity and a high thermal conductivity value, so that heat may penetrate through the wall during the heat charging and discharging periods.

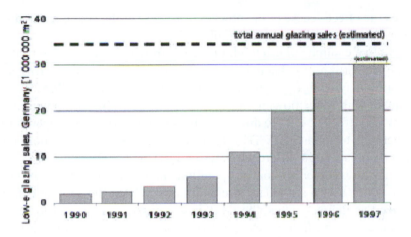

Figure 2. Development of low-e glazing sales in Germany (Reinhart C.F. et al, 2001).

The distribution of thermal mass is based on the orientation of the given surface and the desirable time lag (Lechner, 1991). During the night when outdoor temperatures are lower than indoor temperatures, it is possible to cool the structural mass of the building by natural ventilation. Such a technique, known as 'night ventilation' may contribute in decreasing the cooling load of air conditioned buildings up to 60%, or decreasing the overheating hours of "free-floating buildings," i.e., of buildings not using a cooling system (no mechanical air conditioning) up to 75% (Geros et al., 1999).

Phase change materials incorporated in plaster increase the heat storage capacity in the building and thus contribute in decreasing the average indoor temperatures.

3.5 Heat Dissipation and Hybrid Cooling

As previously mentioned, heat dissipation techniques are based on the transfer of a buildings' excess heat to a low temperature environmental sink. Main sinks are the ambient air, water, the ground and the sky. When heat is dissipated to the ambient air, the technique is known as 'convective cooling;' when water is used the process is known as 'evaporative cooling;' when the ground or the sky are the sinks, the techniques are known as 'ground and radiative cooling' respectively.

Convective cooling by (ventilation is a very effective method to improve indoor comfort, indoor air quality and reduce temperature. Higher air speeds inside the building may enhance thermal comfort when they do not exceed certain values. The technique is usually limited to night time ventilation however daytime ventilation may be used when ambient temperature is lower than indoor temperature.

Convective cooling may be natural, mechanical or hybrid. Natural ventilation is due either to wind forces, temperature differences, or both (Allard, 1998). A serious reduction of the cooling potential is observed in dense urban environments as a result of the dramatic decrease of the wind speed in cities (Geros et al., 2001). Careful positioning of the openings in naturally ventilated buildings is a crucial parameter that determines the effectiveness of the process. A review of the sizing methodologies is given by Anthienitis and Santamouris (2002).

Evaporative cooling applies to all processes in which the sensible heat in an air stream is exchanged for the latent heat of water droplets or wetted surfaces. Evaporative cooling may be direct or indirect. In direct evaporative coolers, air comes in direct contact with water flowing through fibrous pads. The air temperature is thus reduced by about 70-80% of the difference between the dry bulb temperature (DBT) and the wet bulb temperature (WBT). Therefore, direct evaporative cooling is effective when there is a large difference between DBT and WBT.

When the air is cooled without any addition of moisture by passing through a heat exchanger which uses a secondary stream of air or water, the cooling equipment is characterized as indirect. Thus, the DBT is decreased without any increase of the air's moisture content. Indirect evaporative coolers are based on the use of a heat exchanger where the indoor ventilated air passes through the primary circuit where evaporation occurs while the fresh air passes through the secondary circuit. Energy savings of up to 60% compared to mechanical A/C may be achieved in hot dry regions (Santamouris and Assimakopoulos, 1996).

During the summer the soil temperature at a certain depth is considerably lower than the ambient temperature (Mihalakakou et al., 1992). Therefore, the ground offers an important opportunity for dissipation of the buildings' excess heat. There are two strategies for the use of the ground:

- Direct Earth contact cooling techniques, and

- Buried pipes cooling

Earth-contact buildings offer various advantages, i.e., limited infiltration and heat losses, solar and heat protection, reduction of noise and vibration, fire and storm protection and improved security. However they are not free of disadvantages. Inside condensation, slow response to changing conditions, poor daylighting and poor indoor air quality are frequent problems (Carmody et al., 1985).

The concept of buried pipes involves the use of metallic or PVC pipes buried at 1 to 4m in depth (Sinha and Goswami, 1981). Ambient or indoor air is delivered via the tubes where it is precooled and then delivered to the building. When outdoor air is circulated into the pipes the system is characterized as an 'open loop system'; when the indoor air is circulated from the building through the tubes the system is known as a 'closed loop system'(Goswami and Ileslamlou, 1990). The performance of the buried pipes is a function of the inlet air temperature, the ground temperature, the thermal characteristics of the pipes and soil as well as of the air velocity, the pipe dimension and the pipes' depth (Goswami and Dhaliwal, 1985). Application of buried pipes cooling techniques in buildings has shown that the energy benefits are very important (Tombazis et al., 1990). Advanced modeling techniques have been developed quite recently that permit a very accurate sizing of the system under different boundary conditions (Mihalakakou et al., 1994, 1994b, 1995, 1996).

Radiative cooling is based on heat loss by long wave radiation emission from a building to night sky at a much lower temperature.

There are two methods of applying radiative cooling in buildings: direct, or passive radiative cooling, and hybrid radiative cooling. In the first, the building envelope radiates towards the sky and gets cooler, producing a heat loss from the interior of the building. For physical reasons, the part of the building envelope that radiates the most is a flat roof. In the second case, the radiator is not the building envelope but usually a metal plate. The operation of such a radiator is the opposite of an air flat-plate solar collector. Air is cooled by circulating under the metal plate before being injected into the building. Other systems are combinations of these two configurations.

4. Urban Microclimate and Its Impact on the Cooling Needs of Buildings

Cities or urban areas are defined as the physical environment that is composed of:

"a complex mix of natural elements including air, water, land, climate, flora and fauna, and the built environment that is constructed or modified for human habitation and activity, encompassing buildings, infrastructure and urban open spaces (Hardy et al., 2001)."

The last 50 years was a period of the most intensive urbanization that our planet has ever experienced. Urban population has increased from 160 million to about 3 billion in just 100 years, and it is expected to increase to about 5 billion by 2025.

An increase of the urban population by 1% increases the energy consumption by 2.2% — thus the rate of change in energy use is twice the rate of change in urbanization (Jones, 1992).

Increasing urbanization has deteriorated the urban environment. Deficiencies in development control have seriously impacted the urban climate and environmental performance of urban buildings. As reported by Akbari, New York City has lost 175,000 trees, or 20% of its urban forest in the last ten years (Akbari et al, 1992).

As a consequence of heat balance, air temperatures in densely built urban areas are higher than the temperatures of the surrounding rural country. This phenomenon known as 'heat island', exacerbates electricity demand for air conditioning of buildings and increases smog production, while contributing to increased emission of pollutants from power plants.

4.1 Temperature Increase in Cities–Heat Island Phenomenon – Canyon Effect

The heat island phenomenon is mainly due to the following parameters (Oke et al, 1991):

a. Complex radiative exchange between buildings and the screening of the skyline contributes to decreasing the long wave radiation from within street canyons,

b. Large thermal mass of the buildings that stores sensible heat,

c. the anthropogenic heat released from transport, air conditioning, industry, other combustion processes and animal metabolism,

d. the urban greenhouse due to the polluted and warmer urban atmosphere,

e. the canyon radiative geometry that decreases the effective albedo of cities because of the multiple reflection of short wave radiation between the canyon surfaces,

f. the reduction of evaporating surfaces in the urban areas.

g. the reduced turbulent transfer of heat from within streets.

Besides the temperature increase, cities affect many other climatological parameters. As mentioned by Landsberg, solar radiation is seriously reduced because of increased scattering and absorption, while the sunshine duration in industrial cities is reduced by 10 to 20% in comparison with the surrounding countryside (Landsberg, 1981).

Wind speed and direction in cities is seriously decreased compared to the undisturbed wind speed. This is mainly due to the specific roughness of a city, channeling effects through canyons and also because of the heat island effect (Santamouris, 2001). Estimation of the wind speed is of vital importance for passive cooling applications and especially in the design of naturally ventilated buildings (Papadopoulos, 2001). Routine wind speed measurements above buildings or at airports differ considerably from the speed at an urban monitoring site. Quite recently, appropriate algorithms to estimate wind characteristics in urban canyons for naturally ventilated buildings have been developed (Georgakis, 2003).

4.2 Documentation of Heat Island in Major Cities

Urban heat island studies usually refer to 'urban heat island intensity,' which is the maximum temperature difference between the city and the surrounding area. Data compiled by various sources shows that heat island intensity can be as high as 15°C (Santamouris, 2001).

Studies on the intensity of heat island have been performed for many European cities. The work of Watkins et al. (2002) was based on measurements performed in summer of 1999, reported a heat island intensity for London close to 7°C. This value is much higher than the values given by Lyall (1977) and Chandler (1965). Lyall (1977) reported that the magnitude of the nocturnal heat island averaged over June-July 1976 was of the order of 2.5°C . This is not much lower than a daily upper decile limit of 3.1°C found by Chandler (1965), in a comparison of Kensington and Wisley from 1951-60.

Multiyear measurements performed in the Athens (Greece) area have shown that during summer the maximum heat island intensity in the very central area is close to 16°C, and the mean value for the major central area of Athens is close to 12°C. In parallel, cooling degree hours in the central area of the city is about 350% higher than in the suburban areas (Santamouris and Georgakis, 2003).

Data for Göteborg and Malmo in Sweden are reported by Eliasson (1996), and Barring et al. (1985), respectively. The Göteburg data shows an urban heat island ranging from 3.5°C in winter to 6°C in summer, while in Malmo, a mean heat island intensity close to 7°C has been found.

Escourrou (1990/91), reported data on the heat island intensity in Paris, France. As stated, a horizontal thermic close to 14°C has been recorded gradient between Paris and the suburbs.

Heat island data are available for three German cities. Swaid and Hoffman (1990) have reported limited data on the heat island intensity in Essen, Germany, for September 1986. The observed heat island intensity was between 3-4°C for both the day and night period. Kuttler et al. (1996) reported a night time difference of 6°C temperature between the urban and rural areas for Stolberg.

Data on the heat island for various Swiss cities are reported by Wanner (1983). For Bale and Berne the heat island intensity was close to 6°C, while for Biel and Freiburg it was 5°C, and for Zurich it was close to 7°C.

Finally, Abbate (1998), using satellite data for Rome, Italy, reports important temperature differences between high density urban areas and low density urban and agricultural areas.

In the United States, Akbari et al. (1992) presented trends in absolute urban temperatures in several cities. The overall analysis is based on the use of average annual and maximum annual temperatures. The observed increase of temperatures in North American cities becomes clear when the cooling degree days corresponding to urban and rural stations are compared. Taha (1997) reported the increase of the cooling degree days due to urbanization and heat island effects for selected North American locations (see Table 10). As shown, the difference of the cooling degree days can be as high as 92%, while the minimum difference is more than 10%.

Heat island studies are available for three Canadian cities, Montreal, Edmonton and Calgary. Oke and East's 1971 study of heat island characteristics in Montreal showed a maximum intensity during winter nights close to 10.5°C. Hage (1972) studied heat island conditions at Edmonton, Alberta and found a heat island intensity of about 6.5°C.

Table 10. Increase of the cooling degree days due to urbanization and heat island effects. Averages for selected locations for the period 1941-1970 (Taha, 1997).

Location	Urban	Airport	Difference (%)
Los Angeles	368	191	92
Washington DC	440	361	21
St. Louis	510	459	11
New York	333	268	24
Baltimore	464	344	35
Seattle	111	72	54
Detroit	416	366	14
Chicago	463	372	24
Denver	416	350	19

Numerous studies on heat island intensity of tropical cities have been presented by the World Meteorological Organization (WMO, 1986; 1994), and Givoni (1989). Cities presenting a very high intensity of the phenomenon include Bombay, 9.5°C, Delhi, 10°C, Kuala Lumpur, 7°C, and Buenos Aires, 7.5°C.

4.3 Impact of Heat Island on the Cooling Demand of Buildings (Recent Data)

While increased urban temperatures increase electricity demand for cooling in the summer, they may reduce the heating load of buildings during winter.

Cooling load variation calculations of a reference building in the major Athens area, based on experimental data from 20 stations was reported by Hassid et al. (2000). They reported that the cooling load of the reference buildings was about double at the centre of the city than in the surrounding Athens area. Hassid et al. also reported that almost a double peak-cooling load was calculated for the central Athens area than the surrounding area. It is known that higher ambient temperatures result in lower efficiency air conditioners. Hassid et al. reported that the minimum COP values were lower by about 25% in the central Athens. The lower COP necessitates an increase in the size of the A/C systems resulting in increased peak electricity demand and energy consumption for cooling.

Studies on the Tokyo area reported by Ojima (1991) conclude that between 1965 and the year 2000 the cooling load of existing buildings has increased by as much as 50% on average due to the heat island phenomenon.

In the USA, comparisons of high ambient temperatures to utility loads for the Los Angeles area have shown that an important correlation exists. It is found that the electricity demand increases by almost 540 MW per °C increase in ambient temperature (Akbari et al., 1992). There has been a 5°F (2.8°C) peak temperature increase in

Los Angeles since 1940, resulting in an additional 1.5 GW electricity demand due to the heat island effect. Similarly, it has been calculated that summer electricity costs for the USA due to the heat island alone could be as much as $1 million per hour, or over $1 billion per year. It is estimated that 3% to 8% of the current urban electricity demand is used to compensate for the heat island effect alone.

4.4 Impact of Canyon Effect on the Cooling Demand of Buildings. Recent Experimental Data

As already mentioned wind speed in the canopy layer is much lower than the undisturbed wind speed above buildings. The specific wind and temperature regime in canyons dramatically affect the potential for natural ventilation of urban buildings and thus the possibility to use passive cooling techniques instead of air conditioning.

Specific studies investigating the reduction of flow rate in naturally ventilated buildings because of the canyon effect are reported in Santamouris et al. (1998). They noted that the natural ventilation potential in single and cross ventilation configurations is seriously decreased inside the canyon. For single side ventilation configurations the air flow is reduced up to five times, while in cross ventilation configurations the flow is sometimes reduced up to 10 times.

Based on the previously defined potential for natural ventilation in urban canyons, Geros (1998) calculated the reduction of the performance of night ventilation techniques when applied to naturally ventilated buildings located in urban canyons. The study was performed for 10 different urban canyons where detailed meteorological and energy data were measured.

It was found that in single side configurations the cooling load of buildings located in canyons is 6 to 89% higher than the load of unobstructed buildings, while in cross ventilation configurations the cooling load increases by 18 to 72%. Thus, canyon effect has a very considerable impact on the performance of passive cooling techniques located in dense detrimental canyons.

4.5 Increase of the Ecological Footprint of Cities Because of the Heat Island Phenomenon

A population group's ecological footprint is defined by Rees as:

"the area of land and water required to produce the resources consumed, and to assimilate the wastes generated by the population on a continuous basis, wherever on Earth that land is located (Rees, 2001)."

Calculations have shown that the average ecological footprint of every person on Earth corresponds to 1.5 hectacres (ha) of ecologically productive land and about 0.5 ha of productive ocean (Rees, 2001b).

Studies have shown that the eco-footprint of residents of high income countries ranges from 5 to 10 ha per capita (Rees and Wackernagel, 1996; Wackernagel and Rees, 1996). People in the less developed countries have footprints of less than one hectare (Wackernagel et al., 1997; Wackernagel et al., 1999).

Table 11. The energy cost, CO_2 emissions and ecological footprint of the Athens heat island.

Year	1997	1998
Athens' heat island energy cost (kWh/m²)	38.2	29.0
Total Athens' heat island energy cost (GWh)	1772.5	1345.6
CO_2 emissions tons	5317440	4036800
Ecological footprint of the Athens' heat island (ha)	1036901	787176

Data from Paraponiaris et al, (2004).

Buildings' energy consumption for cooling purposes increases the eco-footprint of cities. The higher the consumption the higher the eco-footprint. Degradation of the urban environment and in particular an increase of urban ambient temperatures has a great impact on building's energy consumption, thus increasing the ecological footprint of a city.

Based on heat island data from many urban stations in Athens, Greece, the energy demand increase and the increased ecological footprint have been calculated. The results show that if all buildings of the Athens Municipality were fully air conditioned, then almost one million ha should be reserved annually just to compensate for the extra CO_2 emissions caused by the heat island effect. The energy penalty because of the heat island effect varies between 1,340 and 1,770 GWh per year (see Table 11).

5. Technologies to Improve the Urban Microclimate

Techniques to improve the urban microclimate are receiving increasing attention. Recent research has shown that the use of more appropriate materials, increased use of green areas, use of cool sinks for heat dissipation, appropriate layout of urban canopies, etc., may be effectively used to counterbalance the effects of temperature increase.

The first priority for utilities seems to be adoption of measures to decrease the electricity demand of air conditioners, thus avoiding unnecessary costs of new power plants operating for limited periods. Such a strategy, adopted by the Sacramento Municipal Utility District (SMUD), has proven very effective and economically profitable (Flavin and Lenssen, 1995). It was calculated that a megawatt of capacity is actually eight times more expensive to produce than to save it. This is because energy saving measures have low capital and no running costs, while construction of new power plants involves high capital and running costs.

5.1 The Role of Materials. Impact on the Temperature Regime of Cities

The thermal and optical characteristics of the materials used in the urban environment define at large its thermal balance. Reflectivity of the materials to solar radiation as well as their emissivity is the more important of the optical parameters.

Important research has been performed to better understand the thermal and optical performance of materials as well as their impact on the city climate. The US EPA published a detailed guide on light colored surfaces (Akbari et al., 1992), while Yap (1975) reported that systematic differences between urban and rural surface emissivities are responsible for a portion of the heat island effect. However, Oke et al. (1991) found the role of emissivity to be minor. On the contrary, the effect of thermal properties of the materials, especially admittance, is more important. Admittance is defined as the reciprocal of the thermal resistance or impedence of an element to cyclic heat flow from the environmental temperature point and has the same units as the U value. For a flat land, it is reported that if the urban admittance is 2,200 $J/m^2/K$ and the rural one is 800 units lower, a heat island of about 2°C develops during the night, even when the urban admittance is decreased to 600 $J/m^2/K$, a cool island of over 4°C may be formed during night. Finally, Berg and Quinn (1978), reported that during mid-summer white painted roads with an albedo close to 0.55 have almost the same temperature as the ambient environment, while unpainted roads with albedo close to 0.15 were approximately 11°C warmer than the air.

Research investigations have been carried out recently regarding the thermal performance of various materials used in the urban fabric and mainly in pavements and streets. Asaeda et al. (1996) conducted an experimental study of summertime performance of various pavement materials used commonly in urban environments. As expected, they found that the surface temperature, heat storage and its subsequent emission to the atmosphere were significantly greater for asphalt than for concrete and bare soil. Taha et al. (1992), measured the albedo and surface temperatures of a variety of materials used in urban fabric. They reported that white elastomeric coatings having an albedo of 0.72 were 45°C lower than black coatings with an albedo of 0.08. They also reported that a white surface with an albedo of 0.61 was only 5°C warmer than ambient air whereas conventional gravel with an albedo of 0.09 was 30°C warmer than the air. Doulos et al. (2003), performed comparative measurements of 93 different materials commonly used for pavements in the urban environment. Surface temperature differences of more than 25°C were measured. Analysis of the building materials according to them showed that tiles made of marble, mosaic and stone were cooler than other construction materials. Finally, analysis based on the material's textures showed that tiles with smooth and flat surfaces were cooler than tiles with rough surfaces.

The so-called "cool-materials" are characterized by a high reflectivity factor to the short wave radiation and a high emissivity. They reduce the amount of the absorbed solar radiation by the building envelopes and urban structures and keep their surfaces cooler. Respectively, they are good emitters of long wave radiation and release the energy that has been absorbed as short wave radiation. Using "cool materials" in urban environment planning contributes to lower surface temperatures which affect thermal exchanges with the air, improve comfort in outdoor areas, and decrease the ambient temperature (Akbari et al., 1997; Bretz and Akbari, 1997). Research shows that important energy gains are possible when light color surfaces are used in combination with the plantation of new trees. For example computer simulations by Rosenfeld et al. (1998), showed that white roofs and shade trees in Los Angeles, USA, would lower the need for air conditioning by 18% or 1.04 billion kilowatt-hours, equivalent to a financial gain of close to $100 million per year.

Industrial research has succeeded in developing paints with excellent optical characteristics for the urban environment. Comparative measurements of white marble tiles covered with these new highly reflective paints, shows an almost 6°C lower surface temperature during summertime noon (Santamouris, 2003a).

Large scale changes in urban albedo may have important indirect effects on the city scale. Using meteorological simulations Taha et al. (1988) showed that afternoon air temperatures on summer days can be lowered by as much as 4°C by changing the surface albedo from 0.25 to 0.40 in a typical mid-latitude warm climate. Taha (1994) also used more advanced simulations while investigating the effects of large scale albedo increases in Los Angeles. Taha reported an average decrease of 2°C and up to 4°C may be possible by increasing the albedo by 0.13 in urbanized areas. Further studies by Akbari et al. (1989), showed that a temperature decrease of this magnitude could reduce electricity load from air conditioning by 10%.

5.2 The Role of Green Spaces. Impact on the Temperature Regime of Cities

Trees and green areas greatly contribute in reducing temperatures in our cities and save energy. Trees can provide solar protection to buildings during the summer period while evapotranspiration can cool our cities. Evapotranspiration is defined as the combined loss of water to the atmosphere by evaporation and transpiration. Transpiration is the process by which water in plants is transferred as water vapor in the atmosphere. In addition, trees absorb sound and block erosion-caused by rainfall, filter dangerous pollutants, reduce wind speed and stabilize soil. The American Forestry Association (1989), estimated that the financial value of an urban tree is around $57,000 for a 50 year-old mature specimen. As mentioned by Akbari (1992), this estimate includes a mean annual value of $73 for air conditioning, $75 for soil benefits and erosion control, $50 for air pollution control and $75 for wildlife habitats.

Duckworth and Sandberg (1954), reported that temperatures in San Francisco's heavily vegetated Golden Gate Park average about 8°C cooler than nearby areas that are less vegetated. Bowen (1980) reported 2-3°C temperature reduction due to evapotranspiration by plants. Oke (1977) reported that in Montreal, urban parks can be 2.5°C cooler than surrounding built areas. In Tokyo, vegetated zones in summer have been found to be 1.6°C cooler than non-vegetated spots (Tatsu, 1980; Gao et al. 1994). Saito et al. (1990; 1991) studied the effect of green areas in the city of Kumamoto in Japan. They found that the maximum temperature difference between inside and outside the green area was 3°C. Jauregui (1990; 1991), reported that the park in Mexico city was 2-3°C cooler with respect to its boundaries. Taha et al. (1989; 1991) reported that evapotranspiration can create oases that are 2-8°C cooler than their surroundings. Measurements have shown that evapotranspiration from plants at the National Park of Athens create oases of 1-5°C during the night (Santamouris, 2001).

Givoni (1989) advised spacing trees and public parks throughout the urban area rather than concentrating them in a few spots. This is supported by the measurements of Lindqvist (1992), in Göteborg, Sweden, where the air temperature increased 6°C from 100 m inside the park to a point within the built-up areas 150 m outside the park. More frequently, the air temperature gradient in the transition zone was 0.3-0.4°C per 100 m outside the park.

Numerical simulations have been used to evaluate the effect of additional vegetation to the urban temperature. The computer simulations of Huang et al. (1987) predicted that increasing the tree cover by 25% in Sacramento and Phoenix, USA, would decrease air temperatures by 10°F (5.6°C) 2:00 p.m. in July. Taha (1988), reported simulation results of the effect of canopy on daytime and night time temperature for Davis, California. Taha showed that a vegetative cover of 30% could produce a noon-time oasis of up to 6°C in favorable conditions and a night time heat island of 2°C. Gao (1993) reported that green areas decrease the maximum and average temperatures by 2°C, while vegetation can decrease the maximum air temperatures in streets by 2°C.

Simulations by Sailor (1995, 1998) revealed a potential for reducing peak summertime temperatures in Los Angeles by more than 1.3°C through the implementation of a 14% increase of fractional vegetative cover. He also evaluated the impact of added vegetation on the heating degree days, and cooling degree days, of cities located in the US. He reported that increasing the vegetative cover by 15% over only the residential neighborhoods, reduces the number of cooling degree days by 2-5% and increases the number of heating degree days by 0.5-3.5%. According to the author, one would expect a city-wide savings of up to 5% of summertime air conditioning energy use.

The National Academy of Sciences of the United States (NAS, 1991), reported that the planting of 100 million trees combined with the implementation of light surfacing programs could reduce electricity use by 50 billion kWh per year. This is equivalent to 2% of the annual electricity use in the US. Computer simulations by Akbari et al. (1992) highlight the combined effect of shading and evapotranspiration of vegetation on the energy use of typical one-story buildings in various US cities. It is found that by adding one tree per house, the cooling energy savings range from 12-24%, while adding three trees per house can reduce the cooling load between 17-57%, with shading accounting for only 10-35% of the total cooling energy savings. Simpson and McPherson (1998), calculated the magnitude of tree shade in 254 residential properties in Sacramento, California. They used 3.1 trees per property which reduced annual and peak cooling energy use by 153 kWh, (7.1%), and 0.08 kWh, (2.3%), per tree respectively.

5.3 The Role of Heat Sinks. Impact on the Temperature Regime of Cities

Decrease of the ambient temperature can be achieved by dissipation of the ambient heat to a lower environmental sink, like the ground, water and air.

The use of buried pipes for passive cooling can be traced back as far as 3,000 BC when Persian architects used wind towers and underground air tunnels for cooling (Bahadori, 1976). From 1943-1963, Perrill (1999) used a 120m long, 5m deep air tunnel to air condition entire critical care facilities of a hospital in India. He later constructed a larger system containing more than 1,700 meters of underground air tunnels to air condition another hospital. Later on, some of the most extensive work in this area was done by Goswami and his co-workers (Sinha, Goswami and Klett, 1981; Goswami and Dhaliwal, 1985; Goswami and Ileslamlou, 1990; Arzano and Goswami, 1997). Sinha et al. (1981) showed experimentally that for thin diameter pipes ranging from 8 inches to 24 inches, the material of the pipe does not affect performance much. Goswami and Dhaliwal (1985) confirmed this result by an analytical study showing that the limiting factor in heat transfer is the thermal conductivity of soil. The moisture content of soil has a great impact on the thermal conductivity of soil. Goswami and Ileslamlou (1990) studied the increase in soil temperature around the pipe, the decrease in its thermal conductivity, and effect of both on the performance of this system. They showed that for a long term operation, a closed-loop air tunnel system's COP is as high as 12 and stays more than 7.5 even after operating for 90 days. Arzano and Goswami (1997) showed the effectiveness of this method in dehumidification of indoor air, which is extremely important for hot and humid climates. They used a 0.6 m diameter, 53 m long pipe to cool a 100m^2 house in Florida, which provided 82% of the design cooling load and 47% of the design dehumidification.

Buried pipes have also been used in different projects to supply cool air in an open environment. Monitoring has shown that in 30m long pipes buried at about 3m depth the achieved temperature decrease is close to 12°C (Santamouris et al, 1982). The possible temperature decrease of the ambient air depends on the global thermal balance of the air and the supply of the cool air through the pipes. Alvarez et al. (1992) proposed to irrigate the soil around the pipes in order to counterbalance the low conductivity of the soil and the continuous heating of the ground around the pipes.

Pools, ponds, sprays and fountains are the main components based on evaporative processes. Calculations based on mean summer climatological conditions give an evaporation rate between 150-200W per square meter, which defines the cooling potential of these techniques. Since water surfaces increase air humidity they are very beneficial in dry climates, but they may create some problems in very humid climates. In hot climates, their cooling effect should be maximized through design strategies which prevent diffusion of the cooled air and direct it to inhabited spaces.

A cooling tower system was proposed and applied by Givoni (1994). The system consisted of an open shaft where fine drops of water were sprayed vertically downward. A wind catcher could be placed above the shaft to increase the air flow. Alvarez (1990) showed that when the inlet temperature was close to 36°C, the corresponding exit temperature was around 23°C.

5.4 Recent Case Studies Demonstrating the Reduction of Cooling Needs of Buildings Through Microclimatic Improvements

As previously mentioned techniques aiming to improve microclimate around buildings have gained an increasing acceptance recently. Several big projects have been successfully designed and realized using such techniques.

The first important example of the designs deals with Expo 92 in Seville, Spain, (d'Asiain, 1997; Alvarez et al., 1992). A number of known techniques were used to improve microclimate in environs of Expo 92. To decrease the surface temperature of the outdoor spaces extensive shading was applied in almost all zones where visitors were circulating. This was achieved by using plants, pergolas, etc. Lower surface temperatures decrease the emitted infrared radiation by the materials, while shading protects the visitors from solar radiation and thus improve thermal comfort. Fountains, pools, ponds and sprinklers were used to evaporate water in the ambient air and thus decrease its temperature. Earth-to-air heat exchangers were also used to circulate the ambient air through the ground and thus decrease its temperature. Finally, cooling towers, as described previously, were installed in various parts of the area. Extensive monitoring showed that the application of these passive cooling techniques contributed in decreasing ambient air by up to 5°C.

Another example of microclimatic improvement techniques was Expo 98 in Lisbon, Portugal. Specific air flow studies were performed in order to optimize the air flow through the area. To decrease surface temperature of the materials, decrease ambient temperature and improve outdoor thermal comfort, extensive use of plants, pergolas, pools, ponds and fountains were employed. Monitoring of the area showed that because of the application of passive cooling techniques the ambient temperature was reduced by 3-4°C.

A third important application is the 2004 Olympic Village, in Athens, Greece (Fintikakis and Santamouris, 2002). In order to improve microclimate, cool materials, pergolas and plants, external shading, pools, ponds and fountains as well as air flow enhancement techniques have been used. It is calculated that the ambient temperature has been reduced by 4°C, the maximum wind speed has been reduced by 3m/sec, while the period inside the comfort zone has been increased by 65%.

6. Recent Progress on Solar Control and Heat Protection of Buildings

Important industrial and scientific research has been carried out recently, resulting in new, highly efficient components. Thus, new advanced shading devices that allow a better integration of daylight quality, glare control and efficient solar control are available. In parallel, new advanced types of glazing have been developed that present a much lower U value while able to provide efficient solar control. In real terms the progress achieved by the glazing industry is really impressive. Intelligent windows have been designed and applied in many buildings. Intelligent windows provide solar and heat protection, optimization of daylight and natural or hybrid ventilation in an integrated way. Finally, heat protection techniques, like planted roofs, have gained an increased research interest for optimum use.

In the next sections, the recent progress achieved on the previously mentioned systems and techniques will be presented.

6.1 Recent Industrial Developments on Advanced Solar Control Devices and Systems

Transparent components appropriate for warm climates are characterized by a low transmittance to the solar radiation and a high transmittance to the visible spectrum, thus providing solar control and adequate daylight. The very rapid development of the solar control coatings technology has permitted to use single- or double-layer silver coatings that present very good g and U values. The g-value is the sum of the solar transmittance and a factor for heat gains resulting from sunlight absorbed in the glazing unit. More advanced solar control systems and techniques deal with thermotropic and thermochromic coatings, gasochromic and photochromic glazings and angle selective coatings. All these techniques are on the verge of entering the market (Hutchins, 2003).

Thermotropic and thermochromic coatings are characterized by a variable transparency as a function of their temperature. Thermotropic layers are composed of a mixture of a polymer and water or two polymers. As both components have different refractive indices, when the temperature is above a threshold value they segregate and the material becomes opaque. According to Wilson (1999) a polymer blend in combination with a low-e coating can have a visible transmittance range of 73% at 30°C to 31% at 50°C.

A gasochromic glazing is composed of two glazed panes with a coating on one of the two inner surfaces. The coating consists of a WO_3 film whose color changes to blue when exposed to a low concentration of hydrogen. When the coating is exposed to oxygen it resumes its initial color. Experiments have shown that the transmittance varies from 11-74%. Additional research aimed at changing the color of the coating from blue to gray.

Photochromism is based on the effect of ultraviolet radiation on a photochromic material. A major problem is that solar radiation is absorbed by the glazing surface and not reflected.

The basic idea behind angle-selective coatings is to develop a window structure that presents a high reflectance to incident light at high solar altitudes and a high transmittance at low angles. Thus, the visual contact with the external environment is achieved while solar gains and glare are reduced.

Finally, the development of nanoparticle-doped polymeric solar control glazing seems a promising solution (Hutchins, 2003). Experiments show that the visible transmittance of such a glazing can be close to 0.87, while the solar transmittance close to 0.62.

6.2 Advanced Glazings for Heat Control

Thermal losses or gains from glazing elements represent a significant part of the thermal balance of buildings. Recently the glazing industry has been developing glazing units with a high transmittance to the visible spectrum and a low heat loss coefficient, U. Low e-glass, triple glazing and use of noble gases in a sealed glass unit can contribute in dramatically decreasing heat losses or gains through the glazings.

Actual developments as well as the trend concerning glazing characteristics are shown in Figure 3 (Adapted from Reinhart et al., 2001). Table 12 gives the main optical and thermal characteristics of some reference and advanced glazings (Hutchins, 2003). As shown in Figure 3, the U value of commercially available windows may be as low as 0.6 $W/m^2/K$.

The impact of high insulating glazing on economy and energy consumption is important. According to EUROACE (1998), a Europe-wide domestic window upgrading program could save 94 million ton of carbon dioxide emissions. More than 110,000 jobs could be created in the process, while additional savings of 25 million tons may be achieved through similar measures in the commercial, public and industrial building sectors.

To support the use and application of advanced windows, specific tools able to calculate their visual and thermal characteristics, have been recently prepared (Dijk and Hutchins 2002; Mitchell et al., 2001; Window 1993; Dijk and Baker, 1995). These tools calculate the thermal and solar properties of commercial and innovative window systems on the basis of known component properties (glazings, shading devices, frames and edges, gases, etc.) and are also suitable in calculating the performance of complex facades.

6.3 Other Developments

The use of planted roofs undoubtedly present numerous energy, environmental and social benefits especially in non-insulated buildings. They act positively on the climate of a city counteracting heat island, and contribute in improving the thermal performance of buildings and indoor environmental conditions. In fact, planted roofs

Table 12. Visible and solar energy transmittance, U-value of insulating double-glazed units using low emittance, and solar control coatings for heating- and cooling-dominated applications (Hutchins, 2003).

Glazing	Gas Fill	τ_v	G	U (Wm^{-2}K^{-1})
Clear float single glazing	-	0.90	0.86	6.4
Clear float double glazed unit.	Air	0.81	0.76	2.9
Double Glazed Unit low-e pyrolytic heat mirror for use in heating-dominated applications	Argon	0.75	0.72	1.9
Double Glazed Unit low-e sputtered solar control for use in cooling-dominated applications	Argon	0.66	0.34	1.2

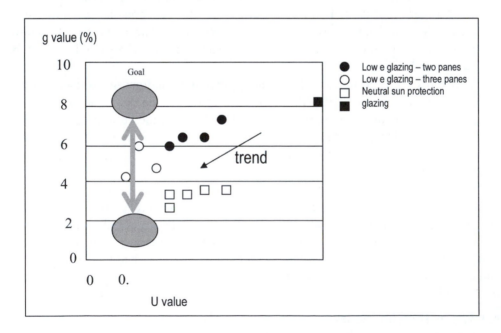

Figure 3. Total solar transmittance against U value for modern glazing, The g-value is the sum of the solar transmittance and a factor for heat gains resulting from sunlight absorbed in the glazing unit (Adapted from Reinhart et al, 2001).

offer increased solar protection and reduction of thermal losses through the fabric, while increasing the thermal capacity of buildings.

Recent research has developed advanced design tools and solutions permitting a better understanding of the performance of planted roofs (Haifeli et al., 1998; Eumorfopoulou et al., 1994, 1995; Cappeli et al., 1998; Palomo, 1999; Domingez and Lozano, 1998; Good, 1990; Niachou et al., 2001).

Niachou et al. (2001) showed that the planted roof systems greatly contribute in decreasing the cooling load of non-insulated buildings close to 50%. This benefit reduces to just 2-5% for buildings with moderate or good insulation levels.

7. Heat Amortization Techniques–Night Ventilation

Night ventilation techniques are based on using cool ambient air, decreasing the indoor air temperature as well as the temperature of the building's structure. The cooling efficiency of night ventilation is mainly based on the relative difference between indoor and outdoor temperatures during the night, the air flow rate, the thermal capacity of the building, and the efficient coupling of air flow and thermal mass.

Important theoretical and experimental research has been carried out to better understand the phenomena and also to develop design tools and computational codes. The specific impact of the urban environment on the efficiency of night cooling techniques is also the subject of recent research.

Extended experimental work on night ventilation techniques is reported by Santamouris and Assimakopoulos (1996), Givoni (1994), Blondeau et al. (1995), Van der Maas and Roulet (1991), Kolokotroni et al. (1997), Zimmerman and Anderson (1998), Geros et al. (1998), and Dascalaki and Santamouris (1998). As Geros et al. (1998) performed measurements on non air conditioned buildings and night ventilated A/C, and showed that the use of this technique decreases peak indoor temperature, up to 3°C. Sensitivity analysis showed that the expected reduction of overheating hours varies between 39% and 96% for air flow rates between 10-30 ACH respectively. For air conditioned buildings, early morning indoor air temperature can be reduced by 0.8-2.5°C depending on the set point temperature. Sensitivity analysis for air conditioned buildings shows that the expected energy conservation varies between 48%-94% for air flow rates 10-30 ACH respectively.

Givoni (1994) reported the development of an empirical formula to predict the indoor maximum temperature as well as the cold storage and the diurnal cooling capacity of the building. Blondeau et al. (1995) developed indices characterizing the energy gain and comfort improvements due to night ventilation.

A detailed methodology to calculate the performance of air conditioned as well as of non air conditioned night ventilated buildings was presented by Santamouris et al. (1995). The method is based on the principle of Modified Cooling Degree Days and is extensively evaluated against experimental data. Other methods which calculate the cooling potential of night ventilation are proposed by Roulet et al. (1997) and Kolokotroni et al. (1997).

As previously mentioned, Geros (1998) performed measurements in 10 urban canyons and calculated the reduction of the performance of night ventilation techniques applied to non air conditioned buildings in urban canyons. Geros found the reduction in performance to be 2.6ºC for single sided naturally ventilated buildings, and 0.2ºC to 3.5ºC for buildings with cross ventilation.

8. Advances in Thermal Comfort Studies

Thermal comfort of buildings has received a lot of attention during the last few years. Existing knowledge covers methods predicting thermal comfort under steady-state conditions. The most well-known and widely accepted methods are the "Comfort Equation" proposed by Fanger (1972), and the J.B. Pierce two-node model of human thermoregulation (Gagge, 1973; Gagge et al., 1986). Several steady state thermal comfort standards have been established based on these models (ISO, 1984; ASHRAE, 1981; Jokl, 1987).

As a result of the thermal interaction between the building shell, occupants, and the cooling system, steady-state conditions are rarely encountered in practice in many types of buildings. Indoor temperatures in "free-floating buildings" vary widely while fluctuations of 0.5ºC and 3.9ºC are found in passive solar buildings with constant set-point control system (Madsen, 1987). Thus, knowledge of thermal comfort under transient conditions is necessary.

Basic thermal comfort research concludes that there is an important discrepancy between steady-state models and those where no mechanical conditioning is applied (Humphreys, 1976). This is mainly due to the temporal and spatial variation of the physical parameters in the building (Baker, 1993).

Experimental comfort surveys have shown significant discrepancies between thermal comfort in real buildings conditions versus laboratory conditions. Nicol and Humphreys (1973) concluded that this discrepancy could be the result

"...of a feedback between the thermal sensation of subjects and their behavior and that they consequently 'adapted' to the climatic conditions in which the field study was conducted (Nicol and Humphreys, 1973)."

Based on data collected in many field studies, Humphreys and Nicol proposed an adaptive comfort model (Humphreys, 1992). They demonstrated that for a group of people the comfort temperature is close to the average temperature they experience. The fundamental assumption of the adaptive approach is expressed by the adaptive principle:

"If a change occurs such as to produce discomfort, people react in ways which tend to restore their comfort (Nicols, 2003)."

To allow for the discrepancies in relating PMV, (Predicted Mean Vote), that is the comfort index of the Fanger's theory, to field measurements, Fanger and Toftum (2002) presented a correction for free-floating buildings in warm climates. Based on the Humphreys work, the Comfort Group of the European PASCOOL research project, carried out field studies to understand the mechanisms by which people make themselves comfortable at higher temperatures (Baker, 1995). They concluded that people are comfortable at much higher temperatures than expected, while finding that people make a number of adaptive actions to make themselves comfortable including moving to cooler parts of the room. Specifically this research found that in 864 monitored hours of more than 500 persons, there were 273 adjustments to building controls and 62 alterations to clothing.

Based on surveys carried out in the frame of the European research project SCATS (Smart Controls and Thermal Comfort), scientists tried to determine the rate of change of comfort temperature using surveys conducted over a period of time (Nicol and Raja, 1996; McCartney and Nicol, 2002). Evidence was found that the comfort temperature in free-running buildings depends on the outdoor temperature. The proposed equation for comfort temperature T_c is given by:

$$T_c = 13.5 + 0.54\, T_o \qquad\qquad\qquad \text{(Eq. 1)}$$

Where, T_o is the monthly mean of the outdoor air temperature.

As Humphreys' and Nicol's 2000 research shows, there is a remarkable agreement between their work on "free-floating buildings" and the 1998 ASHRAE database (Humphreys, 1978; deDear, 1998) (Figure 4). It is also found that the relationship between the two databases for air conditioned buildings is more complex, showing a 2°C difference in indoor comfort temperatures (Humphreys, 1978; deDear, 1998).

Variable indoor temperature comfort standards for air conditioned buildings may result in remarkable energy savings (Auliciems, 1990; Milne, 1995; Wilkins, 1995; Hensen and Centrenova, 2001). Energy savings of about 18% are estimated over that from using a constant indoor temperature in Southern Europe, (Stoops et al., 2000), while the corresponding energy savings for UK conditions have been estimated close to 10%.

9. Heat Dissipation Techniques

Heat dissipation techniques involving ground, evaporative, radiative and convective cooling present a very high cooling potential. The overall efficiency of such systems depends on the specific climatic conditions in the area, the cooling needs and patterns in the building, as well as on the efficiency of the technology used.

The specific performance of heat dissipation techniques has been extensively studied during the last decade. In the frame of the PASCOOL research program, an atlas on the potential of heat sinks was developed (Alvarez et al., 1997), while the Sink research project of the European Commission (Alvarez and Tellez, 1996) studied and

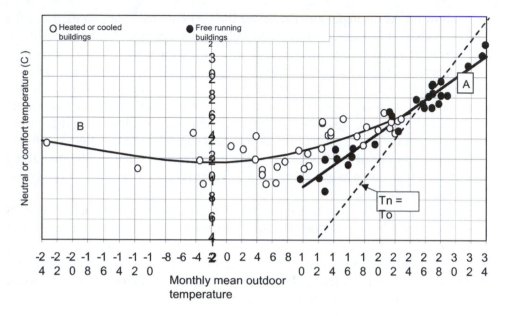

Figure 4. The change in comfort temperature with monthly mean outdoor temperature for free floating and air conditioning buildings, (From Nicols, 2003).

documented the performance of almost all known heat dissipation technologies for the different climatic zones of Southern Europe.

9.1 Natural Ventilation, Technological and Social Developments and Needs

Natural ventilation is an important and simple technique that when appropriately used may improve thermal comfort conditions in indoor spaces, decrease the energy consumption of air conditioned buildings, and contribute to fight problems of indoor air quality by decreasing the concentration of indoor pollutants.

It must be recognized that natural ventilation is not just an alternative to air conditioning. Instead, it is a more effective instrument to improve indoor air quality, protect health, provide comfort, and decrease unnecessary energy consumption.

Given the great inequalities in terms of income and energy use in the world, what natural ventilation may offer is a function of the actual needs, the characteristics of the building as well as the type of energy used, and the services and systems employed. Thus, three main clusters of possible uses/contributions may be defined as a function of income and the corresponding energy use (Figure 5).

- In very poor households natural ventilation can greatly contribute in avoiding indoor air pollution problems caused by combustion processes. Almost 2 billion people are living under these specific conditions, without access to electricity and modern fuels. High concentrations of indoor pollutants pose a tremendous health threat to the population of the less developed countries. World-

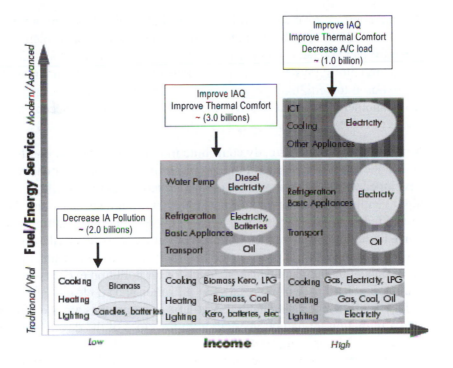

Figure 5. Household fuel transition and possible contribution of natural ventilation (From United Nations Council for Human Settlements (UNCHS): The State of the World Cities, 2001).

wide, close to 2 million deaths per year are attributable to indoor air pollution from cooking fires (Birol, 2002). Recent studies of the World Health Organization (WHO) have shown that 30-40% of 760 million cases of respiratory diseases world-wide are caused by particulate air pollution alone. "Mostly, these health effects are caused by indoor air pollution due to open stove cooking, and heating in developing countries (World Bank, 2000)." Efficient components designed to enhance ventilation in these settlements is a simple task for a low or negligible cost. This must be considered in association with other policies, such as the design of more efficient stoves and the use of cleaner fuels.

- Natural ventilation may greatly contribute in improving indoor air quality and indoor thermal conditions for about 3 billion people of low and medium income. Most of these people live in poorly designed buildings suffering from high indoor temperatures during summer. This population doesn't have the means to use any cooling equipment and relies fully on natural systems and techniques. Design and integration of efficient natural ventilation systems and components like wind and solar towers, can greatly assist in improving indoor thermal comfort. As it concerns indoor air quality, high levels of outdoor air pollution and its impact on the indoor environment, is a major problem for this part of urban population. According to the United Nations Global Environmental monitoring system an annual average of 1.25 billion urban inhabitants are exposed to very high concentrations of suspended particles and smoke, while another 625 million urban citizens are exposed to non-acceptable SO_2 levels (UNCHS, 1993).

- Natural ventilation can greatly contribute in improving thermal comfort, decrease the needs for air conditioning, and improve indoor air quality in the developed world. As previously explained, a serious limitation of natural and night ventilation application in dense urban environments has to do with the severe reduction of wind speed in urban canyons. Outdoor pollution is also a grave limitation for natural ventilation in urban areas. As reported by Stanners and Bourdeau (1996), it is estimated that in 70-80% of European cities with more than 500,000 inhabitants, the levels of air pollution exceeds the WHO standards at least once in a typical year. Filtration and air cleaning is possible only when flow-controlled natural ventilation components are used. Noise is also a limitation for natural ventilation in the cities of the developed world. Stanners and Bourdeau (1996) reported unacceptable noise levels of more than 65 dB(A), affecting between 10-20% of urban inhabitants in most European cities. OECD has calculated that 130 million people in OECD (Organization Of Economic Cooperation and Development) countries are exposed to noise levels that are unacceptable(OECD, 1991).

Extensive experimental and theoretical studies investigating techniques and methods to improve natural ventilation efficiencies as applied in dense urban areas, were conducted by the European research project URBVENT (URBan VENTilation) (Ghiaus et al., 2003). Algorithms to calculate the wind speed in urban canyons as a function of the undisturbed wind speed above the buildings, as well as methodologies to assess the potential of natural ventilation in some hundreds of thousands configurations of urban buildings have been developed and experimentally validated (Santamouris, 2003b). Intelligent software tools based on neural networks that permit estimates of required opening areas required in order to achieve a defined air flow, have been developed and are currently available.

Finally, in the PASCOOL research project a new method (CPCALC+) to calculate the pressure difference around certain building configurations has been developed (Grosso, 2002). The method is based on experimental data collected through wind tunnel experiments as well as on the existing knowledge. These results were used to obtain regression curves for determining the pressure coefficient distribution over flat and tilted roofs.

9.2 Recent Developments on the Field of Hybrid Ventilation

Hybrid ventilation use both natural ventilation and mechanical systems with different features of these systems at different times of the day or season of the year and within individual days to provide a comfortable internal environment (Heiselberg, 2002). Hybrid ventilation systems are based on an intelligent control system that allows switching between natural or mechanical modes in order to minimize energy consumption.

The main advantages of the hybrid ventilation systems are summarized by Heiselberg (2002):

- Hybrid ventilation results in higher user satisfaction as it permits a greater degree of individual control of the indoor climate.

- It optimizes the balance between indoor air quality, thermal comfort, energy use and environmental impact and fulfilling the needs for a better indoor environment and reduced energy consumption.

- It accesses natural and mechanical ventilation modes and exploits the benefits of each mode the best way.

- It is a very appropriate solution for complex buildings as it is associated with more intelligent systems and control.

Different hybrid ventilation systems have been proposed and applied in various types of buildings. An extended review of these systems and their existing applications is provided by Delsante and Vik (2001).

Results from first generation hybridly ventilated buildings show that such techniques have great potential. Hybrid systems were found to be quite effective in providing good IAQ and thermal comfort, while the energy performance was good but not excellent. Hybrid ventilation has proved to be suitable for schools and cellular offices while in open-plan offices there were some complaints.

Hybrid ventilation is a very new and promising technology but several problems still remain. Design tools and methodologies as well as more robust control strategies and sensors for demand control ventilation must be developed in the future.

9.3 Recent Developments in Evaporative Cooling

The efficiency of mechanical evaporative systems has significantly improved recently. Improvements mainly deal with the type of system cooling media and their performance. Efficiencies close to 90% have been achieved in industrial products. Developments in passive evaporative cooling systems are important as well. Improvements of the Passive Downdraft Evaporative Cooling systems (PDEC), and extended testing of this system are among the more important developments.

Cooling towers are traditionally used in Middle East and Persian architecture (Bahadori, 1978). New developments and in particular coolers where the air was forced by a blower through wetted pads have been used in the past in desert areas of the USA. In natural downdraft evaporative coolers, the air is not forced through the pads but provided just by gravity flow (Cunningham and Tompson, 1986). The performance of natural downdraft evaporative coolers have been studied in detail by various authors and design tools have been proposed (Thompson et al., 1994; Givoni, 1991; Sodha et al., 1991; Chalfoun, 1992).

An innovative development of the system is achieved through the PDEC research program of the European Commission (PDEC, 1995). Their improvement consists of replacing the wetted pads with rows of atomizer nozzles that produce an artificial fog by injecting water at high pressure through minute orifices. This feature produces much better regulation of the system, as there is a significant reduction of pressure losses and smaller equipment can be used.

10. Conclusions

The building sector energy consumption is quite high and is expected to further increase because of improving standards of life and swelling world population. Air conditioning use has increasingly penetrated the market during the last few years and greatly contributes to the upsurge of absolute energy consumption. In parallel, urbanization increases city-wide temperatures and thus increases the cooling load as well as the required peak electricity load. Several passive techniques have been proposed to decrease buildings' energy consumption and improve environmental quality, involving techniques aiming at bettering the urban environment and corresponding thermal conditions, and improving buildings' thermal characteristics by using passive and hybrid cooling systems and techniques. Theoretical studies have shown that the application of all the above techniques in buildings may decrease their cooling load up to 70%.

During the last few years important basic and industrial research has been carried out that has resulted in the development of new high-efficiency materials, systems, tools and techniques. However, the continuing increase of energy consumption of air conditioning suggests a more profound examination of the urban environment and the impact on buildings as well as to an extended application of passive cooling techniques. Appropriate research should aim at better understanding micro-climates around buildings, and to understand and describe comfort requirements under transient conditions during the summer period. Also of importance are improving quality aspects, developing advanced passive and hybrid cooling systems, and finally, developing advanced materials for the building envelope.

11. References

Abbate, G., 1998. Heat island study in the area of Rome by integrated use of ERS-SAR and LANDSAT™. Report available from Earthnet online, service provided by the European Space Agency. http://earth.esa.int/workshops/ers97/papers/abbate/.

Adnott, J., (Ed.) 1999. Energy Efficiency of Room Air-Conditioners (EERAC) Study. Directorate General for Energy (DGXVII) of the Commission of the European Communities.

Advanced Windows Information System (WIS), 1994. European Commission DGXII Joule Programme, DG XII Contract JOU2-CT94-0373.

Akbari, H., Rosenfeld, A., Taha, H. 1989. Recent developments in heat island studies: Technical and policy. In H. Akbari, (Ed)., Proceedings of Workshop on Saving Energy and Reducing Atmospheric Pollution by Controlling Summer Heat Islands, Berkeley, CA,. pp. 14-20, 23-24.

Akbari, H., Davis, S., Dorsano S., Huang, J., Winett, S., 1992. Cooling our Communities—A Guidebook on Tree Planting and Light Colored Surfacing. U.S. Environmental Protection Agency, Office of Policy Analysis, Climate Change Division.

Akbari, H., Bretz, S., Kurn, D.M., Hanford, J., 1997. Peak power and cooling energy savings of high-albedo roofs. Energy and Buildings, 25, pp. 117-126.

Allard, F., (Ed.) 1998. Natural Ventilation in Buildings—A design handbook. James and James Science Publishers, London, UK.

Alvarez, S., 1990. Natural cooling techniques. In: T. Steemers, (Ed.), Proceedings of the Workshop on Passive Cooling Techniques, Ispra, Italy, pp. 2-4.

Alvarez, S., Guerra, J., Velazquez, R., 1992. Climatic control of outdoor spaces at EXPO '92. In: Architecture and Urban Space: Proceedings of the Ninth International Plea Conference, Seville, Spain, September 24-27, 1991, pp. 189-194.

Alvarez, S., Téllez, F. 1996. Natural cooling technology assessment. Atlas for Southern Europe countries. Final report of the EU-XVII/4.1030/A/94-88. Altener-SINK project, Brussels, Belgium.

Alvarez, S., Maestre, I.R., Velásquez, R., 1997. Design methodology and cooling potential of environmental heat sinks. International Journal of Solar Energy, 19, pp. 179-197.

American Forestry Association, 1989. Save Our Urban Trees: Citizens Action Guide. Washington D.C.

AFEAS (Alternative Flourocarbons Environmental Acceptability Study). 2001. Issue areas. Production and sales of fluorocarbons. http://www.afeas.org/production_and_sales.html.

Argiriou, A., Santamouris, M., Balaras, C., Jeter, S., 1993. Potential of radiative cooling in southern Europe. International Journal of Solar Energy, 13,189-203.

Arzano, L., and Goswami, D.Y., "Performance Analysis of a Closed-loop Underground Air Tunnel for Residential Housing in a Hot and Humid Climate," Proceedings of the International Solar Energy Society's Solar World Congress, Taejon, Korea, 1997, pp. 239-251. Also in, Proceedings of the American Solar Energy Society's "Solar Forum," 1996, Asheville, NC.

Asaeda T., Vu Thanh, C.A., Wake, A., 1996. Heat storage of pavement and its effect on the lower atmosphere. Atmospheric Environment, 30, 3, pp. 413-427.

ASHRAE, 1981. Thermal environmental conditions for human occupancy. ANSI/ASHRAE Standard 55-1981, ASHRAE, Atlanta, GA, USA.

Athienitis, A., Santamouris, M., 2002. Thermal Analysis and Design of Passive Solar Buildings. James and James Science Publishers, London, UK.

Auliciems, A. 1990. Air-conditioning in Australia III; thermobile controls. Architectural Science Review, 33, 2.

Bahadori, M.N., 1978. Passive cooling in Iranian architecture. Scientific American. 238, No. 2. pp. 144-154.

Baker, N., 1993. Thermal comfort evaluation for passive cooling - a PASCOOL task. In: Foster, N., Scheer, H., (Eds.), Solar Energy in Architecture and Urban Planning: Proceedings of the 3rd European Conference on Architecture, Florence, Italy, pp. 313-319.

Baker, N., 1995. Final Report of the Thermal Comfort Group. PASCOOL Research Program, European Commission, Directorate General for Science Research and Technology, Brussels, Belgium.

Der Petrossian, B., 2001. Conflicts between the construction industry and the environment. Habitat Debate, 5:2, pg. 5.

Barring L., Mattsson, J.O., Lindovist, S., 1985. Canyon Geometry, street temperatures and urban heat island in Malmo, Sweden. Journal of Climatology, 5, pp. 433-444.

Berg, R., Quinn, W., 1978. Use of light colored surface to reduce seasonal thaw penetration beneath embankments on permafrost. In: Proceedings of the 2nd International Symposium on Cold Regions Engineering. University of Alaska, pp 86-99.

Besant-Jones, J.B., Tenenbaum, B., 2001. The California power crisis: Lessons for developing countries. Energy and Mining Sector Board Discussion paper series. The World Bank, Washinghton, D.C. http://rru.worldbank.org/Documents/e_calexp0400.pdf.

Birol, F., 2002. Energy and poverty, In: Proceedings of World Outlook 2002, Energy Forum, World Bank, Washington, D.C., pp.1-49.

Bivens, D.B., 1999. Refrigeration and air conditioning with reduced environmental impact. In: Proceedings of Joint IPCC/TEP Expert Meeting on Options for the Limitation of Emissions of HFCs and PFC's' WHO–UNEP, Petten, The Netherlands, pp.121-132.

Blondeau, M., Sperandio, F.A., 1995. Night accelerated ventilation for building cooling in summer. In: Santamouris, M. (Ed.), Proceedings of the International Conference on Passive Cooling of Buildings, Athens.Greece, pp.34-42.

Bowen, A., 1980. Heating and cooling of buildings and sites through landscape planning. Passive Cooling Handbook, AS/ISES, Newark, DE.

Bretz S.E., Akbari, H., 1997. Long-term performance of high-albedo roof coatings. Energy and Buildings, 25 pp.159-167.

Burton, S., 2001. Energy Efficient Office Refurbishment, James and James Science Publishers, London, UK.

Cappeli D'Orazio, M., Cianfrini, C., Corcicone, M., 1998. Effects of vegetation roof shielding on indoor temperatures. International Journal of Heat & Technology, 16, No. 2, pp. 85-90.

Carmody, J.C., Meixel, G.D., Labs, K.B., Shen L.S., 1985. Earth contact buildings: Applications, thermal analysis and energy benefits. In: Böer, K.W., Duffie, J.A. (Eds.), Advances in Solar Energy, vol. 2, pp. 287-347.

Chalfoun, N., 1992. CoolT, V. 1.4.© Copyright Cool Tower Performance Program, Environmental Research Laboratory, University of Arizona, Tuscon, Arizona.

Chandler, T.J., 1965. City growth and urban climates. Weather, 19, pp 170-171.

Confederation of International Contractors Association, 2002. Industry as a partner for sustainable development: construction. Report developed through a multi-stakeholder process facilitated by the United Nations Environment Programme (UNEP). http://www.uneptie.org/outreach/wssd/docs/sectors/final/construction.pdf.

Cromvall, J., 2001. Final report of AIRLIT-PV research project. European Commission, Directorate General for Research, Brussels, Belgium.

Cunningham, W., Thompson, T., 1986. Passive cooling with natural draft cooling towers in combination with solar chimneys, In: A. Zold (Ed.) Proceedings of PLEA '86, Passive and Low Energy Architecture in Housing, Budapest, Hungary.

Dascalaki, E., Santamouris, M., 1998. The Meletitiki case study building. In: F. Allard (Ed), Handbook on Natural Ventilation of Buildings, James and James Science Publishers, London, UK., p.p. 276-288.

deDear, R.J. 1998. A global database of thermal comfort field experiments. ASHRAE Transactions 104:1b, pp. 1141-1152.

Delsante, A., Vik, T.A., 2001. Hybrid ventilation–State of the art review. IEA Energy Conservation in Buildings and Community Systems Programme. Annex 35: Hybrid Ventilation and New and Retrofitted Office Buildings, Paris, France.

Dijk, H.A. van, Bakker, L., 1995. Development of a European advanced window information system (WIS). Proceedings of the workshop: Window Innovations '95, Ministry of Supply and Services, Canada, pp 337-346.

Dijk H.A. van, Hutchins, M.G., 2002. WinDat. European RTD thematic network on windows as renewable energy sources for Europe — Window energy data network, In Proceedings 4th International Conference on Coatings on Glass (ICCG 4), Braunschweig, Germany, p.p. 56-63.

Dominges, J., Lozano, A., 1998. Green Roof Systems. Computational. Mechanical. Publications. 1, pp. 615-624.

Doulos, L., Santamouris, M., Livada, I., 2003. Passive cooling of open urban areas — the role of materials. Solar Energy Journal, In Press, 2004.

Duckworth, E., Sandberg, J., 1954. The effect of cities upon horizontal and vertical temperature gradients. Bulletin of the American Meteorological Society, 35, pp. 198-207.

Eliasson, I., 1996. Urban nocturnal temperatures, street geometry and land use. Atmospheric Environment, 30:3, pp. 379-392.

Energy Information Administration, 1978-1982, 1984, 1987, 1990, 1993, and 1997. Residential Energy Consumption Surveys; National Oceanic and Atmospheric Administration; Air-Conditioning and Refrigeration Institute, NOAA, Washington, D.C.

Energy Information Administration. 1995. Commercial Buildings Energy Consumption Survey, NOAA, Washington D.C.

Escourrou, G., 1990/1991. Climate and Pollution in Paris. Energy and Buildings, 15-16, pp. 673-676.

Eumorfopoulou K., Economides, G., Aravantinos, D., 1994. Energy efficiency of planted roofs. In E. O. Fernandes (Ed). Proceedings of PLEA '94, Portugal, pp 390-397.

Eumorfopoulou, K., Aravantinos, D. 1995. Numerical approach to the contribution of the planted roof to the cooling of buildings. In M. Santamouris, (Ed.) Proceedings of the International Symposium of Passive Cooling of Buildings, Greece.

EUROACE. 1998. Assessment of potential for the saving of carbon dioxide emissions in European building stock, Report Published By EUROACE, Brussels, Belgium.

Fanger, P.O., 1972. Thermal comfort. Analysis and applications in environmental engineering. Mc Graw Hill, New York.

Fanger, P.O., Toftum, J., 2002. Extension of the PMV model to non-air-conditioned buildings in warm climates. Energy and Buildings, 34:6, pp. 533-536.

Fintikakis, N., Santamouris, M., 2002. Design and results of the outdoor spaces in the Olympic Village Athens 2004. Olympic Village 2004 Company, Athens, Greece (in Greek).

Flavin, K., Lenssen, H.R., 1995. Power Surge: A guide to the coming energy revolution. Earthscan Publishers, London.

Gagge, A.P., 1973. Rational temperature indices of man's thermal environment and their use with a 2-node model of his temperature regulation. Proceedings. Federation. American. Society. Experimental. Bioliology. 32, pp. 1572-1582.

Gagge, A.P., Fobelets, A.P., Berglund, L.G. 1986. A standard predictive index of human response to the thermal environment, ASHRAE Transactions, 92:2B, pp. 709-731.

Gao, W., 1993. Thermal effects of open space with a green area on urban environment, Part I: A theoretical analysis and its application. Journal. Architecture. Planning. Environmental. Engineering. AIJ:488.

Gao, W., Miura, S., Ojima, T., 1994. Site Survey on formation of cool island due to park and inner river in Kote-ku, Tokyo. Journal. Architecture. Planning. Environmental. Engineering, AIJ:456.

Georgakis, C., Santamouris, M., 2004. Air flow in urban canyons. In: Allard, F. (Ed.), Natural Ventilation in Urban Areas, James and James Science Publishers, London, UK, In Press.

GEPVP (Groupent Europeen des Producters de Verre Plat), 2001. Low e- glass in buildings. Impact on the environment and on energy savings. Report published by GEPVP, Brussels, Belgium.

Geros, V., 1998. Ventilation nocturne. Contribution a la reponse thermique des batiments, Ph.D. Thesis, INSA, (Institut National des Sciences Appliquees), Lyon, France.

Geros, V., Santamouris, M., Tsangrassoulis, A., Guarracino, G., 1999. Experimental evaluation of night ventilation phenomena. Energy and Buildings, 29. pp. 141-154.

Geros, V., Santamouris, M., Papanikolaou, N., Guarraccino, G., 2001. Night ventilation in urban environments. In: Proceedings AIVC Conference, Bath, UK.

Ghiaus, C., Allard, F., Mansouri, Y., Axley, J., 2003. Natural ventilation in urban context, In: Santamouris, M., (ed), Solar Thermal Technologies–The State of the art. James and James Science Publishers, London, UK.

Givoni, B., 1989. Urban design in different climates. WMO (World Meteorological Organisation), Technical Report 346.

Givoni, B., 1991. Modeling a passive evaporative cooling tower. In: Proceedings of the 1991 Solar World Congress, International Solar Energy Society, Denver, Colorado, USA.

Givoni, B., 1994. Passive and low energy cooling of buildings. Van Nostrand Reinhold Publishers, Los Angeles.

Good, W., 1990. Factors in planted roof design. Construction. Specifications. 43:11, pg. 132-138.

Goswami, D.Y., Dhaliwal, A.S., 1985. Heat Transfer Analysis in Environmental Control Using an Underground Air Tunnel. Journal of Solar Energy Engineering, Vol. 107, pp. 141-145.

Goswami, D.Y., Sinha, R.R., Klett, D.E., 1981. Theoretical and Experimental Analysis of Passive Cooling Using Underground Air Pipe. In: Proceedings of ISES Meeting, Brighton, England, Aug. 23-28.

Goswami, D.Y., Ileslamlou, S., 1990. Analysis of a Closed Loop Climate Control System Using Underground Air Tunnel. Journal of Solar Energy Engineering, Vol. 112, pp. 76-81, May.

Grosso, M., 2002. CPCALC+: A tool to calculate pressure differences. Final Report of the PASCOOL research project, European Commission, Directorate General for Research, Brussels.

Hage, K.D., 1972. Nocturnal temperatures in Edmonton, Alberta. Journal. Applied Meteorology. 11, pp. 123-129.

Haifeli, P., Lashal, B., Weber, W., 1998. Experiences with green roofs in Switzwerland. In: Fernandes, E.O., (Ed.) Proceedings of the PLEA '98, Portugal, pp. 365-368.

Hardy, J.E., Mitlin, D., Satterthwaite, D., 2001. Environmental problems in an urbanizing world. Earthscan Publishers, London.

Hassid, S., Santamouris, M., Papanikolaou, N., Linardi, A., Klitsikas, N., 2000. The effect of the heat island on air conditioning load. Energy and Buildings, 32:2, pp. 131-141.

Heiselberg, P., (Ed.) 2002. Principles of hybrid ventilation. IEA (international Energy Agency), Energy Conservation in Buildings and Community Systems Programme. Annex 35, Hybrid Ventilation and New and Retrofitted Office Buildings, Paris, France.

Hensen, J.L.M., Centnerova, L., 2001. Energy simulations of traditional versus adaptive thermal comfort for two moderate climate regions. In: Nicols, F., (Ed.) Proceedings of the Moving Comfort Standards into the 21st Century Conference, pp. 78-91.

Huang, Y.J., Akbari, H., Taha, H., Rosenfeld, A.H., 1987. The potential of vegetation in reducing cooling loads in residential buildings. Journal of Climate and Applied Meteorology, 26, pp. 1103-1116.

Humphreys, M.A., 1976.Field studies on thermal comfort compared and applied. Building Services Engineer, 44, 122-129.

Humphreys, M.A., 1978. Outdoor temperatures and comfort indoors. Building Research and Practice (J CIB) Journal Committee International Batiment, 6:2, pp. 92-105.

Humphreys, M.A.,1992. Thermal comfort requirements climate and energy. In: Sayidh A. (Ed.), Proceedings of the World Renewable Energy Congress, Reading, UK.

Humphreys, M.A., Nicol, J.F., 2000. Outdoor temperature and indoor thermal comfort: raising the precision of the relationship for the 1998 ASHRAE database of field studies. ASHRAE Transactions, 206:2 pp. 485-492.

Hutchins, M., 2003. Spectrally selective materials for efficient visible, solar and thermal radiation control. In: Santamouris, M., (Ed.) Solar Thermal Technologies for Buildings–The State of the Art James and James Science Publishers, London, UK, p.p. 123-158.

International Institute of Refrigeration, 2002. Industry as a partner for sustainable development–refrigeration. Report, Paris, France.

International Institute of Refrigeration (IIR), 2003. Information on the efficiency of air conditioners given through its web site, http://www.iifir.org.

ISO (International Standards Organization), 1984. Moderate thermal environments-determination of the PMV and PPD indices and specification of the conditions for thermal comfort. International ISO 7730, Brussels, Belgium.

Japan Air Conditioning and Refrigeration News (JARN), Japan Refrigeration and Air Conditioning Industry Association (JRAIA). 2002. Air Conditioning Market, Tokyo, Japan.

Jauregui, E., 1990-91. Influence of a large urban park on temperature and convective precipitation in a tropical city. Energy and Buildings, 15-16, pp 457-463.

Jokl, M.V., 1987. A new COMECON standard for thermal comfort within residential and civic buildings. In: Proc. Conference Indoor Air 87, Vol 3, 457-460, Berlin.

Jones, B.G., 1992. Population growth, urbanization and disaster risk and vulnerability in metropolitan areas: A conceptual framework. In: Kreimer, A., Mohan, M., (Eds.) Environmental Management and Urban Vulnerability, World Bank Discussion Paper, No 168.

Kolokotroni, M., Tindale, A., Irving, S.J., 1997. NiteCool: Office night ventilation pre-design tool. In: Proceedings of the 18th AIVC Conference-Ventilation and Cooling, Athens, Greece pp. 124-129.

Kuttler, W., Antreas–Bent, B., Robmann, F., 1996. Study of the thermal structure of a town in a narrow valley. Atmospheric Environment, 30:3, pp. 365-378.

Landsberg, H.E., 1981. The urban climate. Academic Press, London.

Lechner, N., 1991. Heating, cooling, lighting. John Wiley, New York.

Limb, M., 2000. Ventilation air duct cleaning–An annotated bibliography. Report published by the Air Infiltration and Ventilation Center (AIVC), Warwick, U.K.

Lindqvist, S., 1992. Local climatological modeling for road stretches and urban areas. Geografiska Annaler, 74A, pp. 265-274.

Lloyd, S., 1992. Ventilation system hygiene–A review. Published by BSRIA (Building Services Royal Institute), Technical Note 18/92, London, UK.

Lopez De Asiain, J., 1997. Espacios Abiertos. Open Spaces, Seville 92, Report published by the University of Seville, Seville, Spain.

Lyall, I.T., 1977. The London heat-island in June-July 1976. Weather, 32:8, pp. 296-302.

Madsen, T.L., 1987. Measurement and control of thermal comfort in passive solar systems. In: Faist, A., Fernandes, E.O., Sagelsdorff, R., (Eds.) Proceedings of the 3rd International Congress on Building Energy Management (ICBEM '87), IV, pp. 489-496, Lausanne, Switzerland.

McCartney, K.J., Nicol, J.F., 2002. Developing an adaptive control algorithm for Europe: Results of the SCATS project. Energy and Buildings 34:6, pp. 623-635.

Mihalakakou, G., Santamouris, M., Asimakopoulos, D.N., 1992. Modeling the earth temperature using multiyear measurements. Journal of Energy and Buildings, 19, pp. 1-9.

Mihalakakou, G., Santamouris, M., Asimakopoulos, D.N., 1994. Modeling the thermal performance of the earth to air heat exchangers. Solar Energy, 53, 3 No 2, pp. 301-305.

Mihalakakou, G., Santamouris, M., Asimakopoulos, D.N., Papanikolaou, N., 1994. The impact of the ground cover on the efficiency of Earth to air heat exchangers. Journal Applied Energy, 48, pp. 19-32.

Mihalakakou, G., Santamouris, M., Asimakopoulos, D.N., Tselepidaki, I., 1995. Parametric prediction of the buried pipes cooling potential for passive cooling applications. Solar Energy, 55:3, pp. 163-173.

Mihalakakou, G., Lewis, J.O., Santamouris, M., 1996. The influence of different ground covers on the heating potential of earth-to-air heat exchangers. Journal of Renewable Energies, 7:1, pp. 33-46.

Milne, G.R., 1995. The energy implications of a climate-based indoor air temperature standard. In: Nicol, J.F., Humphreys, M.A., Sykes, O., Roaf, S., (Eds.) Standards for thermal comfort: indoor air temperature standards for the 21st century. E & FN Spon Publishers, London, 123-145.

Mitchell, R., Kohler, C., Arasteh, D., 2001. Windows 5, Windows and Daylighting Group, Building Technologies Department, Environmental Energy Technologies Division, Lawrence Berkeley National Laboratory, Berkeley, California 94720, http://windows.lbl.gov/software/software.html.

National Academy of Science (NAS), 1991. Policy implications of greenhouse warming. Report of the Mitigation Panel, National Academy Press, Washington, D.C.

Nicol, J.F., Humphreys, M.A., 1973. Thermal comfort as part of a self-regulating system. Building Research and Practice, J. CIB, 6:3, pp. 191-197.

Nicol, J.F., 2003. Thermal comfort state of the art and future directions. In: Santamouris, M., (Ed.) Solar thermal technologies–The state of the art. James and James Science Publishers, London, UK.

Nicol, J.F., Raja, I.A., 1996. Thermal comfort, time and posture: exploratory studies in the nature of adaptive thermal comfort. Report published by the School of Architecture, Oxford Brookes University, Oxford, UK.

OECD (Organization Economic Cooperation and Development), 1996. Fighting noise in the 1990's. Published by OECD, Paris, France.

Ojima, T., 1990-91. Changing Tokyo metropolitan area and its heat island model. Energy and Buildings, 15-16, pp. 191-203.

Oke, T.R., East, C., 1971. The urban boundary layer in Montreal. Boundary Layer Meteorology, 1, pp. 411-437.

Oke, T.R., 1977. The significance of the atmosphere in planning human settlements. In: Wilkin, E.B., Ironside, G., (eds), Ecological Land Classification in Urban areas, Ecological Land Classification Series, No. 3, Canadian Government Publishing Center, Ottawa, Ontario.

Oke, T.R, Johnson, G.T., Steyn, D.G., Watson, I.D., 1991. Simulation of Surface Urban Heat Islands under 'Ideal' Conditions at Night-Part 2: Diagnosis and Causation, Boundary Layer Meteorology 56:339-358.

Palomo, E., 1999. Roof components models simplification via statistical linearization and model reduction techniques, Energy and Buildings, 29, 3, pp. 259-281.

Papadopoulos, A., 2001. The influence of street canyons on the cooling loads of buildings and the performance of air conditioning systems, Energy and Buildings, 33, 6, 601-607.

Papadopoulos, A., Oxizidis, S., Kyriakis, N., 2003. Perspectives of solar cooling in view of the developments in the air-conditioning sector, Renewable Sustainable Energy Reviews, Vol. 7, No. 5.

Paraponiaris, K., Santamouris, M., Mihalakakou, P., 2004. On Estimating the Ecological Footprint of Heat Island Effect over Athens, Greece. Submitted for Publication to Journal of Climate Change.

PDEC: EU DGXII JOR3CT950078, 1995. Final Report of the PDEC project, European Commission, Brussels, Belgium.

Perrill, C.V., "Performance of Passive Cooled Hospital Complexes in India Over a Period of 55 Years," Proceedings of the American Solar Energy Society's "Solar 99" Conference. Also in, the Proceedings of the 23rd National Passive Solar Conference, June 1999, Portland, Maine.

Rees, W.E., Wackernagel, M., 1996. Urban ecological footprints: Why cities cannot be sustainable and why they are a key to sustainability. EIA Review 16: 223-248.

Rees, W., 2001. The Conundrum of Urban Sustainability. In: Devuyst, D., Hens, L., de Lannoy, W., (Eds), How Green is the City, Columbia University Press.

Rees, W., 2001. Global Change, Ecological Footprints, and Urban Sustainability, In: Devuyst, D., Hens, L., de Lannoy, W., (Eds), How Green is the City, Columbia University Press.

Reinhart, C.F., Luther, J., Voss, K., Wittwer, V., van der Weiden, T.C.J., Santamouris, M., Weiß, W., 2001. Energy Efficient Solar Buildings, Position Paper of EUREC published by James and James Science Publishers, London, UK.

Rosenfeld, A., Romm, J., Akbari, H., Lloyd, A., 1998. Painting the Town White and Green, Paper available through the web site of Lawrence Berkeley Laboratory, http://eetd.lbl.gov/HeatIsland/.

Sailor, D.J., 1995. Simulated urban climate response to modifications in surface albedo and vegetative cover. Journal of Applied Meteorology, 34, 1694-1704.

Sailor, D.J., 1998. Simulations of Annual Degree Day Impacts of Urban Vegetative Augmentation, Atmospheric Environment, 32, 1, 43-52.

Saito, I., Osamu, I., Tadahisa K., 1990-91. Study of the Effect of green areas on the thermal environment in an urban area. Energy and Building, 15, 3-4, pp. 493-498.

Santamouris, M., Yianoulis, P., Rigopoulos, R., Argiriou A., Kesaridis, S., 1982. Use of the Heat Surplus from a Greenhouse for Soil Heating. In: Proc. Conference Energex 82, Regina, Canada.

Santamouris, M., Geros, V., Klitsikas, N., Argiriou, A., 1995. Summer: A Computer Tool for Passive Cooling Applications. In: Santamouris, M., (Ed), Proceedings of the International Symposium: Passive Cooling of Buildings, Athens, Greece, June.

Santamouris, M., Argiriou, A., 1993. The European project PASCOOL. In: Proc. 3rd European Conference on Architecture, Florence, May.

Santamouris, M., Argiriou, A., 1997. Passive Cooling of Buildings - Results of the PASCOOL Program. International Journal of Solar Energy, 18, pp. 231-258.

Santamouris, M., Assimakopoulos, D., (Eds), 1996. Passive Cooling of Buildings, James and James Science Publishers, London, UK.

Santamouris, M., Daskalaki, E., 1998. Case Studies. In: Allard, F., (Ed), Natural Ventilation, James and James Science Publishers, London, UK.

Santamouris, M., 1997. Rehabilitation of Office Buildings, Final Report of the SAVE project. European Commission, Directorate General for Energy, Published by CIENE, University of Athens, Athens, Greece.

Santamouris, M., Papanikolaou, N., Koronakis, I., Georgakis, C., Assimakopoulos, D.N., 1998. Natural Ventilation in Urban Environments. In: Proc.1998 AIVC Conference, Oslo.

Santamouris, M., (Ed), 2001. Energy and Climate in the Urban Built Environment, James and James Science Publishers, London, UK.

Santamouris, M., Georgakis, C., 2003. Energy and indoor climate in urban environments: Recent trends. Building Services. Engineering Research Technology, 24, 2.

Santamouris, M., 2003a. Experimental Evaluation of high reflecting paints for the urban environment, Report prepared for Intermat S.A, Published by the University of Athens, Athens, Greece.

Santamouris, M., 2003b. Interim Report of the WP2 of the Urbvent Project, European Commission, Directorate General for Energy, Brussels.

Simpson, J.R., McPherson, E.G., 1998. Simulation of Tree Shade Impacts on Residential Energy Use for Space Conditioning in Sacramento. Atmospheric Environment, 32, 1, 69-74.

Sinha, R.R., Goswami, D.Y., Klett, D.E., 1981. Theoretical and Experimental Analysis of Cooling Technique Using Underground Air Pipe. In: Proceedings of the International Solar Energy Society's Solar World Congress, Brighton, U.K., August.

Sodha M. et al, 1991. Thermal Performance of a room coupled to an evaporative cooling tower, In: Proc. of the 1991 Solar World Congress, Denver, Colorado, USA, pp. 3095-3100.

Stanners, D., Bourdeau, P., (Eds.), 1995. Europe's Environment-The Dobris Assessment. Published by the European Environmental Agency, Denmark.

Stoops, J., Pavlou, C., Santamouris, M., Tsangrassoulis, A., 2000. Report to Task 5 of the SCATS project (Estimation of Energy Saving Potential of the Adaptive Algorithm), Available by European Commission, Brussels, Belgium.

Swaid, H., Hoffman, M.E., 1990. Climatic Impacts of Urban Design Features for High and Mid Latitude Cities, Energy and Buildings, 14, p.p. 325-336.

Taha, H., 1988. Site specific heat island simulations: Model development and application to microclimate conditions. LBL Report No 26105, M. Geogr. Thesis, University of California, Berkeley, CA.

Taha, H., Akbari, H., Rosenfeld, A., 1989. Vegetation microclimate measurements: the Davis project, Lawrence Berkeley Laboratory Rep. 24593.

Taha, H., Akbari, H., Rosenfeld, A., 1991. Heat island and oasis effects of vegetative canopies: Microclimatological field measurements. Theoretical. Applied Climatology, 44, 123-134.

Taha, H., Sailor, D., Akbari, H., 1992. High albedo materials for reducing cooling energy use. Lawrence Berkeley Lab Rep. 31721, UC-350, Berkeley, CA.

Taha, H., 1994. Meteorological and photochemical simulations of the south coast air basin, In: Taha, H., (ed), Analysis of Energy Efficiency of Air Quality in the South Coast Air Basin - Phase II, Rep. No LBL-35728, Lawrence Berkeley Laboratory, Berkeley, CA, Ch. 6, p.p. 161-218.

Tatsou, O., 1980. Thermal environment in urban areas' Report D7 : 1980, Report published by the Swedish Council for Building Research, Stockholm, Sweden.

Thompson, T., Chalfoun, N., Yoklic, M., 1994. Estimating the thermal performance of natural down-draft evaporative coolers, Energy Conversion and Management, Vol. 35, 11, pp. 909-915.

Tombazis, A., Argiriou, A., Santamouris, M., 1990. Performance Evaluation of Passive and Hybrid Cooling Components for a Hotel Complex.International. Solar Energy, 9, 1-12.

United Nations Council for Human Settlements, UNCHS, 2001. The State of the World Cities, 2001, Published by UNCHS, Nairobi, Kenya.

UNCHS, 1993. Development of National Technological Capacity for Environmental Sound Construction, HS/293/93/E, ISBN 92-1-131 214-0, Published by UNCHS, Nairobi, Kenya.

Van der Maas, J., Roulet, C.-A., 1991. Night time ventilation by stack effect. ASHRAE Technical Data Bulletin, Vol. 7, No. 1, Ventilation and Infiltration, pp. 32-40.

Wackernagel, M., Rees, W.E., 1996. Our Ecological Footprint: Reducing Human Impact on the Earth. Gabriola Island, BC and New Haven, CT: New Society Publishers.

Wackernagel, M., Onisto, L., Linares, A.C., Falfan, I.S.L., Garcia, J.M., Guerrero, A.I.S., Guerrero, M.G.S., 1997. Ecological Footprints of Nations. Report to the Earth Council, Costa Rica.

Wackernagel, M., Onisto, L., Bello, P., Linares, A.C., Falfan, I.S.L., Garcia, J.M., Guerrero, A.I.S., Guerrero, M.G.S., 1999. National natural capital accounting with the ecological footprint concept. Ecological Economics 29, 375-390.

Wanner, H., Hertig, J.A., 1983. Temperature and Ventilation of Small Cities in complex terrain (Switzerland). Study supported and published by Swiss National Science Foundation, Berne, 1983.

Watkins, R., Palmer, J., Kolokotroni, M., Littlefair, P., 2002. The London Heat Island: Results from summertime monitoring. Building Services Engineering Research Technology. 23,2, pp. 97-106, 2002.

Wilkins, J., 1995. Adaptive comfort control for conditioned buildings. Proceedings CIBSE National Conference, Eastbourne. Part 2, 9-16. Published by Chartered Inst. of Bldg Serv. Engrs, London.

Wilson, H., 1999. Solar Control Coatings for Windows, Proceedings of the Conference EUROMAT 99, Sep 27-30 Munich, Germany, 1999.

Wilson, H., Gombert, A., Georg, A., Nitz, P., 2000. Switchable Glazing with a Large Dynamic Range in Total Solar Energy Transmittance (g-value). In: Sayidh, A., (Ed), Conf. Proceed. World Renewable Energy Congress- VI (WREC 2000), July 1-7, Brighton, England.

Window 4.1, 1993. A PC Program for analyzing window thermal performance, Lawrence Berkeley Laboratory, Windows and Daylighting Group, LBL Report 32091.

WMO: (World Meteorological Organization), 1986. Proceedings, Technical Conference in Mexico City: Urban Climatology and its Applications with Special Regard to Tropical Areas, Published by WMO - No 652, Geneva.

WMO, (World Meteorological Organization), 1994. Report of the Technical Conference on Tropical Urban Climates. Published by WMO, Dhaka.

World Bank, 2000. Indoor Air Pollution, Fighting a massive health threat in India, Washington, DC., Published by World Bank.

Yap, D., 1975. Seasonal Excess Urban Energy and the Nocturnal Heat Island – Toronto. Archives Meteorology Geophysics Bioclimatology, Series B 23, p.p. 68-80.

Zimmerman, M., Andersson, J., 1998. Low Energy Cooling – Case Study Buildings, International Energy Agency, Energy Conservation in Buildings and Community Systems Programme, Annex 28 Final Report, Published by the International Energy Agency, Paris, France.

Chapter 9

Renewable Solar Energy for Traveling: Air, Land, and Water

by

Paul B. MacCready
AeroVironment Inc.
825 S. Myrtle
Monrovia, California, 91016-3424 USA

Abstract

The goal is to explore transportation needs that can be accomplished with renewable and economical solar energy (plus non-solar tidal energy): direct radiation, and radiation consequences such as temperature, wind, rain, streams, waves, currents, turbulence and shears, and growing plants and trees. The transportation subject encompasses natural creatures from a fraction of a kilogram on up, human powered devices that go to 100 or 200 kg or so, and powered devices that involve human control that include cars, trucks, trains, aircraft, and boats and submarines. Only representative samples of the major categories are presented, types that illuminate the mechanisms of acquiring or using solar energy.

The investigation of land vehicles is split into two parts: the effects obtainable with zero wind, and the additional effects that present themselves when wind plays a significant role. The zero wind cases include ways to improve the use of human muscle to avoid tiring, and the use of battery or other power to achieve vehicular efficiency. The battery powered car appears very attractive because of the continually improving high capability of lithium batteries. Lithium-gasoline hybrids are indicated for early use now, before battery prices decrease enough for them to be relied on to do the whole task. The lithium powered vehicles not only permit the full benefit of regenerative braking, but also, with two-way charging from and discharging to the grid, can provide the utility with a valuable resource for quickly adjusting current to balance the utility's needs, and also for providing extra power for large air conditioning loads.

---

* Corresponding Author. Email: maccready@aerovironment.com

The use of wind to help with travel over land or water is discussed. The effects can be strong with regular sailboats, and will permit faster vehicles (catamarans, land sailers and ice boats) to move upwind or downwind faster than the wind speed as well as move laterally much faster still. A propeller or wind turbine on the vehicle can power the same motions without needing the sail – and permits the vehicle to be operated straight into or with the wind.

Next, flying air vehicles and creatures are considered, not using the already understood upcurrents of slopes, thermals, and waves, but the dynamic power from winds, shears, turbulence and wakes. Such atmospheric effects are evident as we note the flying of albatrosses and vultures in wind without flapping. These lead to major flight benefits for planes, called dynamic soaring.

Movement on and under water means ordinary boats, slender boats, boats elevated above embedded hydrofoils, and submarines operated completely under the water. The surface water vehicles are affected greatly by the wind and wave motions which can be helpful if carefully utilized.

Finally, cases are presented where wind effects on flying are utilized more efficiently by connecting a cable between devices widely separated (air-air, air-ground, or air-water) to operate in different wind regions. The energy effects are maximized if the aerial device is controlled to dash about like a windmill blade rather than staying in one spot.

Illustrating such principles of nature and technology on air, land, and sea can help lead humans to handle much of their transportation and movement of goods while relying on non-polluting, renewable energy.

1. Introduction

The earth is powered by the sun, which provides virtually all earth's energy. For eons, living plants and creatures have lived off the energy arriving from the sun (plus a tiny bit from moon-induced tides): from immediate radiation, to the wind and streams and waves, to the energy in plants and trees from even a hundred years earlier. Ingenious humans initially adapted to using that available renewable energy. However, in the last several centuries, humanity has rocketed upward in population while turning toward the use of oil, gas, and coal which are renewable only in terms of millions of years and which have pollutants associated with their burning or processing.

Dramatic improvements in efficiency are available. As an example, consider the huge percentage of energy that could be saved in powering typical ceiling fans. Modern design lets the fans move the same amount of air for about one sixth the usual electric energy consumed (a factor of two for aerodynamic efficiency improvements in the blades, a factor of three for electronic efficiency in the motors). The economic consequences are significant, but the total impact on society is not large. Smaller percentages, but with a far larger total impact, could now be achieved with battery powered or even hybrid cars: efficiency improvements in the use of battery power, the use of regenerative braking, *and* with the batteries being reverse-coupled to the charging grid so they can also be discharged back to permit the utility to utilize these numerous small car electricity "generators" to handle utility adjustments and longer-term needs for home air conditioning on hot days.

As we look at nature's uses of energy for natural creatures, we begin to get a few clues about efficiency. The albatross soars for days without flapping in windy conditions in the wind gradient of the lowest 30 m above the ocean. The vulture, with a different mechanism, in wind soars at 15-30 m over the tops of trees in a forest, with much maneuvering but no flapping. We look at the human ability for walking and running, but note the higher speed and efficiency of bicycle riding. The clues we get give us insights about achieving great efficiencies in vehicles for air, land, and sea that draw on the energy of animals and humans, of motors, and of the bonuses we can achieve from the mean motions, and turbulence and shears in air and water. In summary, we can benefit by using some of the concepts displayed by small and large natural creatures, and using some of the other concepts we can provide for ourselves. We should, and in the future we must.

Therefore this analysis forces us to explore some efficient movements in air, on land, and through or on water — movement of humans or their products. There are huge improvements available, but also practical constraints. In the future we still will want travel, for work or recreation, but we can travel with *much* more efficiency and *much* less pollution consequences than is presently the case. Here we will talk about a few of the alternatives, especially for small vehicles, and in circumstances of wind and waves, turbulence and shears. This discussion cannot be complete, but at least it will introduce the sorts of potentials that are available.

2. The Special Challenge

Figure 1 shows all the energy used on earth (except nuclear, tidal, and geothermal). The top six categories are usable within a human life span, but the bottom category, fossil fuel (here including coal and natural gas) requires millions of years to renew and so for humans must be considered non-renewable.

Fossil fuel has been ubiquitous during our lifetimes to the point where we have become accustomed to thinking of it as permanently available. However, it is only a finite source. In the U.S., peak production of oil took place in 1970. The peak in global production may have already occurred or will in this decade or the next. Also, there appear to be new large global weather effects occurring, such as temperature changes in the surface atmosphere, humidity effects, more cirrus level cloudiness due to release of water in airplane exhaust, and the enlarged ozone holes at stratospheric altitudes at the poles.

Humans would be wise to comprehend the global weather changes that our industrial activities are altering. There have been major programs to quantify this multidimensional challenge. The challenge could grow to warmer temperatures, temperature

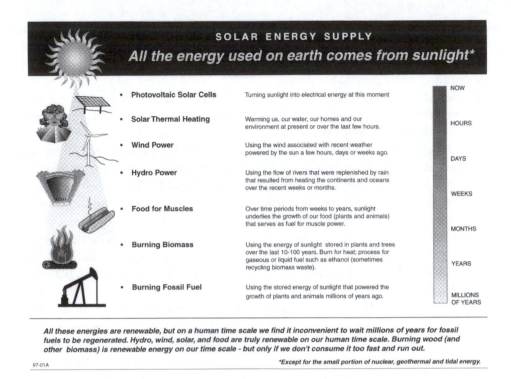

Figure 1. Solar energy supply.

variations by latitude, extreme rainfall distributions, etc. Much weather now seems better and much seems worse, and we struggle to understand the reasons. Are the apparent changes real? And should they be considered natural or artificial? We have previously made considerable improvements in the living conditions of most people in advanced countries, considering the decrease of huge effects from urban pollution. However, as populations grow, the pollution consequences still need further understanding and treatment.

Figure 2, showing the global mass of vertebrate life on land and in the air, provides a dramatic warning concerning human activities. In 2000 when this figure was prepared, the line indicating humans + livestock + pets stood at 98%! Ten thousand years ago, at the early start of civilization, the same line was at less than a negligible 0.1%. The biggest changes in humans + livestock + pets started occurring about 300 years ago as we began seriously digging for coal, and the changes grew faster beginning 100 years ago as technological civilization began using oil. The change from 1925, when I was born and the global population was about 1.7 billion, has been huge in the number of humans (3.7x) and even more when including our increasing livestock and pets. The figures show that we humans have truly taken over the planet, both by the much greater numbers of humans + livestock + pets and the decrease of natural creatures (especially the large animals). As the primary owners it is up to us to create an effec-

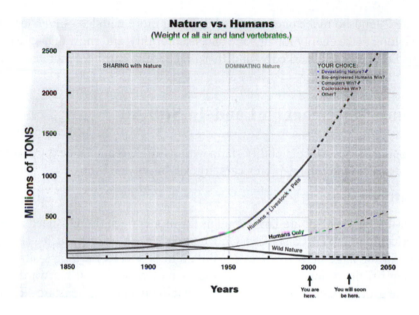

Figure 2. Global mass of vertebrate life on land and in the air.

tive future. There is no way to decrease population. The growth rate in the developed countries is at least leveling off, but a global population of eight or ten billion people seems inevitable.

There seems to be no viable substitute for jet fuel for powering airliners. Hydrogen, even when liquefied, is so voluminous that there would not be room for most passengers. Ethanol derived from cellulosic plants could do the job without a net CO_2 change, but such fuel is presently not cheap enough, and probably could not ever be produced in adequate quantity.

Hoffman (2003) and O'Mahony (2000) provide general articles on the aviation developments at AeroVironment. The cars, bikes, human powered land vehicles, and the water transportation devices are treated in specific articles. To summarize, we are involved in all aspects of transport of unoccupied and occupied devices — with an emphasis both on conserving energy and on using renewable energy.

A decade ago I fashioned the following statement:

"Over billions of years, on a unique sphere, chance has painted a thin covering of life — complex, improbable, wonderful and fragile. Suddenly we humans (a recently arrived species no longer subject to the checks and balances inherent in nature), have grown in population, technology, and intelligence to a position of terrible power: we now wield the paintbrush."

Figure 2 tends to make one interested in doing more with less–*much* less. In this report we look at the energy for air, land and sea mobility, and show that at least much of it can be decreased markedly, and that much of it can be generated and used without pollutants.

3. Surface Movement on Land–In Still Air

A broad look at surface mobility starts with natural creatures, next treats human-powered devices, and then explores cars, trucks, buses and trains.

3.1 Creatures

Natural creatures that are large enough for moderate and high speeds are adept at fast movement for short distances. Only certain large creatures, such as kangaroos and some of the large herbivores, can maintain high speeds for long. Animals must move a leg off the ground and forward and then start moving it rearward before it touches the ground: leg motion that is inefficient even when the legs are slender.

3.2 Human Power

Human-powered devices can move steadily much faster than ordinary human running motion. Except for specially trained and talented athletes, humans tend to be limited in the ability of the legs to maintain power when the legs are fully loaded. A top cyclist can maintain a power of about 200-300 watts output and keep up a speed of around 12-13 m/sec, but only run on legs continuously at less than half that speed. This subject of human power for travel in air, on land, and on water is well covered in a book edited by A.V. Abbott and D.G. Wilson (1995).

Through studying such characteristics we have started to develop Technalegs™ to support smoothly the weight of the upper body while walking or jogging (Figure 3). The goal has been to create apparatus to help standing and walking, so that the operator could carry weight without much effort just as one does on a bicycle. One has the added benefit, in contrast to riding on a bicycle, of only connecting to the ground intermittently. This means that the continuous drag of the bicycle tires is substituted for by the much lower average drag of the shoes which arises only intermittently. The concept works, and you can carry an extra 50 kg on your back without getting tired the way you would ordinarily expect doing normal walking or jogging.

This way of getting rid of wheel friction on the ground can also be made more effective if you walk with high stilts. The connection to the ground is more sparse. Efficient but not really practical. We have also tried to design another approach: making a 6-legged vehicle to substitute for a car, somehow designed so the vehicle could move quickly and almost effortlessly over rough ground. The complexity of the device that could accomplish this feat (and deal with turning and operating in rough terrain) would have been so complex that the goal was deemed unachievable.

Figure 3. Technalegs™ letting you easily carry a heavy weight.

3.3 Solar Powered Vehicles

Solar powered land vehicles suddenly became real when the first solar car race was established in 1987. The rules for the trip from Darwin to Adelaide in Australia specified solar power, augmented as needed by batteries charged by the sun. The strategy of coping with occasional cloudiness was quite complex because the timing of the clouds could not be predicted accurately. The race was won by the GM Sunrayer (Figure 4), for which the main development and testing was done by AeroVironment Inc.

The Sunrayer finished the race in 5-1/4 days, and was 50% faster than the runnerup. The race has continued every two or three years, with the vehicles becoming even more streamlined and the electric motors being designed for efficiency over 95%. This has been an ideal training program for people to focus on the potentials for land vehicles – low drag, efficient power system, and the need for reliability and versatility.

Sunlight is not adequate for normal highway vehicles, which must do their transportation jobs even at night or when clouds continually obstruct the sun. However, these competitions have stimulated the technology of efficiency. Energy from PV cells tends to be expensive, say 20–25 cents per kwhr, satisfactory where it provides

Figure 4. The GM Sunraycer.

a unique benefit but rather expensive compared to the 4 or 5 cents for energy obtainable from wind turbines.

3.4 Automobiles

Cars are so integrated into our lives and expectations that we find it very difficult to consider new power sources for vehicles or substitutes for vehicles. However the U.S. cars which gained such great miles per gallon (mpg) improvements 20-30 years ago have actually not increased in their efficiency in the last two decades. Even ignoring the CO_2 releases, this problem is likely to become a real economic challenge as gasoline prices rise when we move past the global peak of production of gasoline and start into the descending portion of the production

It is instructive to look at final solutions for powering cars, rather than considering how long we can get by on using fossil fuel (gasoline or diesel), because we must realize that we are already long past the time when we should have been decreasing such fuel use substantially. There is now a great deal of study of hydrogen/fuel cell vehicles because of government interest and financing. Enthusiasm for such vehicles is increasingly being tempered because of the inefficiency of the use of hydrogen (which, like electricity, is a carrier, not a fundamental energy source) and the huge costs and limited ranges associated with its use as a fuel for cars. We at AeroVironment have produced hydrogen/fuel cell systems, large and small, for special purposes when the cost was not a consideration. (Our tiny Hornet airplane, flown last March, was probably the first aircraft to be propelled solely by a fuel cell. This is a 15" span airplane that weighs 6 ounces.)

Therefore, our interests at AeroVironment have grown to include battery power for surface vehicles. We have analyzed this to be an inevitable part of the future, and thus have been surprised to find zero interest at present in significant battery power by any major car company. The potential has been demonstrated by the use of small lithium batteries (lithium ion and lithium polymers) for cell phones and microcomputers (and incidentally for small model or operational aircraft). The best of these batteries now put out about 3 times the energy/kg of the Nickel Metal Hydride (NiMH) batteries used in the no-longer-available GM EV-1 and the Toyota RAV 4EV cars, and six times the energy/kg available for lead acid cells. The lithium cells also have very high power outputs. Lithium batteries now offer about 200 Whr/kg. Most are 160-180 Whr/kg, while a few exceed 215 Whr/kg and some 130 Whr/kg cells are also being used. In any case, the energy and power offered is very high.

Cars, in still air, can do an amazing job of traveling smoothly on wheels, with low net energy if we focus on that ability. As a starting reference it is useful to consider the ultimate: an electric car that has 100% efficiency of electricity utilization, zero tire drag and aerodynamic drag, and 100% recovery from braking. Its propulsion energy would be perfectly conserved. Except for the electricity used for control, instruments, radio and lights the range would be infinite.

Next consider actual efficiency which might be obtainable with an ultimate electric vehicle. Recharging efficiency for lithium batteries is already high, typically over 93% for one-hour charge-discharge rates if battery heating and cooling to ambient temperature is assured. Tire drag is low, say 0.7% of the weight – the value for the EV-1 and probably similar for the RAV4-EV. Aerodynamic drag can be low.

The EV-1 of Figure 5 has a small cross section area and a drag coefficient of only 0.19, and eventually it may be possible to maintain a substantially lower drag with laminar flow over most of the body as research is showing with suction on sailplane wings to maintain a laminar flow boundary layer. Regenerative braking energy recovery is likely to be above 80%.

This sort of ultimate reasoning focuses ones attention on the battery powered car as an element of future transportation. Alan Cocconi of AC Propulsion Inc. recently developed a two-person lithium cell vehicle. It used 6800 of the small 18650 cells that have been developed and used for cell phones and microcomputers. The car did a demonstration trip of 250 miles (much of it at 70 mph) leaving a 30-40 mile range in the batteries. The batteries provided acceleration of 0-60 mph in under 5 seconds. The batteries are individually small and efficient but expensive at the moment.

This suggests that the most practical competitive vehicle now would be a hybrid having just a 75-100 mile range on battery only, but when needed, an additional 200-250 miles from an internal combustion engine (ICE) burning gasoline or ethanol. About 90% of the total annual mileage would typically be handled solely by the batteries. As lithium cells improve in capability and cost, the batteries-only range can be economically increased, eventually to the 250-300 mile range car that is 100% electric.

Another important feature of the battery powered car is the use of a bi-directional charging system. This permits the vehicle to provide extra services which are likely to

Figure 5. Our 1989 "Impact," shaped like the EV-1.

pay for much of the cost of electricity supplied to it. One factor of the back-charging is the stabilization of the current supplied by the utility. This is for small amounts, varying + and − for seconds to a few minutes. With proper coordination with the utility, the returned portion of the current could be at a much higher value on price rate than the portions used in the regular direction. Brooks (2002a) demonstrated and described the technique. Brooks (2002b) describes the overall benefit from the range of the lithium battery. Another great benefit comes from using the batteries for the utility to handle the giant loads arising from the use of air conditioners on hot days. The excess can be handled by a number of battery electric cars hooked up to the system, delivering their charge as needed, but recovering the charge later, say from 1 a.m. to 6 a.m.

The use of lithium cells for general car transportation seems inevitable. The economics will steadily improve as the larger batteries become available for cars, the dangers of the battery use will be handled by continued design features, and the battery potential will inevitably grow. (Incidentally, some enthusiasts predict 300 Wh/kg in a year and 400 Wh/kg in two years, and steadily decreasing costs/kWh. Such improvements have a chance of taking place, but probably at a much slow pace.)

During the last half year a number of people have looked more critically at the economics of car transportation in the U.S. A report by the International Center for Technology Assessment (2003) came out with $5.60 to $15.14/gallon. Such reports illuminate the increasing need for change in fueling our U.S. travel habits.

This all is a direct connection of car travel to present solar energy, with the energy arriving through the utility system. The energy in utility grids varies widely by source, and in the U.S. is at present primarily from coal (a polluting source). However the amount from non-oil-gas-coal is increasing steadily, and the energy from wind will likely grow as rapidly as the use of battery electric cars grow. This electric vehicle power also can be employed for small and large trucks, buses and trains – and is steadily being used more in the latter two.

4. Land Surface Vehicles in Wind

All surface creatures and vehicles have air drag, and normally this puts them at an energy disadvantage when moving into the wind and a (somewhat lesser) energy advantage when moving with the wind. For cars and trucks, which ordinarily operate at speeds greater than the wind, the main technology strategy is to streamline the vehicle so that at least the air drag is kept low. For cars this is usually a compromise between a vehicle's efficiency and its appearance. Most airplanes are designed strictly for low drag; appearance means nothing for it turns out that we have been accustomed to the streamlined airplane shapes and find them beautiful. Cars in general are more stylish. Their undersides, seen by the wind but not by buyers, are mostly high drag. For station wagons and trucks there is high drag because of the lowered pressure on the rear. Various techniques have been tried for cutting the drag, such as exterior plates behind the rear, or vortex generators on the rear corners. The results are encouraging, but have not yet been groomed for practical application. For large trucks, the shape of the front end is important, and there are various techniques and devices that eliminate the projection of sharp corners into the wind. In summary, there are well established ways of cutting the aerodynamic drag of cars and trucks, the methods have considerable impact when driving into the wind, and improved methodologies deserve continued attention.

There are some amazing land vehicles that get their energy for movement from the wind. The standard method is to use a winged device, such as the Sirocco Land Sailer (2003). A more dramatic way uses a propeller-windmill rather than a sail. The propeller land sailer (Figure 6) was developed by Andrew Bauer long ago to show that the wind can drive a vehicle downwind at a speed faster than the wind. Bauer demonstrated its success in actual operation, moving downwind with a flag attached showing that it was actually moving faster than the wind because the flag showed the relative wind moving rearward. Bauer made a small, improved model which he operated on a table that had a moving surface installed. In a room with zero wind, the model started moving with the moving surface, but soon began moving in the opposite direction to the surface's movement and reached the table edge. This demonstration was definitive, with the moving surface and no wind corresponding to the case in the field with a fixed surface and some wind.

The reason this works is because of the rotating propeller. As we will see in the later section treating sailboats, an efficient sailboat can move downwind faster than the wind as it moves somewhat crosswind powered by its sail. Here we have the propeller substituting for the moveable sail.

Drawing further on the adaptability of the sort of vehicle shown in Figure 6, we realize that the wind can drive an efficient sand sailer across wind at 4 or even 5 times the wind speed, and that it can even go in a direction giving it a downwind component speed or an upwind component speed exceeding the wind speed. Bundling the sail area into the area of a rotating propeller is a convenience because it permits the

Figure 6. The Propeller Land Sailer of Andrew Bauer.

efficient motion of a sand sailer to operate for straight downward or upwind cases. The propeller is convenient but not very effective as the vehicle moves crosswind, although it could be angled so as to head into the relative wind. Probably this propeller technique could decrease the power required to move a vehicle in any direction when there is a significant wind, but the side forces it creates would make it difficult to apply.

To summarize, the wind can help or hinder the motion of a vehicle. By careful design, the helping function can predominate, but the saving of fuel is probably not worth the complexity and danger of the opportunity. As recreation on dry lakes, the wind can be an attractive contributor. The best rule for serious work is to omit sails or propeller blades and design the vehicle so the drag is as small as possible in any wind speed or direction, and then power the vehicle.

5. Flight

5.1 Effective Flight in Still Air

Still air means air that does not have turbulence or shears in it, although it may be moving horizontally with respect to the earth. This is easy to contemplate if the earth below is covered by an extensive sheet; whether moving fast or slow, the earth's movement has no effect on the aircraft flying above in still air.

Figure 7. Helios flying from Kauai, August 13, 2001.

Aircraft, from giant airliners to lightplanes, are designed for efficient movement in still air. Our special interests have been in getting the efficiency high so that the power can be low and therefore the flights long, and sometimes in deriving the power from other energy sources than gasoline or diesel. The special interests include light weight aircraft, and a number of non-piloted aircraft from 20 grams to over 100 kg. We also have paid special attention to the launch/land modes, including in-flight conversion of helicopter modes to standard winged flight modes.

Our air vehicles have tended to fit into two separate categories, both for unoccupied vehicles. One is the extremely high altitude stationkeeping aircraft (Pathfinder, Pathfinder Plus, Centurion, and Helios) which are to fly for multi-day missions at heights around 20 km. The indicated airspeeds are generally slow, 20-25 mph, and the true airspeed is the same at sea level but 3-1/2 times higher at 20 km. (Higher speeds can be used at altitude to cope with the rare higher wind speeds there, but these higher speeds can decrease the duration.)

Figure 7 shows our Helios shortly after takeoff on its record altitude flight in August 2001. It reached 96,863 feet (29,524 m), over two miles higher than any plane had ever maintained level flight.

The dedication to large, solar-powered aircraft emerged from AeroVironment's initial 1980-81 solar flight program that had evolved from our first human-powered Gossamer Condor (Figure 8). For solar-powered flight, first we put a number of available PV cells onto a panel above the center of the wing of the Gossamer Penguin (Figure 9), a ¾ size human-powered airplane which had emerged on our 1976-79 developments of the Gossamer Condor/Gossamer Albatross human-powered flight program. This was first flown in solar-powered flight on May 18, 1980 by 80 lbs. (13-

Figure 8. The Gossamer Condor.

Figure 9. The Solar Powered Gossamer Penguin.

year-old) Marshall MacCready. Then we turned out the Solar Challenger, 47' span, low aspect plane powered by solar cells on the wing and stabilizer (Figure 10). In the summer of 1981 it performed its main mission, a solar-powered flight of 163 miles from Paris to an airfield in eastern England. The flight was made at 11,000 feet, and got us thinking much more about the altitude potential of solar power. We estimated the climb rate would have peaked at about 30,000 feet, the power available growing faster with altitude than the increased power needed for flight. A still higher altitude would have achieved the highest climb rate except for the cells heating in the turbulent flow region on the wing and tail. Much had evolved from our human-powered airplane projects. The 1979 Gossamer Albatross, 96' span, 70 pound weight, is shown in Figure 11.

Such learning led to an unmanned military program for a stratospheric plane, but the initial flight tests convinced us the design and flight challenges were too great for

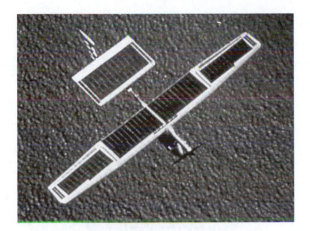

Figure 10. The Solar Challenger.

Figure 11. The Gossamer Albatross.

real success at the time. The project was then discontinued but later restarted as a Star Wars program out of Lawrence Livermore National Lab. That program had an early political ending and the project was subsequently taken over by NASA-Dryden where it remains today. The most thorough documentation of the stratospheric solar planes is presented in a NASA-Dryden report (NASA X-Press, 2002) by Dryden Flight Research Center. In the summer of 2003 the Helios, flown from Kauai in the Hawaiian Islands, suffered a crash due to pitch instability as it was initiating its flight

Figure 12. Black Widow, 6" square, 29 minute flight duration.

with a supply of compressed hydrogen gas (5000 psi, 345 standard atmospheric pressure). Our Pathfinder Plus will be flown in 2004 while plans for the future of Helios get resolved. Tentative possibilities are: 1) to use a lithium battery instead of a 2-way fuel cell system to store solar energy to keep the vehicle flying at night, and 2) to use liquid hydrogen as the energy source for a vehicle, without solar power, that could fly at altitude for 10-14 days to handle the brief stronger wind cases in high northern latitude in winter when solar power may not be adequate.

The other sorts of pilotless aircraft of AeroVironment have been the relatively small drones. The lightest is the 15 gm ornithopter, while the largest have expanded wings for launch from large aircraft. The modern lithium batteries have greatly improved duration, to 25 minutes for the ornithopter, 29 minutes for the 6" Black Widow (Figure 12), 1 hour for the Dragon Eye (Figure 13), 1 hour 30 minutes for the Raven (Figure 14), and over 4 hours for the 8.3 lb. (4 kg) Pointer (Figure 15). Flight duration can be materially increased by vehicle efficiency improvements, but these vehicles emphasize landing and takeoff that are very practical. All the small vehicles except the ornithopter carry along at least a telemetry video camera. Now that IR cameras have developed in weight to under 0.2 kilograms, they are used instead of or in addition to normal view cameras.

Black Widow was developed for DARPA primarily to see how well something this tiny, with video telemetry, could operate on batteries. It was a difficult size/shape to get operating without any waver in its flight, but in final configuration it worked perfectly. It was small and rather quiet and so generally unobservable. However, the

Figure 13. Dragon Eye. 6 lbs; 1 hour flight duration.

Figure 14. Raven. 4 lbs; 1 hr. 30 minute flight duration.

Figure 15. Pointer. 8.3 lbs; flight duration of over 4 hours.

box to carry the vehicle, and several charged batteries, the ground station with a suitable directional antenna, and a suitable shielded viewing screen for recording/ observing what it detected, all greatly increased the system size.

Therefore the Pointer and the smaller/lighter Raven and Dragon Eye vehicles have emerged as the primary small drones for scouting and monitoring, day and night. These all feature easy hand launch, and the landings can be made in small spots.

Our development of various small drones continues actively, with special emphasis on heavy vehicles which can fly fast and slow and even take off and land virtually like a helicopter. Flight within a small room is becoming more important.

5.2 Flight in Turbulence, Shears, and Wakes

Orville Wright made a 9 minute 45 second gliding flight in 1911. Sailplanes with humans aboard have been exploiting atmospheric motions since 1921, starting in Germany. The flight research that led to genuine soaring started with upcurrents on a slope, then thermals, and finally waves in the late 1930s. All such soaring was self explanatory: just sink more slowly in air rising faster than you are sinking and you will climb. For sailplanes and personal aircraft, the books compiling many of the articles by

Bruce Carmichael are worth exploring (Carmichael 2000a, 2000b). Record-aiming sailplanes and light aircraft are the basic topics, and also both documents include a series of articles on dramatic low drag motion through water. This sort of soaring offers huge potential, but need not be discussed further here.

Only in the last seven years or so have models and sailplanes stayed aloft by really concentrating on the turbulence, shears and wakes in the atmosphere. The stage was set early. In 1883 the first paper on the soaring of an albatross over a windy ocean was published (Rayleigh, 1883). In 1921 Dr. Wolfgang Klemperer did a study of dynamic soaring. His English translation was presented much later in Soaring magazine (Klemperer; 1943-45). Over the years several people have written theses or put papers into soaring magazines and technical journals on the topic. In just the past few years the field has received serious attention. Dr. Gary Osoba has studied the flight of both slow and fast sailplanes in turbulence and shears, organized contests for hang gliders that benefit from the turbulence and shears, and set amazing records. Simultaneously, Taras Kiceniuk has worked on the theory and made presentations and written articles on the topic (Kiceniuk, 2001a,b).

Also in the past few years model glider enthusiasts have been doing amazing flight demonstrations associated with the wakes of small hills–staying aloft at fantastic speeds, but limited to operating close to the hill. An excellent reference to these flights is in a videotape (Naton, 2002). The theoretical basis of the technique is described by Sachs and Mayrhofer (2001).

Thus in summary the field of dynamic soaring has been known for over a century, there was some understanding of the fundamentals, but it has taken experiments during the last few years to demonstrate, validate, and expand some of the concepts.

The techniques involve accepting high gust loads on the airplane. Pilots are generally trained to minimize the gust loads in all airplanes/sailplanes, but the most effective techniques involve accepting and exaggerating the gust loads, doing a bit of high frequency manipulation of the lateral and pitch controls. The mechanism for obtaining lift, or thrust, from the turbulence can be best understood by considering an airfoil at zero angle of attack operating across a wind tunnel. With smooth flow it has drag, but in turbulent flow, with the wind hitting it from above and below, it gets some thrust components, or at least less drag, for each upcurrent or downcurrent. The rule in dynamic soaring is to benefit from turbulence rather than minimizing it. In the atmosphere where the scale of turbulence is far larger than in the wind tunnel, the beneficial results can be much greater. The goal is always to move with the turbulent or shear flows rather than minimizing their effects.

The scale of the effects go from birds to sailplanes and small power planes, and could even be useful for much larger aircraft–drones and piloted.

Exploiting shears is the characteristic flight of the albatross. It occasionally soars on the slope of air moving up over waves, but its primary soaring is in the vertical shear zone of horizontal wind in the lowest 30 m above the waves. By gliding down roughly in the direction of the wind the bird encounters slower moving air and thus acquires more airspeed. Then it turns into the wind and, while gliding upward, again acquires some airspeed. Repeating the pattern lets it stay aloft without flapping. Ordinarily the maneuvers are first a left turn, then right, then left, etc. so it covers major distances. Note that it continually turns with the volume where airspeed increases.

Various soaring birds use thermals and slope currents in an understandable way. However, the turkey vulture often does a classic job of soaring low over the forest on a windy day with no flapping, just a great deal of maneuvering. The stability and controllability of the vulture has been studied by Robert Hoey (Hoey, 2002). He has made and adapted models of vultures and other natural soarers (see Figure 16).

For the vulture Hoey finds the spiral stability is satisfactory when the leading edge feathers on the spread tip has a negative 27° angle of attack, with the feathers behind it being less negative. The lift distribution of the whole wing is not elliptical. Rather it includes a negative lift in the region of the tips. This was characteristic of some of the Horton gliders made during World War II, gliders which did not need fins and rudders and which rolled without adverse yaw. The effective soaring of the vulture over the trees needs further research to cope with the continuous flight theme, but the evaluation of their soaring maneuverability at least sheds some light on the challenge.

There are several primary mechanisms involved in dynamic soaring: exploiting vertical gusts, and lateral horizontal gusts. The calculation results are amazing.

Figure 17 is from the recent report by Kiceniuk (2001a) which looks at the consequence of using high G loads for control in turbulent conditions. For example, a person controlling in turbulence to stay as close as possible to 1 G will fare poorly in air that is turbulent but does not have a mean upcurrent. For this small sailplane model of Fig. 13 (70 lbs weight, best L/D = 25 at Induced Drag = 1.4 lbs and Friction Drag = 1.4 lbs) operating in turbulent conditions with 10 ft/sec up for 100 feet, 10 ft/ sec down for 100 feet, then 10 ft/sec up for 100 feet, etc. If instead of operating at G = 1 at 80 ft/sec the plane is controlled to operate at G = 2 for the first 100 ft and G = 0 for the next 100 ft, the average climb turns out to be 5 ft/sec. At 120 mph, the climb is 2.5 ft/sec, but at G = 3 and G = -1 the climb is 9 ft/ sec while for G = 4 and –2 the vehicle climbs at an average of 14 ft/sec. It turns out that this sort of soaring is a strong example of the benefits of accepting (or actually promoting) large G load variation when the gusts have some continuity. The actual rate of climb achieved is a function of the square of the ± vertical velocity of the air. This means that as the pilot goes faster and accepts larger variations of G load, the average climb can exceed the upward velocity of the portion of the rising air.

Another direct part of the dynamic soaring picture is the use of lateral gusts. The pilot should turn away from the gust, not try to minimize its effect by turning into it. Again, the pilot exaggerates rather than minimizes the gust effect, and benefits accordingly. This effect and the previous effect on vertical gusts are pictured and discussed by Kiceniuk (2001a).

Figure 16. Bob Hoey with two RC models of soaring birds.

Dynamic Soaring UAV Calculations

Simplified Square Wave Air Block System Flown at Different Speeds and Gee Loads

Aircraft Specifications:

Best L:D =	25	to one	Drag @ Best L:D speed & 1 gee			
@ Speed of	80	feet/second	Induced Drag =	1.4	lbs	
All Up Weight =	70	lbs	Friction Drag =	1.4	lbs.	
Fastest Aircraft Speed =	240	feet/second	Stall Speed +&- G=	55	55	feet/sec
Speed Increment=	40	feet/second	Neg G drag factor=	1	ratio	

Air Conditions:

Air Block Type 1			Air Block Type 2		
Up Air Motion =	10	feet/second	Upward Air Motion =	-10	feet/second
Hrz, Size =	100	feet	Horizontal Block Size =	100	feet
G Load C1 =	1	x 32 f/s/s	Gee Load Case #1 =	1	x 32 f/s/s

G increment =	1	x 32 f/s/s

Aircraft Energy Gain Expressed as Average Vertical Velocity

(Graph: Vertical Speed, f/s on y-axis from -20.0 to 20.0; Case # 1 to 8 on x-axis)

Forward Speed feet / second
- 80
- 120
- 160
- 200
- 240

Gee Forces in the Different Cases								
Case #	1	2	3	4	5	6	7	8
Gee, Block 1	1.00	2.00	3.00	4.00	5.00	6.00	7.00	8.00
Gee, Block 2	1.00	0.00	-1.00	-2.00	-3.00	-4.00	-5.00	-6.00

Figure 17. Dynamic soaring UAV calculations.

6. Movement On or Under Water

Efficiency for moving on the surface of water has received a great deal of attention, starting with motion on the surface of still water, then expanding to make use of the energy of wind and waves. Rowboats, sailboats, ferries and ocean liners all operate on the surface. In windy conditions they try to go as fast as possible by minimizing their drag and by coping with waves and, in some cases, utilizing the waves to help in their propulsion. Various water vehicles have partially escaped the water's drag by raising the hulls above the water while resting on hydrofoils which are underwater wings. Then there are also underwater craft which do not have to deal with lift but which must cope with their own drag in water. In turbulently-agitated water they can benefit, or be adversely affected, by turbulence and waves. There is a presumption that fish can actually receive thrust from turbulence.

The drag of hulls is a rather complex subject, for it depends on viscous drag (skin friction), inertial drag, and wave drag. The subjects are well reviewed for high speed sailing craft in water by Smith (1963), by Hoerner (1965), and by chapters in Human-Powered Vehicles (edited by Abbott & Wilson, 1995), written by Abbott, Brooks, and Wilson (Human-Powered Watercraft), by Edward S. Van Dusen (Rowing Shells), and by Alec N. Brooks (The 20-Knot Human Powered Watercraft). The wave drag is especially limiting. Smith wrote that for ordinary ships it places a limit to practical "speeds (in knots) of $1.6\sqrt{water\ line\ length\ in\ feet}$," "only 8 knots or 9 mph for a 25 foot displacement hull" (which is about 4 m/sec for a 7.6 m hull). Those of us interested in much higher speeds must use a radically different design concept: very slender hulls that tend to avoid exercising the wave drag (but which may provide inadequate buoyancy); multi slender hulls; and small hydrofoils that remain underwater (and are but slightly influenced by wave drag) while the basic vehicle is supported above the surface by their lift.

Skulls and racing shells deal with these conflicts by suitable compromises. However, the hydrofoils are particularly effective. Three or four small foils can control a sailboat's tipping movement as well as its pitch. As the speed increases the need for foil area decreases. This causes a designer's dilemma. The basic vehicle must be pedaled with pontoon drag at a speed where the pontoons that keep the vehicle afloat at low speed get lifted above the water surface. Then as speed is increased further, the hydrofoils are oversize and cause undesired drag. Some hydrofoils use foils with large dihedral, so the amount of foil area in the water can be adjusted to be appropriate for the speed.

The oars for propulsion are moderately efficient, say 70%, whereas a propeller drive, carefully designed to avoid cavitation, could have an efficiency of around 90%.

Birds floating on water propel themselves by paddling with their legs, a not very efficient travel mechanism. Water strider bugs can walk across quiet water rather efficiently, but not at high speed. They stay above the surface by surface tension.

When a hydrofoil is used under the water to elevate the hull(s) to be completely above the water, the drag of the hydrofoil, and its supporting struts, can be very small

Figure 18. The Flying Fish of Alec Brooks.

even when it is producing enough lift to keep the hull and passenger(s) aloft. A fast, elegant embodiment is shown in Figure 18.

Smith (1963) considered the advantages of using an aerohydrofoil, a sailboat set on small angled fins, that could traverse the water in various directions. Table 1 shows information he derived for his water craft in wind.

In this table $\frac{V_b}{W}$ is the actual speed V_b relative to the wind speed W. When going 90° relative to the wind the vehicle goes 2.37 times the wind speed. His theory agreed roughly with his experiments on a wind driven boat he built for his experiments. Sirocco Land Sailer (2003) reports that a land sailer "can sail at up to five times the speed of the wind. Smaller boats run at two to three times wind speed."

Thus this description by Smith of a water surface vehicle which establishes the mechanism of sailing motions also serves to describe the mechanism for land vehicles, the "land sailers." The difference is that the vehicles on land generally go faster than on water because rolling friction is so much lower than water friction. The ultimate speed in wind comes from ice boats with sails. They can even go faster than land sailers. To summarize, extreme sailed vehicles on water or land can go downwind faster than the wind, crosswind much faster, and go upwind faster than the wind (but not directly into the wind). The mechanism is the same when the wing is transferred to the rotating blades on a properly twisted propeller powered device. Such a propeller device works best heading roughly into or with the wind. With propeller reorientation it can work well in a considerable off-axis direction.

Table 1. Theoretical water craft.

Angle Relative to the Wind	$\dfrac{V_b}{W}$
28°	1.00
60°	2.05
90°	2.37
120°	2.00
150°	1.05

7. Other Techniques With Coupled Devices

So far all the discussion has been about various single vehicles. If we now consider a kite on a string, we realize the kite can be aloft and interact with a stronger wind than is available were it to be affixed onto a surface vehicle. The kite can be used to tow a ground or water vehicle. The kite can also be guided to pull at an angle nearly 80° from the relative wind at the kite altitude, and can provide a good force to the ground vehicle that is pulled within say 80° of the kite line direction. Energy for the ground vehicle comes from the pull component in the direction of motion of the vehicle. Alternatively, the kite can extract energy by a propeller mounted on it. This decreases its performance, but provides an independent source of energy (which presumably is brought to the ground via two insulated wires that comprise the line to the kite).

A further advantage of the kite is that it can be controlled to move fast: left and right, up and down. Its speed becomes much faster than the wind speed, and the power it derives can be far larger than with the motionless kite. To some extent it operates like the whirling blade of a wind turbine, its effectiveness relating to the swept area. Further, when the wind becomes too light to support the stationary kite, the kite moving to and fro can provide line tension and thus remain aloft.

Kites are used rarely for research purposes, but deserve more attention. Giant kites have been experimented with for carrying a significant energy generator aloft. In many locations the wind aloft is 3-4 times stronger than that near the ground, so a wind turbine aloft could in such circumstances obtain 27 to 64 times the energy than it would at the surface. If the generating kite were programmed to dash to and fro, it could extract even more energy. There have been energy projects proposed for stationary kites. The challenges come from operating in air space that must be avoided by aircraft, and dealing with the natural variations of wind.

In summary, there is huge energy aloft, derivable from kite generation, but not yet activated.

Another technology makes use of a pair of kites, or sailplanes, separated by a long line but not connected to the ground. If the two vehicles are operating at different altitudes in air with different speeds and directions, they can extract energy – perhaps enough energy to keep the two vehicles aloft, and possibly enough additional to provide power for immediate or delayed fast propulsion.

8. The Future

The subject here is the movement of devices, occupied or unoccupied, through the air, on land, or on or under water, using only the energy from the sun (and waves) which has arrived recently. Such solar energy manifests itself by radiation, heat, and motion of the air or water. The energy can be stored for subsequent general use as heat or electricity, and for immediate use as the kinetic energy of motion.

The challenge we face was creatively illuminated by Earl Cook (1971). He presents a plot that illustrates humans' increasing use of energy per person. In units of 1000k cal, he plots the number for Primitive Man as 2, Advanced Agricultural Man as 25, Industrial Man as 71, and Technological Man as 220. Direct energy consumption only grows 5-fold from Primitive Man to Technological Man. It is Home and Commerce, Industries and Agriculture, and Transportation that represent the huge increases. We do not know what to expect in 2004, but the numbers will be larger still.

The overall energy concept was well covered by Hermann Scheer (2001), in his classic book "A Solar Manifesto." The message was that *every* country could get by satisfactorily on the energy received within its borders from the sun. (The challenge of fuel energy for aviation was not addressed.) Scheer has been a significant architect of German's rise to prominence in the production and use of wind turbines.

We are finding more ways to come to grips with Hermann's challenge, even economical ways if there were not so much financial support for our oil, gas, and coal industries. Nevertheless, the use of the emerging natural energy technologies is increasing. The inventiveness of people is increasing even more. The change to needing less total energy and getting more of it from solar-related sources seems inevitable, but will take a good bit of modern technological innovation and change.

9. References

Abbott, A.V., Wilson, D.G., 1995 (Editors), Human Powered Vehicles, Human Kinetics. Champaign, IL 61825-5076.

Brooks, A.N. by AC Propulsion, Inc., 2002a. Vehicle-to-Grid Demonstration Project: Grid Regulation Ancillary Service with a Battery Electric Vehicle. Final Report prepared for the California Air Resources Board and the California Environmental Protection Agency, December 10.

Brooks, A.N., 2002b. Perspectives on Fuel Cell and Battery Electric Vehicles. Report presented at the California Air Resources Board ZEV Workshop, December 5.

Carmichael, B.H., 2000a. Personal Aircraft Drag Reduction. Third Edition, Privately Printed, Capistrano Beach, CA.

Carmichael, B.H., 2000b. The Collected Sailplane Articles and Soaring Misadventures of Bruce Carmichael. Privately Printed, Capistrano Beach, CA.

Cook, E., 1971. The Flow of Energy in an Industrial Society. Scientific American. 9, September, pp. 136.

Hoerner, S.F., 1966. Fluid Dynamic Drag. Hoerner Fluid Dynamics, Brick Town, NJ. pp. 11-26 through 11-32.

Hoey, R., 2002. Design and Build a RC Bird Model. Model Airplane News, June.

Hoffman, C., 2003. Aero Boy. Popular Science, 262, No. 6, June, pp. 52-59.

International Center for Technology Assessment, 2003. The Real Price of Gas. http://www.icta.org/projects/trans/rlprexsm.htm.

Kiceniuk, T., 2001a. Dynamic Soaring and Sailplane Energetics. Technical Soaring, 25, No. 4, pp. 221-227.

Kiceniuk, T., 2001b. Calculations on Soaring Sink. Technical Soaring, 25, No. 4, pp. 228-230.

Klemperer, W.B., 1943-1945. Theory of Soaring Flight. Soaring, (March-April 1943, May-June 1943, July-August 1943, Nov-Dec 1943, March-April 1944, May-June 1944, March-April 1945, May-June 1945, Nov-Dec 1945).

NASA X-PRESS, 2002. Special Missions to the Stratosphere. Special Helios Prototype Edition, 44, Issue 2, May 8.

Naton, P., 2002. Endless Lift (videotape). Available from http://www.radiocarbonart.com.

O'Mahony, C., 2000. Pursuing Soaring Visions. Soaring, 62, No. 2, February, pp. 12-16.

Rayleigh, L., 1883. The Soaring of Birds. Nature, XXVII, pp. 534-535.

Sachs, G., Mayhofer, M., 2001. Shear Wind Strength Required for Dynamic Soaring at Ridges. Technical Soaring, 25, No. 4, pp. 209-215.

Scheer, H., 2001. A Solar Manifesto. Second Edition, James & James Science Publishers Ltd., printed in the UK by The Cromwell Press.

Sirocco Land Sailer, 2003. Introduction to Land Sailing. http://www.landsail.org/introduction_to_land_sailing.htm.

Smith, B., 1963. The 40-Knot Sailboat. Gosset & Dunlap Inc., NY, pp. 98-105.

Chapter 10

Modeling Solar Hydrogen Fuel Cell Systems[1]

by

Peter H. Aurora
Engineering Reliability and Product Development
RWE Schott Solar Inc., 4 Suburban Park Drive
Billerica, Massachusetts 01821 USA

John J. Duffy*
Energy Engineering Program
University of Massachusetts/Lowell,
One University Avenue, Lowell, Massachusetts 01854 USA

Abstract

In this study a *solar hydrogen fuel cell system* (SH$_2$FCS) was modeled to help understand the behavior of, and to aid in the optimal design of, such systems at different load and ambient conditions. The main components of this system are photovoltaic modules, solid polymer electrolyte (SPE) water electrolyzers, proton exchange membrane (PEM) fuel cells, and hydrogen and oxygen storages devices. All the models are based on physical and electrochemical principles, as well as empirical relationships. The thermal model of SPE electrolyzer, PEM fuel cell, and tank storage are developed using the lumped parameter approach. A high-pressure electrolyzer is used instead of a compressor to reduce the parasitic loads. A unique approach of placing the electrolyzer in the hydrogen storage tank to prevent leaks from the electrolyzer was modeled (patent applied for). The power management of the system is achieved using an on-off controller; and the control of the temperature for the electrochemical devices is performed using a proportional integral controller strategy. Other actual solar hydrogen systems (i.e., Schatz Solar Hydrogen Project at Humboldt University and the PHOEBUS plant at FZ-Julich in Germany) were simulated, and results were compared to help verify the model. Some laboratory test data was also used to formu-

[1] *A portion of this chapter (resident in Aurora's thesis) resides with the University Microfilms (umi.com) U Mass/Lowell, under a non-exclusive copyright held by the author. Reprinted with permission.*
* *Corresponding Author Email: John_Duffy@uml.edu*

late and verify the electrolyzer model. A system was sized and simulated with actual weather and load data monitored at a remote medical clinic in the mountains of Peru. Good temperature control and energy management performance resulted. Water recirculated through the electrolyzer was found important to keep the electrolyzer cool enough. The energy needed to compress hydrogen using the electrolyzer as a compressor is less than the mechanical energy required for a compressor. The simulated overall energy efficiency of the regenerative fuel cell system (electrolyzer, gas storage, and fuel cell) is relatively low at about 42%. This model will aid in the optimization of the sizing/design of the SH_2FCS to improve the energy efficiency and reduce costs.

1. Introduction

Hydrogen is often referred to as the energy carrier of the future because it can be used to store intermittent renewable energy sources such as solar and wind energy (Ulleberg, 2002). Stand-alone solar energy systems can be integrated with other systems such as hydrogen generators, storage and fuel cells. Recent developments in proton exchange membrane (PEM) fuel cells and electrolyzers are beginning to make them a promising alternative to batteries for storage of energy from solar electric power systems (Cisar et al., 1999). Several solar hydrogen projects were developed over the past decade around the world like WE-NET in Japan, PHOEBUS and HYSOLAR in Germany, and Humboldt University in USA (Schatz Energy Research Center). U Mass Lowell and ElectroChem Inc. have combined efforts to develop an improved energy production/storage system based upon the concept of a regenerative fuel cell and high-pressure electrolyzers (Shapiro, 2002 and Das, 2002). Shapiro, Duffy, and ElectroChem (2002) designed an electrolyzer fuel cell system, which includes a high-pressure PEM electrolyzer, high-pressure hydrogen and oxygen storage tanks, and a PEM fuel cell. They found that the additional expenditure of energy of a high-pressure electrolyzer was 1.5 % to achieve gas pressures of 200 psi from atmospheric pressure.

2. Hydrogen Energy Technology

Hydrogen is one of the most promising alternative fuels for the future because it has the capability of storing energy of high quality and can be produced and "recycled" from a number of abundant renewable sources, including water. Hydrogen has therefore been visualized to become the cornerstone of future energy systems based on solar energy and other renewable energy sources. It can be produced chemically from hydrocarbons (e.g., renewable fuels like methane, ethanol, or methanol). If it is produced from natural energy resources, the cycle is an environmentally friendly process. Because solar energy for all practical purposes can be regarded as an infinite source of energy, its use to produce hydrogen is one of the best options for a sustainable future (Ulleberg, 1998).

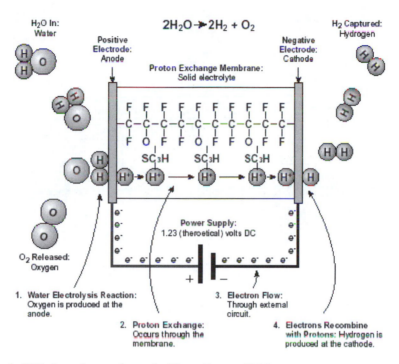

Figure 1. SPE electrolyzer schematic (Home Power, 2001).

2.1 Hydrogen Production

There are basically three pathways—and their combinations— for producing hydrogen with solar energy: electrochemical, thermochemical, and photochemical.

2.1.1 Electrochemical: Water Electrolysis

An attractive and easy way to produce hydrogen is via *water electrolysis*, simply because of the abundance of water on earth. The basic chemical reaction to break the water down to hydrogen and oxygen is given by:

$$H_2O + electric\ energy \rightarrow H_2 + \frac{1}{2}O_2 \qquad \text{(Eq. 1)}$$

For this reaction to occur, an amount of energy must be added (endothermic reaction). An electrolyzer is the device used for water electrolysis, and because it works at relatively low temperature (~80°C) solid polymer electrolyzers (SPE, or proton exchange membrane, PEM, electrolyzer) is widely used. A schematic of an SPE electrolyzer is shown in Figure 1.

2.1.2 Solar thermochemical production of hydrogen

Thermochemical production of hydrogen is based on the use of concentrated solar radiation as the energy source of high-temperature process heat, which drives

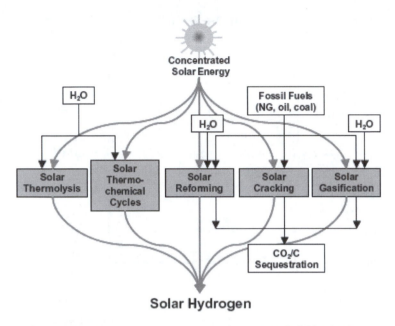

Figure 2. Five thermochemical routes to produce hydrogen (Steinfeld, 2004).

the endothermic chemical transformation. Five thermochemical routes for solar hydrogen production are depicted in Figure 2. Indicated in the figure are the chemical sources of H2: water for the solar thermolysis and the solar thermochemical cycles, fossil fuels for the solar cracking, and a combination of fossil fuels and H2O for the solar reforming and solar gasification. All of these routes involve endothermic reactions that make use of concentrated solar radiation as the energy source of high-temperature process heat (Steinfeld, 2004).

Unlike electrolysis, thermochemical cycles for splitting water can convert low-level thermal energy directly into chemical energy by forming hydrogen and oxygen, and with overall first-law efficiencies exceeding 50% (Huang and T-Raissi, 2004).

a) Solar thermolysis

The single-step thermal dissociation of water is known as water thermolysis,

$$H_2O + thermal\ energy \rightarrow H_2 + \frac{1}{2}O_2 \qquad \text{(Eq. 2)}$$

The water splitting reaction is simple, but it has been impeded by the need of a high-temperature heat source at above 2,500 K for achieving a reasonable degree of dissociation, and by the need of an effective technique for separating H2 and O2 to avoid ending up with an explosive mixture. Effusion separation and electrolytic separation are the methods proposed to separate H2 from the products. The very high temperatures needed by the process (e.g. 3,000 K for 64% dissociation at 1 bar) give severe material problems and can lead to significant re-radiation from the reactor,

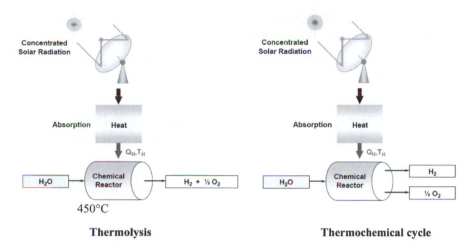

Figure 3. Thermolysis and thermochemical cycles (Steinfeld, 2004).

thereby lowering the absorption efficiency of the solar reactor (Steinfeld, 2004). Figure 3 shows the simplified diagram of a solar thermolysis process.

b) Solar thermochemical cycles

A large number of water splitting cycles have been conceptualized since 1960. Among almost 2,000-3,000 possible thermochemical cycles Mark 1, Mark 15, ZnO/Zn, Fe3O4/FeO and sulfur-iodine have been studied extensively (Advances in Solar Energy, 2003). The well known sulfur-iodine (S-I) cycle proposed by General Atomics Company is presented in this work, the characteristics of this cycle are: all fluids are in continuous process, all the chemicals are recycled and no effluent, the hydrogen is produced at high pressure (22-84 atm), and high cited projected efficiency about 50%. One of the challenges is that it requires temperatures higher than 800°C; this is because higher temperatures give higher efficiencies (the process follows the rules of chemistry and thermodynamics, Carnot cycle).The S-I cycle consists of three steps (Huang, 2004):

$$H_2SO_4(g) \rightarrow SO_2(g) + H_2O(g) + \tfrac{1}{2}O_2(g) \qquad\qquad 850°C \qquad \text{(Eq. 3)}$$

$$I_2(l) + SO_2(aq) + 2H_2O(l) \rightarrow 2HI(l) + H2SO_4(aq) \qquad 120°C \qquad \text{(Eq. 4)}$$

$$2HI \rightarrow I_2(l) + H_2(g) \qquad\qquad\qquad\qquad 450°C \qquad \text{(Eq. 5)}$$

The schematic diagram of the solar thermochemical cycle is shown in Figure 3.

c) Decarbonization of fossil fuels

Three solar thermochemical processes for H_2 production using fossil fuels as the chemical source are considered: cracking, reforming, and gasification. These routes are shown schematically in Figure 2.

The solar cracking route refers to the thermal decomposition of natural gas (NG), oil, and other hydrocarbons, and can be represented by the simplified reaction:

$$C_xH_y \rightarrow xC(gr) + \frac{y}{2}H_2(g)$$
(Eq. 6)

The steam-reforming of NG, oil, and other hydrocarbons, and the steam-gasification of coal and other solid carbonaceous materials can be represented by the simplified net reaction:

$$C_xH_y + H_2O \rightarrow \left(x + \frac{y}{2}\right)H_2(g) + xCO$$
(Eq. 7)

Using solar energy for process heat offers a threefold advantage: (1) the discharge of pollutants is avoided; (2) the gaseous products are not contaminated; and (3) the calorific value of the fuel is upgraded by adding solar energy in an amount equal to the ΔH of the reaction (Steinfeld, 2004).

2.1.3 Photo-electrochemical Hydrogen Generation

One promising method for producing hydrogen using a renewable source is that based on photo-electrochemical water decomposition using solar energy. These electrochemical cells are called photo-electrochemical cells (PEC). The energy conversion efficiency of water photo-electrolysis is determined principally by the properties of the materials used for photo-electrodes.

The materials for the photo-electrodes of PECs should perform two fundamental functions:

- Optical function required to obtain maximal absorption of solar energy;

- Catalytic function required for water decomposition.

The photo-electrochemical water decomposition is based on the conversion of light energy into electricity within a cell involving two electrodes, immersed in an aqueous electrolyte, of which at least one is made of a semiconductor exposed to light and able to absorb light. This electricity is then used for water electrolysis (Bak et al., 2002). There are three options for the arrangement of photo-electrodes in the assembly of PECs:

- Photo-anode made of n-type semiconductor and cathode made of metal

- Photo-anode made of n-type semiconductor and photo-cathode made of p-type semiconductor

- Photo-cathode made of p-type semiconductor and anode made of metal.

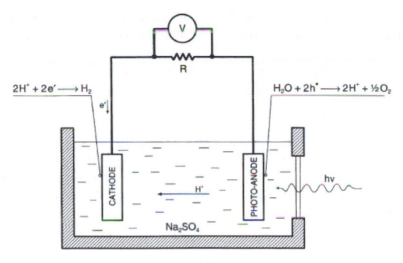

Figure 4. Schematic showing the structure of a PEC cell (Urade, 2004).

A photo-electrochemical cell for the photo-electrolysis is shown in Figure 4. The energy efficiency of the PECs reported varies depending on the cell structure, the experimental conditions, and light energy, light spectrum, photo-anode material, photo-anode processing conditions, cathode, and electrolyte. Fujishima and Honda reported 10% energy efficiency using TiO_2 as anode, Pt-black as cathode and xenon lump as a light source (Balk et al., 2002). Licht et al. (2001) reported 18.3% efficiency using $AlGaAs/SiRuO_2$ as anode, Pt as cathode and, a tungsten-halogen lamp.

2.2 Hydrogen Storage

In ambient conditions hydrogen only exists as a gas. Hydrogen has the highest energy density per unit weight as compared to any other common fuels but it has low density per unit volume. Due to this issue, hydrogen has to be stored at high pressure in order to have small tank storage volumes. Hydrogen can be stored mechanically and/or chemically; some of the hydrogen storage options are shown below.

2.2.1 Pressurized Hydrogen

The most traditional way to store hydrogen in gaseous form is in pressure vessels. Gaseous hydrogen can be stored either above ground in portable or stationary containers or in earth caves. Above ground hydrogen storage vessels vary in size but are typically at a standard pressure of 200 bars (even though in some countries safety regulations limit the maximum pressure to 165 bars). There are also, for example, stationary high-pressure containers (>200 bar) and large low-pressure spherical containers (>15000 m^3 and 12-16 bar). The underground caves are an easy and relatively cheap method for large-scale storage of hydrogen. There are several different kinds of caves that can be used, such as salt caverns, mined caverns, natural caves, and aquifer structures. The pressure in earth caves varies between 80–160 bar and thus the volumetric energy density is about 250–465 kWh/ m^3 (Hottinen et al., 2004). For vehicle applications hydrogen can be stored in fiberglass-wrapped aluminum cylinders up to 3,500 psi (238 atm). Hydrogen can also be stored in a high-pressure tank of 3,000 to 10,000 psi (205-680 atm) pressure. High-pressure hydrogen storage tanks can be composite-wound reinforced cylinders with metal liner (Ananthachar, 2002).

2.2.2 Liquid Hydrogen

Liquid hydrogen has been used as a fuel in space technology for several years. The liquefaction of hydrogen can only be achieved cryogenically by mechanical compression and cooling (-253°C) (Ananthachar, 2002). The total energy required to produce LH$_2$ from gas H$_2$ at standard state is about16,000kJ/kg, where about 25% is cooling energy and 75% is compression work (Ulleberg, 1998).

2.2.3 Metal Hydride

Certain materials absorb hydrogen under moderate pressure (less than 68 atm) at low temperatures, forming unstable hydrogen compounds called hydrides. The storage tank contains powered metals that absorb hydrogen and release heat when the hydrogen is forced into the tank under pressure (Ananthachar, 2002).

Metal hydrides (MH) systems store hydrogen in the inter-atom spaces of a granular metal. Various metals can be used. The hydrogen is released by heating. MH storage devices are reliable and compact, but can be heavy and expensive. Varieties now under development can store about 7% hydrogen by weight. The disadvantage of using MH storage is that it needs energy to release the fuel (Dincer, 2002).

2.2.4 Absorber Storage

A carbon adsorption technique stores hydrogen under pressure on the surface of highly porous superactivated graphite. Some varieties are cooled, while others operate at room temperature.

Graphitic nanofibers (GNFs). Grown by the decomposition of hydrocarbons or carbon monoxide over metal catalysis, the fiber consists of graphene sheets aligned in a set direction. Three distinct structures may be produced: platelet, ribbon and herringbone. The capacity of these GNFs has been found to be 4-7 wt%, but these values have to be verified to provide a reliable benchmark for nanofibers (Atkinson et al., 2001).

Carbon nanotubes (CNTs). There are a number of options for hydrogen storage in carbon nanotubes: the nanotubes can exist as single-walled nanotubes (SWNT diameter 1-2nm) or as multi-walled nanotubes (MWNT, diameter 5-50 nm), and can be utilized in their pristine state or in a doped state. None of the experimental results have been confirmed by independent research groups. Theoretical calculations predict that a range of 4-14 wt% hydrogen adsorption in carbon-based materials is possible (Atkinson et al., 2001).

2.3 Hydrogen Utilization (Fuel Cells)

A fuel cell is an electrochemical device that generates electricity by consuming chemical fuel in a non-combustion process. In the case of hydrogen fuel cell the only product is pure water. Such devices differ from conventional batteries in that their ability to supply electricity is not limited (electricity will be generated as long as fuel and oxidant can be supplied).

$$H_2(g) + \frac{1}{2}O_2(g) \quad \rightarrow \quad H_2O(l) + electric\ energy \qquad \text{(Eq. 8)}$$

Hydrogen fuel is fed into the "anode" of the fuel cell. Oxygen (or air) enters the fuel cell through the cathode. Encouraged by a catalyst, the hydrogen atom splits into a proton and an electron, which take different paths to the cathode. The proton passes through the electrolyte. The electrons create a separate current that can be utilized before they return to the cathode, to be reunited with the hydrogen and oxygen in a molecule of water. The ideal standard potential of an H_2/O_2 fuel cell is 1.229 V at STP with liquid water product and 1.18 V water with gaseous product (DOE, 2000). Based on the DOE classification, the types of fuel cells are:

Polymer Electrolyte Fuel Cells: The electrolyte is an ion-exchange membrane (fluorinated sulphonic acid polymer or other similar polymer). This type of fuel cell works at a relatively low temperature (60-90°C). This type of fuel cell is also called a proton exchange membrane (PEM) fuel cell. A schematic PEM fuel cell is shown in Figure 5 and an actual picture of a fuel cell stack in Figure 6.

Alkaline Fuel Cells: The electrolyte is concentrated (85 wt%) KOH, with the fuel cell operated at relatively high temperatures of about 250°C. The electrolyte can be less concentrated for lower temperature operation, around 120°C.

Phosphoric Acid Fuel Cells: Phosphoric acid concentrated to 100% is used as electrolyte, in fuel cells operated at 150 to 220°C.

Molten Carbonate Fuel Cells: The electrolyte is a combination of alkali carbonates, retained in a ceramic matrix of $LiAlO_2$. The fuel cell operates at a temperature of 600 to 700°C, at which the alkali carbonates form a highly conductive molten salt.

Tubular Solid Oxide Fuel Cells: The electrolyte in this fuel cell is a solid, nonporous metal oxide, usually Y_2O_3-stabilized ZrO_2. The cell operates at 1,000°C where ionic conduction by oxygen ions takes place.

Intermediate Temperature Solid Oxide Fuel Cells: The electrodes and electrolyte used are the same as in Tubular Solid Oxide Fuel Cells, except that the temperature of operation is lowered to 600 to 800°C using thin film technology to promote ionic conduction.

Direct Methanol Fuel Cell: The electrolyte of this fuel cell is also a proton exchange membrane, but in this case methanol is directly supplied to the anode. This kind of fuel cell typically operates at a temperature between 50 to100°C.

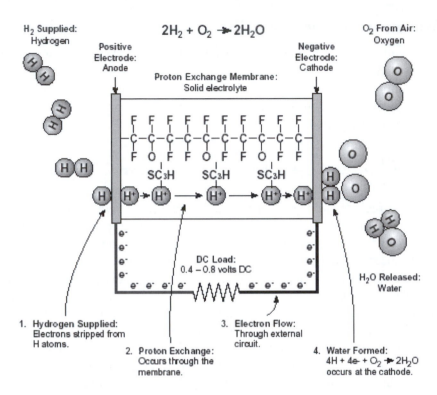

Figure 5. Schematic of a PEM fuel cell (Home Power, 2001).

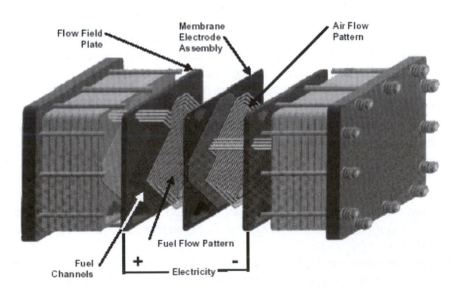

Figure 6. Schematic of a Ballard Mark V PEM fuel cell stack (Ballard Power Systems, Inc., 2003).

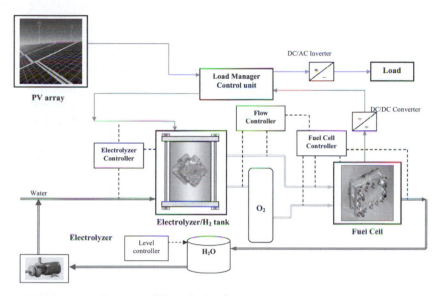

Figure 7. Schematic diagram of the solar hydrogen fuel cell system.

3. System Description

In a *solar hydrogen fuel cell system* water is separated into hydrogen and oxygen and the gases are stored in pressurized tanks. And when there is no solar irradiation and electricity is needed, the hydrogen and oxygen are fed into the fuel cell, which generates electricity, heat and water. The water is then recirculated back to the hydrogen production system. For this work a high-pressure PEM electrolyzer was chosen among all the other hydrogen generation options, because nowadays electrolyzers are well developed and are relatively easily available in the market. The use of a high-pressure electrolyzer will decrease the auxiliary energy consumption and volume storage in stand-alone energy systems. Finally, the main components of this kind of solar hydrogen system include a PV-array, a high-pressure solid polymer electrolyte electrolyzer, hydrogen storage (compressed gas), and a proton exchange membrane fuel cell. The simplified diagram of system under study is shown in Figure 7.

3.1 PEM or SPE Devices: Electrolyzer and Fuel Cell

A PEM or SPE fuel cell (electrolyzer) consists of an anode, a cathode and a hydrated polymer membrane as electrolyte (Figure 1 and Figure 5).The use of a solid membrane reduces the corrosion and electrolyte management problems (You and Liu, 2002). Thus, the PEM device may operate at low temperatures. They also have high power density, short response time and quick startup capability.

An ordinary electrolyte is a substance that dissociates into positively and negatively charged ions in the presence of water, thereby making the water solution electrically conducting. The electrolyte in a polymer electrolyte membrane is a type of

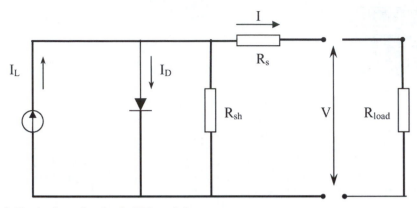

Figure 8. Equivalent circuit of a PV module.

plastic, a polymer, and is usually referred to as a membrane. The appearance of the membrane varies depending on the manufacturer, and the most prevalent membrane is Nafion™ produced by DuPont (Los Alamos National Laboratory, 1999).

4. System Modulation

4.1 Mathematical Modeling: PV Module

The photovoltaic device is modeled based on the voltage-current relationship of the cells under different levels of radiation and at various cell temperatures. The relatively simple idealized one diode model can be used for system design purposes; Figure 8 shows the equivalent circuit of a photovoltaic cell. It can be used either for an individual cell or for a module consisting of several cells or for an array consisting of several modules (Duffie and Beckman, 1991).

At fixed temperature and solar radiation, the current-voltage characteristics of a PV module can be modeled using,

$$I_{PV} = I_L - I_D - I_{sh} = I_L - I_o \{e^{\frac{V+IR_s}{a}} - 1\} - \frac{V_{PV} + I_{PV} R_s}{R_{sh}} \qquad (Eq.\ 9)$$

The power is given by:

$$P = IV \qquad (Eq.10)$$

In most cases the shunt current can be ignored because the shunt resistance is so high that the term goes to zero, particularly for monocrystalline solar cells, so Equation 9 becomes:

$$I = I_L - I_D = I_L - I_o \{e^{\frac{V+IR_s}{a}} - 1\} \qquad \text{(Eq.11)}$$

With the shunt resistance taken as infinity, the model remains with three measured points on the I-V curve to determine the four parameters. These three conditions can be found from manufacturer's data, which usually provide values for V and I at short circuit, open circuit and the maximum power for a given set of reference conditions. At short circuit conditions, the diode current is very small and the light current is equal to the short circuit current.

$$I_L = I_{sc} \qquad \text{(Eq.12)}$$

At open circuit conditions the current is zero and the exponential term in Equation 11 is much larger than 1. Because of this the diode current is given:

$$I_{o,ref} = I_{L,ref} e^{\frac{-V_{oc,ref}}{a_{ref}}} \qquad \text{(Eq.13)}$$

The measured values for Imp and Vmp at the maximum power point, given by the manufacturer, can be substituted into, along with the diode and light current to find the series resistance Rs.

$$R_s = \frac{a \ln\left(1 - \frac{I_{mp}}{I_L}\right) - V_{mp} + V_{oc}}{I_{mp}} \qquad \text{(Eq.14)}$$

The following equations are good approximations for the temperature effects on many PV modules (Duffie and Beckman, 1991):

$$I_L = \frac{G_T}{G_{T,ref}} \left[I_{L,ref} + \mu_{I_{sc}} (T_c - T_{c,ref}) \right] \qquad \text{(Eq.15)}$$

$$a = a_{ref} \frac{T_{c,ref}}{T_c} \qquad \text{(Eq.16)}$$

$$I_o = I_{o,ref} \left(\frac{T_c}{T_{c,ref}} \right)^2 e^{\left[\frac{\varepsilon N_s}{a_{ref}} \left(1 - \frac{T_{c,ref}}{T_c} \right) \right]} \qquad \text{(Eq.17)}$$

$$V_{oc} = V_{oc,ref} + \mu_{V_{oc}} \left(T_c - T_{c,ref} \right)$$ (Eq.18)

$$a_{o,ref} = \frac{\mu_{V_{oc}} T_{c,ref} - V_{oc,ref} + \varepsilon N_s}{\dfrac{\mu_{I_{sc}} T_{c,ref}}{I_{L,ref}} - 3}$$ (Eq.19)

The following two equations are approximations made by Arkin (2001):

$$I_{mp} = \frac{G_T}{G_{T,ref}} \left[I_{mp,ref} + \mu_{I_{sc}} (T_c - T_{c,ref}) \right]$$ (Eq.20)

$$V_{mp} = V_{mp,ref} + \mu_{V_{oc}} \left(T_c - T_{c,ref} \right)$$ (Eq.21)

4.2 Mathematical Modeling: Electrolyzer

The decomposition of the water into hydrogen and oxygen can be achieved by passing an electric current (DC) between two electrodes separated by an electrolyzer. For the reaction to occur a minimum electric voltage must be applied to the two electrodes. To induce current to flow through an electrolytic cell and bring about a non-spontaneous cell reaction, the applied voltage difference must exceed the zero-current voltage (open circuit voltage) by at least the cell overvoltage (or overpotential). The cell overvoltage is the sum of the overvoltage at the two electrodes and the ohmic drop due to the current through the electrolyte. The additional potential needed to achieve a detectable rate of reaction may need to be large when the exchange current density at the electrodes is small (Atkins, 1998). The cell potential for water electrolysis is a sum of four terms:

$$E_{EL} = E_r + \eta_{act} + \eta_{conc} + \eta_{ohm}$$ (Eq.22)

The electrode kinetics of an electrolyzer can be modeled using semi-empirical current-voltage relationships (Ulleberg, 2002). The basic form of the *I-V* curve is given by:

$$V_{EL} = E_r + \frac{r}{A} I_{EL} + s \log \left(\frac{t}{A} I_{EL} + 1 \right)$$ (Eq.23)

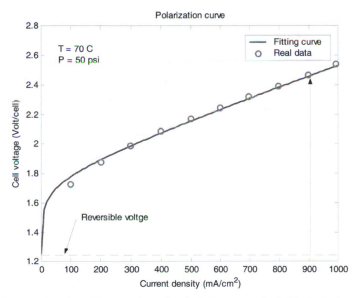

Figure 9. Measured and predicted values for the I-V curve of a SPE electrolyzer.

The *emf* for reversible electrochemical process, called the reversible potential (E_r), is expressed using the Nernst Equation for the split water process (Atkins, 1998):

$$E_r = E^0 - \frac{RT}{nF} \ln[\frac{P_{H_2} P^{1/2}_{O_2}}{P_{H_2O}}]$$
(Eq.24)

In equation (14) P_{H_2}, P_{O_2} are the partial pressures of the gases in atmospheres. The values of r, s and t in (13) are found using experimental data, and they vary for different electrolyzers. Placing (14) into (13) results in the following expression for the electrolyzer voltage:

$$V_{EL} = E^o - \frac{RT}{nF} \ln\left[\frac{P_{H_2} P^{1/2}_{O_2}}{P_{H_2O}}\right] + \frac{r}{A} I + s \log\left(\frac{t}{A} I + 1\right)$$
(Eq.25)

Experimental data gathered by Shapiro (2001) at Electrochem, Inc. (8-cell electrolyzer with 50-cm² active area) is used in this work to find the values of the s, r and t parameters. The parameters in equation (19) were adjusted until the resulting curve fit the actual data reasonably well, as illustrated in Figure 9.

4.3 Mathematical Modeling: Fuel Cell

The maximum possible theoretical voltage (emf) that can be produced across the electrodes of a fuel cell is called the reversible voltage, E_r. The actual cell voltage is

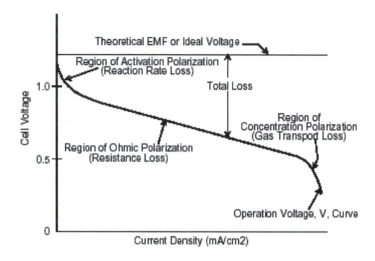

Figure 10. Ideal and actual fuel cell I-V characteristic (From Fuel Cell Handbook DOE, 2000).

lower than its equilibrium voltage because of irreversible losses as shown in Figure 10 (Fuel Cell Handbook DOE, 2000). The actual output voltage of a simple cell can be defined with the following expressions:

$$V_{FC} = E_r + \eta_{act,a} + \eta_{act,c} + \eta_{conc} + \eta_{ohm} \tag{Eq.26}$$

The reversible voltage, E_r, is obtained in open circuit thermodynamic balance. This is calculated using the modified version of Nernst's equation, with a term that takes into account changes in the temperature in relation to the standard reference temperature, and partial pressure of the gases at 25°C (Amphlett et al., 1995).

$$E_r = \frac{\Delta G}{nF} + \frac{\Delta S}{nF}(T - T_{ref}) + \frac{RT}{nF}\ln[\frac{P_{H_2} P^{1/2}_{O_2}}{P_{H_2O}}] \tag{Eq.27}$$

The other three overvoltage terms in equation (20) represent reductions from the reversible voltage to give the useful cell voltage.

The activation overvoltage is directly related to the rates of electrochemical reactions. There is a close similarity between electrochemical and chemical reactions in that both involve an activation barrier that must be overcome by the reacting species. In the case of an electrochemical reaction with $\eta_{act} > 50\text{-}100$ mV the total activation overvoltage (cathode and anode) can be represented by the following expression (Mann et al., 2000).

$$\eta_{act} = \xi_1 + \xi_2 T + \xi_3 T[\ln(c^*_{O_2})] + \xi_4 T[\ln i] \tag{Eq.28}$$

For *Ballard Mark V* PEM fuel cell, the following values are proposed (Fowler *et al.*, 2002):

$$\xi_1 = -0.948(\pm 0.004) \tag{Eq.29}$$

$$\xi_2 = k_{cell} + 0.000197 \ln A + 4.3x10^{-5} \ln c_{H_2}^* \tag{Eq.30}$$

$$\xi_3 = 6.8 \pm 0.2x10^{-5} \tag{Eq.31}$$

$$\xi_4 = -1.97 \pm 0.05x10^{-4} \tag{Eq.32}$$

Where ξ_1 is in volts and ξ_2, ξ_3 and ξ_4 are in volts/K. The parameter k_{cell} includes rate constants for the anode and cathode reactions, as well as some properties specific to the cell design such as effective catalyst surface area, and the concentration of protons and water at the interface.

The ohmic polarization results from the resistance to electron transfer in the graphite collector plates and graphite electrodes plus resistance to proton transfer in the solid polymer electrolyte membrane. According to Ohm's law the overvoltage is given by (Fowler et al., 2002):

$$\eta_{ohmic} = \eta_{ohmic}^{electronic} + \eta_{ohmic}^{proton} = -i(R^{electronic} + R^{proton}) = -iR^{internal} \tag{Eq.33}$$

If the purity of the machined graphite plates is high, the electronic resistance should be insignificant in comparison to the proton resistance; then the ohmic overvoltage is given by iR^{proton}. The general expression for resistance is given by:

$$R^{proton} = \frac{r_M l}{A} \tag{Eq.34}$$

Since Nafion is considered for this work, the following empirical expression is taken from Mann et al. (2000):

$$r_M = \frac{181.6\left[1 + 0.03i + 0.062\left(\dfrac{T}{303}\right)^2 i^{2.5}\right]}{\left[\lambda - 0.634 - 3i \exp\left[3.25\left(\dfrac{T-303}{T}\right)\right]\right]} \tag{Eq.35}$$

As a reactant is consumed at the electrode by electrochemical reaction, there is a loss of voltage due to the inability of the surrounding material to maintain the initial concentration of the bulk fluid. That is, a concentration gradient is formed. At practical current densities, slow transport of reactants/products to/from the electrochemical reaction site is a major contributor to concentration overvoltage. The concentration losses can be expressed by:

$$\eta_{con} = \frac{RT}{nF} \ln\left(1 - \frac{i}{i_L}\right)$$

(Eq. 36)

4.4 Flow Rates and Efficiencies

4.4.1 Flow Rates

According to Faraday's law, the flow rate of hydrogen produced, in the electrolyzer, or consumed, in the fuel cell, is proportional to the current given or produced respectively. Then the total hydrogen flow rate in the stack, which consists of several cells connected in series, can be represented by:

$$\dot{N}_{H_2} = N_c \frac{I}{nF}$$

(Eq. 37)

From the stoichiometry it is found that the flow rate of oxygen is half the flow rate of the hydrogen and the water consumed is the same as the hydrogen flow rate in moles.

$$\dot{N}_{H_2} = \dot{N}_{H_2O} = 2\dot{N}_{O_2}$$

(Eq. 38)

4.4.2 Faraday Efficiency

The Faraday efficiency is the ratio between the actual and theoretical maximum amount of hydrogen produced or consumed. This is also called current efficiency and is less than 100% due to: multiple reactions, hydrogen permeation through the membrane, and shunt currents. The permeation of hydrogen affects more the current efficiency of a PEM device in general. An empirical equation to calculate the Faraday efficiency is given by (Ulleberg, 2002):

$$\eta_F = \frac{\dot{N}_{H_2}(actual)}{\dot{N}_{H_2}(ideal)}$$

(Eq. 39)

4.4.3 Cell Efficiency

The thermal efficiency of an energy conversion device is defined as the amount of useful energy produced relative to the change in stored thermal energy (chemical energy) that is released when a fuel is reacted with an oxidant (Fuel Cell Handbook D.O.E. , 2000).

$$\eta = \frac{UsefulEnergy}{\Delta H} \tag{Eq. 40}$$

For the ideal case of an electrochemical conversion, such as a fuel cell, the change in Gibbs free energy, ΔG, of the reaction is available as useful electric energy at the temperature of the conversion. The ideal efficiency (thermodynamic efficiency) of a fuel cell is then given by:

$$\eta = \frac{\Delta G}{\Delta H} \tag{Eq. 41}$$

At standard conditions of 25°C and 1 atmosphere, the thermal efficiency of an ideal fuel cell operating reversibly on pure hydrogen is $h=83\%$. The actual cell efficiency in a fuel cell can be represented by:

$$\eta_{cell} = \frac{E_{FC}}{U_{tn}} \tag{Eq. 42}$$

Where U_{tn} is the thermoneutral voltage which is a function of the change in enthalpy, and it is given by:

$$U_{tn} = \frac{\Delta H}{nF} \tag{Eq. 43}$$

The cell efficiency of an electrolyzer is given by:

$$\eta_{cell} = \frac{U_{tn}}{V_{EL}} \tag{Eq. 44}$$

In theory, cell efficiency greater than 100%, in the case of an electrolyzer, can be achieved by keeping the cell voltage V in the region between the reversible E_r and the thermoneutral voltage U_{tn}. In practice, this could be done by adding thermal energy to the system (endothermic process).

4.4.4 Energy Efficiency

In the case of the electrolyzer the electric energy supplied is not totally used to split the water into hydrogen and oxygen; a part of this energy is converted into heat. The generation of heat in an electrolyzer is mainly due to electrical inefficiencies. For the same reason the output electric energy produced in the fuel cell is lower than it should be. In order to find the energy efficiency of an electrolyzer, the heating value of the hydrogen must be used.

The heating value or calorific value of a fuel is the magnitude of the heat of reaction at constant pressure or at constant volume at a standard temperature for the complete combustion of a unit mass of fuel. For fuel containing hydrogen, the water in the products affects the value of heat of reaction whether it is liquid or vapor. The term higher heating value, *HHV*, (or gross heating value) is used when the water formed is all condensed to liquid phase; the term lower heating value, *LHV*, (or net heating value) is used when the water formed is all in the vapor phase (Heywood, 1988).

The energy efficiency of a fuel cell is the relation between the energy given by the flow rate of hydrogen and oxygen into the fuel cell and the actual electric energy produced by the fuel cell stack. And because the hydrogen is used in the fuel cell and the product of "combustion" is water, almost all of which is in the liquid phase, the HHV is used in the expression for cell efficiency:

$$\eta_{E_{FC}} = \frac{N_c I_{FC} V_{FC}}{N_{H_2} HHV_{H_2}}$$

(Eq. 45)

And the energy efficiency of an electrolyzer is found dividing the energy produced (hydrogen energy) by the total electric energy consumed.

$$\eta_{E_{EL}} = \frac{N_{H_2} HHV}{N_c I_{EL} V_{EL}}$$

(Eq. 46)

4.5 Water Transport and Management

With PEM electrochemical cells (electrolyzer or fuel cell) high power densities can be obtained at relatively low temperatures. One of the factors that is crucial for high power density is the water transport in the membrane electrode assembly (MEA). The power density can be adversely affected by too little water (drying out), as well as by too much (flooding) (Janssen et al., 2001). The conductivity of the membrane, which consists of a fluorocarbon polymer backbone with chemically bonded sulfonic acid groups as side chains, is a very strong function of its water content. Severe dehydratation of the membrane can result in significantly high ohmic losses in performance (Natarajan et al., 2003).

A PEM cell consists of a MEA placed between two so-called flow plates in which gas channels have been machined. The MEA consists of two electrodes separated by a proton conducting membrane, usually a Nafion® membrane (Figure 6). The proton conductivity of Nafion membrane, and therefore the performance of the electrochemical cell, decreases rapidly when the water content of the membrane decreases (Pukrushpan et al., 2002). Since the membrane easily loses water to non-saturated gases, it is essential that sufficient water is present in the gases as well as in the membrane.

PEM fuel cells need an external humidification subsystem that injects water into the gases (H_2 and O_2). The amount of water that is fed to the cell by humidification of the inlet anode and cathode gas is equal to:

$$n_{w\_a}^{in} = \frac{p_{sat}(T_{hum})}{p - p_{sat}(T_{hum})} n_{H_2}^{in} \qquad \text{(Eq. 47)}$$

$$n_{w\_prod} = \frac{I}{nF} \qquad \text{(Eq. 48)}$$

In steady state, the amount of water fed and produced per unit time is equal to the amount leaving the cell.

The water transport across a membrane is achieved through two distinct phenomena (Pukrushpan et al., 2002). First, the electro-osmotic drag phenomenon is responsible for the water molecules dragged across the membrane from the anode to cathode by the hydrogen proton. The amount of water transported is represented by the electro-osmotic drag coefficient, n_d, which is defined as number of water molecules carried by each proton. In Nafion membranes, the value of this coefficient is between 1 (gas vapor equilibrated membranes) and 2.5 (liquid water equilibrated membrane) (Janssen et al., 2001). The electro-osmotic drag, together with the electrochemical production of water, results in an accumulation of water at the cathode/membrane interface.

Second, the gradient of water concentration across the membrane due to the difference in humidity in anode and cathode gases causes "back-diffusion" of water from the cathode to the anode. The flux towards the anode compensates, at least in part, for the electro-osmotic drag.

The ratio of cathode and anode water fluxes is determined by the driving forces for both ways of water transport, i.e. the gradients in the chemical potential of water, and by the permeability of the materials. The gradients in the chemical potential of water are determined by the thickness of the components, by the water content of the membrane and the humidity of the gases (Janssen et al., 2001). As it was seen before, the humidification of the gases depends on the gas inlet humidification and on the temperature and pressure in the gas channel.

The total water flow across the membrane, assuming positive in the direction from anode to cathode) is given by (Rowe et al., 2001):

$$N_{w\_m} = N_c \left[n_d \frac{i}{F} - D_w \frac{(c_{w,c} - c_{w,a})}{t_m} \right]$$

(Eq. 49)

The coefficients n_d and Δ_w vary with membrane hydration (water content in the membrane) λ_m, which depends on the water content in the gas next to the membrane.

The hydration at both sides of the membrane is determined by the activity of water, which is an average of water activities in the gas in the anode and the cathode (Pukrushpan et al., 2002):

$$\lambda = 0.03 + 17.81a - 39.85a^2 + 36a^3, \qquad 0 < a \leq 1$$
$$\lambda = 14 + 1.4(a - 1), \qquad\qquad\qquad 1 < a \leq 3$$

$and:$

$$a = \frac{x_w P}{P_{sat}} = \frac{P_w}{P_{sat}}$$

(Eq. 50)

The electro-osmotic drag and diffusion coefficients are calculated by:

$$n_d = 0.0029\lambda_m^2 + 0.05\lambda_m - 3.4x10 - 19$$

$$D_w = (2.563 - 0.33\lambda_m + 0.0264\lambda_m^2 - 0.000671\lambda_m^3) \exp\left[2416\left(\frac{1}{303} - \frac{1}{T_{fc}} \right) \right]$$

(Eq. 51)

The water concentration at the membrane surfaces, c_w, is a function of water content on the surface, λ.

$$c_w = \frac{\rho_{m\_dry}}{M_{m\_dry}} \lambda$$

(Eq. 52)

The measured values of drag coefficient for Nafion exhibited $n_d = 2.5$ for $\lambda_m = 22$, and $n_d = 0.9$ for $\lambda_m = 11$. Zawodzinski et al. (1993) have measured drag coefficient as high as 4.0. The partial pressure of the saturated water vapor (P_w) as a function of the temperature follows in good approximation the equation of Clausius–Claperyon (Dohle et al., 2002).

$$P_{H_2O} = p * \exp \chi \qquad \text{(Eq. 53a)}$$

with

$$\chi = \frac{\Delta Hv}{R}\left(\frac{1}{T} - \frac{1}{T*}\right) \qquad \text{(Eq. 53b)}$$

In general equation (53) is applicable for calculating the saturated vapor pressure of liquids in a specific temperature range if:

- The evaporation enthalpy ΔH_v is nearly constant in the relevant temperature range;

- The specific volume of the liquid phase is negligible compared to that of the gas phase;

- The saturated vapor phase can be treated as an ideal gas.

All conditions are fulfilled for water in the temperature range from 60 to 130°C with an evaporation enthalpy of 40k*J/molK*.

4.6 Heat and Power Management: Thermal Model

4.6.1 Energy Balance and Heat Transfer

The principle of energy conservation determines the energy balance equation, which can be generally expressed by:

energy input - energy output + energy production = accumulation of energy (Eq. 54)

Electrolyzer: At steady state, the electric energy and the energy entering the electrolyzer in the water would equal the energy used to split the water (chemical energy), total energy leaving the electrolyzer with gases and water, and through other means, such as circulating coolants, cooling fins, or water vaporization, if present. The energy balance, at steady state can be written as:

$$Q_{mass\_in} + Q_{electric} = Q_{rxn} + Q_{mass\_out} + Q_{heat\_convection} \qquad \text{(Eq. 55)}$$

The energy input into the cell is associated with the energy content of the inflowing process fluid (water), and the major contribution is given by the electrical energy given by the external supply (PV module). The rate of energy outflow is given by the energy of the product fluids, the energy used for the water decomposition reaction, and the joule heat convection from the stack surfaces to the environment. Figure 11 shows the

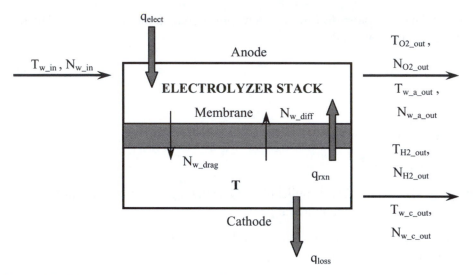

Figure 11. Schematic of the inlet and outlet streams in an electrolyzer system.

schematic of the inlet and outlet flows (mass and energy) for the electrolyzer control volume used to develop the steady state model.

$$Q_{mass\_in} = Q_{water\_in} = N_{w\_a\_in} Cp_w (T_{a\_in} - T_r) \quad \text{(Eq. 56)}$$

$$Q_{electric} = N_c I_{EL} V_{EL} \quad \text{(Eq. 57)}$$

The reaction energy released during the electrochemical reaction is given by:

$$Q_{rxn} = N_c \frac{\left[\Delta H_{H_2O} - \Delta H_{H_2} - \frac{1}{2} \Delta H_{O_2} \right]}{nF} I_{EL} \quad \text{(Eq. 58)}$$

The energy going out, Q_{mass_out}, considers the water outflow (liquid and vapor) on both the hydrogen and oxygen electrodes, in addition to the outflow of oxygen and hydrogen.

$$Q_{mass\_out} = Q_{H_2} + Q_{O_2} + Q_{w\_out} + Q_{v\_out} \quad \text{(Eq. 59)}$$

The difference between the energy given to the electrolyzer, $Q_{electric}$, and the energy needed for the electrochemical reaction, Q_{rxn}, represents the heat generated by the electrolyzer. This heat is generated due to irreversibility, such as the heat resulting from the anodic and cathodic overvoltages. Part of this heat is transferred to the fluid leaving the electrolyzer (hydrogen, oxygen, and water). Some of this heat is consumed during water vaporization; and if there is extra heat, it is released to the ambient due to convection heat transfer to the environment. But since the split water reaction is endothermic, sometimes under low flow conditions the electrolyzer will need to absorb energy from the surrounding. The energy carried out by the hydrogen, oxygen, liquid water and water vapor are expressed using the following equations:

$$Q_{H_2} = N_{H_2} Cp_{H_2} (T_{c\_out} - T_r)$$ (Eq. 60)

$$Q_{O_2} = N_{O_2} Cp_{O_2} (T_{a\_out} - T_r)$$ (Eq. 61)

$$Q_{w\_out} = N_{w\_a\_out} Cp_w (T_{a\_out} - T_r) + N_{w\_c\_out} Cp_w (T_{c\_out} - T_r)$$ (Eq. 62)

A small amount of energy is consumed during the vaporization of water and is incorporated into the energy removed by the water vapor.

$$Q_{v\_out} = N_{v\_a\_out} Cp_v (T_{a\_out} - T_r) + N_{v\_c\_out} Cp_v (T_{c\_out} - T_r) +$$
$$N_{v\_a\_out} \Delta H_{wv} + N_{v\_c\_out} \Delta H_{wv}$$ (Eq. 63)

The heat transfer due to convection between the stack surface and the environment is given by:

$$Q_{loss} = Q_{convection\_ambient} = hA(T - T_\infty)$$ (Eq. 64)

Fuel Cell: The major contribution of the input energy is associated with the reaction enthalpy of the fuel. The rate of energy outflow is given by the energy content of the outgoing process fluids, and the joule heat losses from the stack surfaces. Figure 12 shows the schematic of the control volume and the inlet and outlet flows in a fuel cell system.

$$Q_{mass\_in} - Q_{rxn} = Q_{electric} + Q_{mass\_out} + Q_{loss}$$ (Eq. 65)

The gases entering the anode and cathode of the fuel cell are saturated with water vapor. The energy inflow associated with the fluid coming into the fuel cell is given by:

$$Q_{mass\_in} = Q_{H_2\_in} + Q_{O_2\_in} + Q_{v\_a\_in} + Q_{v\_c\_in}$$ (Eq. 66)

$$Q_{H_2\_in} = N_{H_2\_in} Cp_{H_2} (T_{a\_in} - T_r)$$ (Eq. 67)

$$Q_{O_2\_in} = N_{O_2\_in} Cp_{O_2} (T_{a\_in} - T_r)$$ (Eq. 68)

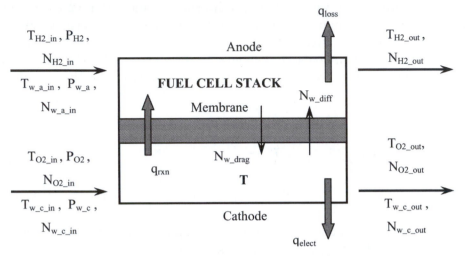

Figure 12. Schematic of the inlet and outlet streams in a fuel cell system.

$$Q_{rxn} = Nc \frac{\left[\Delta H_{H_2} - \frac{1}{2} \Delta H_{O_2} - \Delta H_{H_2O} \right]}{nF} I_{FC} \qquad \text{(Eq. 69)}$$

The vapor injected to the hydrogen and oxygen inflow also carries energy to the fuel cell stack, this energy can be quantified using:

$$Q_{v\_in} = Q_{v\_a\_in} + Q_{v\_c\_in} =$$
$$N_{v\_a\_in} Cp_{wv} (T_{a\_in} - T_r) + N_{v\_c\_in} Cp_{wv} (T_{c\_in} - T_r) \qquad \text{(Eq. 70)}$$

The electric energy generated during the water formation chemical reaction in the fuel cell is given by:

$$Q_{electric} = N_c I_{FC} V_{FC} \qquad \text{(Eq. 71)}$$

The energy outflow is related to the water produced, water vapor and excess hydrogen and oxygen molar flow rate leaving the fuel cell.

$$Q_{mass\_out} = Q_{w\_out} + Q_{v\_out} + Q_{H_2\_out} + Q_{O_2\_out} \qquad \text{(Eq. 72)}$$

$$Q_{H_2\_out} = N_{H_2\_out} Cp_{H_2} (T_{a\_out} - T_r) \qquad \text{(Eq. 73)}$$

$$Q_{O_2\_out} = N_{O_2\_out} Cp_{O_2} (T_{c\_out} - T_r)$$ (Eq. 74)

Part of the water produced inside the fuel cell is vaporized. Thus, the water leaving the fuel cell is a mixture of two water phases. The energy outflow related to the liquid water outflow is given by:

$$Q_{w\_out} = N_{w\_a\_out} Cp_w (T_{a\_out} - T_r) + N_{w\_c\_out} Cp_w (T_{c\_out} - T_r)$$ (Eq. 75)

And the energy associated with the vapor outflow is:

$$Q_{v\_out} = N_{v\_a\_out} Cp_v (T_{a\_out} - T_r) + N_{v\_c\_out} Cp_v (T_{c\_out} - T_r) + N_{v\_a} \Delta H_{wv} + N_{v\_c} \Delta H_{wv}$$ (Eq. 76)

The heat losses are due to convection heat transfer from the surface of the fuel cell stack.

$$Q_{convection\_ambient} = hA(T - T_\infty)$$ (Eq. 77)

4.6.2 Transient Model

An electrochemical system is an open system. For this kind of system, the thermal energy balance is based on the difference between the energy flow into the cell and the flow out of the cell. An accumulation of thermal energy generally occurs during the changes from one operating condition to another. The time variation of energy balances is usually manifested in temperature changes:

$$mCp \frac{d}{dt} T = -\Delta H + Q_{mol} - W_s + V\Delta P$$ (Eq. 78)

An important issue of an electrochemical system (electrolyzer and fuel cell) is the fact that, in contrast to the usual chemical combustion, the complete reaction occurs at two different electrodes. This may become a complication, especially in the estimation of the local entropy sources, which are usually known only for the reaction as a whole.

Electrolyzer: If we consider an electrolyzer as a whole, it is possible to formulate the following general energy balance equation for accumulation of energy in the cell in the form:

$$mCp \frac{dT}{dt} = Q_{mass\_in} + Q_{electric} - (Q_{rxn} + Q_{mass\_out} + Q_{loss})$$ (Eq. 79)

The estimated value of mCp in the accumulation term can be determined by using the summed $\Sigma m_i Cp_i$ values of individual components of the electrolyzer such as the graphite, membrane, and stainless steel.

The $\Sigma m_i Cp_i$ can be called stack thermal capacitance, C, thus Equation 79 can be rewritten as:

$$\frac{dT}{dt} = \frac{1}{C} \left[Q_{mass\_in} + Q_{electric} - (Q_{rxn} + Q_{mass\_out} + Q_{loss}) \right] \qquad \text{(Eq. 80)}$$

Fuel Cell: The model that predicts the transient responses in the fuel cell operation is obtained using the steady-state electrochemical model and the energy equation to form an overall transient model. This lumped thermal capacitance model is given by:

$$\frac{dT}{dt} = \frac{1}{C} \left[Q_{mass\_in} + Q_{rxn} - (Q_{electric} + Q_{mass\_out} + Q_{loss}) \right] \qquad \text{(Eq. 81)}$$

Consideration: To develop these lumped thermal capacitance models, the following assumptions were made:

1. Complete saturation of the gases with water vapor takes place at the anode and cathode because the membrane needs to be humidified.

2. Temperature gradients inside the stack, especially between the anode and cathode, are negligible.

3. The vaporization enthalpy (ΔH_v) of water for the temperature range of 60-130°C is assumed to be constant.

4. Hydrogen reaching the cathode is completely oxidized. This statement means that there is no hydrogen or oxygen going out of the fuel cell stack.

5. In this model the anode and cathode streams are well mixed, and they exit the stack at the same temperature of the fuel cell.

6. The reference temperature in all the cases is zero Kelvin.

4.7 Mathematical Modeling: Gas Storage

The storage of sufficient amounts of hydrogen is one of the most challenging tasks on the way to introducing and establishing hydrogen as an alternative fuel. The use of hydrogen as a fuel requires a storage system that has inherent safety as well as volumetric and gravimetric efficiency. High-pressure hydrogen gas storage is simple, reasonably light and compact, commercially available and relatively safe. Compressed hydrogen gas storage offers a very simple design and high safety due to inherent strength of the pressure vessel (Ananthachar and Duffy, 2004).

4.7.1 Real Gas Theory

The pressure gas storage model was developed using the real gas theory (van der Walls equation). According to the van der Walls equation of state, the pressure p of a real gas in a storage tank can be determined from:

$$P = \frac{nRT}{V - nb} - a\frac{n^2}{V^2} \tag{Eq. 82}$$

$$a = \frac{27R^2T_{cr}^2}{64P_{cr}} \quad \text{and} \quad b = \frac{RT_{cr}}{8P_{cr}} \tag{Eq. 83}$$

The last term in Equation 82, involving a, accounts for the intermolecular attraction forces, while b accounts for the volume occupied by the gas molecules.

4.7.2 Thermal Model

Hydrogen and oxygen are constantly varying in molar content and temperature in the storage tanks. Due to this fact, the pressure inside of the tank varies as well. In order to have a better understanding of what happens inside the tank storage, a thermal model was developed. To this end, the energy balance and mass balance equations are used. Figure 13 shows the tank storage schematic and also the water separator placed before the tank in order to remove all the water (liquid and vapor) before the gases enter the tank. The expansion chamber is used to reduce the pressure of the gases down to the working pressure of the fuel cell (1- 4 atm). The energy equation is given by:

$$\frac{d}{dt}E = -\Delta H + Q_{mol} - W_s + V\Delta P \tag{Eq. 84}$$

$$E = mCpT \qquad \rightarrow \qquad \frac{d}{dt}E = \frac{d}{dt}(nMCpT) \tag{Eq. 85}$$

Placing (85) into (84) and taking the derivative of the right hand-side give:

$$MC_pT\frac{d}{dt}n + nMC_p\frac{d}{dt}T = N_{H_2\_in}C_{pH2}(T_{H2\_in} - T_r) - \\ - N_{H_2\_out}C_{pH2}(T_{H2\_out} - T_r) - hA(T - Ta) \tag{Eq. 86}$$

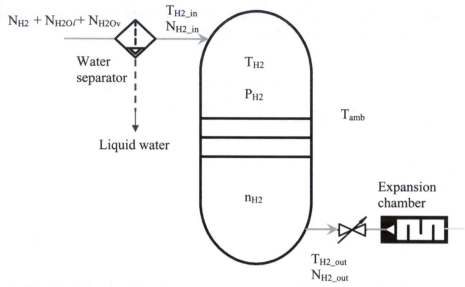

Figure 13. Tank storage schematic.

The mass balance equation is used to find the number of moles inside of the tank.

$$\frac{d}{dt}n = N_{H2\_in} - N_{H2\_out} \qquad\qquad \text{(Eq. 87)}$$

5. System Simulation

5.1 System and Data Used for the Simulation

The system has been designed for use as a stand-alone system for remote applications (with some modifications it could be used for *solar home systems;* in cities, it can even be used as an energy backup in hybrid grid-connected systems). This stand-alone system was designed for Cochapeti (Peru, 3400 m elevation, Latitude 10° S). The components of the proposed system are: photovoltaic module array, SPE high-pressure electrolyzers in parallel, high-pressure tank storage, PEM fuel cell, charge controller, electrolyzer and fuel cell controller, DC/DC converters and DC/AC inverters, temperature controllers, water pump and distribution system. Ambient temperature and solar radiation data from the Andes (Cochapeti) of Peru (Figure 14) is used in this work. The worst solar supply, approximately, in Cochapeti is found during the Peruvian summer days (March). This is because there is a lot of fog in the afternoon.

In order to size the system developed here, the worst-case scenario in Cochapeti (Peru) was used along with load of two vaccine refrigerators and a couple of lights

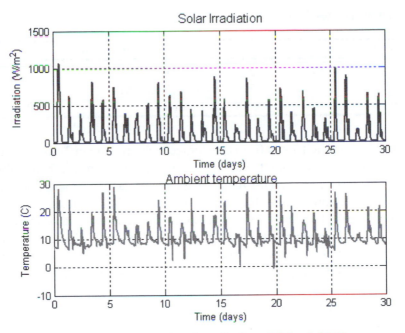

Figure 14. Irradiation and temperature data from Cochapeti (March 2001).

and a nominal voltage of 12 V. The data for Cochapeti was used in order to size the PV array; the equivalent solar hour (ESH) method (Florida Solar Energy Center, 1997) was used to find the energy requirement for the user load in a medical center. The total current needed was considered, including the parasitic and auxiliary currents. The irradiation readings were taken with a SPLite pyranometer from Kipp and Zonen. Temperatures were read with AD590 integrated circuit sensors, currents with precision resistors, and voltages with voltage divider circuits. Sensors were scanned every second, and averages recorded for each hour. The average daily irradiation in March was 3.6 kWh/m^2, which is 3.6 ESH. And the total user load current is 75.7 Ah/day. Finally seven 100-W PV modules were chosen. Table 1 summarizes the parameters used in sizing.

A 200 W SPE electrolyzer is used in this system. In order to get the correct voltage eight cells are connected in series. This electrolyzer is a customized device that uses MEA from ElectroChem, Inc., but should be representative of typical PEM units. The active area of the MEA is 50 cm^2; the membrane used in each MEA is Nafion™112.

The gases going out from the electrolyzer are stored in two pressurized tanks. The standard operating pressure for the PEM fuel cell is 50 psi (~ 3.4 atm), while the maximum planned storage tank pressure is 2000 psi (~ 137 atm). The hydrogen and oxygen are compressed using a high-pressure electrolyzer. Alternatively, the electrolyzer can be placed inside a nitrogen vessel at the pressure of the oxygen and hydrogen, or oxygen vessel itself as proposed by Shapiro, Duffy, Kimble, and Pien (2004).

Table 1. Summarized results of the PV array sizing.

Daily data		PV module	
Irradiation (kWh/m^2)	3.6	Power (W)	100
ESH (h)	3.6	Max. current (A)	5.4
Total Load (Ah)	75.7	# Modules	7.0

Gas will be regulated down from the storage tank pressure to 3.4 atm for use; this means that in normal operation the tank pressure will not be drawn below this level. The tanks must be sized so that the required amount of gas can be withdrawn and still leave the tank pressure at 3.4 atm. As a result, the volumes of the hydrogen and oxygen tanks are 50 liters and 25 liters respectively. The volume of the H_2 tank is twice the O_2 in order to obtain the same pressure of the gases. All the water that goes out of the electrolyzer is removed from the gases, so the gases stored are completely dried. Phase separators (dryers) are used to remove the water from the hydrogen and oxygen pipeline.

A 100 W Ballard Mark V fuel cell is used; the MEA active area is 50.6 cm^2 and it uses a Nafion™ 117 membrane. The stack uses 14 cells connected in series to get adequate voltage. With this number of cells in series the voltage of the fuel cell stack can be up to 16 V when the stack operates at low current and down to 12 V when it works at relatively high current.

5.2 Model of the Solar Hydrogen Fuel Cell System

The model of the system was developed in Simulink using electrochemical and physical processes of the system that are transient and non-linear. The time steps used in the simulations are setup automatically by Simulink. Hourly weather data are interpolated through the simulation time using Matlab's *spline* function. The time steps used in the simulations are setup automatically by Simulink. The Simulink system diagram is shown in Figure 15.

All the simulations have been performed for a system voltage the same as the electrolyzer (maximum voltage17V). The set point temperature for the fuel cell and electrolyzer was 75°C. The values of the heat convection thermal coefficient for both electrolyzer and fuel cell were calculated using the properties and dimensions of component of each PEM device. The temperature control of the electrolyzer and fuel cell stack are performed using proportional controllers. The temperature of the electrolyzer stack is kept at the set point by supplying extra water to the anode, and forced air convection over the stack fins is used in the case of the fuel cell stack, following the approach of Das (2002).

5.3 High Pressure Electrolyzer

In cases where a high-pressure gas storage vessel is used and the hydrogen is produced in a low-pressure electrolyzer, the hydrogen must obviously be compressed

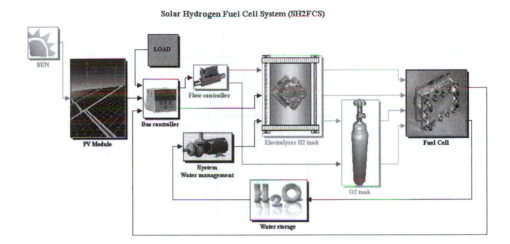

Figure 15. Simulink diagram of the system modeled.

into the pressure vessel. The energy required to compress the hydrogen is taken from the PV modules. One of the unique features of an electrolyzer as a gas generator is that it can generate gases at high pressures with little increase in the energy input required to generate them at low pressure (Cisar et al., 1999). The energy demand in a high-pressure electrolyzer is related to the thermodynamics of the water electrolysis. Using the model developed in this work, it was found theoretically that, the energy demand in a 100 atm (1470 psi) electrolyzer system will increase by about 7.5% compared to a 1 atm system (14.69 psi). This energy needed is less than or equal to the mechanical energy required for a compressor to compress hydrogen from 1 to 100 atm, typically 10% of the total electrolyzer energy demand (Ulleberg, 1998).

The gases produced through water electrolysis are very pure and should not require additional processing. It is advantageous therefore to match electrolysis pressure with that required for further processing or storage. One high-pressure solution adopted for the SPE electrolysis stacks is to enclose the entire assembly in a pressure vessel with a blanket of inert gas for pressure management. This approach assigns the job of providing for both electrolysis stack and system safety to the system designer.

Based on studies cited and on Shapiro et al. (2001), the high-pressure hydrogen subsystem used in this study is the one shown in Figure 16. The electrolyzer will go inside of the hydrogen vessel, to reduce the differential pressure between the inside and outside of the electrolyzer and thereby allow the use of a less expensive electrolyzer. The main components of such a system are:

a) accumulator,

b) differential-pressure relief,

c) electrolyzer,

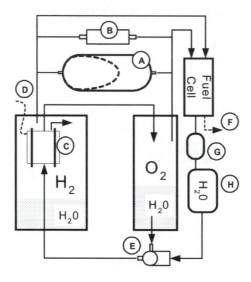

Figure 16: High-pressure electrolyzer system (Shapiro, 2002).

d) external power lead,

e) water circulating pump,

f) fuel cell power out,

g) auxiliary water reservoir,

h) main fuel cell water reservoir.

As apparent in Figure 16, the excess of water is also kept inside of the electrolyzer/ hydrogen vessel; this could be a safety issue. Because of that and in order to prevent any other problem with the water and hydrogen stored in the same tank, the gases (even the oxygen) are dried using a phase separator. To do this, the hydrogen will pass through a dryer and then will go back to the hydrogen pressure vessel. This change would require a slight modification of the system in Figure 16.

5.4 Control System

The system under study is essentially a dynamic system with closed-loop control. In the Simulink block diagram of the system shown above, different blocks represent the different components ("plants").

Power management: The power controller is a simple on-off controller. The current from the PV that goes into the bus controller supplies first the load and, if there is any remaining energy, it goes to the electrolyzer to produce hydrogen. When the energy available from the PV is less that the load, the difference is given by the fuel cell. The idle current for the electrolyzer and the standby current for the fuel cell are the protective currents to ensure the humidification of the PEM devices when the electrolyzer and fuel cell are not working. The PV is directly coupled to the electrolyzer, which means the load is connected in parallel to the electrolyzer.

The power at the bus is distributed to the load and the electrolyzer according to the following control strategy:

$$if \;\; I_{PV} < I_{LOAD} \;\; \rightarrow \;\; I_{FC} = I_{LOAD} - I_{PV} + I_{EL\_idle} \;\; and \;\; I_{EL} = I_{EL\_idle}$$
$$if \;\; I_{PV} = I_{LOAD} \;\; \rightarrow \;\; I_{FC} = I_{EL\_idle} \;\; and \;\; I_{EL} = I_{EL\_idle}$$
$$if \;\; I_{PV} > I_{LOAD} \;\; \rightarrow \;\; I_{FC} = I_{FC\_sb} \;\; and \;\; I_{EL} = I_{PV} - I_{LOAD}$$

(Eq. 88)

Temperature controller: The temperature control of the electrolyzer is accomplished through adding extra water to the electrolyzer stack (Das, 2002); this extra water flow is proportional to the difference between the actual stack temperature and the set point temperature.

The electrolyzer temperature is controlled using a proportional controller, which supplies extra water to the anode; this extra water flow varies according to:

$$If \;\; \Delta T < 0 \;\;\;\;\;\;\; then \;\;\;\;\; N_w = N_o$$
$$If \;\; 0 \le \Delta T \le BW \;\;\;\;\;\;\; then \;\;\;\;\; N_w = K_p \Delta T + N_o$$
$$If \;\; \Delta T > BW \;\;\;\;\;\;\; then \;\;\;\;\; N_w = N_{w,max}$$

(Eq. 89)

The required water flow as a function of the temperature difference is shown in Figure 17, where $N_{w,max}$ represents the maximum water flow that can be pumped to the electrolyzer, which is a function of the operating conditions. The tuning of the controller, setting of gains and bandwidth (range between the highest and lowest ΔT of the device, where $\Delta T = T - T_{sp}$, T_{sp} is the set point temperature), must assure that $N_{w,limit} < N_{w,max}$. If the proportional gain K_p is higher than $N_{w,\,limit}/BW$, then the effective bandwidth is reduced, so $BW_{effective} = N_{w,limit}/K_p$ (Trelles, 2003).

The fuel cell temperature is regulated using an external airflow over the fins or parts of the carbon plates extending out of the stack, which increases or decreases proportionally to the variation of the temperature of the fuel cell stack and the set point. Since the PEM (SPE) devices work in the range of 50-90°C, the set point temperature is set at 75°C. This temperature will ensure that the temperature inside the devices is not higher than the boiling point temperature of the water and will maintain the hydration of the membrane.

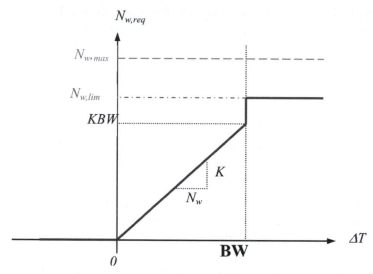

Figure 17. *I-N_w* curve of the electrolyzer temperature controller.

The fuel cell temperature controller is a proportional-integral controller. It supplies airflow to the fuel cell fins at certain velocity, this velocity varies according to:

$$
\begin{aligned}
&\text{If} \quad \Delta T < 0 \qquad\qquad\; \text{then} \quad V_{air} = 0 \\
&\text{If} \quad 0 \le \Delta T \le BW \quad \text{then} \quad V_{air} = K_p \Delta T \\
&\text{If} \quad \Delta T > BW \qquad\quad \text{then} \quad V_{air} = V_{air,max}
\end{aligned}
\qquad\text{(Eq. 90)}
$$

The required air velocity as function of the temperature difference is shown in Figure 18, where $V_{air,max}$ represents the maximum air velocity that the fan can supply. The required airflow can be considered as a signal that goes to the temperature controller and is used for the mathematical model of the system. The simulation system allows the determination of suitable values for K_p.

5.5 Water Management and Humidification

Water management and humidification is a big issue in PEM devices. Depending on the operating conditions, a membrane can have highly non-homogenous water (and therefore ionic conductivity). The membrane beneath a long channel may be dried by hot inlet flow, ideally saturated near the middle of the cell, and flooded near the exit (Mench et al., 2001). Since the reduction and electro-osmotic drag result in water transport and generation at the cathode surface, the flux of water by diffusion can be either from or to the anode surface, depending on the flow and humidity conditions.

Membrane performance suffers without water, because PEM ion conductivity is directly related to the degree of water content and temperature. Alternatively, excessive water at the cathode can cause flooding; this last phenomenon is most likely near

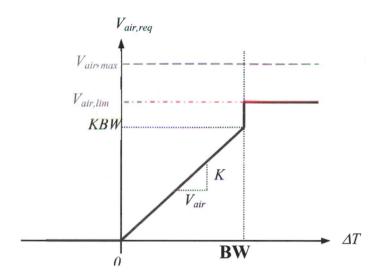

Figure 18. V_{air}–DT curve of the fuel cell temperature controller.

the cathode exit under high current density, high humidification, low temperature, and low flow rate conditions.

System design, modeling and control require knowing the amount and location of water in all parts of the system. ElectroChem's test of its electrolyzer (Shapiro, 2002) indicated that approximately 6 molecules of water are dragged across the membrane for each molecule of hydrogen.

At the anode side of the electrolyzer the water is broken down and oxygen is created and stored. A pump is used to supply the water inside of the electrolyzer; this assures that the electrolysis reaction will not be limited by lack of water. Furthermore, excess water is pumped into the stack as well; this water will cool the electrolyzer down, which also ensures that the membrane is properly hydrated. Localized lack of water, due to gas bubbles on the MEA surface, can cause spot drying, which will reduce efficiency and may lead to membrane failure (Shapiro 2002). Elimination of gas bubbles also minimizes back-diffusion of water through the membrane. All the water going out of the electrolyzer is removed before it goes to the tank storage using a phase separator. This water is stored in the water tank and pumped to the electrolyzer to close the circuit (regenerative fuel cell). Because part of the total water is consumed in this process (to produce H_2 and O_2) an auxiliary water supply is needed.

The membrane of the fuel cell easily loses water to non-saturated gases; because of this it is essential to have sufficient water present in the gases as well as in the membrane. Water vapor is injected to the inlet stream gases to keep the gases humidified and help the membrane to function. All of the water that goes out of the fuel cell stack is circulated back to the water tank; and from this, some portion of the total

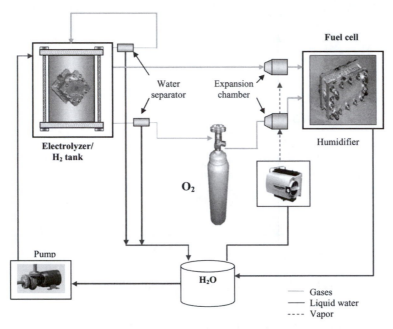

Figure 19. Water management schematic.

volume of water is supplied to the humidifier (Figure 19). Therefore the water in the system is:

- Circulated through the electrolyzer
- Split up in the electrolyzer
- Transported across the electrolyzer membrane
- Separated from the gases before they enter the tank storage
- Introduced in the fuel cell as a vapor
- Transported across the membrane of the fuel cell
- Produced in the fuel cell

6. Results

The model was run for two cases of summer and winter scenarios. For the first scenario it was assumed that the system is turned on for the first time (beginning of the day), which means the storage tanks are empty and the load is bigger than the solar input. All the current received from the PV array goes to the electrolyzer and the load is disconnected (blackout) while the fuel cell is off. When the current generated

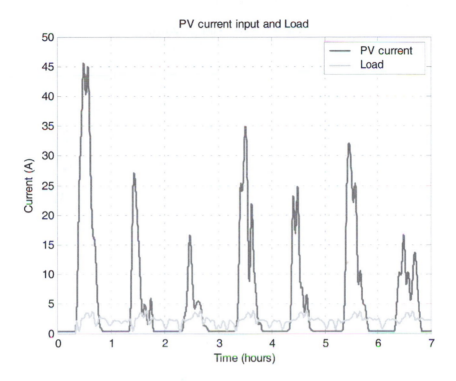

Figure 20. PV output current and load current over time.

by the PV array is greater than the load current, the busbar controller supplies current to the load; and if there is some extra current, this goes to the electrolyzer (Figure 20). After the electrolyzer is switched on, it has to remain on or in the idle stage, so it always has to receive at least the idle current (I_{idle}, protective current). The electrolyzer and fuel cell stack voltage are shown in Figure 21. A DC voltage regulator is used to keep the system voltage at 12 DC V to protect the fuel cell. Figure 22 shows that when the electrolyzer works, the fuel cell remains at the standby stage and vice versa. Most of the load is supplied by the PV array, and the fuel cell does not start its operation until 6 pm (first day) and it continues working until 10 am of the second day when it goes back to the standby. The variation of the gas pressure inside of the hydrogen and oxygen tanks is shown in Figure 23, as it was mentioned before, the volume of the hydrogen storage tank is twice the volume of the oxygen tank; and because the number of moles of hydrogen is also twice the number of oxygen moles, the pressures inside of these two tanks were the same most of the time. The fuel cell temperature controller did not work for this period of time because the temperature of the fuel cell never exceeds the set point temperature (75°C, Figure 24). Table 2 summarizes the volume of water used and the amount of hydrogen consumed and produced during this simulation. The total amount of hydrogen produced was 0.22 kg; 32% of this hydrogen was used by the fuel cell to produce energy. The total volume of water supplied to the electrolyzer was 26.63 *l*, but only 1.97 *l* of water was used to

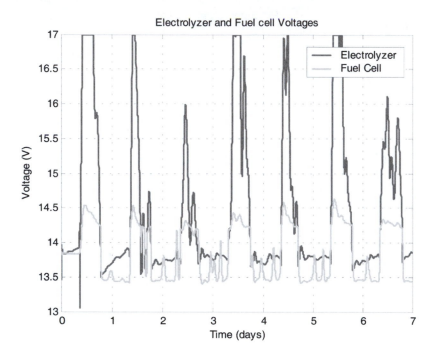

Figure 21. Instantaneous voltage of the electrolyzer and fuel cell.

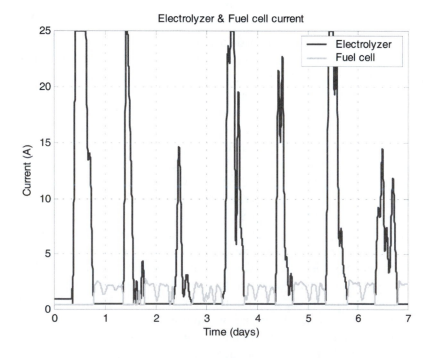

Figure 22. Electrolyzer input current and fuel cell output current over time.

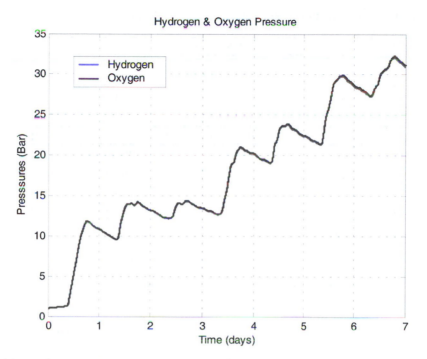

Figure 23. Hydrogen and oxygen pressure variation inside the tank storage.

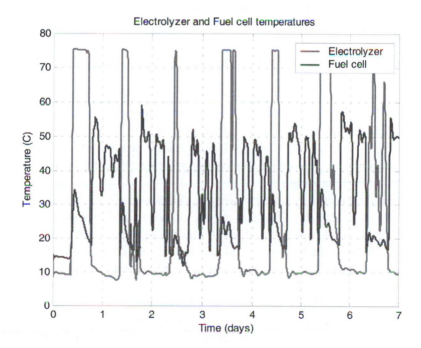

Figure 24. Electrolyzer and fuel cell temperature variation over time.

Table 2. Amounts of hydrogen and water used during the start-up case.

Hydrogen	
	Mass (kg)
H₂ produce	0.22
H₂ consumed	0.07

Water management	
	Volume (L)
Water supplied [1]	26.63
Water consumed [2]	1.97
Water removed [3]	24.65
Water produced [4]	0.94
Water cooling [5]	24.65
Water humidification [6]	0.03

[1] Into the electrolyzer   [4] At the fuel cell

[2] To produce hydrogen   [5] To cool the electrolyzer

[3] From the H_2 and O_2 pipeline   [6] To humidify the fuel cell MEA

produce hydrogen and oxygen; the rest of the water was used to cool the electrolyzer and was recycled. The fuel cell produced 0.94 *l* of water that was circulated back to the water tank. A summary of the first case simulation energy consumption/production and efficiencies of the various components and subsystems of the simulation scenario discussed above (Start-up) is given in Table 3. The results show that about 21% of the energy produced by the PV-arrays went to cover the user load, and the rest went to produce hydrogen in the electrolyzer. The overall system efficiency was about 31.8%; this efficiency was calculated by dividing the energy to the user load by the output energy from the PV. The hydrogen system efficiency was 41.8%. The total efficiency of the system (from the solar irradiation input to the energy consumed in the load) was 4%. The energy stored in the hydrogen tank can supply the load for almost 5 days without any solar input.

The second scenario was winter; during this season the irradiation in Cochapeti is high (> 800 W/m² peak daily). This high irradiation is obtained almost every day during the winter time, even though the ambient temperature can go below 4°C during the night. Figure 25 shows the PV output and user load current. In this case both temperature controllers did work, and the temperature variation over time is shown in Figure 26.

The results obtained using the model developed were compared with other simulation results from actual solar hydrogen system such as the Schatz Solar Hydrogen Project at Humboldt University (Schucan and Scherrer, 2000) and the PHOEBUS plant at FZ-Julich in Germany (Ulleberg, 1998).

Table 3. Summary of the results for one-week simulation winter and summer case.

Component/System		
	March	**July**
	Energy (kWh)	Energy (kWh)
Solar	130.97	216.45
PV	16.39	27.71
User Load	5.22	6.79
Electrolyzer	12.98	23.35
Fuel cell	2.71	3.5
Storage [a]	6.17	10.83
Auxiliary subsystems	*Energy (Wh)*	*Energy (Wh)*
FC cooling [b]	0	152.11
EL cooling [c]	31.25	49.18
	Efficiency (%)	*Efficiency (%)*
Electrolyzer [d]	67.29	66.74
Fuel cell [e]	62.08	64.18
PV	12.51	12.80
PV-load [f]	31.85	24.50
H_2 system [g]	41.78	42.83
Total [h]	3.99	3.14

[a] Hydrogen energy remaining in the tank storage

[b] Fan and accessories

[c] Water pump and accessories

[d] Input electric energy over hydrogen energy produced

[e] Hydrogen energy supplied over electric energy produced

[f] From PV output to user load

[g] Combined electrolyzer storage and fuel cell

[h] From solar irradiation input to user load

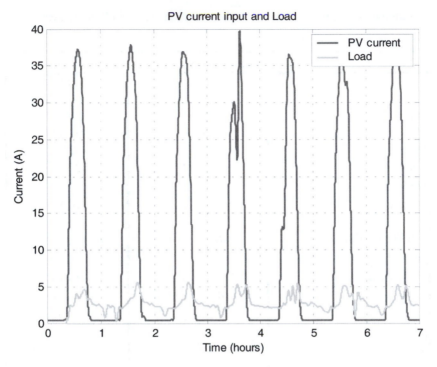

Figure 25. PV output current and load current over the time for the winter case.

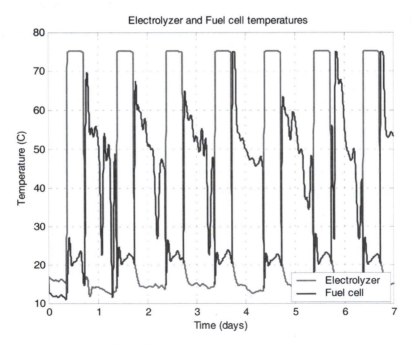

Figure 26. Electrolyzer and fuel cell temperature versus the time, winter case.

Table 4. Comparison of the system modeled in this work with two other real systems.

System	System description	Efficiencies (%)				
		Electrolyzer	Fuel Cell	PV array	Energy storage	System
Schatz Solar Hydrogen Project (Schucan, 2000)	PV array, battery (buffer between PV array and inverter), medium pressure alkaline electrolyzer, H_2 tank storage (no O_2 tank, the fuel cell uses compresed air), PEM fuel cell	79.2	43.1	6.7	34	5.7
Model of the PHOEBUS plant at FZ–Jülich in Germany (Ulleberg, 1998)	PV array, battery, low pressure alkaline electrolyzer, H_2 and O_2 tank storage, compressor,PEM fuel cell	81.8	31.3	10.4	20.7	6
Solar hydrogen fuel cell system modeled (Aurora, 2004)	PV array, high pressure SPE electrolyzer, H_2 and O_2 tank storage,PEM fuel cell	67.3	62.8	12.5	41.78	6.5

Table 4 shows the description of each system and the different important efficiencies. The following definitions are used for the efficiency calculations.

$$Electrolyzer = \frac{\text{Hydrogen produced (Wh)}}{\text{Energy supplied to the electrolyzer (Wh)}} \qquad \text{(Eq. 91)}$$

$$Fuel\ cell\ = \frac{\text{Energy supplied by the fuel cell (Wh)}}{\text{Hydrogen consumed by the fuel cell (Wh)}} \qquad \text{(Eq. 92)}$$

$$PV\ = \frac{\text{PV energy supplied to the system (Wh)}}{\text{Solar energy on the array (Wh)}} \qquad \text{(Eq. 93)}$$

$$Storage\ =\ \text{Electrolyzer efficiency x Fuel Cell efficiency} \qquad \text{(Eq. 94)}$$

$$System\ = \frac{\text{User load + Hydrogen stored (Wh)}}{\text{Solar energy on the array (Wh)}} \qquad \text{(Eq. 95)}$$

The values presented in Table 4, for the case of the PHOBOUS and Schatz System, are average values found using data gathered for at least four months. The system efficiency obtained using the model of this work is the highest among the three systems, because the fuel cell and the PV array efficiencies are also high. The fuel cell is working at low current density and high cell voltage (Figure 10) given a relatively high output power which finally gives high energy efficiency. Even though, the systems compared are not the same size, the results given in Table 4 serve as good comparison to see how accurately the simulation is performed using the mathematical models developed in this work.

7. Summary and Conclusions

Detailed descriptions of the individual component models required to simulate a SH2FCS were presented. These models were based on electrical, electrochemical, thermodynamic, and heat and mass transfer theory. In addition, semi-empirical models were used to model the voltage-current behavior of the electrolyzer and fuel cell. A lumped-parameter, dynamic model of the *solar hydrogen fuel cell system* was developed. The electrolyzer semi-empirical model showed good agreement between simulated and the limited experimental data that was available.

Based on the results of this study, the following conclusions and observations are made:

- The energy needed to compress hydrogen using the electrolyzer as a compressor is less than the mechanical energy required for a compressor.

- A lumped parameter model approach for the electrolyzer, fuel cell and tank storage gives a good understanding of the energy flows and thermal behavior of these components and their influence in the overall performance of the system.

- A relatively simple on-off controller can manage the power flows in the system.

- A proportional controller can maintain the optimal temperature of the electrolyzer with excess water flow through the MEA and subsequent cooling with a heat exchanger of tubing and cooling fins. In the fuel cell, sufficient cooling is obtained through forced air convection over enlarged graphite fins.

- The model developed here can help the designer to determine the necessary specifications of a fuel cell system prior to actual fabrication and testing.

- The overall energy efficiency of the regenerative fuel cell system (electrolyzer, gas storage, and fuel cell) is relatively low at about 42%.

- However, this model will aid in the optimization of the sizing/design of the SH2FCS to improve the energy efficiency and reduce costs.

8. Acknowledgement

The authors acknowledge the helpful discussions with Mike Kimble, now with MicroCell Technologies, MA and the help of John White, Prof., U Mass Lowell, for assistance with MATLAB and Simulink coding.

9. References

Amendola, S., Sharp-Goldman, S., Janjua, S., Kelly, M., Petillo, P., Binder, M., 2000. An Ultra safe Hydrogen Generator: Aqueous, Alkaline Borohydride Solutions and Ru catalyst. Journal of Power Sources, 76, 66-80.

Amphlett, J., Baumert, R., Mann, R., Peppley, A., 1995. Performance Modeling Of The Ballard Mark IV Solid Polymer Electrolyte Fuel Cell. Journal Electrochemical Society, v. 142, n.1, pp. 9-15.

Ananthachar, V., 2002. Efficiencies of Hydrogen Storage Systems Onboard Fuel Cell Vehicles. University of Massachusetts Lowell, MS thesis.

Ananthachar, V., Duffy, J., 2004. Efficiencies of Hydrogen Storage Systems Onboard Fuel Cell Vehicles. Solar Energy, in press.

Arkin, A., 2001. Modeling of PV, Electrolyzer and Gas Storage in a Stand-Alone Solar-Fuel Cell System. University of Massachusetts Lowell, MS thesis.

Atkins, P., 1998. Physical Chemistry 5th Ed. W.H. Freeman & Company, New York, pp 132-133.

Atkinson, K., Roth, S., Hirscher, M., Grunwald, W., 2002. Carbon Nanostructures: An Efficient Hydrogen Storage Medium for Fuel Cell. Fuel Cell Bulletin No. 38, pp 9-12.

Bak, T., Nowotny, J., Rekas, M., Sorrell, C., 2002. Photo-Electrochemical Hydrogen Generation from Water Using Solar Energy, Materials-Related Aspects. Int. Journal of Hydrogen Energy 27, 991-1022.

Cisar, A., Clarke, E., Salinas, C., Murphy, J., 1999. PEM Energy Storage for Solar Aircraft, Lynntech, Inc. Society of Automotive Engineers, Inc.

D.O.E. (U.S. Department of Energy), 2000. Fuel Cell Handbook, 5th Edition. Office of Fossil Energy, Washington, DC, pp 2-22 to 2-30

Das, A., 2002. Modeling, Heat Transfer, and System Integration for a 4-Kw Regenerative PEM Fuel Cell System. University of Massachusetts Lowell, MS thesis.

Dohle, H., Mergel, J., Stolten, D., 2002. Heat and Power Management of a Direct-Methanol-Fuel-Cell (DMFC) System. Journal of Power Sources 111, 268-282.

Duffie, J., Beckman, W., 1991. Solar Engineering of Thermal Processes, second ed. Wiley Interscience, New York, pp. 576-586.

Florida Solar Energy Center, 1997, PV sizing handbook.

Goswami, Y., Böer, K., 2003. Advances in Solar Energy, Vol. 13.American Solar Energy Society (ASES), Boulder, CO, pp. 405-458.

Heywood, J., 1988. Internal Combustion Engine Fundamentals, McGraw-Hill, Inc, Massachusetts, pp. 78-79.

Hottinen, T., Mikkola, M., Lund, P., 2004. Evaluation of Planar Free-Breathing Polymer Electrolyte Membrane Fuel Cell Design. Journal of Power Sources, 129, 68-72.

Huang, C., T-Raissi, A., 2004. Analysis of Sulfur-Iodine Thermochemical Cycle for Solar Hydrogen Production. Part I: decomposition of sulfuric acid. Solar Energy, article in press.

Janssen G.J.M., Overvelde M.L.J, 2001. Water Transport in the Proton-Exchange-Membrane Fuel Cell: Measurements of the Effective Drag Coefficient. Journal of Power Sources 101, 117-125.

Licht, S., Wang, S., Mukerji, S., Soga, T., Unemo, M., Tributsch, H., 2001. Over 18% Solar Energy Conversion to Generation of Hydrogen Fuel: Theory and Experiment for Efficient Solar Water Splitting, International Journal of Hydrogen Energy, 26, 653-659.

Mann, R., Amphlett, J., Hooper, M., Jensen, H., Peppley, B., Roberge, P., 2000. Development and Application of a Generalized Steady-State Electrochemical Model for a PEM Fuel Cell. Journal of Power Sources, 173-180.

Mench, M., Wang C., Thynell S., 2001. An Introduction to Fuel Cell and Related Transport Phenomena, Electrochemical Engine Center, Department of Mechanical and Nuclear Engineering. The Pennsylvania State University.

Natarajan, D., Van Nguyen, T., 2003. Three-Dimensional Effects of Liquid Water Flooding in the Cathode of a PEM Fuel Cell. Journal of Power Sources 115, 6-80.

Pukrushpan, T., Peng, H., Stefanopoulou, A., 2002. Simulation and Analysis of Transient Fuel Cell System Performance Based On a Dynamic Reactant Flow Model. Proceedings of IMECE'02 2002 ASME International Engineering Congress & Exposition, New Orleans, Louisiana, USA.

Rowe, A., Li, X., 2001. Mathematical Modeling of Proton Exchange Membrane Fuel Cells, Journal of Power Sources 102, 82-96.

Schucan, T., Scherrer, P., 2000. Integrated Hydrogen Energy Systems, Chapter 5: Schatz Solar Hydrogen Project, Institute Switzerland, International Energy Agency Hydrogen Implementing Agreement Task 11: IEA 2000.

Schultz, K., General Atomics, 2003. Thermochemical Production of Hydrogen from Solar and Nuclear Energy. Presentation to the Stanford Global Climate and Energy Project.

Shapiro, D., 2002. PEM Electrolyzer and Gas Storage for a Regenerative Fuel Cell System. University of Massachusetts Lowell, MS Thesis.

Shapiro, D., Duffy, D., Kimble, M., Pien, M., 2004, Solar Powered Renerative PEM Electrolyzer/Fuel Cell System. Solar Energy, in press.

Steinfeld, A., 2004. Solar Thermochemical Production of Hydrogen - A Review. Solar Energy, in press.

Thomas, S., Zalbowitz, M., 1999. Fuel Cells: Green Power. Los Alamos National Laboratory, New Mexico.

Ulleberg, O., 1998. Stand-Alone Power Systems for the Future: Optimal Design, Operation & Control of Solar-Hydrogen Energy Systems. Norwegian University of Science and Technology Trondheim.

Ulleberg, O., 2003. Modeling of Advanced Alkaline Electrolyzer: A System Simulation Approach. International Journal of Hydrogen Energy, 28, 21-33

Urade, V., 2004. Photoelectrochemical Generation Hydrogen. School of Chemical Engineering, Purdue University. Website http://atom.ecn.purdue.edu/~vurade/PEC%20Generation%20of%20Hydrogen/Main01.htm

Vieltstrich, W., Lamm, A., Gasteiger, H., 2003. Handbook of Fuel Cells: Fundamentals Technology and Applications, Volume 3: Fuel Cell Technology and Applications Part 1, John Wiley & Sons Ltd, England, pp 64-74.

Vieltstrich, W., Lamm, A., Gasteiger, H., 2003. Handbook of Fuel Cells: Fundamentals Technology and Applications, Volume 1: Fundamentals and Survey of Systems, John Wiley & Sons Ltd, England, pp 99-105.

You, L., Liu, H., 2002. A Two-phase Flow and Transport Model for the Cathode of PEM Fuel Cell. International Journal of Heat Transfer and Mass, 45, 2277-2287.

10. Nomenclature

A_{conv}	convective surface area (m^2)
A_{el}	area of the electrode (m^2)
BW	bandwidth
C	mCp average thermal capacity of the stack
C_p	specific heat (J/K mol)
D_w	diffusion coefficient (cm^2/s)
E	cell voltage (V)
$E°$	standard voltage, 1.229 Volts
E_r	reversible voltage (V)
F	Faraday's constant
G_T	solar irradiation (W/m^2)
HHV	high heating value (J/mol)
I	current (A)
I_D	diode current (A)
I_L	light current (A)
I_O	diode reverse saturation current (A)
I_{sc}	short circuit current (A)
I_{sh}	shunt current (A)
K_i	integral constant
K_p	proportional constant
N_c	number of cell in series
N_s	number of cells in series
N	molar flow rate (mol/s)
P	pressure (atm)
Q	mass energy flow
$Q_{electric}$	electric energy
$Q_{heat\_convection}$	energy lost due to convection to the environment
Q_{mol}	molar heat supplied to the system

Q_{rxn}	electrochemical energy released during the reaction
R	gases constant
R_{fc}	total cell resistance (Ω)
R_{sh}	shunt resistance (Ω)
T_{∞}	ambient temperature (K)
T_{sp}	set point temperature (°C)
T_r	reference temperature (°C)
V	voltage (V)
V_{air}	air velocity
V_{oc}	open circuit voltage (V)
W	work done
ΔG	change in Gibbs energy (kJ/mol)
ΔH	change in enthalpy (kJ/mol)
ΔH_{wv}	water vaporization enthalpy (J/mol)
ΔS	change in entropy (kJ/K.mol)
M_{m_dry}	membrane dry equivalent weight (kg/mol)

Lower case

a	curve fitting parameter (V)
s	coefficient for overvoltage on both electrodes (V)
t	coefficient for overvoltage on both electrodes (m²/A)
r	parameter related to ohmic resistance of electrolyte (ohm.m²)
i	current density (A/m²)
i_o	exchange current density (A/m²)
i_L	limiting density current (A/m²)
l	thickness of the polymer membrane (cm)
n_g	mol number of the gas in the tank
v_g	gas volume (m³)
h	convective heat transfer coefficient (W/Km²)
n	number of moles in the reaction
n_d	electro-osmotic coefficient
c_w	water concentration (mol/cm³)
t_w	membrane thickness (cm)

Greek

η	polarization overvoltage (V)
η_E	energy efficiency
ζ	overvoltage parameter
ε	bangap of the silicon (1.12-1.35 eV)
μ	temperature effect on the PV performance
ρ_{m_dry}	membrane dry density (kg/cm³)

Subscript

a	anode
act	activation
c	cathode
$conc$	concentration
cr	critical
EL	electrolyzer
FC	fuel cell
g	gases
H_2	hydrogen gas
$idle$	idle current
in	inflow
mp	values at maximum power point
O_2	oxygen gas
ohm	ohmic
out	outflow
PV	photovoltaic
ref	reference
sb	stand by
v	water vapor
w	liquid water

Chapter 11

Renewable Energy For The Russian Economy

by

Pavel P. Bezrukih

Energy Ministry of Russian Federation,
7 Kitaigorodsky pr., Moscow, 103074 Russia

Dmitry S. Strebkov, Igor I. Tyukhov*

The All-Russian Research Institute for Electrification of Agriculture (VIESH)
1-st Veshnyakovsky proezd, 2, VIESH, Moscow, 109456 Russia

Abstract

While Russia is home to only 2.5% of the world's population, the country has about 30% of the worldwide geological energy resources. There are compelling reasons for using renewable energy (RE) in spite of this apparent abundance of fossil fuels. There is an urgent need and economic expediency to use RE in Russia. There are great economic and technical opportunities for RE sources to provide significant contributions to the Russian energy supply. The situation is getting better, taking into account the Russian defense industry's activities in the production of non-traditional energy equipment. There are enterprises for the production of wind, micro-hydro, and geothermal systems, as well as photovoltaic equipment, and others. Russia uses only 1.5 million tons of oil while it has the potential equivalent of 270 million tons of economically feasible renewable energy. Right now, the main goal for the Russian RE sector is the creation of a new law on renewable and non-traditional sources of energy. The Energy Ministry started to develop this important law so that RE will have a more significant contribution to the Russian economy in the future.

* Corresponding Author. E-mail: ityukhov@yahoo.com

1. General Estimation of Conditions of Renewable Energy Use in Russia

Russia has the largest land area and most varied geography of any country in the world. Russia is the second largest producer of electricity and the largest producer of natural gas of any country in the world. There are many opportunities for RE in Russia. We prefer to use the term *renewable energy* instead of *solar energy* because of the vastly differing Russian climatic conditions. For this reason, *renewable energy* is a more accurate general term, which naturally includes wind and biomass and also reflects a concept of sustainability. For example, the Russian climate ranges from the semi-arid grass-covered steppes in the South through humid continental in much of European Russia, to the sub-arctic in Siberia and to a tundra climate in the polar North.

Russia has an enormous and diverse RE resource potential. One estimate of RE potential suggests that it might be as high as 30% of the total primary energy supply (Bezrukikh et al., 2002). This estimation is based on the concept of gross, technical and economic potential. *Gross Potential* is the average annual energy total amount of different forms of RE available for extraction (or available resources). *Technical Potential* (part of the *Gross Potential*) takes into consideration environmental constraints which can be effectively used with the help of known technologies. The *Economic Potential* (part of the *Technical Potential*) views renewables as economically justified after considering current prices for fossil fuels, heat, electricity, equipment, materials, transportation and wages. We estimate the economic potential of RE in Russia at more than 2.7×10^8 metric tons of coal equivalent (mtce) ,where 1 mtce=29.31 GJ, without large hydro power stations and fire wood. The potential of different RE sources in Russia is presented in Table 1; regional distributions of different types RE resources are shown in Tables 2-6; and potential resources of energy saving from low potential heat are summarized in Table 7. Russia's total primary energy supply was 614 million tons of oil equivalent (1 Mtoe = 41.868 GJ) in 2000 (or 875 Mtce). Thus, while Russia's estimated economic potential was some 30% of total primary energy supply (TPES), only some 1% of TPES was actually derived from non-hydro RE in

Table 1. Potential of renewable energy resources in Russia (million tons of coal equivalent per year) (1mtce $= 29.31 \times 10^6$ GJ $= 29.31 \times 10^{18}$ J $= 8.1417$ TWh).

Resource	Gross potential	Technical potential	Economic potential
Small Hydropower	360	125	65
Geothermal Energy	*	*	115**
Biomass Energy	10×10^3	53	35
Wind Energy	26×10^6	2,000	10
Solar Energy	2.3×10^6	2,300	12.5
Low Potential Heat	525	115	36
Total Renewable Energy Sources	2.3×10^6	**4,583**	**270**

Table 2. Regional distribution of solar energy resources in Russia (million tons of coal equivalent per year) (1mtce = 29.31x10^{6}GJ = 29.31X10^{18}J = 8.1417 TWh).

Economic district	Gross potential	Technical potential	Economic potential
North	0.172 10^6	172	0.12
North-West	0.023 10^6	23	0.06
Central	0.060 10^6	60	0.62
Central-Chernozem (black earth zone)	0.022 10^6	22	0.25
Volgo-Vyatsky	0.033 10^6	33	0.18
Povolzhsky (Volga region)	0.090 10^6	90	1.30
North-Kavkaz	0.060 10^6	60	0.87
Ural	0.110 10^6	110	1,25
West-Siberia	0.320 10^6	320	1,90
East-Siberia	0.560 10^6	560	3,45
Far East	0.850 10^6	850	2,50
Russia (total)	**~2.3 10^6**	**2300**	**12.50**

Table 3. Regional distribution of wind energy resources in Russia (TWh/year).

Economic district	Gross potential	Technical potential	Economic potential
North	11,040	860	4.3
North-West	1,280	100	0.5
Central	2,560	200	1.0
Central-Chernozem (black earth zone)	2,080	160	0.8
Volgo-Vyatsky	1,040	80	0.4
Povolzhsky (Volga region)	4,160	325	1.6
North-Kavkaz	2,560	20	1.0
Ural	4,880	383	1.9
West-Siberia	12,880	100	5.0
East-Siberia	13,520	1,050	5.2
Far East	24,000	1,860	9.3
Russia (total)	**80,000**	**6,218**	**31**

Table 4. Regional distribution of small hydro energy resources in Russia (TWh/year).

Economic district	Gross potential	Technical potential	Economic potential
North-West	81.6	31.5	24.1
Central	8.2	3.0	2.0
Central-Chernozem (black earth zone)	1.5	0.6	0.3
Volgo-Vyatsky	3.4	1.3	0.9
Povolzhsky (Volga region)	21.5	10.4	5.5
North-Kavkaz	37.5	19.3	11.5
Ural	34.6	17.2	11.5
West-Siberia	74.6	24.6	2.5
East-Siberia	390.8	128.4	66.77
Far East	422.0	146.0	65.38
Russia (total)	**1,105.6**	**382.3**	**200.0**

Table 5. Regional distribution of biomass energy resources in Russia (million tons of coal equivalent per year) (1mtce = 29.31x10⁶ GJ = 29.31X10¹⁸ J = 8.1417 TWh).

Economic district	Gross potential	Technical potential	Economic potential
North-West	713	3.8	2.5
Central	1618	8.6	5.6
Central-Chernozem (black earth zone)	586	3.1	2.1
Volgo-Vyatsky	643	3.4	2.2
Povolzhsky (Volga region)	1241	6.6	4.3
North-Kavkaz	1260	6.8	4.4
Ural	1540	8.1	5.4
West-Siberia	1118	5.9	3,9
East-Siberia	687	3.6	2.4
Far East	594	3.1	2.2
Russia (total)	**10,000**	**53.0**	**35.0**

Table 6. Regional distribution of geothermal energy resources in Russia (million tons of coal equivalent per year) (1mtce = 29.31x10⁶ GJ = 29.31X10¹⁸ J = 8.1417 TWh).

Economic district	Gross potential	Technical potential	Economic potential
North-West	9.0×10^6	1.0×10^6	2.0
Central	0.5×10^6	0.06×10^6	0.5
Povolzhsky (Volga region)	4.5×10^6	0.5×10^6	1.0
North-Kavkaz	31.5×10^6	3.5×10^6	35.0
Ural	1.8×10^6	0.2×10^6	0.5
West-Siberia	49.5×10^6	5.5×10^6	35.0
East-Siberia	2.2×10^6	0.24×10^6	1.0
Far East	81.0×10^6	9.0×10^6	40.0
Russia (total)	**180.0×10^6**	**20.0×10^6**	**115.0**

Table 7. Resources of energy saving from low potential heat (million tons of coal equivalent per year) (1mtce = 29.31x10⁶ GJ = 29.31X10¹⁸ J = 8.1417 TWh).

Resources	Gross potential	Technical potential	Economic potential
Aeration stations	160	30	9
Systems of water rotation supply	200	40	12
Faintly mineralized waters	56	10	3
Non-freezing sources of water	100	20	6
Ventilating discharge of industrial buildings	9	5	1.5
Total	**525**	**105**	**31.5**

2000 (Bezrukikh et al., 2002). Only 1.5 Mtoe is actually used from this amount. Russia's use of RE catastrophically lags behind the available capacity of RE.

2. The Reasons Why Developed Countries are Actively Engaged in Using RE

There are several economic, social, and environmental reasons why developed countries are actively engaged in the use of RE:

- **Ensuring energy security.** The Western world's strong dependence on fossil fuels was highlighted by the fuel crisis of 1973. During 1973-1975 renewable energy development plans were established and accomplished with success. Now the urgency to use RE is again raised in connection with the occurrence of a new energy crisis (particularly, the increasing prices of petroleum and gas).

- **Ecology.** The necessity of decreasing greenhouse gas emissions from traditional power plants is being seriously considered by many countries. These concerns were clearly expressed in the Kyoto Protocol in which vast public funds were attracted and economic measures for attracting the private investment were developed. Russia is the third largest source of carbon-dioxide emissions word-wide, behind the USA and China, and Russia's international commitments to reduce greenhouse-gas emissions should also support renewable energy (Martinot, 1999).

- **The gain of the world markets especially for developing countries.** Production of renewable energy can provide economic development and employment opportunities. Renewable energy can thus reduce poverty in rural areas and reduce pressures for urban mitigation.

- **Preservation of stocks of energy resources for the future generations.** The moral aspect of this reason is clear. Also this issue is important for the concept of sustainable development.

- **Increasing availability of raw material for non-energy use of fuel.** The famous Russian chemist D.I. Mendeleev (inventor of the Periodic Table of the Elements) once said: "To burn oil for heating is the same as to burn banknotes in a heating furnace."

3. Why is it Necessary to be Engaged in RE Promotion in Russia?

All the economic, social, and environmental reasons that motivate developed countries to actively work in the RE field are applicable to Russia also. However, specific existing conditions of the Russian economy and society during this transition period must be considered. The main feature is that more active use and development of RE in Russia could be directed toward solving social problems, decreasing unemployment, development of small business, increasing standards of living, and increasing educational levels and the resulting culture. The needs that RE could fill are presented below in priority order according to economic and social criteria.

3.1 Providing an Energy Supply for Rural Areas Away from the Electrical Grid.

Around 6-8 million tons of liquid fuel (diesel fuel, black oil) and 20-25 million tons of solid fuel (coal) are delivered annually to Russia's Far North, Far East and Siberian areas. Because of increasing transport costs, the price of fuel has doubled to about $350 per ton of oil equivalent, in the republics of Tuva and Altai, and on the Kamchatka peninsula. More than half of these territories' budgets are spent for fuel delivery. The shortage of fuel frequently threatens the quality of life of the people and the state is compelled to solve the fuel delivery problem with the help of the Ministry of Extreme Situations. About 10 million people live in these areas.

3.2 Prevention of Restrictions and Shortages in Energy Supply for the Consumers Connected to Utility Networks.

Restrictions occur almost daily for the consumers of Federal wholesale energy market because of either non-payment for the electric power and fuel or the necessity to save fuel resources. In addition, breaks in the power supply due to emergency disconnections disrupt life in cities and regions and cause damage that is estimated in billions of dollars. The damages and long-term losses in agriculture and in the manufacturing industry caused by intermittent delivery of electric energy are 25-30 times more costly than the cost of energy itself. The creation of an open and controllable market of independent energy producers in these areas will avoid losses from unreliable delivery of energy and also lower the network losses. It is especially urgent to create generating capacities at the ends of local electric transmission lines with 6-10 kV voltage. Many consumers are connected to such lines which are subject to frequent emergency shutdowns. The breaks in power supply last many hours, which intensify the damage to the consumers. This damage is not compensated by the energy supplying organizations.

3.3 Development of Regional and Local Industry.

Development of the RE industry in Russia will result in a steady increase in the export of machines and equipment for RE, and creation of additional work opportunities for the great scientific and technical talent available in Russia. Already such opportunities exist in the trade of some kinds of RE equipment (small wind systems of capacity up to 1 kW, small and micro hydroelectric power stations, individual biogas installations, and photovoltaic systems) with developing and European countries (Table 8).

3.4 Reduction of Ecological Damage in a Number of Cities and Recreational Zones Due to Harmful Emissions from Power Installations.

Ecological conditions will become worse in the case of transition from gas to coal. The proven tools, which do not allow deterioration of environment, are application of thermal pumps, solar installations at the boiler-houses, solar collectors, wind energy systems, small and micro hydroelectric power stations.

3.5 Ensuring Energy Supplies for Remote Regions of Russia, such as Kamchatka, Chukotka, Far East Littoral, and Arkhangelsk.

North American and European countries have proven the competitiveness of RE systems in comparison to traditional energy systems in remote isolated areas. More than twenty million people in Russia live in such areas, without a centralized energy system or with a non-reliable one. Some autonomous systems work only a few hours a day because of a lack of fuel supply and the very high cost of fuel. The poor situation of power supply in these regions does not require any further comments. Potential consumers of renewable sources of energy in Russia are shown in Table 9.

4. Opportunities in Russia

4.1 Russia Has Enormous and Diverse Renewable Energy Resources.

The feasible economic potential of RE is equal to 270 Mtoe, with more than 25% from internal energy consumption. Tables 2-6 show data on the distribution of these resources in all federal districts. In each of the federal districts there are at least two or three kinds of RE with potential for use, therefore it is necessary to develop all kinds of RE in Russia. Russia receives almost none of its energy supply from RE even though there is some contribution from all types of RE. Please see Tables 10-15 for actual electric and thermal energy production based on RE in Russia.

4.2 Development and Industrial Base.

There are developments of all kinds of RE equipment internationally, except for large wind systems (capacity of 100 kW and higher). In Russia, there is a large industrial base supporting the defense industry, which can be used more actively for con-

Table 8. Production of RE equipment in Russia.

	Producer	Type of Equipment	Capacity, productivity
1.	GMKB "Raduga" Dubna,	Windmill	8 kW
2.	GNCRR-CNII "Electropribor" Saint Petersburg	Windmills	
		UVE-40	40 W
		UVE-500	500W
		UVE-300	300W
3.	Rybinsk Instrument-making Plant Rybinsk	Windmills	
		VETEN-0,16	160 W
		VTN-8	8 kW
4.	NPK "Vetrotok" Ekaterinoburg	Windmills	
		VEU-16	16 kW
		VEU-5-4	4,2 kW
		Electric boiler	
		EK-30/40	28 kW
5.	ZAO "Elmotron" Novosibirsk	Windmills	
		VES-1	1 kW
		VES-2	2 kW
6.	AO "Dolina" Kuvandyk Orenburgskaya region	Windmills	
		VEU-2	2 kW
		VEU-5	5 kW
7.	Institute "VNIPTIPMESH" Zernograd Rostov region	Windmills	
		UVGE-500	500 W
8.	ZAO "OKB Krasnoe Znamya" Ryazan	Solar PV panels for radioreceivers	
		Solar-1-9	9 V, 1 W
		Solar-1-6	6 V, 1 W
		Solar-2-9	9 V, 2 W
		Solar PV modules	
		FSM-30	12 V, 40 W
		FSM-40	12 V, 50W
		FSM-50	
		Solar battery	
		Solar 30	12 V, 30 W
		Solar 60	12 V, 60 W
		Autonomous solar water pump system	
			12 V, 120 W
9.	VIESH Moscow	Solar PV modules	
		FSM-30-12	12 V, 30 W
10.	AO"VIEN" Moscow	Autonomous PV solar system	
		ASEP-12/20K	12 V, 60 W
11.	NPO Mashinostroeniya Reutov Moscow region	Solar systems	
		SEU-80	24V, 80 W
		SEU-160	24V, 160 W
		SEU-240	24 V, 240 W
		SEU-320	24V, 320 W
12.	OOO "Sovlux" Moscow	Solar portable folding PV panels	
		P-6.5	19 V, 6.5 W
		P-13	19 V, 13 W
		P-26	19 V, 26 W
		Rigid frame modules	
		R-12.5-20	17.5 V, 12.5 W
		R-20	
			17.5 V, 20 W
		Roof solar modules	
		PM-20	14.9 V, 20 W
		PM-80	19.6 V, 80W

Table 8. Production of RE equipment in Russia (continued from previous page).

	Producer	Type of Equipment	Capacity, productivity
13.	NPO "Kvant" Moscow	Solar modules	
		BSR-10	20 V, 11 W
		BSR-20	20 V, 22 W
		BSR-30	20 V, 33 W
		BSR-40	20 V, 44 W
		BSR-50	20 V, 50 W
		BSR-80	20 V, 66 W
		BSR-100	20 V, 88 W
			20 V, 110 W
		Rigid frame modules	
		R-12.5-20	17.5 V, 12.5 W
		R-20	
			17.5 V, 20 W
		Solar switchable module	
		BSP-10	4.5-6-9-10.5-12 V
			10-10-6-7.5-10.5 W
		Solar PV system of electrical energy supply	
		SEFU-250	16.5 V, 200W
		Solar lantern	
		SF-1	12.5 V, 7.5 W
14.	OAO "Satunr" Moscow	Solar modules	
		BS-25/25	16.1 V, 22 W
		BS-50/55	16.7 V, 55 W
		Unified PV system	
			12 -24-36-48-60-110 V
		FES-U	0.06-10-kW
		PV systems	
		FES-0.01/12	12V, 10 W
		FES-0.1/12-10	12 V, 200W
		FES-0.2/24-10	24 V, 200 W
		FES-0.5/24-270	24 V, 200 W
15.	ZAO"Telecom-STV" Zelenograd Moscow region	Solar PV modules	
		TSM-5	16.2 V, 5-7 W
		TSM-10	16.2 V, 10-120 W
		TSM-22	16.2 V, 20-22-25 W
		TSM-30	16.2 V, 30-33-35 W
		TSM-40	16.2 V, 40-45 W
		TSM-50	16.2 V, 50-53 W
		BS-12-7	14.8 V, 6 W
16.	AOOT "Pozit" Pravda Moscow region	Solar PV modules	
			16 V, 10-9-8-5-4.5 W
		Solar PV modules	
		BS-0.5-9P	9 V, 0.5 W
		BS-0.5-6P	6 V, 0.5 W
17.	NPP "Kvant" Moscow	Thermoelectric generators	
		AIP-150	27.5 V, 150 W
		AIP-450	27.5 V, 450 W
			27.5 V, 750 W
18.	OAO "Kovrov Mechanical Plant" Kovrov Vladimirovskaya region	Solar collector	1 m^2, 100 liters per day
19.	NPP "Konkurent" Zhukovski Moscow region	Solar collector	1 m^2
20.	NPP "Vetrotok" Ekaterinoburg	Solar collector	1 m^2, 80 liters per day

Table 8. Production of RE equipment in Russia (continued from previous page).

	Producer	Type of Equipment	Capacity, productivity
21.	NPO Mashinostroeniya Reutov Moscow region	Solar collector	0.9 m²
22.	MNTO INSEP Saint Petersburg	Micro-hydro plants	
		10 Pr	10 kW
		7.5 Pr	7.5 kW
		50 Pr	50 kW
		22 Pr	22 kW
		90 Pr	90 kW
		50D	50 kW
		25 D	25 kW
		100D	100 kW
23.	MNTO INSEP Saint Petersburg	Small hydro plants	
		GA-1	100-500 kW
		GA-8	150-1500 kW
		GA-5	145-620 kW
		GA-10	290-3000 kW
		GA-2	1000 kW
		GA-12	220-9800 kW
24.	"MAGI" Moscow	Small hydro plants	
		FG	100-250-300-400 kW
		T1,T2	50-90-100-200-3000 kW
25.	Center "Ecoros" Moscow	Individual biogas plants	
		IBGU-1	Bioreactor, 2.2 m³
		BIOEN-1	Bioreactor, 5 m³ x 2
26.	VIESH Moscow	Biogas plants	
		BGU-2	Bioreactor, 2 m3
		BGU-25	Bioreactor, 25 m3
		BGU-50	Bioreactor, 50 m3 Bioreactor, 150 m3
		BGU-150	Bioreactor, 4 x 50 m3
		BGU-500	
27.	TOO Energotechnologiya	Gasgenerator	
		UTG-600	600 kW
28.	AOZT	Heat pumps	
		NT-1000	1000 kW
29.	AOZT "Komressor"	Heat pumps	
		NT 280-4-9-08	370 kW
		NT 410-4-9-08	520 kW
30.	AOZT	Heat pumps	
		TNU-10	10.3 kW
		TNU-12	12 kW
		TNU-15	14 kW

Table 9. Potential consumers of renewable sources of energy in Russia. The number of settlements is shown in the cells above the line; the total number of inhabitants is shown in the cells under the line.

Number of people living in settlements	Decentralised zone				Centralised zone		
	Rural population (est. of 1995)	Camps in (est.of 1994)	Population of small North towns	Hunters, fisherman, nomads, geologists, builders of utilities and pipelines	Dachas (summer country houses) and agricultural associations	City consumers of heat (showers, solar heating systems)	Rural inhabitants in unstable energy supply zones
Up to 50	2,940 / 6,780	256 / 658,000	-	8,000 / 52,000	400,000 / 2,000,000	40,000 / 1,000,000	4,000 / 120,000
51-500	10,387 / 2,294,805	520 / 131,600	-	160 / 8,000	80,000 / 2,000,000	80,000 / 2,000,000	4,800 / 1,200,000
501-3000	5,615 / 5,868,160	91 / 65,800	-	-	1,000 / 1,000,000	1,000 / 1,000,000	3,600 / 3,600,000
3001-10,000	388 / 2,082,839	43 / 131,600	146 / 353,000	-	-	250 / 1,000,000	216 / 1,080,000
Total	10,313,600	394,800	353,000	60,000	5,000,000	5,000,000	6,000,000
Total	**11,121,400**				**16,000,000**		

Table 10. Wind mills parameters in Russia in 2000 and 2001.

Specification figures	VESAO "Yantar-energo," Kaliningrad region	VESAO "Kamchatskenergo," Bering island, Nikolskoe vilidge	VESAO "Kalmenergo," city Elista	VESAO "Rostovenergo," VES-300	VES AEK "Komienergo," "Zapolyarnoe"	VESAO "Bashkirenergo," "Tyupkeldy"	Total
2000							
Installed capacity, MW	1.5	0.5	1.0	0.3	1.5*	2.2	7.0
Number and capacity wind mills, kW; company producer	4x225 1x600 kW Vestas Denmark	2x250 "Micon" Denmark	1x1000 MKB "Raduga"	10x30 HSW-30 Germany	6x250 UVE-250 Russia-Ukraine	4x550 HAG Germany	
Produced electrical energy, GWh	0.987	0.827	0.1	0.037	0.142512	0.824	2.917
Produced electrical energy, GWh	0.979	0.826	0.1	0.037	0.107256	0.819	2.868
Auxiliary expenses, MWh / %	8/0.8	1/0.1	-	-	35.2/24.7	5/0.6	49/1.7
Used installed capacity, numbers of hours / %	658/7.5	1,654/18.9	100/1.1	1,233/14	114/1.3	374/4.2	416/4.7
2001							
Installed capacity, MW	1.5	0.5	1.0	0.3	1.5	2.2	7.0
Number and capacity of wind systems, kW; company producer	4x225 1x600 Vestas Denmark	2x250 "Micon" Denmark	1x1000 MKБ "Радура"	10x30 HSW-30 Germany	6x250 UVE-250 Russia-Ukraine	4x550 HAG Germany	
Produced electrical energy, GWh	2.167	0.729	0.185	0.037	0.211929	0.790	4.12
Supplied electrical energy, GWh	2.157	0.728	0.175	0.037	0.173045	0.785	4.056
Auxiliary expenses, MWh / %	11/0.5	1/0.1	10/5.4	-	38.9/18.3	5/0.6	64/1.5
Used installed capacity, numbers of hours / %	1,445/16.5	1,458/16.7	257/2.9	1,233/14	169/1.9	359/4.1	588/6.7

*One wind system does not work.

Table 11. Specification figures of geothermal power plants in Russia in 2000 and 2001.

Figure	Pauzhetskaya	Verkhne-Mutnovskaya	Total
2000			
Installed capacity, MW	11	12	23
Available power, MW	6	8	14
Produced electrical energy, GWh	32,755	25,444	58,199
Supplied electrical energy, GWh	30,766	22,607	53,373
Auxiliary expenses, MWh/ %	1,989/6.1	2,837/11.1	4,826/8.3
Used installed capacity, numbers of hours / %	2,978/34	2,120/24.2	2,530/28.9
Used installed capacity, numbers of hours / %	5,459/62.3	3,180/36.3	4,157/47.4
2001			
Installed capacity. MW	8.5	12	20.5
Available capacity, MW	7	10	17
Produced electrical energy, GWh	33,522	57,674	91.196
Supplied electrical energy, GWh	31,431	53,477	84.908
Auxiliary expenses, MWh/ %	2,078/6.2	4,210/7.3	6.288/6.9
Used installed capacity, numbers of hours / %	3,047/35	4,806/55	4,448/50.8
Used installed available capacity, numbers of hours / %	4,789/54.7	57,68/65.8	5,364/61.2

Table 12. Specification figures of small hydro power plants in Russia in 2000 and 2001.

Specification figures	2000	2001
Small hydro power plants, which belong to AO"Energo"		
Number of power plants	41	41
Total installed capacity, MW	395,580	474,814
Total available capacity, MW		343,310
Produced electrical energy, TWh	1,832,859	1,918,473
Used installed capacity, numbers of hours / %	4,633/52.9	4,040/46.1
Used installed available capacity, numbers of hours / %		5,588/63.8
Auxiliary expenses, MWh/ %		28,188/1.5
Small hydro power plants, which do not belong to AO "Energo"		
Number of power plants	18	18
Total installed capacity, MW	117,375	135,705
Total available capacity, MW		98,359
Produced electrical energy, TWh	0.468312	0.448690
Used installed capacity, numbers of hours / %	3,990/45.5	3,315/37.8
Used installed available capacity, numbers of hours / %		4,560/52.1
Auxiliary expenses, GWh/ %		6,784/1.5
Total		
Number of power plants	59	59
Total installed capacity, MW	512,955	610,519
Total available capacity, MW		441,669
Produced electrical energy, TWh	2,301,171	2,367,163
Used installed capacity, numbers of hours / %	4,486/51.2	3,877/44.3
Used installed available capacity, numbers of hours / %		5,359/60.0
Auxiliary expenses, GWh/ %		34,972/1.5

Commentary: Data of Russian Goskomstat.

Table 13. Specification figures of small thermal power plants using biomass as a fuel in Russia in 2000 and 2001.

Specification figures	2000	2001
Number of power plants	27	27
Total installed capacity, MW	1245.23	1394.5
Produced electrical energy, TWh	4.490202	5.541604
Used installed available capacity, numbers of hours / %	3606/41.2	3974/45.4
Supplied thermal energy, GJ	88.20606×10^6	101.30870×10^6
Total spent fuel, thousand toe	5574.744	6259.843
Including biomass thousand toe / % from total	2353.087/0.422	2515.52/0.402
Produced electrical energy from biomass, TWh	1.895304	2.226895
Equivalent installed electrical capacity from biomass, MW	525.6	560.4
Number of power plants	18	18
Supplied thermal energy, GJ	37.2318×10^6	40.711×10^6

Commentary:
1. At thermal power plants biomass (waste of remaking, forestry, pulp and paper, wood industry) is used in combination with organic fuel (mazut, coal, gas).
2. Date was taken from reports of technical and economic assessment and equivalent fuel consumption (according to year reports of turbine electrical power plants of capacity 500 kW and over, which do not belong to AO "Energo"; Russian Goskomstat 2001, 2002).

Table 14. Production of electrical energy in Russia based on renewable sources of energy including small hydro power plants, GWh.

№	Power plants	2000	2001
1.	Windmills	2.917	4.12
2.	Geothermal electrical power plants	58.199	91.196
3.	Small hydro power plants: general using	1832.859	1918.473
	Which do not belong to AO "Energo"	468.312	448.690
	Total hydro	2301.171	2371.163
4.	Thermal electrical power plants on biomass	1895.304	2226.895
	TOTAL:	**4256.588**	**4693.374**
	Production of electrical energy in Russia	876×10^3	888.4×10^3
	Share of renewables, %	0.5	0.53

Commentary: Date was taken from reports of technical and economic assessment and equivalent fuel consumption at Russian power plants (according to year reports of 2000, 2001; Russian Goskomstat 2001, 2002).

Table 15. Supplied thermal energy based on RE in Russia, million GJ.

№	Type	2000	2001	Commentary
1.	Thermal power plants based on biomass	37.262	40.696	Statistical reports
2.	Small boilers on biomass	188.41	192.59	Author estimation
3.	Solar collectors	0.1256	0.1298	Expert estimation
4.	Thermal pumps	2.01	2.05	Expert estimation
5.	Combustion plants and disposal plants	1.256	1.256	Expert estimation
6.	Biogas systems, aeration stations	8.374	8.374	Expert estimation
7.	Geothermal systems of heat supply	2.093	0.921	Expert estimation
8.	**Total:**	**239.49**	**245.93**	
9.	Supplied thermal energy. Total in Russia.	5947	6029	Statistical reports

sumer needs. The detailed data on the developed and produced equipment are given in a published catalogue:

> At the same time it is necessary to note, that for last years there was a collapse of some centres of science, existing in USSR, and branches of central enterprises and institutions (for solar field tests in Gelendgik and Crimea, Ashkhabad, Yerevan, NPO "Vetroen" etc.). Therefore, it is necessary to begin creation of centres and testing areas for renewable energy field tests in Russia (Ministry of Fuel and Energy, 2000).

A list of RE equipment (with technical characteristics) which was developed in Russia can be found at the Ministry of Industry and Energy of Russian Federation website in the section on Non-traditional energy (http://www.mpe.gov.ru/, http://www.mte.gov.ru/docs/33/113.html, http://www.mte.gov.ru/docs/77/624.html).

4.3 The Availability of Trained Workforce

For many years specialists in RE at different levels (engineers, bachelor's degree holders, and technicians) have been taught and trained at the Moscow Power Engineering Institute (Technical University), Bauman Technical University, Lomonosov Moscow State University, Moscow State University of Environmental Engineering, Saint Petersburg State Technical University, Institutions of Ekaterinburg, Novosibirsk, Khabarovsk, and etc. Academic councils continuously award candidate and doctor's degrees in RE. However, there remains a lack of technicians and workers for practical services and maintenance of RE installations.

5. Renewable Energy Economics

The capital investment required for renewable energy equipment is at about the level of traditional power engineering equipment or a little bit above. The domestic equipment is cheaper than imported equipment by 30-50% or more.

For example, the cost of 1 kW of capacity for small hydro power stations in Russia is $1,000-1,200, and the cost for a micro hydroelectric power station working as a stand-alone unit is about $600-700, whereas in Europe the corresponding cost is $1,500-1,800/kW.

The specific capital investment for photovoltaic installations exceeds that of traditional installations by five times or more. However, there is a steady trend in the growth of specific capital investment in the traditional power equipment and a decrease in the equipment cost of RE.

For example, in Russia, the specific capital investment for wind energy installation has decreased from $4,000 per kW in 1980 to $900 per kW in 1999. Specific cost of photovoltaic modules during this period decreased from $50,000 per kW to $4,000-

5,000 per kW. For the same period the specific capital cost of thermal power stations has increased from $750/kW to $1,000-1,100/kW, and atomic power station costs have increased from $1,500/kW to $2,200/kW.

5.1 Cost of Electric Energy

The cost of electric energy from many kinds of RE power stations is now at the same level as that of traditional power. Photovoltaics is an exception; the cost of electric power is 4-5 times higher. However, the steady trend of decrease in the cost of RE (including solar cells) over the next 5-10 years will bring these costs nearer to the prices of other kinds of energy.

At present, the price for 1 kWh of electric power from RE in Russia is: 3-4 cents for micro and small hydroelectric power stations, 4-5 cents for wind mills, 5-6 cents for geothermal stations, and 6-7 cents for power stations using wood waste.

From traditional power stations: the price for 1 kWh of electric power equals 5.2-8 cents for coal power stations, 5-6.5 cents for gas power stations, and 4-8 cents for atomic power stations.

In Denmark, for example, the price of the electric power from wind mills has decreased from $0.168/kWh in 1980 to $0.045/kWh in 1998, and from coal power stations the price has increased from $0.058/kWh to $0.064/kWh. The price of electric energy in Russia in centralized utilities is 1.5-2 cents/kWh, and 4-30 cents/kWh or more in autonomous power systems.

5.2 Capital Investment and Payback Period

The simple payback time of capital investment in power engineering is 8-10 years on average. Moreover, a thermal power station is under construction for 6-8 years, and a large hydro power station is under construction for 10-12 years. According to the results of our calculations, the payback time of the various RE projects in Russia is 3-15 years. A wind installation of 50 MW abroad is under construction for about 5-6 months beginning from the signing of the contract, and pays for itself in 8-10 years.

In accordance with calculations, the simple payback time of capital investment of the program "Development of non-conventional power of Russia for 2001-2005 years," which will result in the replacement of 2.2 Mtoe fuel is 5 years. The calculation of various combinations of the factors, which have an influence on renewable energy payback times, shows the following results.

In centralized utilities an acceptable payback time (5-10 years) results in a specific capital investment of $1,500/kW or less and the number of hours of the established capacity use are 2,200 or more per year. For autonomous power systems of these sizes the corresponding numbers are $2,000/kW and 1,500 hours.

6. Priority Measures Offered for Promotion of RE use in Russia

The state law "On the Role of the State in Using Renewable Energy" was adopted by the State Duma (Parliament) in 1999, which was later blocked by President B. Yeltsin. The current goal is to create broad legislative support for the Russian RE sector.

Among the priority legislative measures are:

*   Development of the Federal law "About non-conventional renewable sources of energy" and other normative acts promoting the creation of independent manufacturers of energy systems based on non-conventional renewable energy resources.

*   Realization of the subprogram "Energy supply for regions, including northern and territories, on the basis of use of non-conventional renewable sources of energy and local fuels," Federal Target Program "Energy efficient economy" and financing from the federal budget for 2004 at least 200 million roubles for realization of the priority projects (Figures 1-4).

*   Approval of a resolution by the Government of Russian Federation for promoting alternative RE fuels and RE systems for State buildings.

*   Assignment of a federal body responsible for development of RE branch of power engineering.

The important figures are shown in Tables 16-17 and Figure 1.

7. Russian Renewable Energy Technologies (Short Review)

The renewable sources of energy will displace coal, petroleum, gas and uranium in the production of electric power, heat and liquid fuels (Strebkov, 1997). The renewable energy resources are huge and are accessible to each country. The amount of solar energy falling on the territory of Russia for a week exceeds the energy content of all Russian stocks of petroleum, gas, coal and uranium (Strebkov, 1994). Electric power and liquid fuel are of prime significance to the life of mankind, as these two sources of energy represent a Helmholz free energy (Alekseev, 1995).

Mankind could avoid an energy crisis due to the exhaustion of petroleum, gas, and coal, if it uses renewable energy technologies. Use of renewable energy sources would also solve the problems of pollution from power stations and transport, as well as providing healthy food, good education, qualified medical aid, and an improved quality of life.

If renewable energy sources are going to provide for the world's energy needs the present high cost of renewable technology must be reduced. Two main paths to

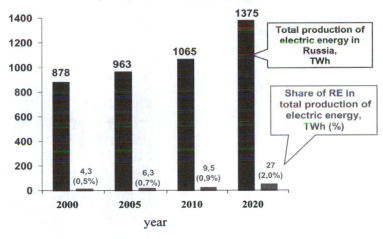

Figure 1. Forecast of using RE in Russia for production of electricity taking into account small hydro power stations up to 30MW.

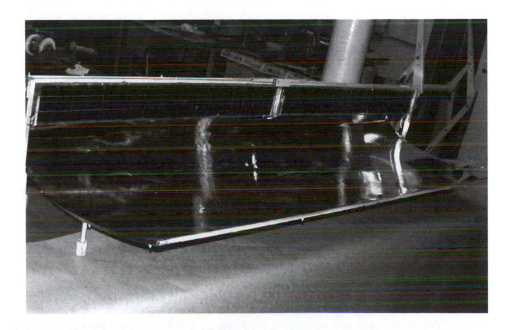

Figure 2. Asymmetric PV concentrator module with aperture angle 36° for 9 months of non-tracking operation. The size is 2,500 x 700 mm. Electric Power is 70 Wp.

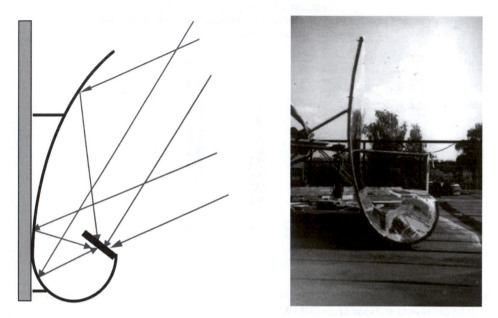

Figure 3. Ray tracing and photo of vertical asymmetrical concentrator.

Figure 4. PV stationary concentrator module. Dimensions: 2,500x350 mm. Electric Power is 50 Wp.

Table 16. Investment needed for the development of RE in Russia*.

	Sum total	From federal budget
2003-2005 years	10.9 billion roubles	1.22 billion roubles
2006-2010 years	45.7 billion roubles	3.2 billion roubles
2011-2020 years	To keep not less than 10 billion roubles	
* Period 2004–2010: investment, predicted on activities of federal target programme «Energy efficient economy»		

Table 17. Share of RE of G8 countries in 1999 (excepting hydro power stations).

	Billion kW hours	% from total production
Canada	7,340	1.24
France	3,500	0.65
Germany	14,820	2.7
Italy	7,290	2.8
Japan	19,740	1.83
Great Britain	8,630	2.3
USA	86,250	2.2
Russia	2,100	0.33

biomass cost reduction are fast pyrolysis liquid fuel technique and cogeneration using gas-steam turbine cycle.

As the cost of fuel in Russia approaches the rising global prices, technologies for producing gas and motor fuel from biomass are becoming economically viable. It has been demonstrated experimentally that farmers who sow rapeseed (Brassica napus) on 10 to 15% of their land do not depend on external sources for motor fuel. In areas rich in peat and wood, ethanol and methanol production would enable the use of gas and synthetic fuels in diesel electric generators and cars.

Russia has 20% of world wood resources. Using the fast pyrolysis technique it is possible to receive 700 kg of liquid fuel from one ton of sawdust (Vainstein and Ksenevitch, 2000; Strebkov, 1998). Connecting this technology with the technology of fast growing energy plantations with productivites of 40 tons of dry mass per hectare, it is possible to decentralize the production of liquid and gaseous fuels in former forestry free regions and to supply each farmer and rural area with motor fuel and electricity. The liquid fuel production requires less than 5% of the available agricultural land. This technology could already be cost effective in countries importing petroleum, for example, in the Czech Republic, where the cost of gasoline is more than $1 US dollar per liter today.

The utilization of biomass results in carbon neutral emission, since photosynthesis removes an equivalent amount of carbon dioxide from the atmosphere. In comparison with coal, biomass does not give off the emission of sulfur, and with the optimum incineration technology it is possible to reduce the emission of nitric oxides.

In Russia, power stations operating on biomass can provide more energy than all of the petroleum stocks of the Komi Republic or all of the nuclear power stations. This could be achieved by using biomass gasification and steam-gaseous combined cycle technologies with efficiencies of 50% instead of direct incineration with efficiencies of 25% (Strebkov, 1998).

In view of such factors as environmental safety, the enormous amount of solar energy available, and 50 years of experience of photovoltaic technologies, it is becoming more and more obvious that solar photovoltaic stations will play a strategic role in future energy production in the world (Strebkov and Koshkin, 1996).

Solar power plants can be used to solve both local and global power production problems. At 15% efficiency, Russia's entire energy needs could be met by solar power plants with an aggregate active area of 4,000 square kilometers (0.024 % of the nation's territory).

Large contributions to the development of photovoltaic energy were made by the following Russian scientists: N.S. Lidorenko, Zh.I. Alferov, V.S. Vavilov, A.P. Landsman, A.M. Vasil'ev, V.M. Evdokimov, V.M. Andreev, M.V. Kagan, M.M. Koltun.

At the present time Russia has a dynamically developing photovoltaic industry and highly skilled staff of scientists and technologists. Key events in Russian photovoltaic technology development are listed in Table 18. Space programs stimulated the development of photovoltaic technology for terrestrial applications. Physical principles for the conversion of solar radiation were worked out and used in developing theoretical models and new designs in the following types of solar cells: cells based on heterostructures; cells with varizone structures having an ultimate efficiency of 93%; silicon cells of a n-p+-p+ structure having a theoretical efficiency of 44% and multi-function high-voltage solar cells. A maximum efficiency of 30% was measured in cascade solar cells based on heterostructures that were developed at St. Petersburg Ioffe Physicotechnical Institute (Strebkov and Koshkin, 1996.).

In an experiment that involved the conversion of concentrated radiation (laser beam), the ultimate electrical capacity of the high-voltage multijunction solar cell (with vertical p-n-junctions) was 3.6 kW/cm^2, the voltage density was 100 V/cm^2, the maximal voltage of the solar array was 32 kV, and the spectral sensitivity at wave length of 1,060 nm was 0.5 Amps/Watts.

Solar power installations are currently sufficient to provide lighting, pump water, and power telecommunications facilities and home appliances in remote areas and in vehicles. The production cost of solar cells is 1.8-2.1 US$/Wp. Modules and solar systems cost 2.5-2.9 US$/Wp and 7-9 US$/Wp respectively (Table 19).

The second phase of mass production and utilization of solar plants in the electric grid is contingent upon the development of technology and materials capable of reducing prices fivefold, to US$ 1-2/Wp, and the cost of electric power to 0.10-0.12 US$/kWh. A basic limitation on cost reduction is the high cost of solar-grade silicon (40-100 US$/kg). Since silicon is the second most abundant element in the earth's crust, sili-

Table 18. Development timeline of photovoltaics in Russia.

Year	Key events
1958	First satellite "Sputnik-3" with solar arrays was launched
1964	0.25 kW solar PV concentrator plant for water pumping was tested in Kara-Kum desert near Ashkhabad (Turkmenistan)
1967	New class of silicon multijunction (with vertical p-n-junctions) and high voltage cell technology was developed
1970	Ion implantation technology was applied to a solar cell manufacturing line
1970	Bifacial silicon solar cell technology was developed and applied to space solar cell production
1975	Testing of 32 kV, 1 m^2 solar arrays for ion plasma space engine
1975	Development of GaAlAs-GaAs solar cell technologies. In 1981 this technology was applied in lunar space program
1980	Multijunction solar cell technology using GaAlAs-GaAs was developed
1981	New types of compound, holographic, and prism concentrators were developed
1983	1 kW PV plant with glass parabolic mirrors and heat tube cooling system was installed in Tashkent (Uzbekistan)
1984	10 kW PV plant with plastic parabolic trough for linear concentration was installed in Ashkhabad (Turkmenistan)
1985	Efficiencies up to 36 % with electric power density of 3,6 kW/cm^2 was demonstrated for laser PV energy conversion
1987	The technology for purification of metallurgical-grade silicon for solar cell applications was developed
1989	500-W solar power plant with glass compound concentrator was developed and 40-kW solar PV village in Krasnodar region was installed
1989	Special technology in production of solar cells for terrestrial applications was initiated
1993	30 % efficiency of cascade GaAlAs-GaGeAs heterojunction with concentrated solar radiation was demonstrated
1996	The technology of 100-square shape bifacial solar cells was developed
1997	1 kW, 10 kW single-wire electric power system for renewable-based electric grid was developed
1998	Chlorine-free solar grade poly silicon technology was developed
2000	Stationary concentrator PV module with ratio concentration 5-30 was developed

Table 19. Production cost structure and annual volume of solar cells and modules production.

Technology Process	Contemporary state		Nearest goal		Long-term goal	
	Production cost, US$ per Wp	Volume of production MWp/year	Production cost, US$ per Wp	Volume of production GWp/year	Production cost, US$ per Wp	Volume production GWp/year
Silicon wafer	1.3-1.5		1.0		0.4	
Solar cell	0.5-0.6		0.4		0.3	
Solar module	0.7-0.8	300	0.6	30	0.5	300
Total module production cost	2.5-2.9		2.0		1.2	

con-based solar power is expected to keep a dominant role in the photovoltaic industry.

An innovative solution in photovoltaics is expected to be low-cost solar grade silicon technology as a long-term goal and with solar concentrator technology as a short-term goal to achieve one GW production volume in two years.

New technology allowing mass production of solar-quality silicon feedstock by a method of direct carbothermic reduction from naturally pure quartzite will provide from 80 to 85% of the necessary silicon. It will cost US$15 per kg, and will reduce the cost of solar electric power. Russian quartzite is among the purest in the world; their available deposits are sufficient for solar photovoltaic plants with a capacity of over 1,000 GW.

Russian scientists have developed a new chlorine-free technology for manufacturing silicon for electric power production, in which one 1 kg of silicon in solar module is equivalent to 75 tons of petroleum in terms of generated electricity during 50 years of operation (Strebkov et al., 2001). For this reason, silicon, frequently referred to as the oil ("black gold") of the 21st century, would be a profitable investment. In the present market, silicon costs twice as much as uranium even though the contents of silicon in the earth's crust exceeds that of uranium by 100,000 times. The new ecologically benign technologies will allow covering roofs and facades of houses with cheap silicon solar cells and to receive electrical energy without lengthy transmission lines the same way as the mobile telephone communications provide service without wires. New classes of holographic, prismatic, and parabolic concentrators have been developed and optical schemes have been designed for the following systems: solar stations with space-based heliostats, focusing heliostats and stationary concentrators and solar stations of the light-guide type with a sliding beam.

Two new solar concentrator technologies recently developed in Russia include quasi-stationary systems and stationary concentrator systems. Quasi-stationary solar concentrator systems require once a day tracking the sun for all modular concentrators of the system. Concentrated solar radiation with concentration ratios of 10 to 70 from each concentrator is collected and directed through a special light- guide optical split to the central receiver. The general idea of this system using modular concentrators is to collect concentrated solar radiation from each module, but without high temperature fluid in the conventional trough solar thermal power system. The area of this concentrator system is can vary from 1-106 m2 and has a capacity of 1 kW or higher.

Bifacial solar cells and modules were initially developed in Russia in 1970 for low orbit satellites to use the direct solar radiation as well as radiation reflected from the earth. Bifacial solar cells have a lower equilibrium temperature (and therefore higher efficiency) because of their transparency for non-active infrared radiation. Some Russian factories have been manufacturing bifacial solar modules for terrestrial applications since 1993.

Using a bifacial solar module as a receiver, we designed and tested stationary solar concentrator modules with concentration ratios of 3 and higher. Such stationary solar concentrator systems can also be used for cogeneration, heating, drying, supplying hot water and for cooking and daylighting illumination of buildings.

Our estimation shows that PV module prices may be reduced to US$ 2-3/Wp from their present value of US$ 4-5/Wp. Constructing such modules as a result of concerted international efforts will substantially reduce the cost of electricity and energy pay-back time for solar PV stations. Russian manufacturing facilities possess the technology and production capacity to manufacture 10 MW of solar cells and modules annually.

A simple and highly efficient technique for producing bifacial solar cells and modules has been worked out at the All-Russia Research Institute for the Electrification of Agriculture. If several Russian plants were to specialize in the production of solar modules, output could rise to over 200 MW by 2005 and to 2,000 MW per annum by the year 2010. To make it happen, however, state and private investment for the support of new energy technologies, especially for solar silicon production technology, must be forthcoming.

The most favorable areas of the Russian Federation for installation of solar power plants producing 200 kWh/m^2 annually, are: Astrakhan and Volgograd regions, Kalmykia, Buryatia, Daghestan, Tuva, Chita Region, the coasts of the Caspian and the Black Seas and the Sea of Japan.

Solar energy production costs are independent of the size or capacity of the generating unit, so that it is often advisable to build a modular system, placing units on the roofs of rural houses or farm buildings. Owners of solar power units can sell excess energy to an electric grid during the day and purchase energy from a utility company at night, using a reversible electric meter. Apart from encouraging small-scale independent energy production, this technique has the advantages of saving on support structures and land, and combining the functions of roof and power source. A modular solar power station with a one million kW capacity can provide electricity for 500,000 rural dwellings.

In view of the high reliability of silicon and solar cells, solar power plants may operate for as long as 50 years or more. However, solar modules using EVA (Ethylene Vinyl Acetate) and other plastics for encapsulation have an operation time of about 25 years. It is essential to increase the operation time by excluding plastics from module structure. We have developed new plastics-free solar module technology with an expected operation time of 50 years. The new technique uses two layers of glass with soldered edges and a special liquid inside for solar cell-glass optical and thermal matching (Strebkov et al., 1999c). The only limitation of the operation time is the necessity to replace old solar cells with more efficient ones. Efficiencies of 25% to 30% are expected to be achieved by the industry in the next 10 to 20 years.

The new technologies of solar energy conversion will be utilized first of all, for off-grid power supply in rural areas. The problems of nighttime and seasonal storage may be solved through utilization of solar-hydrogen systems as well as latitudinal distribution of solar power plants, new storage technologies and low-loss electric power transmission line systems.

It is known that 90% of children are being born in developing countries, where 2 billion people have no access to electric power. At the same time, less developed countries, in contrast to the countries of Europe and Northern America have 1.5-3 times more solar energy resources per unit area of territory.

In the future, developing countries with large solar resources could sell electric power generated by solar plants to Europe, where the solar resource is available mostly from March till September. For this purpose it is necessary to organize electric energy flows in the meridian direction. The electrical power streams in latitudinal direction West-East will enable the use of diurnal variation of solar energy, connected with rotation of the Earth about the axis.

In the year 2000 about 600 power stations in Russia with a total capacity of 207 GW generated 420 billion kWh of electric power in six months. A computer model calculation has shown that if one is to install two solar power stations with a total capacity 187 GW, one on Chukotka and the other in Kaliningrad or in Brest, and connect these power plants by low loss transmission line (Strebkov et al., 2000a), such a solar electric power system would generate enough electricity from April 20 till August 20 to completely meet the energy requirements of Russia. On March 20 and September 20 the solar electric power system will work 22.7 hours per day, and on March 1 and October 1 interruption of operation at night time will be 2 hours.

The electric power production for 6 months will be 420 billion kWh with the area of each solar power station being 25x25 km considering an efficiency of 15% for each power station of 93.5 GW peak power. Small size hybrid PV-wind generators have a capacity of 0.5-1 kW (Table 10). The 1.5 kW solar-wind-diesel hybrid power system is applied for stand-alone small family farms, parking of shepherds and hunters, communications and weather stations, remote from the electric grid and located in areas with an average wind speed not less than 4.0 m/s. The annual saving of diesel fuel is up to 50% compared with a diesel generator set.

One more new technology in Russia is soldered vacuum and gas filled glass panes. The single glass sheet transparent to sunlight is widely used in buildings and in designs of solar power engineering, including optical concentrators, thermal collectors, and photoelectric modules. The thermal characteristics can be improved by depositing special coatings on the glass surface, increasing the long-wave infrared reflection coefficient. The coated glazing (glass panes) consisting of two or three glass layers has better heat-insulating properties. To increase thermal resistance the air layer between the glass panes can be replaced with inert gas with a large molecular mass or a vacuum. In the Table 20 the values of resistance R to heat transfer for miscellaneous glass designs are shown. The IR-coating also reduces the IR emittance of the glass.

Table 20. The main parameters of glass pane.

Type of glass window	Thickness of the gap, mm	Thermal resistance R, m² °C/W
Single layer of a glass	-	0.17
Double layer of glass pane	10 - 16	0.37
Double layer of glass pane with argon fill	10 - 16	0.39
Double layer of glass pane with IR coating	10 - 14	0.56 – 0.58
Double layer of glass pane with argon fill and with IR coating	10 - 14	0.71 – 072
Triple layer of glass pane	6 - 12	0.47 – 0.57
Triple layer of glass pane with argon	6 - 12	0.53 – 0.62
Triple layer of glass pane with IR coating	6 - 12	0.67 – 1.08
Triple layer of glass pane with argon	6 - 12	0.90 – 1.43
Double layer of glass pane with vacuum between the glass layers	0.05	0.44
Triple layer of glass pane with vacuum between the glass layers	0.05	0.57
Double layer with vacuum between the glass layers and with IR coating	0.05	1.2
Triple layer of glass pane with vacuum between the glass layers and with IR coating	0.05	2.0

As it is seen from the table, a triple layer with a vacuum between the glass layers has the highest heat-insulating properties, 10 times that of common plate glass. This considerably improves the performance characteristics for solar power engineering. For comparison, the resistance to heat-transfer of a brick wall 2.5 bricks in depth equals 1-2 m²°C/W. Compared to gas-filled glass panes, the panes separated by a vacuum have negligible gap between the panes. So, for a triple layer glass with a pane thickness of 2.5 mm the vacuum-layered glass has a thickness of 7.6 mm compared with 31.5 mm for the gas-filled. The gap between layers in vacuum glass is set with the help of extra small ceramic gaskets (diameter 0.2 mm, height 40µ) arranged on the surface of the pane with a step of about 20 mm.

The glass panes are sealed around their perimeter with the help of polymer cement bonding. The gap between the panes in a gas-filled system is formed with the help of aluminum gaskets. The aging of polymer under conditions of solar radiation and temperature drops limits the durability of glass panes. Usually they last 5-10 years, because at some time there is a depressurization of connection and the moisture penetrates into the gap. Another disadvantage of gas-filled glass panes is higher heat transfer from the edges of the panes through the aluminum gasket.

A design and a method of manufacturing of vacuum and gas-filled glass panes soldered at the perimeter with glass flux is licensed in Russia. The usage of these glass panes in designs of solar thermal collectors and solar modules is also licensed. The production process of soldered glass panes has also been used for developing flat television screens.

The width of a soldered junction equals 5 mm. The soldering takes place at a temperature about 350°C. New glass panes are 1.5 times more resistant to heat transfer than similar glued glass panes. Their durability is now about 50 years, almost 5 times the original durability. The inert gas is saved inside soldered glass pane during the entire life of service. The available equipment allows the production of soldered glass panes in sizes up to 1m x 2m from plate glass in thicknesses from 2.5 to 10 mm. The gap between glasses in gas-filled glass panes is formed with the help of a glass framework. The cost of soldered glass panes is not that different from the cost of bonded glass panes.

The application of soldered glass panes allows considerably lower thermal losses through glass surfaces that can form the basis for creating energy efficient buildings and improve stability and characteristics of such items of solar power engineering, as solar cookers, thermal collectors for heating of water or air, solar modules, optical concentrators and combinations of these devices. So, for example, the solar cells placed in soldered glass panes and filled with inert liquid can have a service life of 50 years. The same outcome can be achieved in the manufacturing of optical concentrators. The efficiency of flat plate solar collectors with soldered glass panes will not be much different than expensive tubular vacuum collectors.

8. Future Energy Technology

Let us imagine a future energy technology based on Russian and world achievements. Twenty years ago R. Buckminster Fuller proposed interconnecting regional power systems into a single continuous global electrical energy grid. Now this concept is actively supported by Global Energy Network Institute (GENI), registered in California, USA. The president of GENI Peter Meisen participated in the International Solar Congress in Moscow in 1997 and reported that technological advances have made the linking of international and interregional energy networks practicable today (Blakers and Smeltink, 1998; Cable et. al., 1997; May et al., 2000; Geyer and Benemann, 1991).

The electric power systems of South American and Arabian countries have been developed. Baltic and Black Sea power rings, the grid systems of USA and Canada; and the grid systems of Scandinavian and European countries already exist. The existing technologies are capable of transmitting electric power of 10 GW capacity through distance of some thousand kilometers (Alexandrov and Smolovic, 1999).

The Russian scientists have developed a new quasi-superconductive technique using single-wire line with reactive capacitive current for active electric power transmission. The power losses in a single-wire line are reduced 2-2.5 times, and the expenses of non-ferrous metals 50-100 times (Strebkov et al., 2000b,c). A new method of power transmission without wires was proposed using a laser beam as a single-wire (Strebkov et al., 1999a). This technique will bridge producers and consumers of energy in each country in a global power generation. The dream of the famous Yugoslavian engineer Nicola Tesla to receive the electric power practically in any place of the globe will become a reality.

These new technologies will change the concepts of developing transport engineering. Imagine a trolley bus or car with a single wire current lead (only one conductor line), similar to a tram. Trolley buses have two lines for current supply, but trams have one current lead and rails as second current lead (actually two current leads). In our new system we use only single wire current lead. A hybrid car having an internal combustion engine, and a 50 kW electric motor leaves a garage using petrol fuel, and on a highway a telescopic antenna-trolley is hooked up to contact a single-wire and the car uses the electric power for traffic (Strebkov et. al, 1999b; Strebkov et. al, 2000a).

A thorough examination of the great migratory processes happening on the Earth is carried out by the International association "Trajectories of Great Migrations of Mankind" under the leadership of Dr. D.B. Purveev. Since ancient Paleolithic (420 thousand years up to AD) times, the migration of nations from North to Eastern habitats of Asia through the Bering Strait to Northern America began. Ancient nomadic hunters mastered the Great Silk Route during seasonal migrations, which connected the countries of Europe, the Near East and Middle East to China and India. Now a revival of the Great Silk Route is in process. For example, the groundwater transport arteries from South East Asia to Northern Asia and Europe.

Dr. D.B. Purveev with a group of the Russian scientists has offered the international project "Great Continental Ways of Mankind", in which the integration of the European, Asian and American continents in the 21st century is offered (Purveev, 2000). Transcontinental systems consolidating transport and power flows are proposed. The first system is a transport and power line West-to-East: Lisbon - Vladivostok and from the South to North: Australia - Indonesia - Thailand - Vietnam - China - Bering Strait - Alaska - Canada - America. The second system is a meridian stream that will pass through the Great Silk Routes: India - Afghanistan - Kirghistan - Tajikistan - Uzbekistan - Turkmenistan - Kazakhstan, North of Western Siberia. The meridian power and transport streams will be intercrossed in Eastern and Western Siberia with a latitudinal power and transport flow, building up a great cross through Europe and Asia.

A third system will be a meridian transport and power line to connect Cape Town with Oslo through Eastern Africa, Arabian countries, Turkey, countries of the Black Sea, countries of East Europe and Scandinavia. The fourth meridian power line will unite countries of Western Africa, the Mediterranean, Western Europe, Britain and Ireland. A meridian power line will link countries of Southern and Northern America.

The latitude power line in an equatorial zone up to 30° of northern altitude aggregating the countries of Asia, Africa and Latin America will be set up. The equatorial power line, and also latitudinal power line of Lisbon-Vladivostok will be made through the Pacific and Atlantic Ocean, Northern and Central America. A cross meridian and latitude power lines will build up a globally integrated power generation system on the Earth. The future global power system will operate using solar, wind, hydro, geothermal and biomass power stations.

At the Okinawa Summit (G8) in July 2000 the leaders set up a task force to assess the barriers and to recommend actions to better encourage the use of RE in developing countries (Renewables Feature at G8 Summit, 2000). Concrete proposals for action were developed at the Genoa G8 Summit in June 2001.

Market liberalization, high efficiency and the low price of local power cogeneration, reliability and environmental friendliness are the key factors to the competitiveness of small independent renewable energy producers.

The excess electric energy generated by consumers-producers will be billed by means of two-way electric meters. The cost of installing such meters is to be borne by the utility company upon request and the cost will be reimbursed to the utility company by the electricity generated by renewable energy consumers-producers.

The increasing decentralized energy production is a new trend in world power engineering. This new approach to energy generation is considered by some economists in the West as the electric revolution (Gibson, 2000; Luque, 1988). Environmental considerations are beginning to have a significant influence on the energy market. The share of environmentally dangerous large electric power stations will be decreased. Electric market liberalization will lead to connection of millions of small-scale independent electric power producers in every country to the electric grid. Purchasing, selling and controlling energy flows among millions of renewable energy consumers-producers will require the implementation of innovative information technologies, Internet and electronic commerce. Therefore, the development of information technologies and renewable micro power technologies are interrelated and progress in each technology will promote the development of the other. As an example of the symbiosis between information and energy technologies we have demonstrated the possibility of electric power transmission along telecommunication fiber-optic line.

All power stations in the 21st century operating on renewable sources of energy will be automatically controlled by digital computers. The latitudinal cross flows of energy will double the capacity of power stations at the expense of transfer of power, unused in night time, to cover the peak loads on the illuminated part of the Earth. Consolidating the ecologically benign energy resources across the earth and in space can create comfortable living conditions for everyone.

9. Solar Stationary Trough Concentrator Technologies For Stand-Alone and Large-Scale Application

Parabolic trough technologies using the Rankine steam turbine-generator cycle were identified as the most commercially ready and available technology for large-scale solar thermal electric power generation. A solar power plant with a total installed generating capacity of 354 MW$_e$ continues to operate today. Solar efficiency of 12.5% capital cost of 3.39 US\$/Wp and a cost of electricity 0.101-0.166 US\$/kWh were calculated, compared to electricity cost 0.043-0.069 US\$/kWh for a coal-fired Rankine plant (Pitz-Paal, 2003). In the long-term it is reasonable to expect the cost of parabolic trough concentrator technology to continue to decrease and approach the cost of conventional power.

Disadvantages of traditional solar concentrator technology are: the high cost of concentrator field, poor response to diffused light, the need for moving parts and high

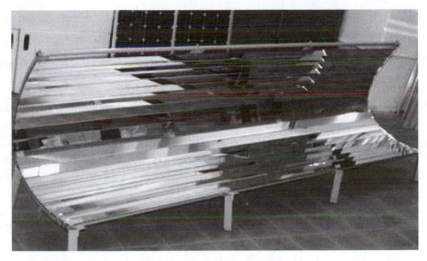

Figure 5. PV stationary concentrator module. Dimensions: 2500x700 mm. Electric Power is 100 Wp.

accuracy tracking, intensive maintenance, low reliability, and the size of the concentrator is limited due to tracking the Sun.

In the past, attempts to commercialize PV solar concentrator technology were not successful. This is due to high cost and poor module design. We developed a solar stationary parabolic trough concentrator module for stand-alone and large-scale applications (Figures 2-5). The advantages of stationary parabolic trough technology are:

1. The cost of the concentrator field is US$1-2/Wp;

2. The cost of electrical energy does not depend much on the capacity installed;

3. No tracking system is needed for the concentrator and receiver;

4. The size of a PV parabolic trough concentrator can be very large, up to 10^{12} m^2; so large scale solar power plants could be constructed;

5. Up to 25% of the diffuse solar radiation can be collected.

6. The concentration ratio for 12 months operation of symmetric stationary concentrator is from 3.5 to 14. For Northern countries application during 9 month of polar summer the stationary symmetric concentrator has a concentration ratio from 5 to 20. A quasi-stationary symmetric concentrator module with a concentration ratio of 7.5-30 needs position correction every 6 months. Asymmetric trough concentrator has a concentration ratio 50% of symmetric trough;

7. Low cost stand-alone PV system;

8. Good opportunities for building integrated design;

9. High temperature solar collector application for steam generation and for solar driven air conditioning system;

10. Combined electric generation and hot water production could be implemented for concentrations of 3.5-30.

The concentration system includes three innovative solar energy technologies developed by Prof. D.S. Strebkov at the All-Russian Research Institute for Electrification of Agriculture (VIESH). Bifacial PV technology modules were originally designed by Prof. D. Strebkov and his colleagues for Russian low orbit satellites in 1970 (Strebkov et al., 1976, Zadde, 1976). Today bifacial solar cells and modules are manufactured in Russia for terrestrial application.

Non-tracking solar concentrator modules appear to show a significant economic effect compared with the conventional flat plate solar panel design (Strebkov, 2000d).

Plastic-free PV module technology has an estimated operating life time of more than 30 years. This new technology uses two sheets of glass with soldered edges and a special liquid filling for solar cell/glass optical and thermal properties matching (Strebkov et al., 1999c; Strebkov, 1998; Strebkov et. al, 2003). Solar module manufacturing companies all over the world use EVA and/or other plastics for solar module encapsulation. These organic materials are to a substantial extent responsible for the relatively low module operation time (20-25 years) and poor cooling conditions, especially during operation at high solar radiation intensity. For a price estimation of PV parabolic trough concentrator modules several assumptions were made:

1. The cost of a PV bifacial receiver is 500 US$/m^2 (the price of VIESH production facility);

2. The cost of a glass mirror reflector is 20 US$/m^2 (the price of glass mirror factory);

3. The cost of the non-tracking metallic support structure is 30 US$/m^2 of module aperture (the price of Russian mechanical factory);

4. The cost of tempered low-iron solar glass protective coating is 20 US$/m^2 (the price of Russian production factory);

5. The peak power output of the concentrator module is 100 W/m^2.

The results of cost calculation of symmetrical PV trough concentrator modules are presented in Table 21.

The cost of the modules for all aperture angles is less than 2US$/Wp. At maximum concentration ratio it is close to 1US$/Wp. For the average concentration ratio of 7-15 the price of the module is less than 1.5US$/Wp. A low concentration ratio of 3.5-7.5 is attractive due to the collection of diffuse radiation in the range of the aperture angle and the possibility of passive cooling. All these prices can be decreased by a factor 1.5-2, if we apply a solar PV receiver with better efficiency and a module power output of 150-200 Wp/m^2.

Table 21. The cost of a symmetric solar stationary trough concentrator module with 10% efficiency, US$/Wp.

Concentration ratio	Aperture angle		
	48°	36°	24°
Low	2.00	1.76	1.46
average	1.49	1.29	1.15
max	1.16	1.06	1.0

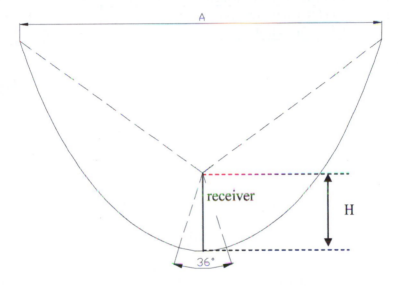

Figure 6. Forming of the compound parabolic trough concentrator with aperture angles and an angle of a turn equal to 36°.

Now let us consider a basic design (Figure 6) consisting of two semi-parabolic cylindrical concentrators with aperture angles of 36° with respect to the general focal axis. Semi-parabolic cylinders are turned around a focal axis so that the dihedral angle between focal planes is equal to the aperture angle. The reflecting surface within the limits of an angle of a turn represents an element of the circular cylinder with radius equals to a focal length of parabolas. The bifacial receiver is installed in a plane of symmetry of the concentrator (Strebkov et al., 2003).

A similar design with an aperture angle of 36° does not demand tracking the Sun for 9 months (for example, from February till November). In northern latitudes the available solar energy for this period makes up 85-90 % of the annual value. A prominent feature of the stationary module is that both semi-parabolic-cylinders work simultaneously resulting in a doubled concentration ratio. The further increase of concentration ratio can be attained by inserting additional rotating semi-cylindrical reflectors with axes in a plane of symmetry of the concentrator (Figure 7) into the base design. These additional semi-cylindrical reflectors should be rotated 180° twice a year.

Table 22. Basic technical and economic characteristics, used in calculation.

The characteristic	Value
Parameter of a parabola, p	3,42 m
Middle width the concentrator, A	8,008 m
Length of forming generatrix (semi-parabolic-cylindrical) of main reflector, loo	12,916 m
Cost of facets of the main mirror reflector, $Cfac$	$20/ m^2
Cost of a skeleton of the main reflector, Cr	$20/ m^2
Cost of additional mirror reflectors, Car	$30/ m^2
Cost of the solar cells Csc	$450/m^2

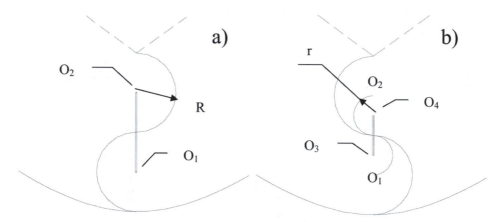

Figure 7. Forming additional rotary semi-cylindrical reflectors.

Let us determine the cost of one kW installed capacity for several variants of the concentrating module design. The initial data and symbols used in calculation are given in Table 22.

Cost of one running meter of concentrating module:

$$C_{rm} = l_{oo} (C_{fac} + C_r) + l_{ar} C_{ar} + H C_{sc} \tag{Eq. 1}$$

where:

l_{ar} - total length of forming additional reflectors;

H - width of the solar cells.

Values l_{ar} and H vary depending on the number and sizes of additional semi-parabolic-cylindrical reflectors. In this paper the following variants of solar concentrator semi-parabolic-cylindrical designs are considered:

Variant 1: The base variant of the design without additional reflectors (Figure 6): $H = p/2$.

Variant 2: The module with two additional semi-parabolic-cylindrical mirror reflectors. The radius of the semi-cylinders equals $R=p/8$ (Figure 7a).

Table 23. Characteristics of constructive elements of the module.

Characteristic	Variant		
	1	2	3
R, m	-	0.43	0.43
r, m	-	-	0.21
lar, m	-	2.68	4.02
H, m	1.71	0.86	0.43
K, relative units	4.7	9.4	18.7

Table 24. Cost of constructive elements one running meter of the module, US$.

Element	Variant		
	1	2	3
Facets of the basic reflector	258.32		
Skeleton of the basic reflector	258.32		
Additional reflectors	-	80.40	120.6
Solar Cells	769.60	384.80	193.50
Total Crm	1286.10	981.80	830.70

$l_{ar} = 2\pi R = \pi p/4$,
$H = 2R = p/4$.

Variant 3: The module with four additional semi-parabolic-cylindrical mirror reflectors. The radii of the semi-cylinders *are* $R = p/8$ and $r = p/16$ (Figure 7b).

$l_{ar} = 2\pi R + 2\pi r$,
$H = 2r = p/8$.

Numerical values of these characteristics are shown in Table 23. For reference, the geometrical concentration ratio G_c (ratio of entrance aperature size to size of receiver) for all three variants of the parabolic-cylindrical concentrator are given in this table.

$G_c = A/H$.

Additional cylinders are rotated along the axes O_1-O_4 depending on the time of the year. For example, in Figure 7b, big cylinders are rotated twice a year, small ones four times a year.

The cost of one running meter of the module, calculated from expression (1), and the cost of its constructive elements are given in Table 24.

Hence, the length L (in meters) of 1 kW concentrating module:

$$L = \frac{1}{A\,\eta} \qquad \text{(Eq. 2)}$$

where η = efficiency of the solar cells (in relative units).

The graph of dependence is represented in Figure 8.
The cost of the installed capacity of one kW will be:

$$C = C_{rm} L \qquad \text{(Eq. 3)}$$

Our results of this calculation for the three variants of the module design are shown in Figure 9. Analysis of the results shows that use of additional reflectors considerably reduces the cost per unit of the installed capacity in comparison with the base variant. Influence of the additional reflectors on cost is not so great.

Further calculations concern the module with a concentration ratio of 18.7 (variants 3) and a 10% solar cell efficiency.

Decreasing the cost of facets or a skeleton in 1 US\$/m^2 results in US\$16 less per installed kWp. To achieve a similar effect the cost of additional reflectors should be reduced by 3.20 US\$/m^2, and the cost of solar cells should be reduced by 29.80 US\$/m^2. It is important to notice that solar cells are the most expensive element, and decreasing their cost can be significant in comparison with the other constructive elements.

Efficiency of solar cells determines the length of one kW module, influencing the cost of all constructive elements, but with increase in efficiency this influence is reduced (Figure 8). A growth of efficiency from 10% to 11% would result in a US\$100 decrease in the cost per installed kWp. An increase from 14% to 15% would give a cost decrease per installed kWp of US\$52.

The main results of developed technology are:

1. The solar stationary concentrator module is a new attractive solar electric conversion technology for stand-alone and large-scale application with a cost of 1-2 US\$/Wp.

2. The promising features of this technology are the low cost for installed kW and the possibility to construct a large scale concentrator field without a tracking system with concentration ratio 3.5-30.

Further development is connected with cogeneration of heat and electric power and the application of 15-20% efficiency solar cells. New solar stationary concentrator technology can be applied for electricity generation, high temperature thermal applications and absorption cooling systems.

10. Comparison of the Characteristics of Solar and Thermal Power Plants

Thermal power plants (TPP) based on fossil fuels dominate the market of the electric power production. The purpose of this section is to compare technical and economic parameters of TPP and fuel-free solar power stations (SPS). The parameters considered include efficiency of conversion of energy, number of annual hours of installed capacity use, lifetime, and cost of capital expenses and the electric power.

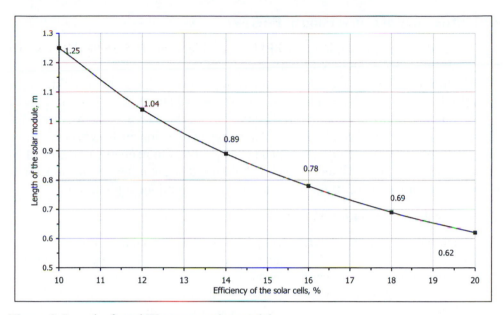

Figure 8. Length of one kW concentrating module.

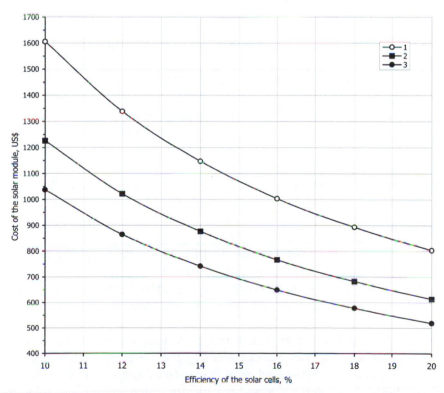

Figure 9. Cost of one kW stationary parabolic-cylindrical concentrating module (explanation in the text).

10.1 Efficiency of Conversion of Energy

Thermal coal power stations (TCPP) have efficiencies of 25-36 %. TPPs with a steam-gas cycle, have the maximal efficiency of 52%.

The limiting efficiency of photoelectric conversion of solar energy is estimated as 93% (Evdokimov, 1985; Evdokimov, 1986). The maximal efficiency of 36.9 % was achieved under laboratory conditions (Spectrolab) for three layer cascade solar cells. Existing photoelectric technologies of solar energy conversion are based on the quantum nature of the photoelectric effect and necessary condition: energy of the incident photons should be greater than or equal to the forbidden gap of the semiconductor. It is possible to further increase the efficiency by using multijunction structures.

10.2. Number of Working Hours

The number of working hours of using TPP (first of all coal power plants) capacity per one-year amounts to 5,000 hours. For SPS at average day length of 12 hours and 300 sunny days in one year SPS achieves operation time 3,600 hours per year. However, in solar power engineering-annual number of working hours of SPS with the peak capacity corresponding to standard peak solar irradiance equal 1,000 W/m^2 at temperature 25°C is used. For average latitudes in Europe this parameter is equal to 1,000 hours. In desert conditions the number of peak capacity hours reaches 2,000-2,900 hours per year. Practically, this means that to provide round-the-clock production of electric power, it is necessary to use three SPS of identical capacity, one of which provides electric energy supply during the day, and the other two SPS generate electric power for storage (for example in hydro-storage or for generation of hydrogen and then in fuel cells) to provide electric energy to consumers at night.

Another approach consists of three solar power stations which are installed along the equator through eight time zones, incorporating an electric main and forming a global solar energy system, working 24 hours per day.

In comparison with TPP capacity, peak SPS capacity should be increased 2-3 times that adversely affects its competitiveness. However, SPS has no fuel component in the price and does not pollute the environment.

10.3. Lifetime

The lifetime of TPP is about 30 years, the lifetime of SPS is equal to 20 years for a tropical climate and 25 years in a temperate climate.

The degradation of the parameters of SPS is caused by the light and thermal aging of basic polymeric materials: EVA and tedlar, which are used for the hermetic sealing of solar cells.

If polymeric materials are excluded from SPS design, the lifetime of SPS will increase to 40-50 years and will be determined by the lifetime of tightness preservation of the glass soldering in the module. Polymer-free design of solar modules consists of two sheets of the tempered glass hermetically sealed on the edges soldering or welding. Solar cells are located inside SPS and placed in an organic-silicon liquid (Strebkov, 1998; Strebkov et al., 2003).

Tests have shown that the tightness of a soldered joint of a solar module glass box lasts up to 50 years.

10.4. Cost of Installed Capacity

The cost of the installed capacity of TPP is 600-1000 US$/kW, and the cost of the electric energy is 0.025-0.05 US$/kW hour. The cost of the electric energy generated by TPP does not take into account the cost of environmental damage from emissions of TPP such as greenhouse gases and acid rains which influence productivity, preservation of forests, reproduction of fish stocks, health of people, and the global and local climate (Strebkov and Koshkin, 1996). In view of this damage the cost (so called external cost) of electric energy is doubled and will be 0.05-0.1 US$/kW hour.

The installed capital cost of SPS now equals 3.5-7 US$/kW, and the cost of electric power is 0.15-0.25 US$/kW hour (Srebkov et. al., 1998). Decreasing the capital costs and the cost of electric energy, generated by SPS, can be achieved by decreasing the solar silicon cost, increasing the efficiency of solar cells and modules, increasing the SPS lifetime and the number of hours of annual use of peak capacity, by using solar energy concentrators.

Let us consider the influence of new solar silicon technologies and the use of concentrators on the cost of SPS.

Silicon makes up 40% of the cost of solar cells and 25% of the cost of solar modules. Currently, solar polycrystalline silicon (poly-Si) at a price of 30 US$/kg is used for manufacturing SPS. Poly-Si is produced using chlorine-containing products (trichlorosilane and hydrochloric acid). This manufacturing process is ecologically dangerous and requires large amounts of electric energy (300 kWh/kg).

New chlorine-free technologies of solar silicon production are being developed in Russia. Energy costs are reduced ten times, and the manufacturing cost is 12 US$/kg (Strebkov et al., 2001b). Considering that silicon (in the form of quartzite) is the second most abundant element in the earth's crust (after oxygen), it is reasonable to expect that solar silicon will continue to keep it's leading role as the basic material for PV and that new technologies will eventually reduce the cost of silicon to that of high-quality glass (20-50 US$/m^2).

Using stationary concentrators developed in Russia helps to reduce the amount of solar silicon in SPS because a part of the area of SPS is replaced by surfaces of glass mirrors. The glass mirrors cost 20 US$/m^2, or 15 times less than 1m^2 of SPS without the concentrator. The concentration ratio of 10-15 corresponds to approximately similar contribution of solar cells' cost and the concentrator's cost into the cost of the concentrator module at a level of 0.25-0.5 US$/W. Thus, the total cost of SPS will be 0.5-1 US$/W.

Stationary concentrators do not demand tracking the Sun, they have a lifetime of mirrors up to 30 years and allow the realization of a principle of thermal power station for the combined production of electric power and heat. Heat can be used for space heating and hot water supply. Stationary concentrators can produce high-temperature heat (100-200°C), which can be used in absorption coolers for air-conditioning buildings, and also in technological installations for drying vegetables and fruits, clearing of drains and conversion of biomass.

The solar concentrator market has two large niches: agriculture, where 35 % of the population in the world has no access to electric power and ecologically clean fuel-free production of energy, which is necessary for sustainable development and the survival of mankind in the climate conditions of global warming.

The following parameters of SPS can be achieved by the year 2010:

- Capital cost of 1 kW peak power as 0.5-1.0 US$/W;

- Cost of electric energy generation 0.04-0.08 US$/kW hour.

- Efficiency of 18-24 %, lifetime up to 40-50 years.

Achievement of indicated cost values of SPS will result in multiple expansion of demand. The increase of annual volume of production SPS from 500 MW at present to 5 GW will be accompanied by further cost reductions and increasing sales volume. Taking into account increasing fossil fuel prices and the introduction of an emissions tax on TPP, including a payment for releasing carbon dioxide, SPS is expected to become competitive in years 2015-2025. As for nuclear stations, it is necessary to remember that the Chernobyl atomic power station cost 1 billion US$, the direct damage of the explosion has cost 10 billion US$, and for past years the total sum of damage from failure has increased up to 100 billion US$ and continues to grow.

11. Conclusion

The utilization of Russia's enormous and regionally diverse renewable energy potential is quite low. Russian specialists already have considerable technical and scientific experience. New domestic policies and regulations should help in overcoming barriers which seriously limit investments in RE technologies. International agency policies, and international cooperation that help to establish and support RE service companies are also very important. Yet "Russia is the sleeping giant of renewable energy potential," said William C. Ramsay. Renewables can contribute to its energy needs in a cost-effective way and yield important economic and social benefits.

12. References

Alekseev, V.V., 1995. Power basis of economy. The bulletin of Moscow State University, 5, No. 2 (in Russian).

Alexandrov, G.N., Smolovic, S.V., 1999. Flexible lines for electric energy transmission over long distances. Fifth Symposium "Electrical-Engineering-2010", October 19 22, Moscow region, pp. 35 42.

Bezrukikh, P.P., Arbuzov, J.D., Borisov, G.A., Vissarionov, V.I., Evdokimov, V.M., Malinin, N.K., Ogorodov, N.V., Puzakov, V.N., Sidorenko G.I., Shpak, A.A., 2002. Resources and efficiency of the use of renewable sources of energy in Russia, Saint-Petersburg, Nauka.

Blakers, W., Smeltink, J., 1998. The ANU PV/Trough concentrator system. In: 2nd World Conference and Exhibition on Photovoltaic Solar Energy Conversion, July 6-10, Hofburg Kongresszentum, Vienna, Austria.

Cable, R., Coen, G., Price, H., Rearney, D., 1997. SEGS Plant Performance1989-1997. In: ASME International Solar Energy conference, Albuquerque, NM, June.

Evdokimov, V.M., 1985. New theoretical models of solar cells and prospects of efficiency increasing. In: Semenov, N.N., Shilov, A.E. (Eds), Conversion of Solar Energy, Nauka (Science), Moscow, pp. 13-19.

Evdokimov, V.M., 1986. Problems of theory and prospects of efficiency increasing of solar cells. In: Alferov, Zh. I., Shmartsev, Yu. V. (Eds), Photodetectors and Solar Cells, Nauka (Science), Leningrad, pp. 148-180.

Geyer, N., Benemann, J., 1991. 600MW Solar Program in California. An American exception or a world-wide breakthrough. Paper R7. United Nations ECE Solar seminar Alushta, USSR, 22-26April 1991, 27 pp.

Gibson, J.S., 2000. Responding to a revolution. Power Engineering Journal, Jan Feb., p.48 51.

Luque, A., 1988. Solar Cells and Optics for Photovoltaic Concentration. Adam Milger, Bristol and Philadelphia, p.7-9.

Martinot, E., 1999. Renewable energy in Russia: market, development and technology transfer. Renewable and Sustainable Energy Reviews 3, 49-75.

May, K., Walker, A., Dominick, J., Westby, B., 2000. Performance of a large parabolic trough solar water heating system at Phoenix Federal Correctional Institution. In: Proceedings ASES Annual Conference, American Solar Energy Society.

Ministry of Fuel and Energy, AO, 2000. Equipment of non-traditional and small power engineering handbook. New and renewable energy sources. Scientific Advisor, Bezrukikh, P.P., Arbuzov, Y.D. (Ed.), Moscow. Currently being translated to English.

Pitz-Paal, R, 2003. Solar Thermal Power Today opportunities and Future Perspectives. Proceedings of ISES World Congress, Goteborg, Sweden.

Purveev, D.B., 2000. International integration and interrelation of continents in 21 century. Diverse Measuring, No. 4, p. 4-6.

Renewables Feature at G8 Summit, 2000. Renewable Energy World, vol.3, No. 5, p.9.

Strebkov, S., Zadde, V.V., Zaitseva, A.K., Lidorenko, N.S., Landsman, N.S., Bordina, N.M., 1976. Semiconductor photoelectric generator. Russian Patent No.434872, priority claimed 16.11.1970. Published in Russian Patent Bullletin No.20, 1976.

Strebkov, D.S., 1994. On development of solar power in Russia. Thermal Engineering, No. 2, p. 53-60.

Strebkov, D.S., Koshkin, N.L., 1996. On development of photovoltaic power engineering in Russia. Thermal Engineering, vol. 43, No. 5, p. 381-384.

Strebkov, D.S., 1997. Problems of development of renewable energy sources. Mechanization and Electrification of Agriculture, No. 6, p. 4-8 (in Russian).

Strebkov, D.S., 1998. Energy utilization of a biomass. Renewable Energy, No. 3, p. 9-12.

Strebkov, D.S., Irodionov, A.E., Tarasov, V.P., Tveryanovich, E.V., Silaeva, A.N., 1998. A method of calculation of technical and economic characteristics of solar power stations in conditions of market economy. Moscow, Russian Academy of Agriculture, VIESH.

Strebkov, D.S., Avramenko, S.V., Nekrasov, A.I., 1999a. The method and apparatus of electric power transmission, Russian Patent No.2143775. Published in Russian Patent Bulletin No. 36, 1999.

Strebkov, D.S., Avramenko S.V., Nekrasov, A.I., 1999b. A method of power supply of the electrical transport and device for its realization: Russian Patent No. 2136515, priority claimed 26.08.1998, Published in Russian Patent, Bulletin No. 25, 1999.

Strebkov, D.S., Kidjashov, Y.K., Zadde, V.V., Bezrukih, P.P., 1999c. The method of fabrication of solar photovoltaic module, Russian Patent No. 2130670, Priority claimed 24.03.1998, Published in Russian Patent Bulletin No. 14.20.05.1999.

Strebkov, D.S., Avramenko, S.V., Nekrasov, A.I., 2000a. Single-wire electrical power system for mobile electrical units. Materials of international technological conference on automation of agricultural production, Minsk, June 6-8, p. 65 66.

Strebkov, D.S., Avramenko, S.V., Nekrasov, A.I., 2000b. Single-wire power electric system for renewable - based electric grid. EuroSun 2000, June 19-22: Copenhagen, Denmark, Book of abstracts, p. 298.

Strebkov, D.S., Avramenko, S.V., Nekrasov, A.I., 2000c, Investigation of a single-wire system of electrical energy transmission. International conference: An ecology and agricultural engineering. St.-Petersburg, vol. 1, p. 50-55.

Strebkov, D.S., Tverjanovich, E.V., Bezrukih, P.P., 2000d. Solar Module with Concentrator, Russian Patent # 2172903, priority claimed 07.04.2000, published in Russian Patent Bulletin #24, 2000.

Strebkov, D.S., Zadde, V.V., Pinov, A.V., Touyryan, K., Murphy, L., 2001. Crystalline Silicon Technologies in CIS Countries. In: 11th workshop on Crystalline Silicon Solar Cells materials and process. Holiday Inn Estes Park, Colorado, NREL, pp. 199-207.

Strebkov, D.S., Irodionov, A.E., Bezrukih, P.P., 2003. Method of solar photovoltaic module fabrication. Russian Patent No.2130670, priority claimed 19.03.2001, published in Russian Patent Bulletin No. 21.

Strebkov, D.S., Tveryanovich, E.V., Bezrukhih, P.P., Irodionov, A.E., 2003. Solar Module with Concentrator (variants), Russian Patent #2209379, priority claimed 19.03.2001, publised in Russian Patent Bulletin #21, 2003.

Vainstein, E.F., Ksenevitch, I.P., 2000. Study of a possibility of receiving of the basic components of alternate fuel for transport from organic residues. Drive Engineering, No. 2 (24), p. 43-50.

Zadde, V.V., Strebkov, D.S., Lidorenko, N.S., Landsman, N.S., Unishkov, V.A., 1976. Semiconductor photoelectric generator, US Patent No.3948682 cl.136/89, 6.04.1976.

Chapter 12

An Innovative, High Temperature and Concentration Solar Optical System at the Turn of the 19th Century: The Pyreheliophoro

by

Manuel Collares Pereira*

INETI, Renewable Energies Department,

Edificio G, Az. dos Lameiros,

1649-038 Lisbon, Portugal

Abstract

One of the greatest achievements at the turn of the 20th century in the field of solar energy was invented by Manuel Antonio Gomes, a Jesuit Priest. The device, a solar concentrator named the "Pyreheliophoro," won the Grand Prize at the 1904 World Fair, in St. Louis, Missouri, USA. Gomes, born in 1868 to a poor family in Northern Portugal (nicknamed Himalaya because of his height), attended Catholic Seminary as a way to get an education. While he never received a high level of academic instruction, his strong aptitude for physics, chemistry, mechanics, and botony has left lasting contributions in those fields. In Paris, Gomes studied under Berthelot, Violle, and Moissan. His motivation for the study of solar energy germinated from an idea for synthesizing fertilizers using nitrogen present in air. With these desirably cheap and abundant fertilizers he wanted to feed the poor and rural Portuguese population and the rest of the world. Gomes felt the fertilizer synthesis could be achieved with high temperatures coming from high solar concentration, as did other researchers of the day. He initiated a long process of devising new ways of concentrating solar radiation. It started with a metallic lens of the Fresnel type, culminating with the Pyreheliophoro, which was capable of producing a sustained temperature of 3,800°C, which was the highest temperature ever produced with solar energy at that time. Unfortunately, his work had little chance to survive in the dawning of the oil era. He went on to invent and develop other remarkable ideas, including an explosive called

---

* *Corresponding Author. Email: collares.pereira@ineti.pt*

Figure 1. Manuel António Gomes, "Father Himalaya", 1902.

himalayite and a rotary steam engine. He became member of The Portuguese Academy of Sciences and his voice was known as truly original, as he proclaimed concepts which made him a true ecologist "avant la lettre," defending concepts like what we today call sustainable development. This truly remarkable man died in 1933, probably a victim of his own botanical experiments. The commemoration of the 100th anniversary of the St. Louis Grand Prize in 2004 is an excellent reason to bring the man and, in particular, his solar energy achievements to the attention of the solar community.

1. Introduction

The ISES initiative of recovering the recent and not so recent history of solar energy and its pioneers has prompted several investigations into the past. Several gems of scientific and technical ingenuity which were way ahead of their time, have been uncovered. The one described in this paper produced quite a stir in its own time. It was soon to be forgotten given that the world was in transition from the nineteenth to the twentieth century and was about to embark in the oil race, and solar energy was not even given half a chance to be "in the race" at that time.

The man behind the work described here was a truly remarkable personality, a self made scientist (without a proper academic scientific training), and a catholic priest. Throughout his life he tried to compensate for his lack of training by traveling the globe to interact and study with the top notch scientists of his day; people like Berthelot, Moissan, and Violle. He visited such scientifically relevant countries as France (mainly Paris), England, Germany and the U.S.

His name was Manuel António Gomes, soon nicknamed by a friend as Himalaya, because he was taller than his colleagues. He added this nickname to his name and became known by it. He was born in 1868 in Cendufe, a small village in the North of Portugal. He was a child in a large family with little economic resources. As usual in those days and in such circumstances, he entered the Seminary as a way to study and to succeed in life. He was ordained a priest and worked as a practicing priest until his death in 1933. In spite of his very controversial lifestyle, unorthodox views of the Church and its dogmas, and the very critical position on items such as the forced celibacy of priests and the constant fight for a more socially responsible and committed Church. In short, he embraced the truly liberal, republican, socialistic and idealistic ideas of the day (Figure 1).

Father Himalaya was quite famous in his lifetime and was respected for his achievements. He became a member of the Portuguese Sciences Academy and had an attentive audience amidst the politicians of his day.

This paper is dedicated to his crowning achievement in the field of optics and solar thermal, but he is also known for many other original contributions. He got a truly large number of patents in Europe and in the U.S. and, most notably, was awarded Grand Prize at the St. Louis World Exhibition of 1904.

Father Himalaya lacked a proper high level training in physics and other basic sciences, which would have been very helpful to shape his enormous qualities as an experimentalist and as a mechanical genius. His training in chemistry was probably more in-depth due to his interaction with Berthelot and other important chemists. Among other things he invented and developed an explosive (a chlorine based, smoke-less-powder, called himalayite—it is said to be more powerful, easier, and safer to use than dynamite) which he put to many peaceful uses in agriculture and in quarries.[1] His explosive was sought after by several armies of the world (U.S., German, Portuguese, etc.) and his involvement with some of them is still more or less shrouded in mystery. Another one of his inventions, deserving a mention in this brief account, is the one of a rotary steam engine, looking very much like the rotary engines first proposed and developed many years later.[2]

He also loved nature and was a self trained biologist, a practitioner of natural medicine, and most remarkably, an ecologist "avant la lettre." He was an explicit and stout advocate of sustainable development through a proper balance of humanity, its needs and nature, which he regarded not just as a provider but also as an important part of the whole scheme of things. He constantly called for Renewable Energies (solar, hydro, tidal, wave) as the means for long term and balanced solutions for the many problems caused by poverty and starvation facing the world of his time and in particular of his own country. In this regard he had a truly modern view of the World and of the place of Man in Nature, a view which is taking another hundred years to affirm itself.

This paper is an effort by the author to reconstruct the thought process that lead to the Pyreheliophoro, based on the information contained in a remarkable biographical book written by Prof. Jacinto Rodrigues, which is now being translated into several languages and being used as the basis for a movie on the extraordinary life of this towering man and personality (Rodrigues, 1999). Most of the other references come from the book by Rodrigues (1999) and serve only as a possible for further research by the interested reader, without access to the book itself.

---

[1] Portuguese patents 8544-1912, 8932-1913, which were preceded by several U.S. French and British patents: U.S. patents 301524 (1906) 853085 (1907) 869158 (1907) French patents 374 656(1907) 374932 (1907), British patents 3179 (1907) 3199, 4439 (1908), 20931 (1908).

[2] Portuguese patent-9719, March 1917.

Figure 2. Solar device of Mouchot-Piffre.

2. The Pyreheliophoro

2.1. The First Steps: The Metallic Lens

From his early days Father Himalaya saw solar energy as a means to provide energy not just for the production of hot water or steam, but as a direct means to provide energy for industrial processes, that are associated with materials production or processing, if high enough temperatures could be achieved. One of his objectives was to produce nitrogen based agricultural fertilizers by extracting the nitrogen directly from the air! While he could never have achieved this with his devices, as we can today well understand, he did manage to achieve perhaps the highest controlled temperatures of the day, about 3,800°C, in the solar furnace of his pyheliophoro which was a truly remarkable achievement.

During his visit to Paris at the turn of century, Father Himalaya probably became quite familiar with the works of A. Mouchot (1869), Louis de Royaumont ("Conquête du Soleil" Ed. Marpon et Flammarion (Royaumont, 1880-1886)), and Charles Metelier (1890). From his correspondence and from documents found among his belongings it is fair to assume that he must have had some degree of familiarity with the works of John Ericson (Rodrigues, 1999; Meinel and Meinel, 1977) W. Adams (Rodrigues, 1999; Meinel and Meinel, 1977), Calver (Rodrigues, 1999; Meinel and Meinel, 1977; Serões, 1901a), and Aubrey Eneas (Rodrigues, 1999; Gazeta Ilustrada, 1901; Serões, 1901b; Meinel and Meinel, 1977).

Critical of the devices produced by Mouchot-Piffre, he soon realized that he needed to modify them to obtain higher temperatures and to break the mechanical coupling between the solar furnace (placed in the "focal zone") and the structure supporting

Figure 3. Solar experimental apparatus of Eneas.

Figure 4. Parabolic trough of Ericson.

the mirrors. If possible he also wanted a stationary solar furnace, in which only the optics would do the tracking of the sun's apparent motion in the sky.

The device developed by Mouchot-Piffre is shown in Figure 2. It was a paraboloid like shaped structure with reflecting inner walls and a furnace placed along its optical axis.

Mirrors in the shape of a truncated cone (Figure 3), (Eneas, Pasadena, California, 1901) and large flat mirrors (Calver, Tucson, 1901) were among the most common solar optics of the day. John Ericson (Rodrigues, 1999; Meinel and Meinel, 1977) proposed a parabolic shaped mirror in 1880 (Figure 4), but Himalaya's ideas went in rather different directions.

Serões (1901a), Gazeta Ilustrada (1901), and Serões (1901b) are explicit instances of Portuguese magazines dedicating space to new inventions and Fr. Himalaya likely read them. It was not possible to consult the referred magazines and therefore their level of technical detail is not known to the author. However these were publications for a general audience and little should be expected beyond some photos or drawings and a reference to the purpose of the inventions.

To the interested reader the author recommends a modern book (Meinel and Meinel, 1977) containing an interesting introductory chapter on the history of solar energy. This book makes an explicit mention of Father Himalaya and his crowning solar achievement—the Pyreheliophoro—at the St. Louis Fair of 1904.

The ideas of Aubrey Eneas were perhaps more relevant to the developments proposed by Father Himalaya, since some of the aspects involving the tracking mechanism were incorporated into the Pyreheliophoro. This is just a guess judging from the referred drawings and the probable knowledge that Father Himalya had of them. However it is really impossible to be more specific given the lack of known documents explaining clearly where he was coming from and what, at each stage, he set out to achieve.

From the way his work progressed and from the available documents (correspondence with one of his brothers, notes, newspaper clips with observations by others and most importantly from the patents he was awarded) it is possible to piece together a probable thought process. That effort is presented below.

Fr. Himalaya realized early on that the very high temperatures he was after, required a very high solar radiation concentration factor which directed his thoughts towards two axis type tracking. With materials processing in mind, he needed solutions that would not have, in his own words, "the furnace between the reflector and the sun." He thus first thought of lenses to do the job, since these could send the concentrated light down and out, towards the target. However the required dielectric lenses (glass, at the time!) were not a practical idea in those days. His first remarkable attempt can be seen in Figure 5 and in the photographs of Figure 6 corresponding to prototypes actually built and tested. It is a very original metallic Fresnel lens type, done with flat-strip mirrors, ring shaped, the whole ingeniously tracking the sun in elevation and compensating for the earth's rotation, by moving together on circular rails. These

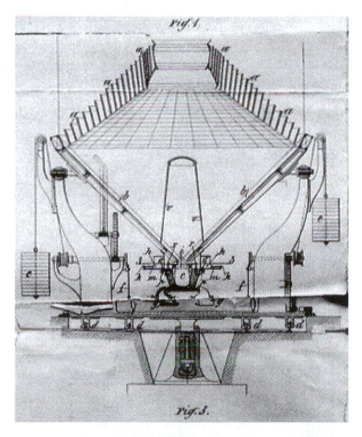

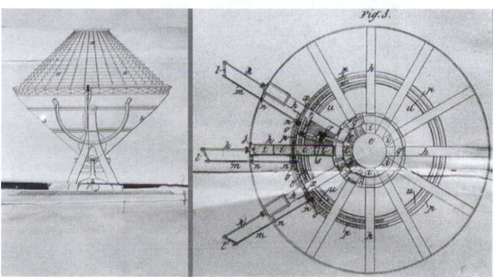

Figure 5. Excerpts from Patent 292,360 (1899).

a.

b.

Figure 6. Photographs of Father Himalya's Prototype.
    a.   Fr. Himalya standing in front of the device in Paris.
    b.   The device at Castel d'Ultrera.

first efforts are the basis of the first patent (Patent 292,360, 1899) awarded to him on the subject of solar energy.

His experiments were carried out in the French Pyrenees, (Castel d'Ultrera) not far from Odeillo and Font-Romeu (of later day fame, for very similar solar reasons!)

The results he obtained were not as good as he expected, but it seems that he was able to achieve temperatures in excess of 900ºC (sufficient to melt iron), a remarkable

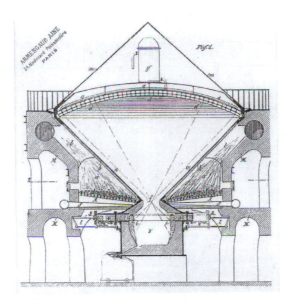

Figure 7. Cross section of radiative furnace, excerpt from patent 293,512, Paris (1899).

achievement, given the choice he had of materials for the mirrors, and a good measure of the mechanical precision with which he was able to produce his device. It should be noted that the solar furnace itself was an object of careful developments, to be able to contain the materials he was melting/processing with it. His temperature measurements were crucially dependent on what he was able to melt.

In fact the furnace itself was the object of patents, perhaps the most important of which being Patent 293,512 (1899).

A radiative type furnace taken from that patent is shown in Figure 7, where the side walls c,c' were to be heated with the burning of fuel and the heat radiated into the triangular shaped cavity was to be concentrated (focused) down onto the hot cavity F, by a paraboloid shaped upper wall d. This furnace was later very easily (and much better) adapted to the solar focussing optics to be described next, with solar radiation coming through an aperture placed in d and the side walls c,c' now serving as what today would be described as a second stage concentrator.

In the process of these developments he also invented a radiometer to measure solar radiation intensity using his metallic lens concept. Unfortunately it has not been possible to find details about this potentially interesting development.

2.2. Designing the Pyreheliophoro for the St. Louis Fair of 1904

2.2.1 Preliminary Work

His next serious attempt was carried out in Lisbon. This second invention, a tracking section of a parabaloid and a solar furnace can be seen in some of the figures

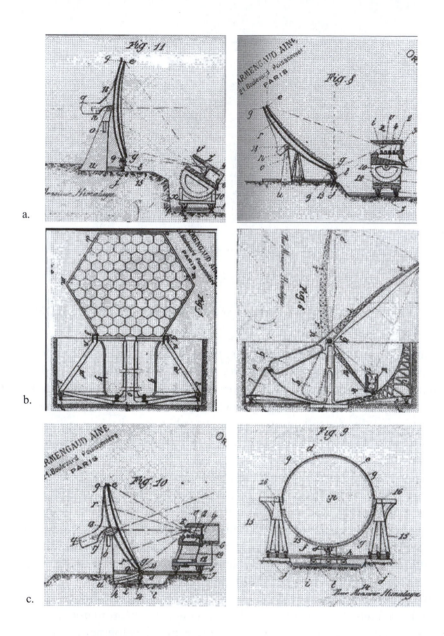

Figure 8.
Excerpts from patent 307,699, Paris (1901).
    a. Cross section of parabolic mirror and solar furnace
    b. Front view of hexagonal parabolic mirror and cross section displaying relative motion capacity and weight balance
    c. Front view of round parabolic mirror and cross-section of another mirror/furnace configuration.

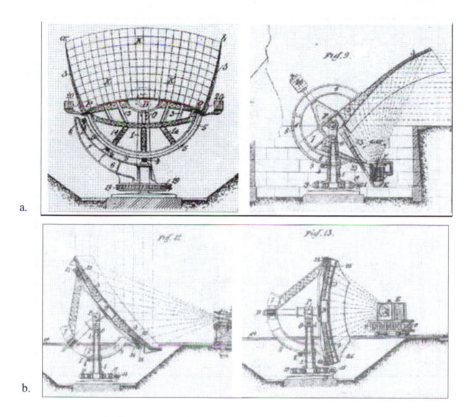

a.

b.

Figures 9a-b. Exceprts from Patent 307,699 (1901).

(Figure 8a-c and Figure 9). There are several patents in different countries corresponding to this invention, the main ones being: patent 307,699, Paris, 1901, 3,746 Lisbon, 1901 or patent 797,891, 1905, Washington, D.C. Archives. In this design there is no longer a lens like concept. Instead, the very high concentration factor achieved is based on a primary mirror concept, now with full separation of the optics from the furnace, in contrast with the Mouchot-Piffre ideas.

A conceptual leap, as explained in the patents, is the fact that in previous 3 dimensional solutions radiation arrived at the focal zone from all sides, never allowing for sufficient concentration to be achieved on its outside walls (see Mouchot, Figure 2). A paraboloidal sector allows for the maximum concentration achievable being redirected into furnace Z for direct effect on the substances to process or heat. The built in flexibility of motion always ensures that reflected rays are directed at all times into the furnace Z. In modern terms (Winston and Welford, 1978) we can see that the conical entrance aperture to the furnace ensures a second stage concentration, taking care of reflection and tracking inaccuracies (spillage).

The complete set of drawings show a large number of novel possible combinations of mirrors and furnaces, their relative motions and sun tracking capabilities. In Figure 8 the concentrating mirror is a full paraboloid (circular or approximated by an

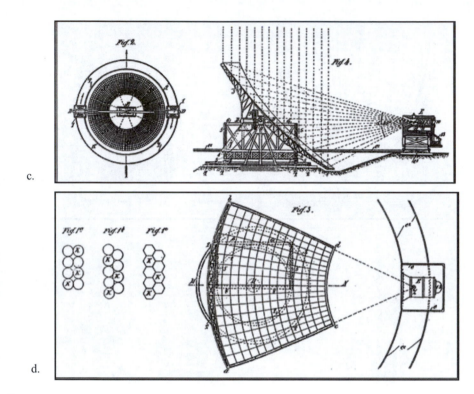

Figures 9c-d. Excerpts from Patent 307,699 (1901).

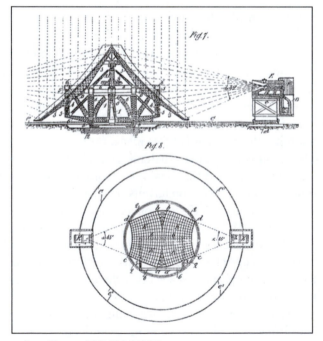

Figure 9e. Excerpts from Patent 307,699 (1901).

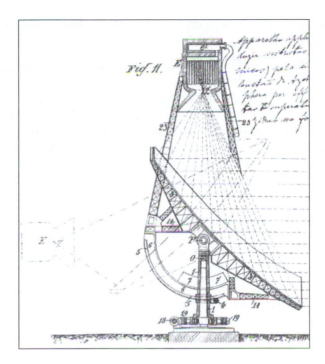

Figure 9f. Excerpts from Patent 307,699 (1901).

hexagon) with its optical axis and axis of symmetry going through its center. It is produced with a collection of small mirrors with square rectangular, hexagonal or circular shapes (in this case the patent talks about different ways to pack the circles) which would be glued or fixed by other means to a supporting frame having the required curvature. Father Himalaya speaks of using flat mirrors but also curved ones to achieve higher concentration. The reflector is associated with a furnace that can be moved in conjunction with the mirror, thus satisfying the constraints imposed by the tracking of the sun's apparent daily motion. The movement of the furnace is achieved in several different ways, on horizontal rails or on circular ones, and also (as in drawings 11 and 8 of Figure 8a) as combination of both possibilities. There are several mechanical solutions to balance the weights and strains on system components (in particular the mirror assembly).

However the entrance to furnace Z, in Figure 8c does not provide the same amount of final concentration/light gathering capability as in Figure 8a-b or as in Figure 9b. In fact, in this case getting light from the top edge of the parabolic mirror does not allow the top edge of the furnace to extend out as much as in other instances. The evolution from Figure 8 to Figure 9 goes in the direction of a greater flexibility for the relative motions of mirror and furnace, some of the configurations being more suitable for higher concentration, in particular that which results from the furnace itself, functioning as a second stage concentrator in a modern sense of the word. One visible change, distinguishing the devices in Figure 9, in general, is that the mirror is now a sector of paraboloid, asymmetric, with its optical axis falling outside the useful mirror surface, allowing-at comparable focal lengths-for a larger separation between mirror and furnace, thereby seemingly increasing the variety of mechanisms to provide for their respective sun tracking motions.

The solution of two concentrating mirrors, back to back, moving on the same tracking structure (for instance, drawing 7 in Figure 9e) is rather interesting, making an efficient use of the rails and other parts of the tracking mechanism, in this instance producing with the same movement, two simultaneous hot spots where the furnace would be located. A quite distinct situation is achieved in the case where the optics and the furnace are combined in a unique set-no rails (for instance drawing 11 in Figure 9f).

A concern about accessibility of the furnace is patent in some of these drawings and the means to achieve it are ingenious indeed. Father Himalaya really wanted to use his invention for industrial applications, in the chemical, agrochemical, metal processing industries and, therefore, he provided solutions for configurations which would be best suited for gases or for solids turning into liquids at the very high temperatures achieved. In this context it is interesting to note the difference between a furnace on a level with the mirror, as in Figure 9b, above it (as in Figure 9f or below it as in Figure 9a. It is also interesting to note the use of the degree of freedom which comes from the different relative heights of the tracks for the furnace and the base of the mirror pedestal which is often placed in the centre of a hole, below ground level.

The case of Figure 9f deserves a further comment since it has a note in Father Himalaya's own handwriting, in Portuguese. It mentions the application to the "production of nitrates (chemical fertilizers) from the simple combustion of the Nitrogen present in the atmosphere at very high temperatures obtained within the focus Z", an old dream of his and perhaps the main objective he had in mind for all his efforts in this field of solar energy. In fact there is a visible arrangement for the collection of gases at the top of the furnace, which distinguishes this configuration from others.

These drawings also show different approaches to tracking the sun in azimuth and elevation. Use is made of rotation around center poles to compensate for the Earth's rotation, with the furnace sometimes moving in a separate fashion, on rails, or as one with the mirror, but always with the possibility of adjusting to the sun's elevation. It is interesting to note that these good and innovative ideas were not put to use in the final Pyreheliophoro (Section 2).

The reason was surely–and this is an important conceptual point–that Father Himalaya knew that none of these movements could be made fully automatic in a modern sense, (i.e, in truly unattended operation) since that would require modern day combinations of tracking motors and sun sensors which were not available at that time.

Experiments with one of the configurations described in these patents (presumably one with the furnace mounted on a circular rail, as can be inferred based on the photographs of Figure 10) were carried out in Lisbon (March/April 1902). Inaccuracies in the design and mechanical problems plagued the prototype. On the day of the

a.

b.

Figures 10a-b. The prototype built for the Lisbon tests with what lookes like the furnace in the background.

public demonstration, the concentrated radiation destroyed the supporting structure and it was a fiasco!

Father Himalaya must have realized that he needed a new idea for a truly practical system able to track the sun, unattended, at maximum concentration–while correcting the faults with the prototype. A simple clock mechanism would do the trick, but that required a radically new design, bringing again the optics and the furnace to a single structure. That became the Pyreheliophoro, to be described next.

For that he needed to bring the optics and furnace back to a single structure and this is where he must have obtained some inspiration from the ideas of Eneas (see Figure 9). He adapted Eneas' ideas to his by moving off axis type paraboloid sectors and furnaces closer to ground level, where they could be easily operated in any industrial application suitable for the high temperatures obtained.

3. The St. Louis Fair of 1904

Father Himalaya worked hard developing his new ideas and in a short time he was ready to show his new concept at the World Fair in St. Louis. In 1901 (Patent 3746) he had already used the word Pyreheliophoro to designate the concept presented above but the invention he presented at the St. Louis world's fair is what eventually became known as the Pyreheliophoro. As before, the explanations that follow result from a careful examination of the photographs that follow and from the educated guesses they allow. In fact, the writings of Father Himalaya that are available today are even more scant about the Pyreheliophoro than about his previous inventions. No final patent was found on the Pyreheliophoro even though it is his crown achievement in this field and he himself was well aware of that. One explanation may be the fact that he fought the U.S. patent office for a long time because he wanted an explicit acknowledgement that this invention was suited for industrial applications, something that the U.S. Patent office never granted him.

A particular reference should be made to the fact that the drawing, Figure 12b, was made for Dr. Rodriques' book (1999) by Prof. João Gabriel da Silva (1998).

The final configuration as it was assembled in St. Louis is shown in Figures 11 and 12. This solution integrates the two distinct motions necessary for the optics to track the sun at each operating latitude, but with the required simplicity, as explained below.

The device was 13 m tall. The total reflector area- a sector of a paraboloid- was 80m^2 giving a mean focal distance of 10 m. There were 6,117 small (123mm x98mm) silvered glass mirrors painstakingly fixed to the underlying structure. The focal area was designed to be no larger than a circle with 150mm diameter. This resulted in a total geometric concentration factor of ~4500 X. Father Himalaya claimed a final concentration factor of 6,117 X, hinting at a smaller focal spot, approximately with the same area as each individual mirror. In truth, the conic entrance to the furnace (even though it is not an ideal optics for that purpose, as we know today) would have been

a.

b.

Figures 11a-b. View of the completed Pyreheliophoro as assembled in St. Louis.

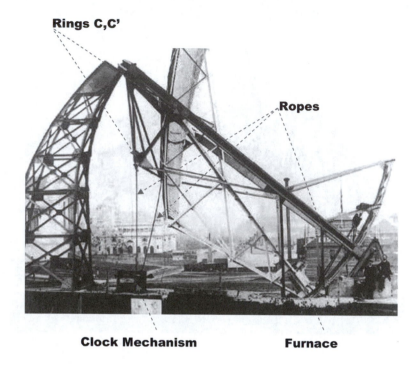

Figure 12a. View of thePyreheliophoro.

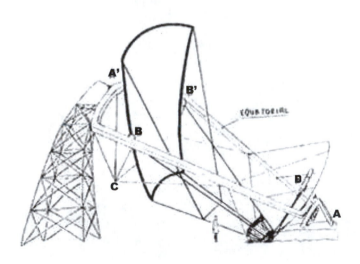

Figure 12b. Schematic drawing of the Pyreheliophoro (Prof. J. Gabriel da Silva).

Figure 13. The medal awarded to the Pyreheliophoro.

instrumental in the recovery of tracking inaccuracies and radiation spillage that might have occurred , and would effectively enhance the final concentration.

The way it operated is as follows: (see Figures 11 and 12).

The system could be set with its axis AA' parallel to the Earth's rotation axis (an equatorial mount type arrangement- presumably each model would have the capability of slight tilt adjustments of the equatorial mount axis to the exact value of the local latitude). Simple tracking around this axis was achieved with ropes pulling on rings C and C', powered by a clock type mechanism, which was fixed to the ground immediately below. This would ensure that the sun would always be on a plane perpendicular to the plane of the equatorial mount structure.

To concentrate its radiation on the furnace entrance, the system would now move the mirror assembly itself together with the furnace around axis BB', making the furnace describe an arc of a circle on some sort of a "rail", with its centre on the line BB'. On bar D there was a pulley-chain mechanism to accomplish this movement.

This daily adjustment ensured that the axis of the paraboloid would point to the sun at all times throughout the day, since this whole system also moved as one with the equatorial mount!

In short, this corresponds to the present day complex two axis tracking with step or variable speed electrical motors, with sun sensors and/or computer assistance used in high concentration solar optics. This is substituted by a very simple yet accurate constant speed clock tracking system. This clever and practical solution was essential for the very high concentration of solar heat which Father Himalaya set out to achieve. There was no other practical way to produce such results at that time, without the devices we have today.

The Pyreheliophoro was operated during the fair, to the amazement of its countless visitors. The best measure of its success is the fact that it got the Grand Prize (Figure 13).

According to Rodrigues (1999) several books (Demy, 1904; Buel, 1904-1905) and newspapers of the day referred to this event, reporting specifically about the Pyreheliophoro. Articles on the device ran in the *Sunday Magazine St. Louis Post Dispatch*, the *St. Louis Republic*, the *Western Watchman* and even the *New York Times/New York Herald*. There was even a mention of the Pyreheliophoro in the October 1904 issue of *Scientific American*. The press also made countless references to the device in Portugal.

Father Himalaya only got three very clear days during which he claims to have obtained at least 3,800°C, a very impressive achievement for the day.

U.S. entrepreneurs wanted to take it (or make copies of it) to display its capabilities for the public in other fairs. The Pyreheliophoro could melt iron easily and would turn a piece of wood placed in the furnace to smoke almost instantaneously. This was a true source of wonder to all who witnessed it.

Father Himalaya would have none of it. He wanted his system to be used in more noble applications, as he put it, namely in industries like those requiring higher temperatures than the ones obtainable at that time through combustion or electric arcs (<3,500°C).

He got nowhere in his fights over what to do with the system. The U.S. patent office never granted him the classification of industrial interest that he so desperately wanted, and he was never able to negotiate a deal with potential buyers. As commented above, this is quite likely why so little was written by him directly about it. In fact the U.S. patent (Patent 797,891, 1905) awarded to Father Himalaya in 1905 was only after the Fair in St. Louis and covers only the work prior to St. Louis and described in 1.2, without any reference to the potential of the invention for industrial use.

Before the Pyreheliophoro was dismantled, Father Himalaya tried to donate it to the local University. He also tried other institutions such as the Carnegie Institute, but was never successful. Citing two major flaws in the systems design (no description was available), a Mr. Shinker, presumably of the fair grounds, made the decision that the Pyreheliophoro could not remain on the fair grounds and would have to be removed.

It is not clear what eventually happened to the Pyreheliophoro, but there are many theories regarding its fate. Some people think that it may have been stolen, others believe it may simply have been dismantled. There are also stories that many of the small mirrors which made up the large reflecting section of the parabaloid eventually made their way into the hands of local school children. Whatever its fate, the Pyreheliophoro enjoyed a bright, but short career right at the beginning of the oil era.

4. Conclusion

In conclusion it is fair to say that the Pyreheliophoro and all the work leading up to it were remarkable achievements for Father Himalaya's time, truly deserving their dissemination among the solar energy scientists of today and even among the general public interested in the history of science and technology. The timing couldn't be better, since it is during the year 2004 when the 100th anniversary of the Grand Prize at the St. Louis fair will take place.

It is a pity that Father Himalaya did not write a detailed account of his efforts or attempt to write down his successive achievements in the form of scientific papers which would have made the task of the author much easier, removing the guess work that was involved in this paper and giving a better account of what he thought and really knew about the work of others. In any case from what was presented here the originality and power of his inventive capacity in the solar energy area is well established.

His preliminary lens ideas were quite interesting, as well as his many attempts at increasing concentration and delivering solar radiation to a furnace in a practical way, for practical applications. His crowning achievement was indeed the Pyreheliophoro, with which he managed to deliver a very high concentration of solar radiation on a continuous, daily, basis through a play of clever optics and simple mechanisms. Quite likely it produced in a sustainable manner the highest temperature ever at that time with a solar device, a world record for the day. Even today it is not easy to obtain the same very high temperatures and the limitations we have today in terms of the knowledge of optics, materials, and sensors are quite different from those at the turn of the 19th century.

For the record it is important to point out that Father Himalaya was not always correct in his interpretations or in his stated goals. That is quite understandable when only later day science could prove him wrong. However in other instances, even when scientists of his day might have been able to correct him, he did not know better. Given his lack of a high level formal training in the basic sciences, he had no way to encompass in depth the very wide range of knowledge that he needed to not make those specific wrong statements or claims. That should not prevent us from admiring his powerful and creative mind, brilliantly complemented by his cunning practical eye, to translate ideas into useful devices.

For instance, after his forays into solar energy, Father Himalaya moved on to other topics which, as commented upon in the Introduction, included among many other remarkable things, explosives (the so called himalayite) and rotary (steam) engines.

After the St. Louis Fair, he continued to work and travel all over the World, and got wide recognition, especially from his fellow country men and in particular in the region he came from, Minho, in the North of Portugal. Throughout the rest of his life he continued on a collision course with many of the established views of the day, in such diverse areas as religion, politics, agriculture, medicine, industry, social development, ecology to a degree which would make him feel quite at home in today's world. Perhaps that really constitutes the best summary of his persona: a man 100 years ahead of its time. He died on December 21, 1933.

5. Acknowledgements

The author wishes to first thank Prof. Jacinto Rodrigues for all the conversations and collaboration received, in particular for the copies of the patents and other documents which were thus possible and easy to consult. Secondly the author wishes to thank Rui Rodrigues from INETI who helped with handling of all photographs and drawings. The author further wishes to thank Prof. J.J. Delgado Domingos from IST for the copies of the documents in his possession, an indispensable complement of the ones obtained form Prof. Jacinto Rodrigues and Ms. Maria Abreu e Lima from Cooperativa de Actividades Artísticas and Mr. Humberto Nelson from Pagella, for the photographs made directly available.

6. References

Most of the other references come from the book by Rodrigues, Jacinto, 1999 and serve only as a possible starting point for further research by the interested reader, without access to the book itself.

Buel, J.W. (Ed.), 1904-1905. Louisiana and the Fair. World's Progress Publishing Co. pp. 3128.

da Silva, João Gabriel (Universidade de Coimbra), 1998. Private communication to Prof. Jacinto Rodrigues, integrated into (Rodrigues, 1999).

Demy, A., 1904. Exposition Universelle. Bibliothèque de St. Geneviève, Paris, pp. 709-711.

Gazeta Ilustrada (Journal), 1901 (drawings with Aubrey Eneas truncated cone).

History of the Louisiana Purchase Exhibition, 1905. Universal Exposition Publishing Company, St. Louis, pp. 281.

Meinel, A.B., Meinel, M.P., 1977. Applied Solar Energy, an Introduction. Addison Wesley Publishing Company, Second Printing, January.

Metelier, C., 1890. La conquête paciphique de l'Afrique par le soleil, Paris, Ed.

Mouchot, A., 1869. La Chaleur Solaire et ses applications industrielles. Paris Gauthier Villar, Imprimeur Libraire.

Patent 3746, Livro 4º, fl.151, August 31, 1901, Rep. Propriedade Industrial, Lisboa.

Patent 292,360, 7 September 1899, INPI, Paris.

Patent 293,512 October 20, 1899, INPI, Paris.

Patent 307,699, January 31, 1901, INPI, Paris.

Patent 797,891, August 22 1905, National Archives Washington D.C, U.S.A.

Rodrigues, Jacinto, 1999. A Conspiração Solar do Padre Himalaya-Edição Cooperativa de Actividades Artísticas (ISBN-972-9089-44-2).

Royaumont, Louis de, 1880-1886. Conquête du Soleil. Ed. Marpon et Flammarion.

Serões (Journal), April 1901a (on Calver's system in Tucson, Arizona).

Serões (Journal), June 1901b, (Solar engine by Aubrey Eneas, Pasadena, California).

Welford, W.T., Winston, R., 1978. The Optics of Non Imaging Concentrators. Academic Press.

Errata

Corrected Figures

from Volume 14, Chapters 10 and 11

Erratum 1

Renewable Energy Technology Mix for Atmospheric Carbon Dioxide Stabilisation by 2050

(Chapter 10 in Volume 14)

by

David R. Mills* and Christopher J. Dey
School of Physics, A28
University of Sydney
2006 Sydney, New South Wales
Australia

* *Corresponding Author. Email: d.mills@physics.usyd.edu.au*

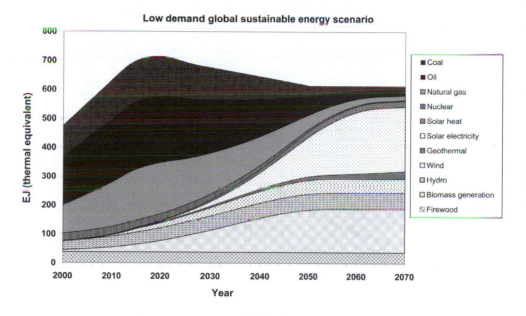

Figure 1. A scenario for fossil fuel replacement prepared by the authors showing how renewable energy sources could be allocated in a reasonable low demand future energy scenario. The first part of the total demand curve boundary is indicated as a solid line based upon IEO projections out to 2010. The second boundary section is a broken line between 2010 and 2050 to highlight the fact that this demand scenario is a nominal load pattern which simply connects the 2010 and 2050 load points (the 2050 load point is based upon assumptions given in the text). The third section assumes a constant load from a constant population from 2050; this may be a pessimistic assumption given that technology would be expected to further improve, but there could also be a continuing increase in energy service demand in less developed countries which counteracts improved technology. In this figure, PV and solar thermal direct solar electricity installations are lumped together and cumulative capacity is assumed to expand at almost 25% per year until 2035, when the annual capacity increase is frozen at the same level as today's natural gas installed energy capacity increase. After the middle of the century, direct solar electricity could become the most important element in reducing fossil fuel usage.

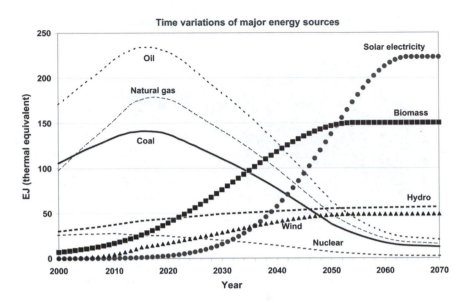

Figure 2. Variations in the annual delivery of major fuels and technologies using the assumed load pattern in Figure 1. Variations in the assumed nominal load drop between 2010 and 2050 in Figure 1, or some increased fuel switching between natural gas and coal, will affect the curvature of the downward slope of the demand of individual fossil fuels. However, for atmospheric stability to be achieved, the general required trend is clear.

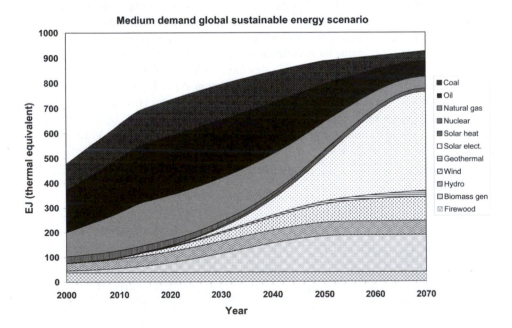

Figure 3. The demand curve for Scenario B in which constant per capita demand is assumed as population increases. A more aggressive renewable energy programme closer to the technical resource limits is required, and it takes 20 years longer than in Scenario A to reduce fossil fuel use to 50% of 1990 levels.

Erratum 2

Development Strategies for Solar Thermal Electricity Generation

(Chapter 11 in Volume 14)

by

David R. Mills* and Christopher J. Dey
School of Physics, A28
University of Sydney
2006 Sydney, New South Wales
Australia

* Corresponding Author. E-mail: d.mills@physics.usyd.edu.au

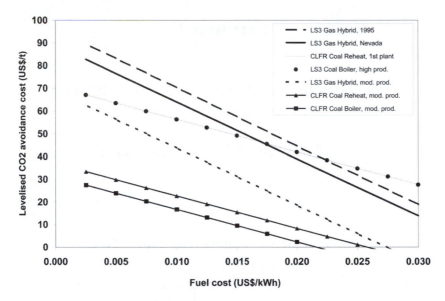

Figure 1. Results of levelized CO_2 avoidance costs versus coal fuel cost for line focus STE technologies. The results for the large production LS3 boiler coal saver and CLFR prototype reheat coal saver are identical. The different slopes of the curves for plant types A-C compared with types D-G demonstrate that the coal savers are less sensitive to the price of the displaced fuel.

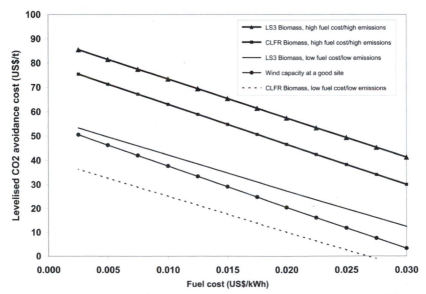

Figure 2. Results of levelised CO_2 avoidance costs versus fossil fuel cost for solar/biomass line focus solar technologies and wind generation. The high CF solar/biomass options replace baseload coal, while the low CF wind is assumed to replace 1/2 coal and 1/2 gas. The line focus technologies are uncompetitive against wind for high biomass prices and emissions, but the CLFR can better wind where biomass has low emissions and low cost. Biomass emissions will also drop as the transport economy becomes cleaner, and collection cost will drop as the industry develops in size. However, because wind offsets a grid mix of fuel, the fuel cost offset by wind may be higher than the cost of coal offset by biomass/solar plant. On this basis, a CLFR biomass plant may be comparable to wind plant in moderate production.

WORD INDEX